Rings, Modules and Codes

CONTEMPORARY MATHEMATICS

727

Rings, Modules and Codes

Fifth International Conference
Noncommutative Rings and their Applications
June 12–15, 2017
University of Artois, Lens, France

André Leroy
Christian Lomp
Sergio López-Permouth
Frédérique Oggier
Editors

AMS AMERICAN MATHEMATICAL SOCIETY

2010 *Mathematics Subject Classification.* Primary 16D40, 16D70, 16W20, 16S36, 11T71, 13P05, 08A05.

Library of Congress Cataloging-in-Publication Data

Names: International Conference on Noncommutative Rings and Their Applications (5th : 2017 : Lens, France) | Leroy, André (André Gerard), 1955- editor. | Lomp, Christian, 1969- editor. | López-Permouth, S. R. (Sergio R.), 1957- editor. | Oggier, Frédérique, editor.

Title: Rings, modules and codes : Fifth International Conference on Noncommutative Rings and Their Applications, June 12-15, 2017, University Of Artois, Lens, France / André Leroy, Christian Lomp, Sergio López-Permouth, Frédérique Oggier, editors.

Description: Providence, Rhode Island : American Mathematical Society, [2019] | Series: Contemporary mathematics ; volume 727 | Includes bibliographical references.

Identifiers: LCCN 2018041947 | ISBN 9781470441043 (alk. paper)

Subjects: LCSH: Noncommutative rings–Congresses. | Noncommutative algebras–Congresses. | Rings (Algebra)–Congresses. | Modules (Algebra)–Congresses.

Classification: LCC QA251.4 .I582 2019 | DDC 512/.46–dc23

LC record available at https://lccn.loc.gov/2018041947

Contemporary Mathematics ISSN: 0271-4132 (print); ISSN: 1098-3627 (online)

DOI: https://doi.org/10.1090/conm/727

∞ The paper used in this book is acid-free and falls within the guidelines
established to ensure permanence and durability.
Visit the AMS home page at `https://www.ams.org/`

10 9 8 7 6 5 4 3 2 1 24 23 22 21 20 19

Contents

Preface

This proceeding contains many of the articles related to talks presented at the international meeting "Noncommutative Rings and their Applications V" that was held in Lens in June 2017. The theme of the meeting was essentially ring theory, module theory, and coding theory but related topics were also present. This is reflected in the present volume. The work and the enthusiasm of the participants created a nice and friendly atmosphere. This encouraged discussions leading to revitalizing cooperations or creating new ones. Many countries were represented and this international aspect was very nice to see. Prof. Oggier gave a marvelous course, a part of which is included in this volume. The four invited speakers Philippe Langevin, Christian Lomp, Sergio López-Permouth and Pace Nielsen enriched the conference with their talks. They all contributed to this volume in one way or the other. For the most part, contributors to this volume delivered related talks at the conference.

All papers were subject to a strict process of refereeing and we would like to thank warmly all the colleagues who served as anonymous referees for the articles in this volume and provided their valuable comments and recommendations. Their expertise and professionalism greatly contributed to the quality of this volume. Without their help, this proceeding would not exist.

The meeting was supported by the Laboratory of Mathematics in Lens (LML) and by different bodies of this university (RI, BQR) as well as the GDR "TLAG".

We would like to thank all the participants for their kindness and enthusiasm during the conference. We would also like to thank Angelique Gérard, secretary of the Laboratory in Lens, for all her work and for making the administrative tasks as light as possible for all of us.

The editorial staff of the AMS, especially the clear and kind emails from Christine Thivierge and her patience, helped me a lot preparing the volume.

The editors

List of Participants

Mehdi Aaghabali
University of Edinburg.

Evrim Akalan
Hacettepe University, Ankara.

Mustafa Alkan
Akademiz University, Antalya.

Meltem Altun-zarslan
Hacettepe University, Ankara.

Dilshad Alghazzawi
Université d'Artois.

María José Arroyo
Universidad Autónoma Metropolitana, Iztapalapa (Mexico).

Pinar Aydoğdu
Hacettepe University, Ankara.

Grzegorz Bajor
Warsaw University of Technology.

Vladimir Bavula
University of Sheffield.

Mhammed Boulagouaz
King Khalid University, ABHA.

Christian Brown
University of Nottingham.

Engin Buyukasik
Department of Mathematics, Urla, Izmir.

Ahmed Cherchem
Université USTHB, Alger.

Yilmaz Mehmet Demirci
Sinop University, Sinop.

Abdelrahman Eid
Université of Lille1.

Youness El Khatabi
Université Moulay Ismail.

Pinar Eroğlu
Dokuz Eylul University, Izmir.

Alberto Facchini
University of Padova.

Jose Gomez-Torrecillas
University of Granada.

Małgorzata Hryniewicka
University of Białystok.

Emil Ilić-Georgijević
University of Sarajevo.

Gizem Kafkas Demirci
Izmir Institute of Technology.

Yeliz Kara
Uludag University, Ankara.

Handan Kose
Ahi Evran University, Kirsehir.

Arda Kör
Gebze Technical University.

Philippe Langevin
Université de Toulon.

André Leroy
Université d'Artois.

Christian Lomp
University of Porto.

Sergio López-Permouth
University of Athens (Ohio).

Jerzy Matczuk
Warsaw University.

Pace Nielsen
Brigham Young University, Salt Lake City.

Frédérique Oggier
Nanyang Technological University.

Salahattin Özdemir
Dokuz Eylul University, Izmir.

Deividi Pansera
University of Porto.

Mehmet UC
Mehemet Akif Ersoy University, Burdur.

Serap Şahinkaya
Gebze Technical University.

Liang Shen
Southeast University, Nanjing.

Felix Ulmer
Université de Rennes.

Michał Ziembowski
Warsaw University.

Contemporary Mathematics
Volume **727**, 2019
https://doi.org/10.1090/conm/727/14620

Rings whose proper images are almost self-injective

Adel Alahmadi, André Leroy, and Surender K. Jain

Abstract. In 1966, L. Levy characterized commutative noetherian rings whose proper homomorphic images are self-injective as motivated by a property of Dedekind domains. The purpose of this paper is to characterize commutative noetherian rings whose proper homomorphic images are almost self-injective, a property that holds for a larger class of rings, including serial rings.

Introduction

The following theorem was proved by L. Levy in [**L**].

Theorem 1. *Let R be a commutative noetherian ring. Then every proper homomorphic image of R is self-injective if and only if R is the direct sum of the following types (not necessarily all): (I) Principal Ideal artinian ring, (II) Dedekind domain, (III) Local ring whose maximal ideal M has composition length 2 and satisfies $M^2 = 0$.*

The main result of our paper is the following theorem:

Theorem 2. *Let R be a commutative noetherian ring. Then every proper homomorphic image of R is almost self-injective if and only if R is the direct sum of the following three types of rings (not necessarily all) (I) Serial ring, (II) Dedekind domain, (III) Local ring with the maximal ideal $M = A \oplus B = Soc(R)$. Furthermore, in each case the Krull dimension is bounded by 2.*

The concept of almost self-injective ring was introduced by Baba in [**B**] and studied, among others, by Harada and Tozaki [**HT**] in connection with serial rings and almost quasi-Frobenious rings.

A commutative ring R is almost self-injective, if for any R-homomorphism $f : I \longrightarrow R$, I an ideal of R, either f extends to a homomorphism $g : R \longrightarrow R$, or there exists a decomposition $R = R_1 \oplus R_2$ and a homomorphism $h : R \longrightarrow R_1$, where $R_1 \neq 0$, such that $hf(x) = \pi(x)$ for all $x \in I$, π is the usual projection of R onto R_1. In other words one of the following two diagrams can be completed:

$$(1)\quad \begin{array}{ccc} 0 \longrightarrow I & \xrightarrow{i} & R \\ \mathrm{f}\downarrow & \swarrow g & \\ R & & \end{array} \qquad (2)\quad \begin{array}{ccc} 0 \longrightarrow I & \xrightarrow{i} & R = R_1 \oplus R_2 \\ \mathrm{f}\downarrow & & \downarrow \pi \\ R & \xrightarrow{h} & R_1 \end{array},$$

2010 *Mathematics Subject Classification.* Primary 16D50.
Key words and phrases. Almost self-injective rings, serial rings.

and we will sometimes say that the map f can be "completed" (by g or h).

R is called a uniserial ring or a valuation ring if each pair of ideals (equivalently principal ideals) can be compared by the inclusion relation. A ring R is called serial if it is a direct sum of uniserial rings.

All rings considered in this paper are commutative and have identity different from zero.

1. Main result

Let us first mention that an almost self-injective ring which is indecomposable must be uniform. Indeed such a ring has no idempotents and hence it is π-injective (=quasi-continuous) (cf. Lemma 1, [**AJ**]). Moreover, an almost self-injective noetherian ring is a finite direct sum of uniform ideals (cf. p. 66, [**DHSW**]).

We start by showing that a finite direct sum of valuation rings is such that every homomorphic image is almost self-injective.

LEMMA 3. *Every homomorphic image of a finite direct sum of valuation rings is almost self-injective.*

PROOF. Let us first consider the case of a single valuation ring. Such a ring is a principal ideal ring and it is obvious that each homomorphic image of a valuation ring is still a valuation ring. So, it is enough to show that a valuation ring is an almost self-injective ring. Since a valuation ring is a local ring, it is indecomposable. Let $f : aR \longrightarrow R$ be an R-module homomorphism. Let $f(a) = b$. Consider the case when $aR \subseteq bR$. Then there exists $c \in R$ such that $a = bc$. We define $h \in End_R(R)$ by $h(1) = c$. Then $h(f(ar)) = h(br) = brc = ar$, for any $r \in R$ and thus $h \circ f = \mathrm{Id}_{aR}$. On the other hand if there exists $d \in R$ such that $b = ad$, we define $g \in End_R(R)$ by $g(1) = d$. Then $g(a) = ag(1) = ad = b = f(a)$. Thus in this case g extends f to R.

Now let us consider the case of a finite direct sum $R = R_1 \oplus R_2 \oplus \cdots \oplus R_l$ of valuation rings. We note that every ideal I of R is of the form $I_1 \oplus I_2 \oplus \cdots \oplus I_l$ where I_j is an ideal in R_j, for $j = 1, \ldots, l$. If $I_1, I_2, \ldots, I_l$ are ideals in these rings and $f : I = I_1 \oplus I_2 \oplus \cdots \oplus I_l \longrightarrow R$ is an R-homomorphism, we define $f_i := \pi_i \circ f : I \longrightarrow R_i$ where $\pi_i : R \longrightarrow R_i$ are the projection maps, i.e. $f = (f_1, \ldots, f_l)$. If each f_i can be extended to $g_i : R_i \longrightarrow R_i$ then the map f can also be extended to $g = (g_1, \ldots, g_l)$ and we are done. If there exists an index s such that f_s does not extend, then there exists a map $h_s \in End(R_s)$ such that $h_s \circ f_s = \mathrm{Id}_{I_s}$. We can then define $h = (0, \ldots, 0, h_s, 0 \ldots, 0)$. Then we have $\pi_s \circ \mathrm{Id}_I = h \circ f$. This means that in both cases we can complete the diagram, showing R is an almost self-injective ring. □

LEMMA 4. *In an almost self-injective indecomposable ring if two elements $a, b \in R$ are such that $ax = 0$ implies $bx = 0$, then either $Ra \subset Rb$ or $Rb \subset Ra$. In particular, if R is a domain then it is a valuation domain.*

PROOF. Since the map $Ra \to Rb$ sending a to b is well defined and can be completed in one of the two ways as given in the definition of almost self-injectivity. These two different ways leads to the fact that either a divides b or b divides a. □

Our next proposition will be useful in proving our main result.

PROPOSITION 5. *Let R be a commutative noetherian ring having the property that every proper homomorphic image is almost self-injective. Then the following hold:*

(1) *If R is domain then it is a Dedekind domain and the Krull dimension of R is 1. Moreover, if R is istelf almost self-injective then R is a valuation ring with a unique nonzero prime ideal (= maximal ideal) and all the ideals are power of this prime ideal.*
(2) *If R is not a domain, there exists a finite number of maximal ideals and the Krull dimension of R is bounded by 2.*

PROOF. (1) We first claim that there are no ideals between M and M^2, where M is a maximal ideal. Since R is a domain, $M^2 \neq 0$ and the ring $S = R/M^2$ is a local almost self-injective ring. If x, y are nonzero elements in M/M^2 then x and y have the same annihilator, namely M/M^2. This implies that $xS \cong yS$. Since S is almost self-injective the map $\varphi : xS \longrightarrow S$ given by $\varphi(x) = y$ can be completed and hence either there exists $u \in S$ such that $y = xu$ or there exists $v \in S$ such that $x = yv$. Since the square of M/M^2 in $S = R/M^2$ is zero, the elements $u, v \in S$ cannot belong to M/M^2. Thus u, v are invertible. This leads to the fact that for every pair of elements $x, y \in S$ we have $Sx = Sy$. This yields the claim. We can then use a theorem of Cohen (cf. [**C**]) that a commutative noetherian domain satisfying the property that there are no ideals between a maximal ideal M and its square must be a Dedekind domain.

If the ring R is itself almost self-injective Lemma 4 above shows that the ring R is a valuation ring. Since in a Dedekind domain every ideal is a product of prime ideals and every nonzero prime ideal is maxima,l the last statement follows.

(2) It is well known that a commutative noetherian ring has only a finite number of minimal prime ideals. If P is such a minimal prime ideal, then R/P is a domain and Lemma 4 shows that it is a valuation domain and hence a local ring. In particular, there is a unique maximal ideal containing P. This shows that the number of maximal ideals in R is fewer then the number of minimal prime ideals. Moreover the above statement (1) shows that every ideal in R/P is a power of the maximal ideal. In particular, the Krull dimension of R is bounded by 2. □

Before proving the main theorem we prove the following crucial result for local ring satisfying our property i.e. every proper homomorphic image is almost self-injective.

LEMMA 6. *Let R be a commutative local ring with maximal ideal M such that each proper homomorohic image is almost self-injective. Then either R is a valuation ring or $M^2 = 0$ with composition length of M equal to 2. Indeed, every proper homomorphic image is a valuation ring in each case.*

PROOF. We claim that every ideal is either minimal or essential. Suppose there exist nozero ideals I and K such that I is not minimal and $I \cap K = 0$. Then for every nonzero ideal C properly contained in I, $(I/C) \cap (K + C)/C = 0$. Since R is local and $C \neq 0$, R/C is local and almost self-injective, and hence uniform. In particular, I/C is essential, a contradiction. This proves the claim. Since every proper homomorphic image of R is uniform, $udim(R) \leq 2$. In particular, the socle of R is a direct sum of at most two minimal ideals. We divide the proof in three cases.

Let the socle of R be zero. Then every ideal of R is essential. Let I and K be two nonzero ideals. Since $R/(I \cap K)$ is uniform and $\frac{I}{I\cap K} \cap \frac{K}{I\cap K} = 0$, it follows that $I \subseteq K$ or $K \subseteq I$. Thus R is a valuation ring.

If the socle or R consists of a single minimal ideal, then all ideals are essential and the proof comes from the above case.

Finally, let $soc(R) = A \oplus B$ where A and B are minimal ideals. Let M be the unique maximal ideal. Then $MSoc(R) = 0$. In particular, for any $x \in M$ we have that $ann(x) \neq 0$ and $Rx \cong R/ann(x)$ is uniform. This means that Rx cannot contain the $Soc(R) = A \oplus B$ and hence Rx cannot be essential. So Rx must be minimal. This shows that $M \subseteq Soc(R)$ and so $M = Soc(R)$. Clearly, $M^2 = 0$, as desired.

Furthermore every proper homomorphic image of R is clearly a valuation ring, completing the proof. □

We now prove the main Theorem 2 as stated in the introduction.

PROOF. We denote the prime radical of R by N.

Suppose first that $N \neq 0$. Then all homomorphic images of R/N are almost self-injective. Invoking the result (8.2, p.66, [**DHSW**]) , we get $R/N = \oplus_{i=1}^{l} e_i R/e_i N$ where $\{e_1 + N,...,e_k + N\}$ is an orthogonal family of idempotents such that $1 + N = (e_1 + N) + ... + (e_k + N)$. Since N is nilpotent, without loss of generality we can assume that e_i are idempotents in R (cf. Theorem 21.28 p.319 [**l**]), and thus the decomposition of R/N can be lifted to $R = \oplus_{i=1}^{l} e_i R$ say.

If $l > 1$, then each R_i is uniform almost self-injective and hence valuation. The conclusion then follows by the . Lemma 6. Now suppose $l = 1$. Then R/N is semiprime uniform almost self-injective and hence a valuation domain by Lemma 4. In particular R/N is local, this implies R is local. The conclusion follows by the above lemma. This completes the proof in the case when $N \neq 0$

Suppose now that $N = 0$. Then R is a semiprime noetherian ring. We prove the result by induction on the uniform dimension of R. If $u.\dim = 1$, then R is domain and hence a Dedekind domain by the Lemma 4

Let us suppose that the result holds for commutative noetherian rings with $udim(R) < n$, for some $n > 1$ We consider two cases. First suppose, there exists a non-essential maximal ideal M. Then $R = M \oplus K$ for some nonzero ideal K. Then by the chinese remainder theorem $R \cong R/M \times R/K$, where R/M is a field. Since $udim(R/K) < udim(R)$, the induction hypothesis gives the desired conclusion.

If on the other hand every maximal ideal is essential, then according to Proposition 5 the minimal prime ideals are also essential. Since there are only a finite number of minimal prime ideals the nilpotent radical N is essential, a contradiction because R is semiprime. This proves the "if" part of the theorem..

"only if part". Let us now show that if a ring belongs to one of the three families given in the statement of the main theorem then every proper homomorphic image is almost self injective. If R is a Dedekind domain then it is well known that every proper image of R is self-injective. For the remaining two types, refer to 6 and Lemma 3. □

We may compare our result with the one obtained by L. Levy in [**L**]. Firstly, we remark that a principal ideal artinian ring is also a valuation ring. The example given below shows that there exists a family of rings of a type obtained by us which does not fall in the class of rings obtained by Levy.

EXAMPLE 7. Let k be a field and consider the product $R = k[[X]] \times k[[Y]]$. This ring is not a domain. It is neither a local ring nor artinian but it is the product of two valuation domains and hence satisfies our property. So this ring doesn't belong to the family obtained by Levy but belongs to the third family we obtained. In other words this ring is restricted almost self-injective but not restricted self-injective.

References

[AJ] A. Alahmadi and S. K. Jain, *A note on almost injective modules*, Math. J. Okayama Univ. **51** (2009), 101–109. MR2482408

[B] Y. Baba, *Note on almost M-injectives*, Osaka J. Math. **26** (1989), no. 3, 687–698. MR1021440

[C] I. S. Cohen, *Commutative rings with restricted minimum condition*, Duke Math. J. **17** (1950), 27–42. MR0033276

[DHSW] N. V. Dung, D. V. Huynh, P. F. Smith, and R. Wisbauer, *Extending modules*, Pitman Research Notes in Mathematics Series, vol. 313, Longman Scientific & Technical, Harlow; copublished in the United States with John Wiley & Sons, Inc., New York, 1994. With the collaboration of John Clark and N. Vanaja. MR1312366

[HT] M. Harada and A. Tozaki, *Almost M-projectives and Nakayama rings*, J. Algebra **122** (1989), no. 2, 447–474, DOI 10.1016/0021-8693(89)90229-9. MR999086

[L] L. S. Levy, *Commutative rings whose homomorphic images are self-injective*, Pacific J. Math. **18** (1966), 149–153. MR0194453

[l] T. Y. Lam, *A first course in noncommutative rings*, Graduate Texts in Mathematics, vol. 131, Springer-Verlag, New York, 1991. MR1125071

[S] S. Singh, *Almost relative injective modules*, Osaka J. Math. **53** (2016), no. 2, 425–438. MR3492807

DEPARTMENT OF MATHEMATICS, KING ABDULAZIZ UNIVERSITY, JEDDAH 21589, SAUDI ARABIA

Email address: adelnife2@yahoo.com

DEPARTMENT OF MATHEMATICS, UNIVERSITÉ D'ARTOIS, RUE JEAN SOUVRAZ, 62300 LENS, FRANCE

Email address: andre.leroy@univ-artois.fr

DEPARTMENT OF MATHEMATICS, KING ABDULAZIZ UNIVERSITY, JEDDAH 21589, SAUDI ARABIA; AND OHIO UNIVERSITY, ATHENS, OHIO 45701

Email address: jain@ohio.edu

Contemporary Mathematics
Volume **727**, 2019
https://doi.org/10.1090/conm/727/14621

Enveloping skewfields of the nilpotent positive part and the Borel subsuperalgebra of $\mathfrak{osp}(1, 2n)$

Jacques Alev and François Dumas

ABSTRACT. We study an analogue of the Gelfand-Kirillov property for some Lie superalgebras. More precisely, we consider in the classical simple orthosymplectic Lie superalgebra $\mathfrak{osp}(1, 2n)$ of dimension $n^2 + 3n$ the positive nilpotent subsuperalgebra $\mathfrak{n}^+$ of dimension $n^2 + n$ and the solvable Borel subsuperalgebra $\mathfrak{b}^+$ of dimension $n^2 + 2n$. We prove that the enveloping algebras $\mathcal{U}(\mathfrak{n}^+)$ and $\mathcal{U}(\mathfrak{b}^+)$ are rationally equivalent to some super-analogues of Weyl algebras over polynomial superalgebras.

Introduction

In the classical context of the seminal paper [**7**], an algebraic finite-dimensional complex Lie algebra $\mathfrak{g}$ is said to satisfy the Gelfand-Kirillov property when its enveloping algebra $\mathcal{U}(\mathfrak{g})$ is rationaly equivalent to a Weyl algebra $A_n(K)$ over a commutative purely transcendental field extension K of $\mathbb{C}$. It was proved that this property is satisfied when $\mathfrak{g}$ is nilpotent ([**7**]), or more generally solvable ([**5**], [**10**], [**12**]). This profound problem gave rise to an abundant literature with many developments within Lie theory itself (see references in [**3**] or [**16**]) or for variants concerning quantum groups, invariant algebras or other classes of interesting noncommutatives algebras.

The study of an analogue of the Gelfand-Kirillov property for Lie superalgebras was explored in [**13**] and [**2**]. In this context, it is relevant to introduce for any integer $n \geq 1$ the noncommutative polynomial superalgebra $\widetilde{\mathcal{O}}(\mathbb{C}^n)$ and an analogue $\widetilde{A}_n(\mathbb{C})$ of the Weyl algebra. By definition $\widetilde{\mathcal{O}}(\mathbb{C}^n)$ is the algebra generated over $\mathbb{C}$ by n indeterminates $y_1, \ldots, y_n$ with relations $y_iy_j + y_jy_i = 0$ for all $1 \leq i \neq j \leq n$. Then $\widetilde{A}_n(\mathbb{C})$ is the algebra generated over $\widetilde{\mathcal{O}}(\mathbb{C}^n)$ by n other indeterminates $x_1, \ldots, x_n$ with relations $x_iy_j + y_jx_i = x_ix_j + x_jx_i = 0$ for all $1 \leq i \neq j \leq n$ and $x_iy_i - y_ix_i = 1$ for any $1 \leq i \leq n$. These algebras are noetherian domains and then they admit a skewfield of fractions. Otherwise the only case in the classification of classical simple finite dimensional complex Lie superalgebras $\mathfrak{g}$ where the enveloping algebra $\mathcal{U}(\mathfrak{g})$ is a domain is when $\mathfrak{g}$ is an orthosymplectic Lie superalgebra $\mathfrak{osp}(1, 2n)$ (see [**4**], [**9**]). This is the case we consider here.

2010 *Mathematics Subject Classification.* Primary 17B35; Secondary 16S30, 16S85, 16K40.
Key words and phrases. Simple Lie superalgebra, enveloping algebra, Gelfand-Kirillov hypothesis, Weyl algebra, skewfield.

For any integer $n \geq 1$, the superalgebra $\mathfrak{g} = \mathfrak{osp}(1, 2n)$ is of dimension $2n^2 + 3n$. In its $\mathbb{Z}_2$-graded decomposition $\mathfrak{g} = \mathfrak{g}_{\overline{0}} \oplus \mathfrak{g}_{\overline{1}}$, the even part $\mathfrak{g}_{\overline{0}}$ is isomorphic to the symplectic Lie algebra $\mathfrak{sp}(2n)$ of dimension $2n^2 + n$, and the odd part $\mathfrak{g}_{\overline{1}}$ is a vector space of dimension $2n$. There is a natural definition of the nilpotent positive part $\mathfrak{n}^+ = \mathfrak{osp}^+(1, 2n)$, which is a nilpotent subsuperalgebra of $\mathfrak{g}$ of dimension $n^2 + n$. In its $\mathbb{Z}_2$-graded decomposition $\mathfrak{n}^+ = \mathfrak{n}^+_{\overline{0}} \oplus \mathfrak{n}^+_{\overline{1}}$, the even part $\mathfrak{n}^+_{\overline{0}} \simeq \mathfrak{sp}^+(2n)$ is just the nilpotent positive part of dimension n^2 in the ordinary triangular decomposition of the Lie algebra $\mathfrak{g}_{\overline{0}} \simeq \mathfrak{sp}(2n)$, and the odd part $\mathfrak{n}^+_{\overline{1}}$ is a subspace of dimension n in $\mathfrak{g}_{\overline{1}}$. Denoting by $\mathfrak{h}$ the Cartan Lie subalgebra of $\mathfrak{g}_{\overline{0}}$, the Borel subsuperalgebra $\mathfrak{b}^+$ of $\mathfrak{g}$ is defined as the subsuperalgebra generated by $\mathfrak{n}^+$ and $\mathfrak{h}$. We have $\mathfrak{b}^+ = \mathfrak{n}^+ \oplus \mathfrak{h} = \mathfrak{b}^+_{\overline{0}} \oplus \mathfrak{n}^+_{\overline{1}}$, where the even part $\mathfrak{b}^+_{\overline{0}} = \mathfrak{n}^+_{\overline{0}} \oplus \mathfrak{h}$ is the positive Borel Lie subalgebra of $\mathfrak{g}_{\overline{0}} \simeq \mathfrak{sp}(2n)$. In particular, $\mathfrak{b}^+_{\overline{0}}$ and $\mathfrak{b}^+$ are of dimensions $n^2 + n$ and $n^2 + 2n$ respectively. The enveloping algebras $\mathcal{U}(\mathfrak{n}^+)$ and $\mathcal{U}(\mathfrak{b}^+)$ can be described as iterated skew polynomial algebras; then they are noetherian domains and therefore we can consider their skewfields of fractions. We prove in the paper:

MAIN THEOREM. *For any $n \geq 2$, the enveloping algebras of the subsuperalgebras $\mathfrak{n}^+$ and $\mathfrak{b}^+$ of $\mathfrak{osp}(1, 2n)$ satisfy the following isomorphisms:*

(i) $\operatorname{Frac}\mathcal{U}(\mathfrak{n}^+) \simeq \operatorname{Frac}\left(A_{n(n-1)/2}(\mathbb{C}) \otimes \widetilde{\mathcal{O}}(\mathbb{C}^n)\right)$, *and its center is a purely transcendental extension of dimension n over $\mathbb{C}$.*

(ii) $\operatorname{Frac}\mathcal{U}(\mathfrak{b}^+) \simeq \operatorname{Frac}\left(A_{n(n-1)/2}(\mathbb{C}) \otimes \widetilde{A}_n(\mathbb{C})\right)$, *and its center is $\mathbb{C}$.*

The method we use to prove this theorem allows to recover as a corollary the classical Gelfand-Kirillov property for the even parts, with already known descriptions of $\operatorname{Frac}\mathcal{U}(\mathfrak{n}^+_{\overline{0}})$ as a Weyl skewfield $D_{n(n-1)/2}(K)$ over a commutative field K which is a purely transcendental extension of dimension n over $\mathbb{C}$, and of $\operatorname{Frac}\mathcal{U}(\mathfrak{b}^+_{\overline{0}})$ as a Weyl skewfield $D_{n(n+1)/2}(\mathbb{C})$ over $\mathbb{C}$. The particular case $n = 2$ of the main theorem was previously proved in Proposition 3.2 and Theorem 3.4 of [**2**]. The natural and probably difficult question of an extension of the main theorem to the enveloping algebra of the Lie superalgebra $\mathfrak{osp}(1, 2n)$ itself must be appreciated by recalling that even for the even part the answer is unknown, the case of $\mathfrak{sp}(2n)$ being one of the situations where the original Gelfand-Kirillov conjecture remains open (see [**16**]).

The paper is in three parts. The first one introduces the analogues of Weyl algebras for polynomial superalgebras and some canonical form of their skewfields of fractions. The second part describes the enveloping algebras $\mathcal{U}(\mathfrak{n}^+)$ and $\mathcal{U}(\mathfrak{b}^+)$ as iterated skew polynomial algebras. The last section contains the proof of the main theorem, based on suitable rational changes of generators reducing step by step the commutation relations.

1. Analogues of Weyl algebras for polynomial superalgebras

1.1. Polynomial superalgebras.

1.1.1. *Notations.* It is usual in quantum group theory to denote by $\mathcal{O}_\Lambda(\mathbb{C}^n)$ the algebra of polynomials in n indeterminates $y_1, \ldots, y_n$ with coefficients in $\mathbb{C}$ and noncommutative multiplication twisted by relations $y_i y_j = \lambda_{ij} y_j y_i$ for any $1 \leq i, j \leq n$, where $\Lambda = (\lambda_{ij})$ is a $n \times n$ multiplicatively skew-symmetric matrix with entries in $\mathbb{C}$. In the particular cases where all $\lambda_{ij} = 1$, or $\lambda_{ij} = -1$ for all $i \neq j$, we use respectively the two following notations:

(i) $\mathcal{O}(\mathbb{C}^n)$ is the commutative algebra of polynomials in n indeterminates $y_1, \ldots, y_n$ with coefficients in $\mathbb{C}$.

(ii) $\widetilde{\mathcal{O}}(\mathbb{C}^n)$ is the noncommutative algebra of polynomials in n indeterminates $y_1, \ldots, y_n$ with coefficients in $\mathbb{C}$ and product twisted by relations $y_i y_j = -y_j y_i$ for any $1 \leq i \neq j \leq n$.

1.1.2. *Superalgebra structure.* The monomials ${y_1}^{\beta_1} \cdots {y_n}^{\beta_n}$ whose total degree $\beta_1 + \cdots + \beta_n$ is even generate a commutative subalgebra R_0 of $\widetilde{\mathcal{O}}(\mathbb{C}^n)$. Denoting by R_1 the R_0-submodule of $\widetilde{\mathcal{O}}(\mathbb{C}^n)$ generated by $y_1, \ldots, y_n$, the $\mathbb{C}$-submodules R_0 and R_1 satisfy $\widetilde{\mathcal{O}}(\mathbb{C}^n) = R_0 \oplus R_1$ and $R_i R_j \subseteq R_{i+j}$ (with indices taken modulo 2). This gives rise to a structure of superalgebra on $\widetilde{\mathcal{O}}(\mathbb{C}^n)$. The algebras $\widetilde{\mathcal{O}}(\mathbb{C}^n)$ being noetherian domains, they admit skew fields of fractions. The following proposition gives an alternative form of $\widetilde{\mathcal{O}}(\mathbb{C}^n)$ in terms of planes $\widetilde{\mathcal{O}}(\mathbb{C}^2)$ up to rational equivalence.

PROPOSITION 1.1.1. *For any integer $n \geq 2$, we have:*

If $n = 2p$, then $\mathrm{Frac}\,(\widetilde{\mathcal{O}}(\mathbb{C}^n)) \simeq \mathrm{Frac}\,(\widetilde{\mathcal{O}}(\mathbb{C}^2)^{\otimes p})$

If $n = 2p+1$, then $\mathrm{Frac}\,(\widetilde{\mathcal{O}}(\mathbb{C}^n)) \simeq \mathrm{Frac}\,(\widetilde{\mathcal{O}}(\mathbb{C}^2)^{\otimes p} \otimes \mathcal{O}(\mathbb{C}))$

PROOF. Suppose that $n \geq 3$ and define from generators $y_1, \ldots, y_n$ of $\widetilde{\mathcal{O}}(\mathbb{C}^n)$ with $y_i y_j = -y_j y_i$ for all $1 \leq i \neq j \leq n$ the monomials: $y_1' = y_1$, $y_2' = y_2$ and $y_i' = y_1 y_2 y_i$ for any $3 \leq i \leq n$. They satisfy $y_1' y_2' = -y_2' y_1'$ and

$$y_1' y_i' = y_i' y_1', \quad y_2' y_i' = y_i' y_2', \quad y_i' y_j' = -y_j' y_i' \quad \text{for all } 3 \leq i \neq j \leq n.$$

It is clear that the subfield of $\mathrm{Frac}\,\widetilde{\mathcal{O}}(\mathbb{C}^n)$ generated by $y_1', y_2', \ldots, y_n'$ is equal to $\mathrm{Frac}\,\widetilde{\mathcal{O}}(\mathbb{C}^n)$. We deduce that $\mathrm{Frac}\,\widetilde{\mathcal{O}}(\mathbb{C}^n) \simeq \mathrm{Frac}\,(\widetilde{\mathcal{O}}(\mathbb{C}^2) \otimes \widetilde{\mathcal{O}}(\mathbb{C}^{n-2}))$, and the result follows by iteration. □

1.2. Weyl algebras for polynomial superalgebras.

1.2.1. *Notations.* The definition of quantum differential calculi on various quantum algebras with adapted de Rham complexes gives rise to suitable versions of quantum Weyl algebras. In the case of quantum spaces $\mathcal{O}_\Lambda(\mathbb{C}^n)$, these algebras $A_n^{\overline{q},\Lambda}$ (introduced in [**1**] and [**19**] and studied in many papers) take in consideration both the quantization parameters $\Lambda = (\lambda_{ij})$ of the space and the quantization parameters $\overline{q} = (q_1, \ldots, q_n)$ of the "differential" operators on this space. We are interested here in two particular cases:

(i) For $\lambda_{ij} = q_i = 1$ for all $1 \leq i, j \leq n$, we find the usual Weyl algebra $A_n(\mathbb{C})$, that is the algebra of polynomials in $2n$ indeterminates $x_1, \ldots, x_n, y_1, \ldots, y_n$ with coefficients in $\mathbb{C}$ and a noncommutative product twisted by relations:

$$(1.1) \qquad \begin{cases} x_i x_j - x_j x_i = y_i y_j - y_j y_i = x_i y_j - y_j x_i = 0 & (1 \leq i \neq j \leq n), \\ x_i y_i - y_i x_i = 1 & (1 \leq i \leq n). \end{cases}$$

We also introduce the subalgebra $U_n(\mathbb{C})$ of $A_n(\mathbb{C})$ generated by $y_1, \ldots, y_n$ and $w_1, \ldots, w_n$ with $w_i = y_i x_i$ for any $1 \leq i \leq n$. We have:

$$(1.2) \qquad \begin{cases} w_i w_j - w_j w_i = y_i y_j - y_j y_i = w_i y_j - y_j w_i = 0 & (1 \leq i \neq j \leq n), \\ w_i y_i - y_i w_i = y_i & (1 \leq i \leq n). \end{cases}$$

(ii) For $q_i = 1$ and $\lambda_{ij} = -1$ for all $1 \leq i \neq j \leq n$, we denote by $\widetilde{A}_n(\mathbb{C})$ the algebra of polynomials in $2n$ indeterminates $x_1, \ldots, x_n, y_1, \ldots, y_n$ with coefficients in $\mathbb{C}$ and a noncommutative product twisted by:

$$(1.3) \qquad \begin{cases} x_i x_j + x_j x_i = y_i y_j + y_j y_i = x_i y_j + y_j x_i = 0 & (1 \leq i \neq j \leq n), \\ x_i y_i - y_i x_i = 1 & (1 \leq i \leq n). \end{cases}$$

We denote by $\widetilde{U}_n(\mathbb{C})$ the subalgebra of $\widetilde{A}_n(\mathbb{C})$ generated by the elements $y_1, \ldots, y_n$ and $w_1, \ldots, w_n$ with $w_i = y_i x_i$ for any $1 \leq i \leq n$. We have:

$$(1.4) \qquad \begin{cases} y_i y_j + y_j y_i = 0 & (1 \leq i \neq j \leq n), \\ w_i w_j - w_j w_i = w_i y_j - y_j w_i = 0 & (1 \leq i \neq j \leq n), \\ w_i y_i - y_i w_i = y_i & (1 \leq i \leq n). \end{cases}$$

1.2.2. *Remarks.* The algebras $\widetilde{A}_n(\mathbb{C})$ are the algebras denoted by $S^{\Lambda}_{n,n}$ and studied in [**17**] and [**18**] when all non diagonal entries of Λ are equal to -1. Section 11 of [**19**] gives a slightly different presentation of these algebras as "generalized enveloping algebras"; with the notations of this article, case (i) corresponds to the values $q_{ij} = p_i = q = 1$, and case (ii) is for $q_{ij} = -1$ and $p_i = q = 1$. They also appear in example 2.1 of [**8**] (taking $q = 1$ and $p_{ij} = -1$). We refer to paragraph 1.3.3 of [**17**] for more ringtheoretical references and a survey, and only mention here that, for any integer $n \geq 1$, the algebra $\widetilde{A}_n(\mathbb{C})$ is simple, has center $\mathbb{C}$, and has the same Hochschild homology and cohomology as the usual Weyl algebra $A_n(\mathbb{C})$.

1.2.3. *Superalgebra structure.*

(i) Let $\mathfrak{g}_0$ and $\mathfrak{g}_1$ be two $\mathbb{C}$-vector spaces of dimension n with respective basis $\{z_1, \ldots, z_n\}$ and $\{y_1, \ldots, y_n\}$. Let $\mathfrak{k}_0$ be a linear extension of $\mathfrak{g}_0$ of dimension $2n$ with basis $\{z_1, \ldots, z_n, w_1, \ldots, w_n\}$. Then we define a Lie superalgebra structure $\mathfrak{g} = \mathfrak{g}_0 \oplus \mathfrak{g}_1$ of dimension $2n$ and a Lie superalgebra structure $\mathfrak{k} = \mathfrak{k}_0 \oplus \mathfrak{g}_1$ of dimension $3n$ by setting for the values of the super-bracket on these generators:

$$\begin{cases} \{y_i, y_i\} = z_i, \ [w_i, y_i] = y_i, \ [w_i, z_i] = 2z_i \ \text{ for any } 1 \leq i \leq n, \\ \text{the brackets are zero for all other pairs of generators.} \end{cases}$$

By construction $\mathfrak{g}_0$ is a subsuperalgebra of $\mathfrak{k}_0$ and $\mathfrak{g} = \mathfrak{g}_0 \oplus \mathfrak{g}_1$ is a subsuperalgebra of $\mathfrak{k}$. It is clear by PBW theorem (see Theorem 6.1.2 in [**14**]) that the enveloping algebra $\mathcal{U}(\mathfrak{k})$ is isomorphic to $\widetilde{U}_n(\mathbb{C})$ of GK-dim $2n$. Moreover the subalgebra $\mathcal{U}(\mathfrak{k}_0)$ is isomorphic to $U_n(\mathbb{C})$ of GK-dim $2n$, and the subalgebra $\mathcal{U}(\mathfrak{g})$ is isomorphic to $\widetilde{\mathcal{O}}(\mathbb{C}^n)$ of GK-dim n.

(ii) We have for the usual Weyl algebra the classical isomorphism $A_n(\mathbb{C}) \simeq A_1(\mathbb{C})^{\otimes n}$. We also have $\widetilde{A}_n(\mathbb{C}) \simeq \widetilde{A}_1(\mathbb{C})^{\widehat{\otimes} n}$ in the tensor category of super-algebras assigning the parity 1 to the generators x_i and y_i.

(iii) All isomorphims and tensor products considered in the following are in the tensor category of associative algebras.

1.2.4. *Remark.* All algebras introduced in 1.2.1 can be described as iterated skew polynomial algebras over $\mathbb{C}$; then they are noetherian domains, and therefore they admit a skewfield of fractions. The skewfield $\operatorname{Frac} A_n(\mathbb{C}) = \operatorname{Frac} U_n(\mathbb{C})$ is the well known Weyl skewfield $D_n(\mathbb{C})$, with center $\mathbb{C}$. Similarly we deduce from (1.4) that $\operatorname{Frac} \widetilde{A}_n(\mathbb{C}) = \operatorname{Frac} \widetilde{U}_n(\mathbb{C})$, and it is proved in Proposition 3.3.1 of [**18**] that its center is also $\mathbb{C}$. In parallel with Proposition 1.1.1, we have for these skewfields the following canonical form.

THEOREM 1.2.1. *For any integer $n \geq 1$, we have the following isomorphisms:*

(i) *If $n = 2p$ is even, then* $\operatorname{Frac} \widetilde{A}_n(\mathbb{C}) = \operatorname{Frac}(\widetilde{A}_2(\mathbb{C})^{\otimes p})$.

(ii) *If $n = 2p+1$ is odd, then* $\operatorname{Frac} \widetilde{A}_n(\mathbb{C}) = \operatorname{Frac}(\widetilde{A}_2(\mathbb{C})^{\otimes p} \otimes A_1(\mathbb{C}))$.

PROOF. We can suppose $n \geq 1$. We consider the subalgebra $\widetilde{U}_n(\mathbb{C})$ defined in 1.2.1 with generators $w_1, \ldots, w_n, y_1, \ldots y_n$ and relations (1.4). As in the proof of Proposition 1.1.1, we introduce: $y_1' = y_1$, $y_2' = y_2$ and $y_i' = y_1 y_2 y_i$ for $3 \leq i \leq n$. Then we define: $w_1' = w_1 - (w_3 + \cdots + w_n)$, $w_2' = w_2 - (w_3 + \cdots + w_n)$ and $w_i' = w_i$ for $3 \leq i \leq n$. Obvious calculations give:

$$(1.5) \qquad \begin{cases} y_1'y_2' = -y_2'y_1', \quad w_1'w_2' = w_2'w_1', \quad y_1'w_2' = w_2'y_1', \quad y_2'w_1' = w_1'y_2', \\ [w_1', y_1'] = y_1', \quad [w_2', y_2'] = y_2'. \end{cases}$$

$$(1.6) \qquad \begin{cases} y_1'y_j' = y_j'y_1', \quad y_2'y_j' = y_j'y_2', \quad y_1'w_j' = w_j'y_1', \\ y_2'w_j' = w_j'y_2', \quad w_1'w_j' = w_j'w_1', \quad w_2'w_j' = w_j'w_2', \end{cases} \quad \text{for } 3 \leq j \leq n.$$

We calculate for $\ell = 1$ or 2 and any $3 \leq j \leq n$ the commutator: $[w_\ell', y_j'] = [w_\ell, y_1y_2y_j] - \sum_{i=3}^{n} [w_i, y_1y_2y_j] = y_1y_2y_j - [w_j, y_1y_2y_j] = 0$ to conclude :

$$(1.7) \qquad y_j'w_1' = w_1'y_j' \text{ and } y_j'w_2' = w_2'y_j' \text{ for any } 3 \leq j \leq n.$$

Moreover:

$$(1.8) \qquad [w_j', y_j'] = y_j' \quad \text{for any } 3 \leq j \leq n,$$

$$(1.9) \qquad y_i'y_j' = -y_j'y_i', \quad w_i'w_j' = w_j'w_i', \quad w_i'y_j' = y_j'w_i', \quad \text{for } 3 \leq i \neq j \leq n.$$

We denote by V the subalgebra of $\widetilde{U}_n(\mathbb{C})$ generated by $y_1', \ldots, y_n', w_1', \ldots, w_n'$, by W the subalgebra of V generated by $y_1', , y_2', w_1', w_2'$, and by W' the subalgebra of V generated by $y_3', \ldots, y_n', w_3', \ldots, w_n'$. It follows from (1.5) that $W' \simeq \widetilde{U}_2(\mathbb{C})$, from (1.8) and (1.9) that $W'' \simeq \widetilde{U}_{n-2}(\mathbb{C})$, and from (1.7) and (1.9) that $V \simeq W \otimes W'$. Since it is clear that $\operatorname{Frac} V = \operatorname{Frac} \widetilde{U}_n(\mathbb{C})$, we deduce that $\operatorname{Frac} \widetilde{U}_n(\mathbb{C}) \simeq \operatorname{Frac}(\widetilde{U}_2(\mathbb{C}) \otimes \widetilde{U}_{n-2}(\mathbb{C}))$. We conclude that $\operatorname{Frac} \widetilde{A}_n(\mathbb{C}) \simeq \operatorname{Frac}(\widetilde{A}_2(\mathbb{C}) \otimes \widetilde{A}_{n-2}(\mathbb{C}))$ and finish by induction on n. □

2. The Lie superalgebra $\mathfrak{osp}(1, 2n)$ and its enveloping algebra

2.1. Generators and relations. We refer for instance to [**6**], [**15**], [**2**]. The basefield is $\mathbb{C}$. For any $n \geq 1$, the odd part $\mathfrak{g}_{\bar{1}}$ of $\mathfrak{g} = \mathfrak{osp}(1, 2n)$ is a vector space of dimension $2n$ and the even part $\mathfrak{g}_{\bar{0}}$ is isomorphic to the symplectic Lie algebra $\mathfrak{sp}(2n)$ of dimension $2n^2+n$. As a Lie superalgebra, $\mathfrak{osp}(1, 2n)$ is generated by the elements $b_i^{\pm}$ $(1 \leq i \leq n)$ of a basis of $\mathfrak{g}_{\bar{1}}$. Then the elements $\{b_j^{\pm}, b_k^{\pm}\}$ $(1 \leq j \leq k \leq n)$ and $\{b_j^+, b_k^-\}$ $(1 \leq j, k \leq n)$ form a basis of $\mathfrak{g}_{\bar{0}}$. The brackets in $\mathfrak{osp}(1, 2n)$ are given by the "para-Bose" relations on the generators:

$$(2.1) \qquad [\{b_j^\xi, b_k^\eta\}, b_\ell^\epsilon] = (\epsilon - \xi)\delta_{j\ell} b_k^\eta + (\epsilon - \eta)\delta_{k\ell} b_j^\xi$$

$$(2.2) \quad [\{b_i^\xi, b_j^\eta\}, \{b_k^\epsilon, b_\ell^\varphi\}] = (\epsilon - \eta)\delta_{jk}\{b_i^\xi, b_\ell^\varphi\} + (\epsilon - \xi)\delta_{ik}\{b_j^\eta, b_\ell^\varphi\} \\ + (\varphi - \eta)\delta_{j\ell}\{b_i^\xi, b_k^\epsilon\} + (\varphi - \xi)\delta_{i\ell}\{b_j^\eta, b_k^\epsilon\},$$

with $1 \le i, j, k, \ell \le n$ and $\epsilon, \varphi, \xi, \eta$ are $\pm$. By PBW theorem, the enveloping algebra $\mathcal{U}(\mathfrak{osp}(1,2n))$ is generated by the $2n^2+n$ elements:

$$b_i^{\pm} \text{ and } k_i := \tfrac{1}{2}\{b_i^-, b_i^+\} \text{ for } 1 \le i \le n, \tag{2.3}$$

$$a_{ij}^{\pm} := \tfrac{1}{2}\{b_i^{\pm}, b_j^{\pm}\} \text{ for } 1 \le i < j \le n, \tag{2.4}$$

$$s_{ij} := \tfrac{1}{2}\{b_i^-, b_j^+\} \text{ and } t_{ij} := \tfrac{1}{2}\{b_i^+, b_j^-\} \text{ for } 1 \le i < j \le n, \tag{2.5}$$

of $\mathfrak{osp}(1,2n)$. In particular, denoting

$$c_i^{\pm} := \tfrac{1}{2}\{b_i^{\pm}, b_i^{\pm}\} \text{ for } 1 \le i \le n, \tag{2.6}$$

we obtain the enveloping algebra $\mathcal{U}(\mathfrak{sp}(2n))$ of the even part $\mathfrak{g}_{\bar{0}}$ as the subalgebra of $\mathcal{U}(\mathfrak{osp}(1,2n))$ generated by the $2n^2+n$ elements $(b_i^{\pm})^2 = c_i^{\pm}, k_i$ for $1 \le i \le n$, and $a_{ij}^{\pm}, s_{ij}, t_{ij}$ for $1 \le i < j \le n$. The commutation relations in the associative algebra $\mathcal{U}(\mathfrak{osp}(1,2n))$ are deduced from (2.1) and (2.2) taking $\{x,y\} = xy+yx$ if $x, y \in \mathfrak{g}_{\bar{1}}$, and $[x,y] = xy - yx$ otherwise. In particular commutation relations between the generators defined by (2.3), (2.4), (2.5), (2.6) are obtained by injecting them in relations (2.1) and (2.2). For instance:

$$b_i^+ b_j^+ + b_j^+ b_i^+ = 2a_{ij}^+ \qquad \text{for all } 1 \le i < j \le n, \tag{2.7}$$

$$[c_i^+, b_j^+] = [c_i^+, c_j^+] = 0 \qquad \text{for all } 1 \le i < j \le n, \tag{2.8}$$

$$[a_{ij}^+, b_\ell^+] = [a_{ij}^+, c_\ell^+] = 0 \qquad \text{for all } 1 \le i < j \le n,\ 1 \le \ell \le n, \tag{2.9}$$

$$[a_{ij}^+, a_{k\ell}^+] = 0 \qquad \text{for all } 1 \le i < j \le n,\ 1 \le k < \ell \le n. \tag{2.10}$$

Similarly with (2.1) and (2.5), we deduce for all $1 \le i < j \le n$ and $1 \le k \le n$:

$$[t_{ij}, b_j^+] = b_i^+ \text{ and } [t_{ij}, b_k^+] = 0 \text{ if } k \ne j. \tag{2.11}$$

The relations involving the t_{ij}'s are obtained from (2.2), (2.4) and (2.5), which give for $1 \le i < j \le n$ and $1 \le k < \ell \le n$:

$$[t_{ij}, a_{k\ell}^+] = \delta_{jk}\tfrac{1}{2}\{b_i^+, b_\ell^+\} + \delta_{j\ell}\tfrac{1}{2}\{b_i^+, b_k^+\}, \tag{2.12}$$

$$[t_{ij}, t_{k\ell}] = \delta_{jk}\,\tfrac{1}{2}\{b_i^+, b_\ell^-\} - \delta_{i\ell}\,\tfrac{1}{2}\{b_k^+, b_j^-\}, \tag{2.13}$$

or in other words:

$$[t_{ij}, a_{k\ell}^+] = 0 \text{ if } j \ne k, j \ne \ell \qquad (1 \le i < j \le n,\ 1 \le k < \ell \le n), \tag{2.14}$$

$$[t_{ij}, a_{j\ell}^+] = a_{i\ell}^+ \qquad (1 \le i < j < \ell \le n), \tag{2.15}$$

$$[t_{ij}, a_{ij}^+] = c_i^+ \qquad (1 \le i < j \le n), \tag{2.16}$$

$$[t_{ij}, a_{kj}^+] = a_{ik}^+ \text{ if } i < k \qquad (1 \le i < j \le n,\ 1 \le k < j \le n,), \tag{2.17}$$

$$[t_{ij}, a_{kj}^+] = a_{ki}^+ \text{ if } k < i \qquad (1 \le i < j \le n,\ 1 \le k < j \le n,). \tag{2.18}$$

$$[t_{ij}, t_{j\ell}] = t_{i\ell} \qquad (1 \le i < j < \ell \le n), \tag{2.19}$$

$$[t_{ij}, t_{ki}] = -t_{kj} \qquad (1 \le k < i < j \le n), \tag{2.20}$$

$$[t_{ij}, t_{k\ell}] = 0 \text{ if } j \ne k, i \ne \ell \qquad (1 \le i < j \le n,\ 1 \le k < \ell \le n). \tag{2.21}$$

The action of the generators k_i follows from (2.1) and (2.3):

$$[k_i, k_j] = 0 \qquad (1 \le i, j \le n), \tag{2.22}$$

$$[k_i, b_i^+] = b_i^+,\ [k_i, b_j^{\pm}] = 0 \qquad (1 \le i \ne j \le n), \tag{2.23}$$

then with (2.4) and (2.5), for all $1 \leq i < j \leq n$ and $1 \leq \ell \leq n$:

$$\begin{cases} [k_i, a_{ij}^+] = a_{ij}^+, \\ [k_j, a_{ij}^+] = a_{ij}^+, \\ [k_\ell, a_{ij}^+] = 0 \text{ if } \ell \neq i, j. \end{cases} \qquad \begin{cases} [k_i, t_{ij}] = t_{ij}, \\ [k_j, t_{ij}] = -t_{ij}, \\ [k_\ell, t_{ij}] = 0 \text{ if } \ell \neq i, j. \end{cases} \tag{2.24}$$

2.2. Chevalley generators and Serre relations. Following [**15**] (up to a renormalization of the n-th generator by $\sqrt{2}$), we introduce:

$$t_i := t_{i,i+1} \text{ for any } 1 \leq i \leq n-1, \text{ and } t_n := b_n^+, \tag{2.25}$$

which satisfy:

$$[t_i, t_j] = 0 \text{ if } |i-j| > 1 \qquad (1 \leq i, j \leq n), \tag{2.26}$$

$$[t_i, t_{i+1}] = t_{i,i+2} \qquad (1 \leq i \leq n-2), \tag{2.27}$$

$$[t_{n-1}, t_n] = b_{n-1}^+. \tag{2.28}$$

It also follows from (2.11) that for any $1 \leq i \leq n-1$ and $1 \leq j \leq n$:

$$[t_i, b_{i+1}^+] = b_i^+ \text{ and } [t_i, b_j^+] = 0 \text{ if } j \neq i+1. \tag{2.29}$$

By straightforward calculations using (2.20), (2.26), (2.27), (2.28), the elements $t_1, \dots, t_n$ defined in (2.25) satisfy the following Serre identities which already appear as (2.11) in [**15**]:

$$t_i^2 t_{i+1} - 2t_i t_{i+1} t_i + t_{i+1} t_i^2 = 0 \qquad (1 \leq i \leq n-1), \tag{2.30}$$

$$t_i^2 t_{i-1} - 2t_i t_{i-1} t_i + t_{i-1} t_i^2 = 0 \qquad (2 \leq i \leq n-1), \tag{2.31}$$

$$t_n^3 t_{n-1} - (t_n^2 t_{n-1} t_n + t_n t_{n-1} t_n^2) + t_{n-1} t_n^3 = 0. \tag{2.32}$$

2.3. Enveloping algebra of the nilpotent subsuperalgebra $\mathfrak{n}^+$.

2.3.1. *Definition of* $\mathfrak{n}^+$. We define $\mathfrak{n}^+ := \mathfrak{osp}^+(1, 2n)$ as the subsuperalgebra of $\mathfrak{osp}(1, 2n)$ generated by Chevalley generators $t_1, t_2, \dots, t_n$. By (2.25), (2.28) and (2.29), $\mathfrak{n}^+$ contains the elements $b_1^+, b_2^+, \dots, b_n^+$. Hence by (2.4), $\mathfrak{n}^+$ contains the elements a_{ij}^+ for $1 \leq i < j \leq n$. It also follows inductively from (2.19) and (2.27) that $\mathfrak{n}^+$ contains the elements t_{ij} for $1 \leq i < j \leq n$. Hence:

$$\mathfrak{n}^+ = \mathfrak{n}_{\bar{0}}^+ \oplus \mathfrak{n}_{\bar{1}}^+,$$

where $\mathfrak{n}_{\bar{1}}^+$ is the $\mathbb{C}$-vector space with basis $(b_i^+)_{1 \leq i \leq n}$, and $\mathfrak{n}_{\bar{0}}^+ \simeq \mathfrak{sp}^+(2n)$ is the Lie subalgebra of $\mathfrak{g}_{\bar{0}} \simeq \mathfrak{sp}(2n)$ with basis $(c_i^+, a_{ij}^+, t_{ij})_{1 \leq i < j \leq n}$. Then:

$$\dim \mathfrak{n}_{\bar{0}}^+ = \dim \mathfrak{sp}^+(2n) = n^2 \quad \text{and} \quad \dim \mathfrak{n}_{\bar{1}}^+ = n.$$

Let us observe that relations (2.8), (2.9) and (2.10) imply that the Lie subalgebra with basis $(c_i^+, a_{ij}^+)_{1 \leq i < j \leq n}$ in $\mathfrak{n}_{\bar{0}}^+$ is abelian of dimension $n(n+1)/2$.

2.3.2. *Description of* $\mathcal{U}(\mathfrak{n}^+)$ *as an iterated skew polynomial algebra.* We denote by U^+ the enveloping algebra $\mathcal{U}(\mathfrak{n}^+)$ of $\mathfrak{n}^+$. By PBW theorem U^+ is generated as an associative algebra over $\mathbb{C}$ by the n^2 elements b_i^+, a_{ij}^+ and t_{ij} for $1 \leq i < j \leq n$. The relations detailed in 2.1 allow us to describe U^+ as an iterated skew polynomial extension. We start with $A := \mathbb{C}[a_{ij}^+]_{1 \leq i < j \leq n}$, which is by (2.10) a commutative polynomial algebra in $n(n-1)/2$ variables. Then we set:

$$B_0 := A, \quad B_1 := A[b_1^+], \quad B_j := B_{j-1}[b_j^+ \,;\, \sigma_j, \delta_j] \text{ for } 2 \leq j \leq n, \tag{2.33}$$

where σ_j is the automorphism of B_{j-1} and δ_j is the σ_j-derivation of B_{j-1} defined by $\sigma_j(a^+_{k\ell}) = a^+_{k\ell}$ and $\delta_j(a^+_{k\ell}) = 0$ for all $1 \leq k < \ell \leq n$, and $\sigma_j(b^+_i) = -b^+_i$ and $\delta_j(b^+_i) = 2a^+_{ij}$ for any $1 \leq i < j$, in order to translate relations (2.7) and (2.9). At the last step we denote:

$$B := B_n = \mathbb{C}[a^+_{ij}][b^+_1][b^+_2\,;\,\sigma_2,\delta_2]\cdots[b^+_n\,;\,\sigma_n,\delta_n]. \tag{2.34}$$

Then we add the generators t_{ij} for $1 \leq i < j \leq n$ inductively:

$$\begin{aligned} &R_1 := B[t_{1,2}\,;\,d_{1,2}][t_{1,3}\,;\,d_{1,3}]\cdots[t_{1,n}\,;\,d_{1,n}],\\ &R_i := R_{i-1}[t_{i,i+1}\,;\,d_{i,i+1}][t_{i,i+2}\,;\,d_{i,i+2}]\cdots[t_{i,n}\,;\,d_{i,n}], \qquad (2 \leq i \leq n-2)\\ &R_{n-1} := R_{n-2}[t_{n-1,n}\,;\,d_{n-1,n}] = U^+, \end{aligned}$$

where the derivations d_{ij} are suitably defined to account for relations (2.11) and (2.14) to (2.21). To sum up we conclude:

$$A = \mathbb{C}[a^+_{ij}]_{1\leq i<j<n}, \tag{2.35}$$

$$B = A[b^+_i\,;\,\sigma_i,\delta_i]_{1\leq i\leq n}, \tag{2.36}$$

$$U^+ = B[t_{ij}\,;\,d_{ij}]_{1\leq i<j\leq n}. \tag{2.37}$$

Let us recall that in this presentation, A is commutative and the b^+_i's commute with the elements of A.

2.3.3. *Restriction to the even part.* Because of relations (2.8) and (2.9), we can consider in B the commutative subalgebra C generated over A by the elements $c^+_k = (b^+_k)^2$ for $1 \leq k \leq n$. It follows from previous results that:

$$C = \mathbb{C}[a^+_{ij}, c^+_k]_{1\leq i<j<n, 1\leq k\leq n}, \tag{2.38}$$

$$\mathcal{U}(\mathfrak{n}^+_{\bar 0}) = C[t_{ij}\,;\,d_{ij}]_{1\leq i<j\leq n}. \tag{2.39}$$

2.4. Enveloping algebra of the solvable Borel subsuperalgebra $\mathfrak{b}^+$.

2.4.1. *Definition of* $\mathfrak{b}^+$. The Cartan Lie subalgebra $\mathfrak{h}$ of $\mathfrak{g}_{\bar 0}$ is the Lie subalgebra with basis $(k_i)_{1\leq i\leq n}$, which is abelian by relation (2.22). The positive Borel subsuperalgebra $\mathfrak{b}^+$ is defined as the subsuperalgebra of $\mathfrak{osp}(1,2n)$ generated by $\mathfrak{n}^+$ and $\mathfrak{h}$. We have clearly:

$$\mathfrak{b}^+ = \mathfrak{n}^+ \oplus \mathfrak{h} = \mathfrak{b}^+_{\bar 0} \oplus \mathfrak{n}^+_{\bar 1}, \quad \text{with } \mathfrak{b}^+_{\bar 0} := \mathfrak{n}^+_{\bar 0} \oplus \mathfrak{h},$$

where the even part $\mathfrak{b}^+_{\bar 0}$ is the positive Borel Lie subalgebra of $\mathfrak{g}_{\bar 0} \simeq \mathfrak{sp}(2n)$.

2.4.2. *Description of* $\mathcal{U}(\mathfrak{b}^+)$ *as an iterated polynomial algebra.* We denote by U^+_b the enveloping algebra $\mathcal{U}(\mathfrak{b}^+)$ of $\mathfrak{b}^+$. The commutation relations of the k_i's with the generators of U^+ given by formulas (2.22) to (2.24) allow an easy description of U^+_b as the following iterated skew polynomial algebra:

$$U^+_b = U^+[k_1,\,\Delta_1][k_2,\,\Delta_2]\cdots[k_n,\,\Delta_n], \tag{2.40}$$

where the derivations Δ_i vanish on all k_j's and satisfy $\Delta_i(b^+_i) = b^+_i$ and $\Delta_i(b^+_j) = 0$ for any $1 \leq j \neq i \leq n$, $\Delta_i(a^+_{ij}) = a^+_{ij}$ and $\Delta_i(t_{ij}) = t_{ij}$ for $1 \leq i < j \leq n$, $\Delta_i(a^+_{ji}) = a^+_{ji}$ and $\Delta_i(t_{ji}) = -t_{ji}$ for $1 \leq j < i \leq n$, $\Delta_i(a^+_{k\ell}) = 0$ and $\Delta_i(t_{k\ell}) = 0$ for $k \neq i, \ell \neq i$.

2.4.3. *Restriction to the even part.* It follows from previous results that, with notations (2.40) and (2.39):

$$\mathcal{U}(\mathfrak{b}^+_{\bar 0}) = \mathcal{U}(\mathfrak{n}^+_{\bar 0})[k_i\,;\,\Delta_i]_{1\leq i\leq n} = C[t_{ij}\,;\,d_{ij}]_{1\leq i<j\leq n}[k_i\,;\,\Delta_i]_{1\leq i\leq n}. \tag{2.41}$$

3. Proof of the main theorem

3.1. Some general technical results. We will use the following easy lemmas.

LEMMA 3.1.1. *Let A be a commutative $\mathbb{C}$-algebra, a an element of A, and R the iterated skew polynomial algebra generated over A by two generators x_1 and x_2 commuting with the elements of A and with relation $x_1x_2 + x_2x_1 = 2a$. Then the element $y = x_1x_2 - a$ satisfies $x_1y = -yx_1$ and $x_2y = -yx_2$.*

PROOF. Obvious calculation. □

LEMMA 3.1.2. *Let A be a commutative $\mathbb{C}$-algebra. Let $(a_{ij})_{1\leq i<j\leq n}$ be a family of elements of A, and R the iterated skew polynomial algebra generated over A by n generators $x_1, x_2, \ldots, x_n$ commuting with the elements of A and with relations $x_ix_j + x_jx_i = 2a_{ij}$ for all $1 \leq i < j \leq n$. Then the $n-1$ elements $y_i = x_1x_i - a_{1i}$ for any $2 \leq i \leq n$ satisfy:*

$$x_1y_i = -y_ix_1 \qquad \text{and} \qquad y_iy_j + y_jy_i = 2(a_{1i}a_{1j} - a_{ij}x_1^2).$$

PROOF. The first identity follows from previous lemma. For the second one, we observe firstly that:

$$\begin{aligned} x_jy_i &= x_jx_1x_i - x_ja_{1i} = (-x_1x_j + 2a_{1j})x_i - a_{1i}x_j \\ &= -x_1x_jx_i + 2a_{1j}x_i - a_{1i}x_j = x_1x_ix_j - 2a_{ij}x_1 + 2a_{1j}x_i - a_{1i}x_j \\ &= (x_1x_i - a_{1i})x_j - 2a_{ij}x_1 + 2a_{1j}x_i = y_ix_j - 2a_{ij}x_1 + 2a_{1j}x_i. \end{aligned}$$

We use this identity and Lemma 3.1.1 to calculate:

$$\begin{aligned} y_jy_i &= (x_1x_j - a_{1j})y_i = x_1(y_ix_j - 2a_{ij}x_1 + 2a_{1j}x_i) - a_{1j}y_i \\ &= -y_ix_1x_j - 2a_{ij}x_1^2 + 2a_{1j}x_1x_i - a_{1j}y_i \\ &= -y_i(x_1x_j - a_{1j}) - 2a_{1j}y_i - 2a_{ij}x_1^2 + 2a_{1j}x_1x_i \\ &= -y_iy_j - 2a_{1j}y_i - 2a_{ij}x_1^2 + 2a_{1j}x_1x_i \\ &= -y_iy_j + 2a_{1j}(x_1x_i - y_i) - 2a_{ij}x_1^2. \end{aligned}$$

We conclude that $y_jy_i = -y_iy_j + 2a_{1j}a_{1i} - 2a_{ij}x_1^2$. □

3.2. Enveloping skewfield of the Lie subsuperalgebra $\mathfrak{n}^+$.

NOTATIONS 3.2.1. Recalling the description (2.35), (2.36), (2.37) of U^+, we introduce the skewfields of rational functions:

$$K := \operatorname{Frac} A = \mathbb{C}(a_{ij}^+)_{1\leq i<j\leq n}, \tag{3.1}$$

$$L := \operatorname{Frac} B = K(b_i^+ \, ; \, \sigma_i, \delta_i)_{1\leq i\leq n}, \tag{3.2}$$

$$F := \operatorname{Frac} U^+ = L(t_{ij} \, ; \, d_{ij})_{1\leq i<j\leq n}. \tag{3.3}$$

NOTATIONS 3.2.2. We define the following families of elements of B:

$$y_1 := b_1^+, \quad x_{1,i} := b_i^+, \quad u_{1,i,j} := a_{i,j}^+ \quad \text{for all } 1 \leq i < j \leq n, \tag{3.4}$$

and more generally for any $1 \leq k \leq n$:

$$\begin{cases} u_{k,i,j} := u_{k-1,k-1,i}\, u_{k-1,k-1,j} - u_{k-1,i,j}\, y_{k-1}^2 & \text{for all } k \leq i < j \leq n, \\ x_{k,i} := x_{k-1,k-1}\, x_{k-1,i} - u_{k-1,k-1,i} & \text{for any } k \leq i \leq n, \\ y_k := x_{k,k}. \end{cases} \tag{3.5}$$

It is clear that we have with notation (2.33):

$$y_k \in B_k,\ x_{k,i} \in B_i,\ u_{k,i,j} \in B_{k-1} \quad \text{for all } 1 \leq k \leq n,\ k \leq i < j \leq n. \tag{3.6}$$

LEMMA 3.2.3.

(i) *The elements* $y_1, y_2, \ldots, y_n$ *of* B *satisfy the relations:*

$$\begin{cases} y_{k-1}y_k = -y_ky_{k-1} & \textit{for any } 2 \leq k \leq n, \\ y_ky_\ell = y_\ell y_k & \textit{for all } 1 \leq k, \ell \leq n,\ \ell \neq k \pm 1, \\ y_ka_{ij}^+ = a_{ij}^+y_k & \textit{for all } 1 \leq k \leq n, 1 \leq i < j \leq n. \end{cases}$$

(ii) *For any* $1 \leq k \leq n$, *the elements* $u_{k,i,j}$ *with* $k \leq i < j \leq n$ *are in the commutative subalgebra* $A[y_1^2, y_2^2, \ldots, y_{k-1}^2]$ *of* B.

PROOF. By iterative application of Lemma 3.1.2, we have for any $1 \leq k \leq n$:

$$x_{k-1,k-1}x_{k,i} = -x_{k,i}x_{k-1,k-1} \quad \text{for any } 2 \leq k \leq i \leq n, \tag{3.7}$$

$$x_{k,i}x_{k,j} + x_{k,j}x_{k,i} = 2u_{k,i,j} \quad \text{for all } k \leq i < j \leq n. \tag{3.8}$$

We deduce by induction on ℓ that, for any $1 \leq \ell \leq n$:

$$\begin{cases} y_\ell\, x_{\ell+1,i} = -x_{\ell+1,i}\, y_\ell & \text{for any } \ell + 1 \leq i \leq n, \\ y_\ell\, x_{k,i} = x_{k,i}\, y_\ell & \text{for any } \ell + 2 \leq k \leq i \leq n, \\ y_\ell\, u_{k,i,j} = u_{k,i,j}\, y_\ell & \text{for any } 1 \leq k \leq i < j \leq n. \end{cases} \tag{3.9}$$

Point (i) of the lemma follows obviously from these relations and point (ii) follows by iterative definition (3.5) of the $u_{k,i,j}$'s. □

PROPOSITION 3.2.4.

(i) *The subalgebra of* B *generated over* $\mathbb{C}$ *by* $y_1, \ldots, y_n$ *is isomophic to* $\mathcal{O}_\Lambda(\mathbb{C}^n)$ *where all entries of the symmetric matrix* $\Lambda = (\lambda_{ij})_{1 \leq i,j \leq n}$ *are equal to 1 except* $\lambda_{i,i+1} = -1 = \lambda_{i+1,i}$ *for any* $1 \leq i \leq n-1$.

(ii) *Its skewfield of fractions* Q *satisfies:*

$$L = Q(a_{ij}^+)_{1 \leq i < j \leq n}, \tag{3.10}$$

$$Q \simeq \operatorname{Frac}(\widetilde{\mathcal{O}}(\mathbb{C}^n)), \tag{3.11}$$

where each a_{ij}^+ *commutes with the elements of* Q.

PROOF. By induction from relations (3.5), we prove that for any $1 \leq k \leq n$,

$$y_k = (y_{k-1}y_{k-2}\cdots y_2y_1)b_k^+ + r_{k-1}, \tag{3.12}$$

where the term r_{k-1} lies in the subalgebra of B_{k-1} generated by $y_1, \ldots, y_{k-1}$ and the a_{ij}^+'s for $1 \leq i < j \leq k$. It follows that the skewfield L defined by (3.2) is also the skewfield of rational functions generated over K by $y_1, \ldots, y_n$. Then the commutation relations of Lemma 3.2.3 imply point (i) and assertion (3.10) of point (ii). In order to prove assertion (3.11), we suppose that $n \geq 3$ and define: $y_1' = y_1$, $y_2' = y_2$, $y_3' = y_1y_3$ and $y_i' = y_i$ for any $4 \leq i \leq n$. It is clear that the subfield of Q generated by the elements $(y_i')_{1 \leq i \leq n}$ is equal to Q. Moreover by straightforward calculations, we have:

$$\begin{cases} y_1'y_2' = -y_2'y_1', \quad y_1'y_i' = y_i'y_1', \quad y_2'y_i' = y_2'y_1', & \text{for any } 3 \leq i \leq n, \\ y_3'y_4' = -y_4'y_3', \quad y_3'y_i' = y_i'y_3', & \text{for any } 5 \leq i \leq n, \end{cases}$$

We deduce that $Q \simeq \operatorname{Frac}(\widetilde{\mathcal{O}}(\mathbb{C}^2) \otimes T_{n-2})$ where T_{n-2} is the subalgebra of Q generated by the elements $(y'_i)_{3 \le i \le n}$. Applying the same reduction on the generators of T_{n-2}, we conclude by iteration and using Proposition 1.1.1 that $Q \simeq \operatorname{Frac}(\widetilde{\mathcal{O}}(\mathbb{C}^n))$. □

LEMMA 3.2.5. *For any $1 \le i \le n-1$ and any $1 \le m \le n$, we have:*

$$[t_i, u_{m,i,i+1}] = x_{m,i}^2 \tag{3.13}$$

$$[t_i, u_{m,k,i+1}] = u_{m,k,i} \qquad \text{for } m \le k < i \le n, \tag{3.14}$$

$$[t_i, u_{m,i+1,\ell}] = u_{m,i,\ell} \qquad \text{for } m < i+1 < \ell \le n, \tag{3.15}$$

$$[t_i, u_{m,i+1,\ell}] = 0 \qquad \text{for } m = i+1 < \ell \le n, \tag{3.16}$$

$$[t_i, u_{m,k,\ell}] = 0 \qquad \text{for } k \ne i+1 \text{ and } \ell \ne i+1, \tag{3.17}$$

$$[t_i, x_{m,i+1}] = x_{m,i} \qquad \text{for } i+1 > m, \tag{3.18}$$

$$[t_i, x_{m,i+1}] = 0 \qquad \text{for } i+1 = m, \tag{3.19}$$

$$[t_i, x_{m,k}] = 0 \qquad \text{for } k \ne i+1. \tag{3.20}$$

PROOF. We study the action on B of the Chevalley generators $t_1, t_2, \dots, t_{n-1}$ defined in 2.2. Let us recall that, by (2.29) we have:

$$[t_i, b_{i+1}^+] = b_i^+ \quad \text{and} \quad [t_i, b_k^+] = 0 \ \text{ if } k \ne i+1. \tag{3.21}$$

Similarly by (2.14) to (2.18), we obtain:

$$[t_i, a_{i,i+1}^+] = (b_i^+)^2, \tag{3.22}$$

$$[t_i, a_{k,i+1}^+] = a_{k,i}^+ \qquad \text{for } 1 \le k < i \le n, \tag{3.23}$$

$$[t_i, a_{i+1,\ell}^+] = a_{i,\ell}^+ \qquad \text{for } i+1 < \ell \le n, \tag{3.24}$$

$$[t_i, a_{k,\ell}^+] = 0 \qquad \text{for } k \ne i+1 \text{ and } \ell \ne i+1. \tag{3.25}$$

Then for fixed i, we prove the assertions of the lemma by induction on m. By (3.4), assertion (3.21) is the case $m = 1$ of assertions (3.18) to (3.20), and (3.22) to (3.25) are the case $m = 1$ of assertions (3.13) to (3.17). Suppose that the eight assertions are satisfied to some rank $m \ge 1$. By (3.5), we have:

$$\begin{aligned}[t_i, u_{m+1,k,\ell}] = [t_i, u_{m,m,k}]u_{m,m,\ell} + u_{m,m,k}[t_i, u_{m,m,\ell}] \\ - [t_i, u_{m,k,\ell}]x_{m,m}^2 - u_{m,k,\ell}[t_i, x_{m,m}^2].\end{aligned}$$

It follows from (3.19) and (3.20) that the fourth term is zero. Then using induction assumptions (3.13) to (3.17), tedious but straightforward calculations prove that these five assertions remain valid at the rank $m+1$. Similarly by (3.5) we have:

$$[t_i, x_{m+1,k}] = [t_i, x_{m,m}]x_{m,k} + x_{m,m}[t_i, x_{m,k}] - [t_i, u_{m,m,k}].$$

It follows from (3.19) and (3.20) that the first term is zero. Then using induction assumptions (3.13) to (3.20), it is easy to check that: $[t_i, x_{m+1,i+1}]$ equals 0 if $m+1 = i+1$ and equals $x_{m+1,i}$ if $i+1 > m+1$, and $[t_i, x_{m+1,k}] = 0$ for any $k \ne i+1$. This finishes the proof. □

COROLLARY 3.2.6. *Each generator t_{ij} for $1 \le i < j \le n$ commutes with each generator y_m for $1 \le m \le n$.*

PROOF. By (3.19) and (3.20), the Chevalley generators $t_1, t_2, \dots, t_{n-1}$ commute with $y_m = x_{m,m}$ for any $1 \leq m \leq n$. By (2.19) and (2.27), each generator t_{ij} for $1 \leq i < j \leq n$ is an iterated bracket of $t_1, t_2, \dots, t_{n-1}$. □

PROPOSITION 3.2.7. *The skewfield of fractions of U^+ satisfies:*

$$F = Q\,(a^+_{ij})_{1\leq i<j\leq n}\,(t_{ij}\,;\,d_{ij})_{1\leq i<j\leq n}, \tag{3.26}$$

where the generators t_{ij} commute with all elements of Q, and the derivations d_{ij} translate the commutation relations between the t_{ij}'s and the a^+_{ij}'s [see relations (2.14) to (2.18)] or the t_{ij}'s with each other [see relations (2.19) to (2.21)].

PROOF. This is just a summary of assertions (3.3) and (3.10), Proposition 3.2.4 and Corollary 3.2.6. □

LEMMA 3.2.8. *Denote by $(q_{i,j})_{1\leq i<j\leq n}$ the elements of F defined by:*

$$q_{i,j} := y_i^{-2} u_{i,i,j} \qquad \text{for any } 1 \leq i < j \leq n. \tag{3.27}$$

Then the $q_{i,j}$'s commute with each other and with all elements of Q, and the subfield of L generated over Q by the $(q_{i,j})_{1\leq i<j\leq n}$ is equal to L.

PROOF. The first assertion follows from Corollary 3.2.3 and third relation in (3.9). To prove the second assertion, we observe directly from definition (3.5) that for any $1 \leq i < j \leq n$, we have:

$$u_{i,i,j} = e_{i-1,j} + (-1)^{i-1} a^+_{i,j} y_1^2 y_2^2 \cdots y_{i-1}^2,$$

where the element $e_{i-1,j}$ lies in the subalgebra E_{i-1} generated by $y_1^2, y_2^2, \dots, y_{i-1}^2$ and the $a^+_{k,\ell}$'s for $1 \leq k \leq i-1$. Then we can write:

$$a^+_{i,j} = (-1)^{i-1} y_1^{-2} y_2^{-2} \cdots y_{i-1}^{-2}\, q_{i,j} + e'_{i-1,j}$$

with $e'_{i-1,j} := (-1)^i y_1^{-2} y_2^{-2} \cdots y_{i-1}^{-2} e_{i-1,j} \in E_{i-1}$. This proves the rest of the lemma by induction on i. □

LEMMA 3.2.9. *For any $1 \leq i < j \leq n$ and any $1 \leq k < \ell \leq n$, we have:*

$$[t_{ij}, q_{i,j}] = 1, \tag{3.28}$$

$$[t_{ij}, q_{k,\ell}] = 0 \qquad \text{if } j \neq \ell, \tag{3.29}$$

$$[t_{ij}, q_{k,j}] = 0 \qquad \text{if } i < k, \tag{3.30}$$

$$[t_{ij}, q_{k,j}] = q_{k,i} \qquad \text{if } i > k. \tag{3.31}$$

PROOF. We proceed by induction on $j - i$. For $j = i + 1$, the identities directly follows from Lemma 3.2.5. For the general case, we deduced from (2.19) and (2.25) that $t_{i,j+1} = [t_{i,j}, t_{j,j+1}] = [t_{i,j}, t_j]$. Then we have $[t_{i,j+1}, q_{k,\ell}] = [[t_{i,j}, t_j], q_{k,\ell}] = [t_{i,j}, [t_j, q_{k,\ell}]] - [t_j, [t_{i,j}, q_{k,\ell}]]$ for any $1 \leq k < \ell \leq n$, and we finish the proof by induction considering successively the different cases for the indices (i, j) and (k, ℓ). □

COROLLARY 3.2.10. *The elements $(q_{i,j})_{1\leq i<j\leq n}$ defined in Lemma 3.2.8 and the elements $(p_{i,j})_{1\leq i<j\leq n}$ defined inductively by:*

$$p_{1,j} := t_{1,j} \quad \text{and} \quad p_{i,j} := t_{i,j} - \textstyle\sum_{m=1}^{i-1} q_{m,i} p_{m,j} \tag{3.32}$$

satisfy $[p_{ij}, q_{ij}] = 1$ and $[p_{ij}, p_{k\ell}] = [q_{ij}, q_{k\ell}] = [p_{ij}, q_{k\ell}] = 0$ for $(i,j) \neq (k,\ell)$.

PROOF. Follows obviously from Lemma 3.2.9 by induction on i. □

PROPOSITION 3.2.11. *We have:* $F = Q(q_{ij})_{1\le i<j\le n}\,(p_{ij}\,;\,\partial_{q_{ij}})_{1\le i<j\le n}$.

PROOF. It's clear by Lemma 3.2.8 that $L = Q(q_{ij})_{1\le i<j\le n}$. By definition (3.32) and Proposition 3.2.7, the subfield of F generated over L by the elements $(p_{i,j})_{1\le i<j\le n}$ is equal to F. Hence the proposition just summaries the results of Lemma 3.2.8 and Corollary 3.2.10. □

COROLLARY 3.2.12. *The skewfield F is isomorphic to* $\mathrm{Frac}\,(A_{n(n-1)/2}\otimes\widetilde{\mathcal{O}}(\mathbb{C}^n))$, *and its center is* $\mathcal{Z}(F) = \mathcal{Z}(Q) = \mathbb{C}(y_1^2,\dots,y_n^2)$.

PROOF. By Corollary 3.2.10, the generators p_{ij} and q_{ij} commute with all elements of Q and the subalgebra W of F they generate is isomorphic to the usual Weyl algebra $A_{n(n-1)/2}$. Then the theorem follows from Proposition 3.2.11 and assertion (3.11) of Proposition 3.2.4. □

CONCLUSION 3.2.13. We have proved the first assertion of the main theorem.

3.3. Enveloping skewfield of the Lie subalgebra $\mathfrak{n}_{\bar 0}^+$.

NOTATIONS 3.3.1. In order to determine the skewfield of fractions of the enveloping algebra of the even part $\mathfrak{n}_{\bar 0}^+$ of $\mathfrak{n}^+$ described in (2.39), we complete the notations (3.1), (3.2), (3.3) and (3.11) by defining:

$$Q_0 := \mathbb{C}(y_1^2, y_2^2,\dots,y_n^2) \subset Q, \tag{3.33}$$

$$L_0 := K((b_1^+)^2,(b_2^+)^2,\dots,(b_n^+)^2) \subset L, \tag{3.34}$$

$$F_0 := \mathrm{Frac}\,\mathcal{U}(\mathfrak{n}_{\bar 0}^+) = L_0(t_{ij}\,;\,d_{ij})_{1\le i<j\le n} \subset F. \tag{3.35}$$

LEMMA 3.3.2. *We have the equality* $L_0 = K(y_1^2,y_2^2,\dots,y_n^2)$.

PROOF. Returning to notations (3.5) and using (3.9) and (3.8), we check by straightforward calculations that: $x_{k,i}^2 = -x_{k-1,k-1}^2x_{k-1,i}^2 + u_{k-1,k-1,i}^2$ for all $2\le k\le i\le n$. Using this identity and point (ii) of Lemma 3.2.3, we prove that $y_1^2 = (b_1^+)^2$, $y_2^2 = -(b_1^+)^2(b_2^+)^2 + a_{1,2}^2$, and :

$$y_i^2 = (-1)^{i-1}y_1^2y_2^2\cdots y_{i-1}^2(b_i^+)^2 + v_{i-1}\quad\text{for any } 2\le i\le n,$$

where the term v_{i-1} lies in the subalgebra generated over A by $y_1^2, y_2^2,\dots,y_{i-1}^2$. Then the result follows by induction on i. □

PROPOSITION 3.3.3. $F_0 = Q_0\,(q_{ij})_{1\le i<j\le n}\,(p_{ij}\,;\,\partial_{q_{ij}})_{1\le i<j\le n}$, *with center* Q_0.

PROOF. It is clear by previous lemma that $L_0 = Q_0(a_{ij})_{1\le i<j\le n}$. Then the calculations in the proof of Lemma 3.2.8 show that the $q_{i,j}$'s are elements of L_0, and moreover that the subfield of L_0 generated over Q_0 by $(q_{ij})_{1\le i<j\le n}$ is equal to L_0. Finally we deduce from (3.32) that the skew subfield of F_0 generated over L_0 by $(p_{ij})_{1\le i<j\le n}$ is equal to F_0. □

CONCLUSION 3.3.4. The previous proposition showed that:

$$\mathrm{Frac}\,\mathcal{U}(\mathfrak{n}_0^+) \simeq \mathrm{Frac}\,(A_{n(n-1)/2}(\mathbb{C})\otimes\mathcal{O}(\mathbb{C}^n)).$$

Hence we retrieve in the context of the classical Gelfand-Kirillov property the usual description of the enveloping skewfield of the nilpotent Lie algebra $\mathfrak{sp}^+(2n)$ as a Weyl skewfield $D_{n(n-1)/2}$ over a purely transcendental extension of dimension n over $\mathbb{C}$.

3.4. Enveloping skewfield of the Lie subsuperalgebra $\mathfrak{b}^+$. Resuming all notations of the previous section and using (2.40) and (3.3), we set:

$$H := \operatorname{Frac} U_{\mathfrak{b}}^+ = F(k_1,\, \Delta_1)(k_2,\, \Delta_2)\cdots(k_n,\, \Delta_n). \tag{3.36}$$

Our first step is to determine the action of the generators $k_1, \ldots, k_n$ of the Cartan subalgebra $\mathfrak{h}$ on the generators $y_i,\ p_{ij},\ q_{ij}$ $(1 \le i < j \le n)$ introduced in the description of F in Proposition 3.2.11.

LEMMA 3.4.1. *For any $1 \le \ell \le n$, we have the following identities:*

$$[k_\ell, y_m] = \begin{cases} y_\ell & \text{if } m = \ell, \\ 2^{m-\ell-1} y_m & \text{if } m > \ell, \end{cases} \quad \text{and} \quad [k_\ell, y_m] = 0 \text{ if } m < \ell. \tag{3.37}$$

$$\begin{cases} [k_\ell, q_{\ell,j}] = -q_{\ell,j} & [k_\ell, p_{\ell,j}] = p_{\ell,j} & \text{for any } j > \ell, \\ [k_\ell, q_{i,\ell}] = q_{i,\ell} & [k_\ell, p_{i,\ell}] = -p_{i,\ell} & \text{for any } i < \ell, \\ [k_\ell, q_{i,j}] = 0 & [k_\ell, p_{i,j}] = 0 & \text{for any } i \neq \ell \neq j. \end{cases} \tag{3.38}$$

PROOF. We start with the action of k_ℓ on b_i^+ and $a_{i,j}^+$ for any $1 \le i < j \le n$ given by relations (2.23) and (2.24). Using the iterative definitions (3.4) and (3.5), we obtain for $1 \le m < \ell$ the identities:

$$[k_\ell, x_{m,i}] = \begin{cases} x_{m,\ell} & \text{if } i = \ell, \\ 0 & \text{if } i \neq \ell, \end{cases} \quad \text{and} \quad [k_\ell, u_{m,i,j}] = \begin{cases} u_{m,\ell,j} & \text{if } i = \ell, \\ u_{m,i,\ell} & \text{if } j = \ell, \\ 0 & \text{if } i \neq \ell \neq j. \end{cases}$$

Similarly for $m = \ell$:

$$[k_\ell, x_{\ell,i}] = \begin{cases} x_{\ell,\ell} & \text{if } i = \ell, \\ 0 & \text{if } i \neq \ell, \end{cases} \quad \text{and} \quad [k_\ell, u_{\ell,i,j}] = \begin{cases} u_{\ell,\ell,j} & \text{if } i = \ell, \\ 0 & \text{if } i \neq \ell \neq j. \end{cases}$$

Finally for $m > \ell$: $[k_\ell, x_{m,i}] = 2^{m-\ell-1} x_{m,i}$ and $[k_\ell, u_{m,i,j}] = 2^{m-\ell} u_{m,i,j}$. Assertion (3.37) follows trivially from these relations since $y_m = x_{m,m}$. Moreover, for $1 \le i, j \le n$, we have by definition (3.27) and (3.32):

$$[k_\ell, q_{i,j}] = [k_\ell, y_i^{-2} u_{i,i,j}] = -2y_i^{-3}[k_\ell, y_i]u_{i,i,j} + y_i^{-2}[k_\ell, u_{i,i,j}], \tag{3.39}$$

$$[k_\ell, p_{i,j}] = [k_\ell, t_{i,j}] - \textstyle\sum_{m=1}^{i-1} [k_\ell, q_{m,i}] p_{m,j} - \sum_{m=1}^{i-1} q_{m,i} [k_\ell, p_{m,j}]. \tag{3.40}$$

We deduce (3.38) applying case by case the previous identities. □

NOTATION 3.4.2. An obvious observation arising from (3.38) is:

$$[k_\ell, p_{i,j} q_{i,j}] = 0 \quad \text{for all } 1 \le \ell \le n,\ 1 \le i < j \le n. \tag{3.41}$$

These relations lead us to introduce the following elements of H:

$$h_\ell := k_\ell + \textstyle\sum_{m=\ell+1}^{n} q_{\ell,m} p_{\ell,m} - \sum_{m=1}^{\ell-1} q_{m,\ell} p_{m,\ell} \quad \text{for any } 1 \le \ell \le n, \tag{3.42}$$

which satisfy the relations described in the following lemma.

LEMMA 3.4.3. *For all $1 \le \ell, m \le n$ and $1 \le i < j \le n$, we have:*

$$[h_\ell, q_{i,j}] = [k_\ell, p_{i,j}] = [h_\ell, h_m] = 0 \quad \text{and} \quad [h_\ell, y_m] = [k_\ell, y_m].$$

PROOF. By (3.42) we have:

$$[h_\ell, q_{i,j}] = [k_\ell, q_{i,j}] + \textstyle\sum_{m=\ell+1}^{n} q_{\ell,m} [p_{\ell,m}, q_{i,j}] - \sum_{m=1}^{\ell-1} q_{m,\ell} [p_{m,\ell}, q_{i,j}].$$

If $i \neq \ell$ and $j \neq \ell$, all brackets appearing in the right hand are zero by Corollary 3.2.10 and relation (3.38). If $i = \ell$, the equality becomes $[h_\ell, q_{\ell,j}] = [k_\ell, q_{\ell,j}] + q_{\ell,j}[p_{\ell,j}, q_{\ell,j}] = 0$. If $j = \ell$, we obtain $[h_\ell, q_{i,\ell}] = [k_\ell, q_{i,\ell}] - q_{i,\ell}[p_{i,\ell}, q_{i,\ell}] = 0$. The calculations are similar for $p_{i,j}$. It is trivial that $[(p_{i,j}q_{i,j}),(p_{i',j'}q_{i',j'})]=0$ for any $1\leq i<j\leq n$ and $1\leq i'<j'\leq n$. Then $[h_\ell,h_m]=0$ follows from the commutativity of the Cartan subalgebra and observation (3.41). By Proposition 3.2.11, the generators $y_1,\dots,y_n$ commute with all $p_{i,j}$'s and $q_{i,j}$'s; then the last identity of the lemma is clear. □

PROPOSITION 3.4.4. *Let W be the subalgebra of F generated over $\mathbb{C}$ by the $n(n-1)$ elements p_{ij} and q_{ij}. Let P be the subalgebra of H generated over $\mathbb{C}$ by $y_1,\dots,y_n$ and $h_1,\dots,h_n$. Then $H\simeq \operatorname{Frac}(W\otimes P)$.*

PROOF. It is clear by (3.36), (3.42) and Proposition 3.2.11 that the subfield of H generated over F by $h_1,\dots,h_n$ is equal to H. Then the result follows from Lemma 3.4.3 and Proposition 3.2.11. □

PROPOSITION 3.4.5. *Let us define $y'_1,\dots,y'_n$ in Q by:*

$$y'_1=y_1 \quad \text{and} \quad y'_m=y_1^{-1}y_2^{-1}\cdots y_{m-1}^{-1}\,y_m \quad \text{for any } 2\leq m\leq n. \tag{3.43}$$

Let P' be the subalgebra of H generated over $\mathbb{C}$ by $y'_1,\dots,y'_n$ and $h_1,\dots,h_n$. Then $P'\simeq \widetilde{U}_n(\mathbb{C})$ and $H\simeq\operatorname{Frac}(W\otimes P')$.

PROOF. It is clear by (3.43) that $\operatorname{Frac}P'\simeq\operatorname{Frac}P$ and $H\simeq\operatorname{Frac}(W\otimes P')$. By tedious but simple calculations that we leave to the reader, we have:

$$y'_\ell y'_m=-y'_m y'_\ell,\quad [h_\ell,y'_m]=0 \quad\text{and}\quad [h_\ell,y'_\ell]=y'_\ell \tag{3.44}$$

for all $1\leq\ell\neq m\leq n$. We conclude with point (ii) of 1.2.1. □

COROLLARY 3.4.6. *We have $H\simeq\operatorname{Frac}(A_{n(n-1)/2}(\mathbb{C})\otimes\widetilde{A}_n(\mathbb{C}))$, with center $\mathbb{C}$.*

PROOF. We have already observed in Corollary 3.2.12 that $W\simeq A_{n(n-1)/2}(\mathbb{C})$ and in Remark 1.2.4 that $\operatorname{Frac}\widetilde{A}_n(\mathbb{C})\simeq\operatorname{Frac}\widetilde{U}_n(\mathbb{C})$. Then the isomorphism $H\simeq\operatorname{Frac}(A_{n(n-1)/2}(\mathbb{C})\otimes\widetilde{A}_n(\mathbb{C}))$ follows from previous proposition. Using the usual notation $D_m(E)$ for the skewfield of fractions of a Weyl algebra $A_m(E)$ over a skewfield E, we have $H\simeq D_{n(n-1)}(E)$ for $E=\operatorname{Frac}\widetilde{A}_n(\mathbb{C})$. Its center satisfies $\mathcal{Z}(H)\simeq\mathcal{Z}(D_{n(n-1)}(E))=\mathcal{Z}(E)$, and $\mathcal{Z}(E)=\mathbb{C}$ by Proposition 3.3.1 of [**18**]. □

CONCLUSION 3.4.7. The second assertion of the main theorem is proved.

3.5. Enveloping skewfield of the Lie subalgebra $\mathfrak{b}_{\bar{0}}^+$.

NOTATIONS 3.5.1. In order to determine the skewfield of fractions of the enveloping algebra of the even part $\mathfrak{b}_{\bar{0}}^+$ of $\mathfrak{b}^+$ described in (2.41), we denote:

$$H_0:=\operatorname{Frac}\mathcal{U}(\mathfrak{b}_{\bar{0}}^+)=F_0(k_1\,;\,\Delta_1)\cdots(k_n\,;\,\Delta_n), \tag{3.45}$$

where F_0 is described by Proposition 3.3.3.

PROPOSITION 3.5.2. *We have $H_0\simeq\operatorname{Frac}(W\otimes P'_0)$, where P'_0 is the subalgebra of P' generated by $(y'_1)^2,\dots,(y'_n)^2,h_1,\dots,h_n$, which is isomorphic to the enveloping algebra $U_n(\mathbb{C})$ defined in point (i) of* 1.2.1.

PROOF. By Proposition 3.3.3 and (3.42), the subfield of H_0 generated over F_0 by $h_1,\dots,h_n$ is equal to H_0. Since $\operatorname{Frac}P'_0=Q_0$ by (3.43), the result follows from Proposition 3.3.3. □

CONCLUSION 3.5.3. Because of the isomorphism $\operatorname{Frac} U_n(\mathbb{C}) \simeq \operatorname{Frac} A_n(\mathbb{C})$ observed in Remark 1.2.4 and the canonical isomorphism $A_{n(n+1)/2}(\mathbb{C}) \simeq A_{n(n-1)/2}(\mathbb{C}) \otimes A_n(\mathbb{C})$, the previous proposition proves that:

$$\operatorname{Frac} \mathcal{U}(\mathfrak{b}_{\bar{0}}^{+}) \simeq \operatorname{Frac} A_{n(n+1)/2}(\mathbb{C}).$$

Then we retrieve in the context of classical Gelfand Kirillov property the usual description of the enveloping skewfield of the solvable Borel subalgebra of the Lie algebra $\mathfrak{sp}(2n)$ as the Weyl skewfield $D_{n(n+1)/2}(\mathbb{C})$.

3.6. Final observation. A thorough reading of the previous proofs shows that the isomorphisms we obtain at the level of the skewfields of fractions are already realized at the level of the localization of the algebras under consideration by the multiplicative set generated by the n elements $y_1, \dots, y_n$. Since these elements are homogeneous, this result is consistent with the notion of skewfield of fractions in the category of superalgebras.

References

[1] J. Alev and F. Dumas, *Sur le corps des fractions de certaines algèbres quantiques* (French, with English summary), J. Algebra **170** (1994), no. 1, 229–265, DOI 10.1006/jabr.1994.1336. MR1302839

[2] J. Alev and F. Dumas, *On enveloping skew fields of some Lie superalgebras*, J. Algebra Appl. **15** (2016), no. 4, 1650071, 16, DOI 10.1142/S0219498816500717. MR3462202

[3] J. Alev, A. I. Ooms, and M. Van den Bergh, *The Gelfand-Kirillov conjecture for Lie algebras of dimension at most eight*, J. Algebra **227** (2000), no. 2, 549–581, DOI 10.1006/jabr.1999.8192. MR1759835

[4] M. Aubry and J.-M. Lemaire, *Zero divisors in enveloping algebras of graded Lie algebras*, J. Pure Appl. Algebra **38** (1985), no. 2-3, 159–166, DOI 10.1016/0022-4049(85)90006-4. MR814174

[5] W. Borho, P. Gabriel, and R. Rentschler, *Primideale in Einhüllenden auflösbarer Lie-Algebren (Beschreibung durch Bahnenräume)* (German), Lecture Notes in Mathematics, Vol. 357, Springer-Verlag, Berlin-New York, 1973. MR0376790

[6] A. Ch. Ganchev and T. D. Palev, *A Lie superalgebraic interpretation of the para-Bose statistics*, J. Math. Phys. **21** (1980), no. 4, 797–799, DOI 10.1063/1.524502. MR565726

[7] I. M. Gelfand and A. A. Kirillov, *Sur les corps liés aux algèbres enveloppantes des algèbres de Lie* (French), Inst. Hautes Études Sci. Publ. Math. **31** (1966), 5–19. MR0207918

[8] A. Giaquinto and J. J. Zhang, *Quantum Weyl algebras*, J. Algebra **176** (1995), no. 3, 861–881, DOI 10.1006/jabr.1995.1276. MR1351367

[9] M. Gorelik and E. Lanzmann, *The annihilation theorem for the Lie superalgebra* osp$(1, 2l)$ (English, with English and French summaries), C. R. Acad. Sci. Paris Sér. I Math. **327** (1998), no. 3, 237–242, DOI 10.1016/S0764-4442(98)80139-0. MR1650257

[10] A. Joseph, *Proof of the Gelfand-Kirillov conjecture for solvable Lie algebras*, Proc. Amer. Math. Soc. **45** (1974), 1–10, DOI 10.2307/2040597. MR0379617

[11] V. G. Kac, *Lie superalgebras*, Advances in Math. **26** (1977), no. 1, 8–96, DOI 10.1016/0001-8708(77)90017-2. MR0486011

[12] J. C. McConnell, *Representations of solvable Lie algebras and the Gelfand-Kirillov conjecture*, Proc. London Math. Soc. (3) **29** (1974), 453–484, DOI 10.1112/plms/s3-29.3.453. MR0357529

[13] I. M. Musson, *On the Goldie quotient ring of the enveloping algebra of a classical simple Lie superalgebra*, J. Algebra **235** (2001), no. 1, 203–213, DOI 10.1006/jabr.2000.8468. MR1807662

[14] I. M. Musson, *Lie superalgebras and enveloping algebras*, Graduate Studies in Mathematics, vol. 131, American Mathematical Society, Providence, RI, 2012. MR2906817

[15] T. D. Palev and J. Van der Jeugt, *The quantum superalgebra* $U_q[\mathrm{osp}(1|2n)]$*: deformed para-Bose operators and root of unity representations*, J. Phys. A **28** (1995), no. 9, 2605–2616. MR1338534

[16] A. Premet, *Modular Lie algebras and the Gelfand-Kirillov conjecture*, Invent. Math. **181** (2010), no. 2, 395–420, DOI 10.1007/s00222-010-0249-8. MR2657429

[17] L. Richard, *Hochschild homology and cohomology of some classical and quantum noncommutative polynomial algebras*, J. Pure Appl. Algebra **187** (2004), no. 1-3, 255–294, DOI 10.1016/S0022-4049(03)00146-4. MR2027905

[18] L. Richard, *Équivalence rationnelle d'algèbres polynomiales classiques et quantiques* (French, with English and French summaries), J. Algebra **287** (2005), no. 1, 52–87, DOI 10.1016/j.jalgebra.2004.12.008. MR2134258

[19] M. Wambst, *Complexes de Koszul quantiques* (French, with English and French summaries), Ann. Inst. Fourier (Grenoble) **43** (1993), no. 4, 1089–1156. MR1252939

Laboratoire de Mathématiques (FRE 2011 - CNRS), Université de Reims, Moulin de la Housse, BP 1039, 51687 Reims cedex 2, France
Email address: jacques.alev@univ-reims.fr

Laboratoire de Mathématiques Blaise Pascal (UMR 6620 - CNRS), Université Clermont Auvergne, 3 place Vasarely, CS 60026, 63178 Aubière cedex, France
Email address: Francois.Dumas@uca.fr

Contemporary Mathematics
Volume **727**, 2019
https://doi.org/10.1090/conm/727/14622

The Krull-Schmidt-Remak-Azumaya Theorem for G-groups

Meltem Altun-Özarslan and Alberto Facchini

ABSTRACT. We present some results related to G-groups and their category G-**Grp** in connection with the category **Grp** of all groups and the category R-Mod of all left modules over a fixed ring R. In particular, we prove that the Krull-Schmidt-Remak-Azumaya Theorem holds for G-groups, where G is a fixed group, for finite families of G-groups. A key role in this setting is played by the class of completely indecomposable G-groups.

1. Introduction

In this paper, we present and give a survey of some properties of G-groups, where G is a fixed group. G-groups show a behavior that is a pleasant mixture of the behavior of groups and that of left modules over a ring. In Section 2, we legitimize the existence of G-groups in the panorama of algebraic structures. Then we present the main properties of G-groups and their category. In Section 4, we prove the Krull-Schmidt-Remak-Azumaya Theorem for G-groups in the finite case. It holds for the direct-product decompositions of G-groups in a suitable subcategory $\mathcal{C}_G$ of G-**Grp**. The G-subgroups of a G-group in $\mathcal{C}_G$ are necessarily normal subgroups. A key role is played by completely indecomposable G-groups, which are the analog in the category of G-groups of modules with a local endomorphism ring in the category R-Mod. Finally, we prove Fitting's Lemma for the G-groups in $\mathcal{C}_G$.

All our groups are multiplicatively written. The proofs of the results we present in this paper that are not original can be found in the articles [**4**], [**1**] and [**2**].

2. G-groups and other algebraic structures

The aim of this Section is to give a motivation for the study of G-groups. We begin with a very general digression on algebraic structures. As far as algebras are concerned, we follow Bourbaki [**3**].

2010 *Mathematics Subject Classification.* Primary 16D70.
Key words and phrases. G-group, Krull-Schmidt-Remak-Azumaya Theorem.
The first author was supported by TUBITAK 2214/A Grant #1059B141600036.
The second author was partially supported by Dipartimento di Matematica "Tullio Levi-Civita" of Università di Padova (Project BIRD163492/16 "Categorical homological methods in the study of algebraic structures" and Research program DOR1690814 "Anelli e categorie di moduli").

K-algebras. Let K be a commutative ring with identity. A (not-necessarily associative) *K-algebra* is any unitary K-module M with a K-bilinear mapping $(x, y) \mapsto xy$ of $M \times M$ into M (equivalently, a K-linear mapping $M \otimes_K M \to M$). Thus all algebra axioms are satisfied except at most for associativity of multiplication. It is possible to construct the *opposite* M^{op} of such an algebra M by defining multiplication in M^{op} via $(x, y) \mapsto yx$.

If M, M' are K-algebras, a *K-algebra homomorphism* $\varphi\colon M \to M'$ is any K-linear mapping such that $\varphi(xy) = \varphi(x)\varphi(y)$ for every $x, y \in M$. A *derivation* of a K-algebra M is any K-linear mapping $D\colon M \to M$ such that $D(xy) = (D(x))y + x(D(y))$ for every $x, y \in M$. If M is any K-algebra, its endomorphisms form a monoid, that is, a semigroup with a two-sided identity, with respect to composition of mappings $\circ$.

An example of non-associative K-algebra is the *Mendel algebra.* Let K be a commutative field of characteristic $\neq 2$ with identity. Let M be the vector space over K with basis $\{S_1, S_2, \dots, S_n\}$. Define a multiplication $*$ on M setting

$$S_i * S_j = \frac{1}{2}(S_i + S_j)$$

for every $i, j = 1, 2, \dots, n$ and extending by K-bilinearity. The algebra M is the simplest example of *Mendel algebra.* It is an easy mathematical model of Mendel's Law of Segregation, which states that every individual organism contains two alleles for each trait. An offspring receives a pair of alleles for a trait by inheriting one allele for each trait from each parent. The Mendel algebra is commutative but not associative.

Associative K-algebras. The main example of associative K-algebra is, for any K-module A, the *endomorphism K-algebra* $\mathrm{End}_K(A)$ of any K-module A. If M is any (not-necessarily associative) K-algebra and $x \in M$, the mapping $\lambda_x\colon M \to M$, defined by $\lambda_x(y) = xy$ for every $y \in M$, is an element of the associative K-algebra $\mathrm{End}_K(M)$.

For any (not-necessarily associative) K-algebra M, there is a canonical mapping λ of M into the associative K-algebra $\mathrm{End}_K(M)$, defined by $\lambda\colon x \to \lambda_x$ for every $x \in M$. This mapping λ is a K-algebra morphism if and only if M is associative. Thus, for any associative K-algebra M, it is natural to define left M-modules as follows. A *left M-module* is any K-module A with a K-algebra homomorphism $\lambda\colon M \to \mathrm{End}_K(A)$. Similarly, we can define *right M-modules* as K-modules A with a K-algebra antihomomorphism $\rho\colon M \to \mathrm{End}_K(A)$. Here by *$K$-algebra antihomomorphism* $\psi\colon M \to M'$ between two K-algebras M, M' we mean any K-linear mapping ψ such that $\psi(xy) = \psi(y)\psi(x)$ for every $x, y \in M$. Clearly, a mapping $M \to M'$ is a K-algebra antihomomorphism if and only if it is a K-algebra homomorphism $M^{\mathrm{op}} \to M'$. It follows that right M-modules coincide with left M^{op}-modules. More precisely, when we say that right M-modules coincide with left M^{op}-modules, we mean that there is a canonical category isomorphism between the category of all right M-modules and the category of all left M^{op}-modules. Similarly, left M-modules coincide with right M^{op}-modules.

If the K-algebra M is commutative, then a mapping $M \to M'$ is a K-algebra antihomomorphism if and only if it is a K-algebra homomorphism $M \to M'$. It follows that right M-modules coincide with left M-modules whenever M is commutative.

Lie K-algebras. The main example of Lie K-algebra is, for any K-algebra M, the *Lie K-algebra of derivations* $\mathrm{Der}_K(M)$ of the K-algebra M. If M is any K-algebra and D, D' are two derivations of M, then the composite mapping DD' is not a derivation of M in general, but $DD' - D'D$ is. Thus, for any K-algebra M, we can define the Lie K-algebra $\mathrm{Der}_K(M)$ as the subset of $\mathrm{End}_K(M)$ consisting of all derivations of M with multiplication $[D, D'] := DD' - D'D$ for every $D, D' \in \mathrm{Der}_K(M)$.

A well known second example of Lie K-algebra is the following. Let L be any associative K-algebra. Define $[x, y] := xy - yx$ for every $x, y \in L$. This is a K-bilinear mapping $L \times L \to L$, and L, with this multiplication, turns out to be a Lie K-algebra.

As a third example, let A be any K-module and L the associative K-algebra $L := \mathrm{End}_K(A)$ of all K-endomorphisms of A. Then L with the operation $[-,-]$ defined in the previous paragraph, is a Lie K-algebra, denoted by $\mathfrak{gl}(A)$.

For any Lie K-algebra M and any element $x \in M$, the mapping $\lambda_x \colon M \to M$, defined by $\lambda_x = [x, -]$, is an element of the Lie K-algebra $\mathrm{Der}_K(M)$, usually called the *adjoint* of x, or the *inner derivation* defined by x, and denoted by $\mathrm{ad}_M\, x$ instead of λ_x.

For every Lie K-algebra M, there is a canonical Lie K-algebra homomorphism $\mathrm{ad} \colon M \to \mathrm{Der}_K(M)$, defined by $\mathrm{ad} \colon x \mapsto \mathrm{ad}_M\, x$ for every $x \in M$.

It is possible to define *left M-modules* for any Lie K-algebra M. Let M be any Lie K-algebra. A *left M-module* is a K-module A with a Lie K-algebra homomorphism $\lambda \colon M \to \mathfrak{gl}(A)$. Similarly, we can define *right M-modules* as K-modules A with a K-algebra antihomomorphism $\rho \colon M \to \mathfrak{gl}(A)$. But any Lie K-algebra M is isomorphic to its opposite algebra M^{op} via the isomorphism $M \to M^{\mathrm{op}}$, $x \mapsto -x$. It follows that the category of all right M-modules is canonically isomorphic to the category of all left M-modules for any Lie K-algebra M. Therefore it is useless to introduce both right and left modules, it is sufficient to introduce left M-modules and call them simply "M-modules". This cannot be done for associative K-algebras, because for an associative K-algebra M the structure of right M-modules can be very different from that of its left M-modules. For instance, it is easy to construct examples of associative K-algebras that are right noetherian, but not left noetherian. Over such an associative K-algebra, the structure of the category of right modules is very different from that of left modules.

Monoids. A *monoid* is a semigroup with a two-sided identity, that is, an element $1_M \in M$ such that $1_M x = x 1_M = x$ for every $x \in M$. The main example of monoid is, for any set X, the monoid X^X of all mappings $X \to X$. Here multiplication is composition of mappings. If M is any monoid and $x \in M$, we have the mapping $\lambda_x \colon M \to M$, that is, a morphism in the category of sets, defined by $\lambda_x(y) = xy$ for every $y \in M$, which is an element of the monoid M^M. We have a canonical injective monoid morphism $\lambda \colon M \to M^M$, $\lambda \colon x \mapsto \lambda_x$.

Correspondingly, we can define "left M-modules", now called *left M-sets*, for any monoid M. A *left M-set* is any set X with a monoid morphism $\lambda \colon M \to X^X$. Similarly, we can define *right M-sets* as sets X with a monoid antihomomorphism $\rho \colon M \to X^X$. Again, right M-sets coincide with left M^{op}-sets, and left M-sets coincide with right M^{op}-sets.

If the monoid M is commutative, right M-sets coincide with left M-sets.

REMARK 2.1. M-sets, where M is a monoid, also have a categorical interpretation. For any category $\mathcal{C}$, the endomorphisms of an object C of $\mathcal{C}$ are the elements of a monoid $\mathrm{End}_{\mathcal{C}}(C)$, the *endomorphism monoid* of C. Conversely, given any monoid M, we can construct the category $\mathcal{C}$ with a unique object $*$ with endomorphism monoid $\mathrm{End}_{\mathcal{C}}(*) := M$ and with composition the multiplication of M. The functors of this category $\mathcal{C}$ into the category **Set** of sets correspond to M-sets and the natural transformations between two functors $\mathcal{C} \to \mathbf{Set}$ correspond to M-set morphisms.

Left modules over an associative ring R have a similar behavior, as follows. The endomorphisms of an object C of a preadditive category $\mathcal{C}$ form an associative ring $\mathrm{End}_{\mathcal{C}}(C)$. Conversely, given any associative ring R, we can construct the category $\mathcal{C}$ with a unique object $*$ with endomorphism ring $\mathrm{End}_{\mathcal{C}}(*) := R$, and with morphism composition the multiplication of R. The additive functors of this preadditive category $\mathcal{C}$ into the category **Ab** of abelian groups correspond to left R-modules, and natural transformations between two additive functors from $\mathcal{C}$ to **Ab** correspond to left R-module morphisms.

Monoids with zero. A *monoid with zero* is a multiplicative monoid with an element $0_M \in M$ such that $0_M x = x 0_M = 0_M$ for every $x \in M$. The zero element in a monoid, when it exists, is unique. By definition, a morphism of monoids with zero must respect multiplication, send the identity to the identity and send zero to zero.

One of the main examples of monoid with zero is the endomorphism monoid of any object in the category of pointed sets. The *category* $\mathbf{Set}_*$ *of pointed sets* has as objects the pairs (X, x_0), where X is a non-empty set and x_0 is a selected element of X, called the *base point* of X. A morphism $(X, x_0) \to (X', x_0')$ in $\mathbf{Set}_*$ is any mapping $f \colon X \to X'$ such that $f(x_0) = x_0'$. For any pointed set (X, x_0), the endomorphism monoid $\mathrm{End}_{\mathbf{Set}_*}(X, x_0)$ of (X, x_0) in the category $\mathbf{Set}_*$ is a monoid with zero. The zero in this monoid is the mapping $X \to X$ that sends all the elements of X to x_0.

If M is any monoid with zero 0_M and $x \in M$, then we have the morphism

$$\lambda_x \colon (M, 0_M) \to (M, 0_M)$$

in the category $\mathbf{Set}_*$, defined by $\lambda_x(y) = xy$ for every $y \in M$. Moreover, we have the canonical injective morphism of monoids with zero $\lambda \colon (M, 0_M) \to \mathrm{End}_{\mathbf{Set}_*}(M, 0_M)$, $\lambda \colon x \mapsto \lambda_x$.

Correspondingly, define *left M-sets* for any monoid $(M, 0_M)$ with zero, as follows. A *left M-set* is any pointed set (X, x_0) with a morphism of monoids with zero

$$\lambda \colon (M, 0_M) \to \mathrm{End}_{\mathbf{Set}_*}(X, x_0).$$

Similarly, define *right M-sets* as pointed sets (X, x_0) with an antihomomorphism of monoids with zero $\rho \colon (M, 0_M) \to \mathrm{End}_{\mathbf{Set}_*}(X, x_0)$. Clearly, right M-sets coincide with left M^{op}-sets, and left M-sets coincide with right M^{op}-sets, and, for M commutative, right M-sets coincide with left M-sets.

Groups and the Cayley representation. A group is a special type of monoid, so that all what we have said about monoids applies to groups. One of the main examples of group is, for any set X, the group $\mathrm{Sym}(X)$ of all bijections $X \to X$. If G is any group and $x \in G$, the mapping $\lambda_x \colon G \to G$ considered above is a bijection. We have a canonical Cayley representation $\lambda \colon G \to \mathrm{Sym}(G)$, $\lambda \colon x \mapsto \lambda_x$, which is an injective group morphism.

Correspondingly, we have "left G-sets". A *left G-set* is any set X with a group morphism $\lambda\colon G \to \mathrm{Sym}(X)$. Similarly, *right G-sets* are sets X with a group antihomomorphism $\rho\colon G \to \mathrm{Sym}(X)$. But any group G is isomorphic to its opposite group G^{op} via the isomorphism $G \to G^{\mathrm{op}}$, $x \mapsto x^{-1}$. This is the mother of all symmetries in groups. Hence the category of all right G-sets is canonically isomorphic to the category of all left G-sets for all groups G, and it is useless to introduce both right and left G-sets.

Since groups are monoids, the categorical interpretation of left M-sets in Remark 2.1 applies directly to left G-sets. Given any group G, we can construct the category $\mathcal{C}$ with a unique object $*$ and with endomorphism monoid $\mathrm{End}_{\mathcal{C}}(*) := G$. The functors of this category $\mathcal{C}$ into the category **Set** of sets correspond to G-sets and the natural transformations between two functors $\mathcal{C} \to \mathbf{Set}$ correspond to G-set morphisms.

Groups G and action of G on G via inner automorphisms. There is another very natural action of a group G onto itself, different from that in the previous subsection. For any group G, we can construct its *automorphism group* $\mathrm{Aut}(G)$. If G is any group and $x \in G$, the mapping $\alpha_x\colon G \to G$, defined by $\alpha_x(y) = xyx^{-1}$ for every $y \in G$, is the *inner automorphism* of G given by left conjugation by x. There is a canonical group morphism $\alpha\colon G \to \mathrm{Aut}(G)$, defined by $\alpha\colon x \mapsto \alpha_x$ for every $x \in G$.

Correspondingly, we have "left G-groups". For a fixed group G, a *left G-group* is any group H with a group morphism $\alpha\colon G \to \mathrm{Aut}(H)$. Similarly, we can define *right G-groups* as groups H with a group antihomomorphism $\beta\colon G \to \mathrm{Aut}(H)$. As we have said above, any group G is isomorphic to its opposite group G^{op}. Hence the category of all right G-groups is canonically isomorphic to the category of all left G-groups for all groups G, and it is therefore useless to introduce both right G-groups and left G-groups.

As above for G-sets, given any group G, we can construct the category $\mathcal{C}$ with a unique object $*$ and with endomorphism monoid $\mathrm{End}_{\mathcal{C}}(*) := G$. The functors of this category $\mathcal{C}$ into the category **Grp** of groups correspond to G-groups and the natural transformations between two functors $\mathcal{C} \to \mathbf{Grp}$ correspond to G-group morphisms.

3. G-groups, their category, and their properties

We are now ready for a presentation of G-groups and their properties. For the proofs of all the properties of the category G-**Grp** we present in this section, see [**1**].

As we have just said in the previous subsection, for any two groups G and H we say that H is a *(left) G-group* if there is a fixed group homomorphism $\alpha\colon G \to \mathrm{Aut}(H)$. That is, a G-group is a pair (H, α), where H is a group and $\alpha\colon G \to \mathrm{Aut}(H)$ is a group morphism. Equivalently, a G-group can be defined as a group H endowed with a mapping $\cdot\colon G \times H \to H$, $(g, h) \mapsto gh$, called *left scalar multiplication*, such that

(a) $g(hh') = (gh)(gh')$,
(b) $(gg')h = g(g'h)$, and
(c) $1_G h = h$

for every $g, g' \in G$ and every $h, h' \in H$.

A *G-group morphism* $f\colon (H, \varphi) \to (H', \varphi')$ is any group homomorphism $f\colon H \to H'$ such that $f(gh) = gf(h)$ for every $g \in G$, $h \in H$. We will denote by

$\mathrm{Hom}_G(H, H')$ the set of all G-group morphisms of H into H'. G-groups form a category G-**Grp**. The category **Grp** of groups and the category 1-**Grp**, where 1 is the trivial group (that is, the group with one element), are isomorphic categories. This is the analogue of the fact that the category **Ab** of abelian groups and the category $\mathbb{Z}$-Mod of modules over the ring $\mathbb{Z}$ of integers are isomorphic categories, because 1 and $\mathbb{Z}$ are the initial objects in the category of groups and the category of rings, respectively.

We will look closely at a number of similarities and differences between the category G-**Grp** of G-groups and the category R-Mod of all left modules over a fixed ring R. The category R-Mod has as objects the pairs (A, λ), where A is an additive abelian group and $\lambda\colon R \to \mathrm{End}_{\mathbb{Z}}(A)$ is a ring morphism, and as set of morphisms of (A, λ) into (A', λ') the set of all group homomorphisms $f\colon A \to A'$ such that $f(rx) = rf(x)$ for every $r \in R$, $x \in A$.

The category G-**Grp** has finite products, finite coproducts and equalizers, and its only injective object is the trivial group. In particular, G-**Grp** is a semi-abelian category. The category R-Mod is an abelian category with enough injectives.

In order to study G-groups in an algebraic perspective, let's begin with an example, the example of the *regular G-group* (G, α). Here $\alpha\colon G \to \mathrm{Aut}(G)$, $g \mapsto \alpha_g$, where $\alpha_g(x) = gxg^{-1}$ for every $g, x \in G$, is the canonical mapping associating to every element g of G the inner automorphism given by left conjugation by g. This regular G-group G is the analog of the regular left R-module ${}_RR$, that is, the ring R viewed as a left module over itself.

Recall that a subgroup H' of a G-group H is said to be a *G-subgroup* of H, denoted $H' \leq_G H$, if $gh' \in H'$ for every $g \in G$, $h' \in H'$. For example, the G-subgroups of the regular G-group G are exactly the normal subgroups of G. If $f\colon (H, \varphi) \to (H', \varphi')$ is a G-group morphism, then $A \leq_G H$ implies $f(A) \leq_G H'$, and $A' \leq_G H'$ implies $f^{-1}(A') \leq_G H$.

As in the categories **Grp** and R-Mod, the monomorphisms in the category G-**Grp** are exactly the morphisms that are injective mappings, and the epimorphisms are the morphisms that are surjective mappings.

REMARK 3.1. We will use the "standard additive notation" for mappings between multiplicative groups. That is, if H and L are multiplicative groups and $\alpha, \beta\colon H \to L$ are any two mappings, we denote by $\alpha + \beta\colon H \to L$ the mapping defined by

$$(\alpha + \beta)(h) = \alpha(h)\beta(h)$$

for every $h \in H$. With respect to this operation $+$, the set L^H of all mappings of H into L turns out to be a group that is the direct product of $|H|$ (cardinality of H) copies of the group L. In case α and β are group morphisms, the mapping $\alpha + \beta$ turns out to be a group morphism if and only if $[\alpha(H), \beta(H)] = 1$, that is, if and only if $\alpha(h_1)$ commutes with $\beta(h_2)$ for all $h_1, h_2 \in H$. As usual, 1 and 0 denote the identity endomorphisms and the trivial homomorphisms, respectively.

Now suppose that G is a group and H is a G-group. The set H^H of all mappings $H \to H$, endowed with the addition $+$ defined above and the composition of mappings $\circ$ as multiplication, turns out to be a right near ring (that is, a group with respect to $+$, a monoid with respect to $\circ$, right distributivity holds, but left distributivity does not in general.) The subset $\mathrm{End}_G(H) := \mathrm{Hom}_G(H, H) \subseteq H^H$ is a partial ring with identity, in the following sense. Set $S := \{\, (\alpha, \beta) \in \mathrm{End}_G(H) \times \mathrm{End}_G(H) \mid [\alpha(H), \beta(H)] = 1\}$, so that the addition $+$ is a mapping $+\colon S \to \mathrm{End}_G(H)$.

Then $\mathrm{End}_G(H)$ is a multiplicative submonoid of H^H, and $\mathrm{End}_G(H)$ also has the partially defined operation $+$. Moreover, the equality $0+\alpha=\alpha+0=\alpha$ always holds, and the identities $\alpha+\beta=\beta+\alpha$, $(\alpha+\beta)+\gamma=\alpha+(\beta+\gamma)$, $\alpha(\beta+\gamma)=\alpha\beta+\alpha\gamma$, $(\alpha+\beta)\gamma=\alpha\gamma+\beta\gamma$ hold, for $\alpha,\beta,\gamma\in\mathrm{End}_G(H)$, whenever both members of the equalities are defined. Notice that both distributivity laws hold in $\mathrm{End}_G(H)$, when the operations are defined. Finally, $\mathrm{End}_G(H)$ is a ring if and only if H is abelian.

Our aim in the next section will be to define when the partial ring $\mathrm{End}_G(H)$ is *local* in a suitable sense, in order to prove, in the case of local partial endomorphism rings, a form of the Krull-Schmidt-Remak-Azumaya Theorem.

REMARK 3.2. Let us go back to the partially defined addition $+$ considered in the previous Remark 3.1 and defined on the set $\mathrm{Hom}(H,L)$ of all group morphisms between two groups H,L. The partial operation $+$ on $\mathrm{Hom}(H,L)$ is related to the group morphisms $H_1\times H_2\to L$ for arbitrary groups H_1, H_2, L. Namely, the set $H_1\times H_2\to L$ is clearly in one-to-one correspondence with the set of all pairs $(\psi_1,\psi_2)\in\mathrm{Hom}(H_1,L)\times\mathrm{Hom}(H_2,L)$ such that $[\psi_1(H_1),\psi_2(H_2)]=1$.

The direct product $H_1\times H_2$ is the product in the category **Grp**. The coproduct of H_1 and H_2 in the category **Grp** is their free product H_1*H_2. Thus, for any group L, we have canonical bijections $\mathrm{Hom}(L,H_1\times H_2)\cong\mathrm{Hom}(L,H_1)\times\mathrm{Hom}(L,H_2)$ and $\mathrm{Hom}(H_1*H_2,L)\cong\mathrm{Hom}(H_1,L)\times\mathrm{Hom}(H_2,L)$.

Correspondingly, let K be a commutative ring with identity and K-**Alg** the category of all associative K-algebras. The direct product $R_1\times R_2$ of two such K-algebras R_1,R_2 is their product in the category K-**Alg**. Their coproduct in the category K-**Alg** is the free product R_1*R_2. Hence, for any associative K-algebra S, there are canonical bijections $\mathrm{Hom}(S,R_1\times R_2)\cong\mathrm{Hom}(S,R_1)\times\mathrm{Hom}(S,R_2)$ and $\mathrm{Hom}(R_1*R_2,S)\cong\mathrm{Hom}(R_1,S)\times\mathrm{Hom}(R_2,S)$. For associative K-algebras R_1,R_2, we can also define their tensor product $R_1\otimes_K R_2$, and it turns out that, for any associative K-algebra S, there is a one-to-one correspondence between $\mathrm{Hom}(R_1\otimes_K R_2,S)$ with the set of all pairs $(\psi_1,\psi_2)\in\mathrm{Hom}(R_1,S)\times\mathrm{Hom}(R_2,S)$ such that $[\psi_1(R_1),\psi_2(R_2)]=0$. This is exactly the same behaviour as the direct product of groups $H_1\times H_2$.

Also, there is a canonical onto mapping $R_1*R_2\to R_1\otimes_K R_2$, whose kernel is the two-sided ideal of the free product R_1*R_2 generated by all elements $r_1r_2-r_2r_1$ ($r_1\in R_1$ and $r_2\in R_2$). Similarly, the kernel of the canonical onto mapping $H_1*H_2\to H_1\times H_2$ is the normal subgroup of the free product H_1*H_2 generated by all elements $h_1h_2h_1^{-1}h_2^{-1}$ ($h_1\in H_1$ and $h_2\in H_2$).

The category G-**Grp** contains some particularly relevant full subcategories:

(1) The category $\mathcal{C}_G$ whose objects are all the G-groups (H,φ) with $\varphi(G)\supseteq\mathrm{Inn}(H)$, where $\mathrm{Inn}(H)$ is the subgroup of $\mathrm{Aut}(H)$ of all inner automorphisms of H. If H is an object of $\mathcal{C}_G$ and H' is a G-subgroup of H, then H' is necessarily a normal subgroup of H, so that it is possible to construct the quotient G-group H/H'.

(2) The category $\mathcal{C}^*_G$ whose objects are all the G-groups (H,φ) with $\varphi(G)\subseteq\mathrm{Inn}(H)$.

(3) Their intersection $\mathcal{C}_G\cap\mathcal{C}^*_G$, whose objects are such that $\varphi(G)=\mathrm{Inn}(H)$. This subcategory $\mathcal{C}_G\cap\mathcal{C}^*_G$ contains the regular object ${}_GG$.

(4) The category $\mathbb{Z}[G]$-Mod of all left modules over the group algebra $\mathbb{Z}[G]$ of G. More precisely, consider the functor $F\colon\mathbb{Z}[G]\text{-Mod}\to G\text{-}\mathbf{Grp}$ which sends any

left $\mathbb{Z}[G]$-module (A, λ), where $\lambda\colon \mathbb{Z}[G] \to \operatorname{End}_{\mathbb{Z}}(A)$ to $(A, \lambda|_G)$, where $\lambda|_G\colon G \to \operatorname{Aut}(A)$ is the restriction of λ. It is easily seen that F is an embedding of categories, so that we can view G-**Grp** as an extension of the category $\mathbb{Z}[G]$-Mod. Since $\operatorname{Inn}(H) = 1$ for any abelian group H, we have the inclusions $\mathbb{Z}[G]\text{-Mod} \subseteq \mathcal{C}_G \subseteq G$-**Grp**.

(5) The category **Grp** of groups can also be embedded as a full subcategory of G-**Grp**, by sending any group H to the G-group (H, λ), where $\lambda\colon G \to \operatorname{Aut}(H)$ is the trivial morphism that sends everything to 1. Thus we have an inclusion of categories **Grp** $\subseteq \mathcal{C}_G^* \subseteq G$-**Grp**.

(6) Finally, notice that the intersection of the two subcategories $\mathbb{Z}[G]$-Mod and **Grp** is the category **Ab** of abelian groups.

Thus we have the following diagram:

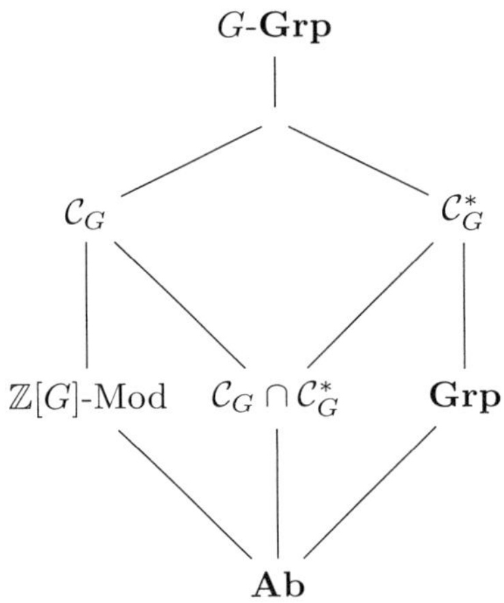

REMARK 3.3. Notice that if H is a G-group in $\mathcal{C}_G$, then all the G-subgroups H' of H are also in $\mathcal{C}_G$. On the contrary, the direct product of two G-groups in $\mathcal{C}_G$ in not necessarily in $\mathcal{C}_G$. More precisely, let us show that:

PROPOSITION 3.4. *If H is a G-group, then $H \times H$ is in $\mathcal{C}_G$ if and only if H is abelian.*

PROOF. Suppose $H \times H$ in $\mathcal{C}_G$. Then, for every $h \in H$, conjugation $\alpha_{(1,h)}$ by the element $(1, h)$ of $H \times H$ is an automorphism of $H \times H$. Assume that $\beta\colon G \to \operatorname{Aut}(H)$ describes the G-group structure on H. As $H \times H$ is in $\mathcal{C}_G$, there exists an element $g \in G$ such that $\begin{pmatrix} \beta(g) & 0 \\ 0 & \beta(g) \end{pmatrix}: H \times H \to H \times H$ is conjugation by $(1, h)$, which is $\begin{pmatrix} \alpha_1 & 0 \\ 0 & \alpha_h \end{pmatrix}$, where α_1 is conjugation by the element 1 and α_h is conjugation by h. Thus $\alpha_1 = \beta(g) = \alpha_h$. It follows that h is in the center of H. Since this holds for every $h \in H$, we have that H is abelian.

Conversely, if H is abelian, then $H \times H$ is abelian, so that $H \times H$ is necessarily in $\mathcal{C}_G$. □

REMARK 3.5. We also have a further parallel between the categories R-Mod and $\mathcal{C}_G$. We know that, for any left module ${}_RA$, there is a one-to-one correspondence φ between the set I of all idempotent elements of $\operatorname{End}({}_RA)$ and the set

$$\{ (B, C) \mid B, C \leq {}_RA,\ {}_RA = B \oplus C \}$$

of all pairs (B, C) of submodules of ${}_RA$ whose sum is direct and equal to ${}_RA$. If $e \in I$, the corresponding pair is $\varphi(e) = (\ker e, e({}_RA))$.

In the category **Grp** of groups, we know that, for any group H, there is a one-to-one correspondence φ between the set of all idempotents of $\mathrm{End}(H)$ and the set of all pairs $\{ (H', H'') \mid H', H'' \le H,\ H = H' \rtimes H'' \}$ (semidirect product), where, again, the pair that corresponds to an idempotent e is the pair $(\ker e, e(H))$ [**4**, Lemma 5.1].

For any G-group H in $\mathcal{C}_G$, this correspondence restricts to a one-to-one correspondence between the set of all idempotents of $\mathrm{End}_{G\text{-}\mathbf{Grp}}({}_GH)$ and the set of all pairs $\{ (H', H'') \mid H', H'' \le H,\ H = H' \times H'' \}$ of direct-product decompositions of the G-group H. When H is in $\mathcal{C}_G \cap \mathcal{C}_G^*$, direct-product decompositions $H' \times H''$ of the group H coincide with product decompositions $H' \times H''$ of the G-group H in the category G-**Grp** [**4**, Lemma 5.2].

Let $\zeta(H)$ denote the *centre* of any group H. Recall that an endomorphism f of a group H is *central* if it induces the identity endomorphism on $H/\zeta(H)$, that is, if $f(h) \in h\zeta(H)$ for every $h \in H$ [**5**, p. 83]. An endomorphism of H is central if and only if it is of the form $1 + f'$ for some endomorphism f' of H with $f'(H) \subseteq \zeta(H)$ [**4**, Lemma 4.2]. More generally, an endomorphism f of a group H is called *normal* if it commutes with every inner automorphism of H. Note that every central endomorphism is normal [**5**, 3.3.6(ii)].

PROPOSITION 3.6. *Let ${}_GH$ be a G-group.*

(a) *If ${}_GH$ is in $\mathcal{C}_G$, then every automorphism of ${}_GH$ in the category G-**Grp** is a central automorphism of H.*

(b) *If ${}_GH$ is in $\mathcal{C}_G^*$, then every central automorphism of the group H is an automorphism of the G-group ${}_GH$.*

PROOF. (a) If $f \colon H \to H$ is an automorphism of H in the category G-**Grp** (equivalenty, in the category $\mathcal{C}_G$), then f is a normal endomorphism of H in the sense of Robinson [**5**, Example 4 on Page 30], that is, f commutes with every inner automorphism of H. By [**5**, 3.3.6(ii)], the automorphism f is central.

(b) Let ${}_GH \in \mathcal{C}_G^*$ and $f \colon H \to H$ be a central automorphism of H. Since any central automorphism is normal and ${}_GH$ is in $\mathcal{C}_G^*$, it is easily seen that $f(gh) = gf(h)$ for every $g \in G$ and $h \in H$. □

Notice that Robert Erich Remak in his dissertation "Über die Zerlegung der endlichen Gruppen in indirekte unzerlegbare Faktoren" ("On the decomposition of finite groups into indirect indecomposable factors", 1911) proved that if a finite group G has two direct-product decompositions $G = G_1 \times G_2 \times \cdots \times G_t = H_1 \times H_2 \times \cdots \times H_s$ into indecomposables, then $t = s$ and there is a *central* automorphism φ of G and a permutation σ of $\{1, 2, \dots, t\}$ such that $\varphi(G_i) = H_{\sigma(i)}$ for all the indices $i = 1, \dots, t$. Now ${}_GG \in \mathcal{C}_G \cap \mathcal{C}_G^*$, hence Remark's result does not concern the direct-product decompositions of the group G in the category **Grp** of groups, but the direct-product decompositions of the regular object ${}_GG$, which is an object of $\mathcal{C}_G \cap \mathcal{C}_G^*$, in the category G-**Grp**.

4. The Krull-Schmidt-Remak-Azumaya Theorem for G-groups

Let H be a G-group. A G-subgroup H' of H is called a *G-direct factor* of H if there exists a G-subgroup H'' of H such that $H = H' \times H''$. If there are no proper non-trivial G-direct factors of H and $H \neq 1$, then H is said to be *G-indecomposable.*

DEFINITION 4.1. A G-group H is *completely indecomposable* if it is $\neq 1$ and, for every pair $\alpha, \alpha' \colon H \to H$ of G-endomorphisms of H, that is, endomorphisms of H in the category G-**Grp**, such that $\alpha + \alpha'$ is defined and is an automorphism of H, either α or α' is an automorphism of H. Recall that $\alpha + \alpha'$ is defined if and only if $[\alpha(H), \alpha'(H)] = 1$.

We leave to the reader the easy proofs of the following Lemmas. The proof of Lemma 4.2 is similar to that of [**4**, Lemma 5.4].

LEMMA 4.2. *Completely indecomposable G-groups are G-indecomposable.*

LEMMA 4.3. *Let H be a completely indecomposable G-group. Suppose that $\alpha_1, \dots, \alpha_n$ are G-endomorphisms of H such that $[\alpha_i(H), \alpha_j(H)] = 1$ for every $i \neq j$. If $\alpha_1 + \dots + \alpha_n$ is an automorphism of H, then there exist an index $i = 1, \dots, n$ such that α_i is an automorphism of H.*

In connection to Remark 3.2, notice that if H_1, H_2, L_1, L_2 are G-groups, the set $\mathrm{Hom}_{G\text{-}\mathbf{Grp}}(H_1 \times H_2, L_1 \times L_2)$ of all G-group morphisms of $H_1 \times H_2$ into $L_1 \times L_2$ is in a canonical one-to-one correspondence with the set of all 2×2-matrices

$$\begin{pmatrix} \psi_{L_1,H_1} & \psi_{L_1,H_2} \\ \psi_{L_2,H_1} & \psi_{L_2,H_2} \end{pmatrix} \in \begin{pmatrix} \mathrm{Hom}(H_1, L_1) & \mathrm{Hom}(H_2, L_1) \\ \mathrm{Hom}(H_1, L_2) & \mathrm{Hom}(H_2, L_2) \end{pmatrix}$$

of G-morphisms such that

$$[\psi_{L_1,H_1}(H_1), \psi_{L_1,H_2}(H_2)] = 1 \qquad \text{and} \qquad [\psi_{L_2,H_1}(H_1), \psi_{L_2,H_2}(H_2)] = 1.$$

Here $\psi_{X,Y}$ denotes a G-group morphism from Y to X. Moreover, this one-to-one correspondence transforms composition of G-group morphisms into row-by-column product of matrices, and gives a monoid isomorphism between $\mathrm{End}(H \times L)$ and the monoid of all 2×2-matrices $\begin{pmatrix} \mathrm{Hom}(H, H) & \mathrm{Hom}(L, H) \\ \mathrm{Hom}(H, L) & \mathrm{Hom}(L, L) \end{pmatrix}$ with

$$[\psi_{H,H}(H), \psi_{H,L}(L)] = 1 \qquad \text{and} \qquad [\psi_{L,H}(H), \psi_{L,L}(L)] = 1.$$

The proof of [**4**, Lemma 5.9] shows that:

LEMMA 4.4. *Let H_1, H_2, L_1, L_2 be G-groups and $\psi \colon H_1 \times H_2 \to L_1 \times L_2$ be a G-group isomorphism. Write ψ as a 2×2-matrix*

$$\begin{pmatrix} \psi_{L_1,H_1} & \psi_{L_1,H_2} \\ \psi_{L_2,H_1} & \psi_{L_2,H_2} \end{pmatrix} \in \begin{pmatrix} \mathrm{Hom}(H_1, L_1) & \mathrm{Hom}(H_2, L_1) \\ \mathrm{Hom}(H_1, L_2) & \mathrm{Hom}(H_2, L_2) \end{pmatrix}.$$

If $\psi_{L_1,H_1} \colon H_1 \to L_1$ is a G-group isomorphism, then H_2 and L_2 are isomorphic G-groups.

Note that the proof of the following Theorem is similar to the proof of the analogous result for completely indecomposable groups due to Facchini and Lucchini [**4**, Theorem 5.10], but we include it here for the convenience of the reader.

THEOREM 4.5 (Krull-Schmidt-Remak-Azumaya Theorem for G-groups). *Let $H_1, \dots, H_n$ be completely indecomposable G-groups with $H_1 \times \dots \times H_n$ in $\mathcal{C}_G$.*

(a) *If $H_1 \times \dots \times H_n = L \times M$ as G-groups, then there is a partition $I_L \dot\cup I_M$ of the set $\{1, 2, \dots, n\}$ such that $L \cong \prod_{i \in I_L} H_i$ and $M \cong \prod_{i \in I_M} H_i$ (isomorphic as direct products of G-groups).*

(b) *If* $H_1\times\cdots\times H_n\cong L_1\times\cdots\times L_m$*, where the* G*-groups* L_j *are indecomposable, then* $n=m$ *and there is a permutation* σ *of* $\{1,2,\dots,n\}$ *such that* $H_i\cong L_{\sigma(i)}$ *for every* $i=1,2,\dots,n$.

PROOF. (a) We will prove the statement by induction on $n\geq 1$. If $n=1$, there is nothing to prove (Lemma 4.2). Suppose $n>1$ and the result is true for direct products of $n-1$ completely indecomposable G-groups. Assume $H:=H_1\times\cdots\times H_n=L\times M$. Define $K=H_2\times\cdots\times H_n$, so that $H=H_1\times K=L\times M$. Let $\pi_L\colon H\to L$, $\pi_M\colon H\to M$, $\pi_1\colon H\to H_1$, $\pi_K\colon H\to K$ denote the corresponding canonical projections and $\varepsilon_L\colon L\to H$, $\varepsilon_M\colon M\to H$, $\varepsilon_1\colon H_1\to H$, $\varepsilon_K\colon K\to H$ denote the embeddings of L,M,H_1,K into H, respectively.

Then $\pi_1\varepsilon_L\pi_L\varepsilon_1$ and $\pi_1\varepsilon_M\pi_M\varepsilon_1$ are G-group endomorphisms of H_1, their sum is defined and is equal to 1_{H_1}. Since H_1 is completely indecomposable, either $\pi_1\varepsilon_L\pi_L\varepsilon_1$ or $\pi_1\varepsilon_M\pi_M\varepsilon_1$ is an automorphism of H_1. It may be assumed without loss of generality that $\pi_1\varepsilon_L\pi_L\varepsilon_1$ is an automorphism with an inverse α.

Now $\beta:=\pi_L\varepsilon_1\alpha\pi_1\varepsilon_L$ is an idempotent G-group endomorphism of L. From Remark 3.5 and the equalities $\beta(L)=\pi_L\varepsilon_1(H_1)$ and $\ker(\beta)=\ker(\pi_1\varepsilon_L)$, we get that $L=\pi_L\varepsilon_1(H_1)\times\ker(\pi_1\varepsilon_L)$ and the projection $\pi\colon L=\pi_L\varepsilon_1(H_1)\times\ker(\pi_1\varepsilon_L)\to\pi_L\varepsilon_1(H_1)$ of L onto $\pi_L\varepsilon_1(H_1)$ with $\ker(\pi)=\ker(\pi_1\varepsilon_L)$. More precisely, $\pi=(\pi_L\varepsilon_1)|^{\pi_L\varepsilon_1(H_1)}\circ\alpha\circ(\pi_1\varepsilon_L)$ where $\pi_L\varepsilon_1|^{\pi_L\varepsilon_1(H_1)}$ is the isomorphism obtained restricting the codomain of $\pi_L\varepsilon_1$ to $\pi_L\varepsilon_1(H_1)$.

If we notice that $\ker(\pi_1\varepsilon_L)=K\cap L$, then the equalities $H=H_1\times K=L\times M$ can be rewritten as $H=H_1\times(H_2\times\cdots\times H_n)=\pi_L\varepsilon_1(H_1)\times((K\cap L)\times M)$. So, the identity morphism of H can be written as a 2×2 matrix in which the $(1,1)$-entry is $\pi\pi_L\varepsilon_1\colon H_1\to\pi_L\varepsilon_1(H_1)$. Thus the $(1,1)$-entry in the 2×2 matrix is $(\pi_L\varepsilon_1)|^{\pi_L\varepsilon_1(H_1)}\circ(\pi_1\varepsilon_L\pi_L\varepsilon_1)^{-1}\circ\pi_1\varepsilon_L\pi_L\varepsilon_1$, which is an isomorphism, because it is the composite mapping of three isomorphisms. By Lemma 4.4, $H_2\times\cdots\times H_n\cong(K\cap L)\times M$, and so by the induction hypothesis, there is a partition $I'\dot\cup I_M$ of $\{2,3,\dots,n\}$ such that that $K\cap L\cong\prod_{i\in I'}H_i$ and $M\cong\prod_{i\in I_M}H_i$. Set $I_L=\{1\}\cup I'$. Then $\prod_{i\in I_L}H_i\cong H_1\times(\prod_{i\in I'}H_i)\cong\pi_L\varepsilon_1(H_1)\times(K\cap L)\cong L$.

(b) We use induction on n. The case $n=1$ is trivial by Lemma 4.2. Thus assume $H_1\times\cdots\times H_n\cong L_1\times\cdots\times L_m$ with the H_i completely indecomposable G-groups, the L_j indecomposable G-groups and $n\geq 2$. By applying (a) to the decomposition $L_1\times(L_2\times\cdots\times L_m)$, we get a partition $I_{L_1}\dot\cup I_{L_2\times\cdots\times L_m}$ of $\{1,2,\dots,n\}$ such that that L_1 and $L_2\times\cdots\times L_m$ are isomorphic to the direct products $\prod_{i\in I_{L_1}}H_i$ and $\prod_{i\in I_{L_2\times\cdots\times L_m}}H_i$, respectively. But L_1 is indecomposable, so that there is a unique element (say i_0) in I_{L_1}. Applying the induction hypothesis to the group $\prod_{i\in I_{L_2\times\cdots\times L_m}}H_i=\prod_{i\neq i_0}H_i$, which is isomorphic to $L_2\times\cdots\times L_m$ yields that $n=m$ and there is a bijection $\sigma'\colon\{1,2,\dots,n\}\setminus\{i_0\}\to\{2,3,\dots,m\}$ such that $H_i\cong L_{\sigma'(i)}$ for every $i=1,2,\dots,n$ different from i_0. Defining $\sigma\colon\{1,2,\dots,n\}\to\{1,2,3,\dots,m\}$ as the map extending σ' such that $\sigma(i_0)=1$ gives the desired result. □

5. Fitting's Lemma for G-groups in $\mathcal{C}_G$

LEMMA 5.1. *Let* H *be a* G*-group in* $\mathcal{C}_G$*. Then there is a one-to-one correspondence between the set of all finite* G*-direct decompositions of* H *and the set of all finite sets* $\{\pi_1,\dots,\pi_n\}$ *of idempotent* G*-endomorphisms of* H *satisfying*

$$\pi_1+\cdots+\pi_n=1$$

$$\pi_i\pi_j = 0 \textit{ for every } i \neq j.$$

PROOF. Let $H = H_1 \times \cdots \times H_n$ be a G-direct decomposition of the G-group H. Since every element h of H can be written in a unique way as a product $h = h_1 \dots h_n$ of elements $h_i \in H_i$, $i = 1, \dots, n$, we can define the canonical idempotent G-endomorphisms $\pi_i \colon H \to H$, $\pi_i(h) = h_i$. Clearly, $\pi_i^2 = \pi_i$. Moreover, for every $h = h_1 \dots h_n \in H$ with $h_i \in H_i$, $(\pi_1 + \cdots + \pi_n)(h) = \pi_1(h) \dots \pi_n(h) = h_1 \dots h_n = h$ and, for $i \neq j$, $\pi_i(\pi_j(h)) = \pi_i(h_j) = 1$. Thus the set $\{\pi_1, \dots, \pi_n\}$ of idempotent G-endomorphisms of H satisfies the desired conditions.

Conversely, suppose that there is a finite set of G-endomorphisms $\{\pi_1, \dots, \pi_n\}$ of H satisfying the two conditions above. Set $H_i = \pi_i(H)$, so that all the subgroups H_i are normal, because $H \in \mathcal{C}_G$. From $\pi_1 + \cdots + \pi_n = 1$, it follows that $H = H_1 \dots H_n$. Moreover, if $h \in H_i \cap \prod_{j \neq i} H_j$, then $h = \pi_i(h')$ for some $h' \in H$ and $\pi_i(h) = h$. But $\pi_j(h) = 1$ since $h \in \prod_{j \neq i} H_j$ and $\pi_i\pi_j = 0$ for $i \neq j$. So, $h = 1$. Hence $H = H_1 \times \cdots \times H_n$. $\square$

LEMMA 5.2. *Let H be a G-group in $\mathcal{C}_G$. If ${}_GH$ satisfies the ACC and the DCC on G-subgroups, and f is a G-group endomorphism of H, then:*

(1) *There exists an integer $n \geq 1$ such that $f^n(M) = f^{n+1}(M)$ and $\ker(f^n) = \ker(f^{n+1})$. Moreover, $H = \ker(f^n) \times f^n(H)$ for any such integer $n \geq 1$.*
(2) *A G-endomorphism f of H is an automorphism if and only if it is injective, if and only if it is surjective.*

PROOF. (1) Let f be a G-group endomorphism. Then the chains of G-subgroups

$$\ker f \subseteq \ker f^2 \subseteq \ker f^3 \subseteq \dots$$

and

$$f(M) \supseteq f^2(M) \supseteq f^3(M) \supseteq \dots$$

are stationary, and therefore both of them become constant at a sufficiently large index n by the ACC and the DCC. Thus $f^n(M) = f^{n+1}(M)$ and $\ker(f^n) = \ker(f^{n+1})$. This proves the existence of such an n. Now assume that $n \geq 1$ is such that $f^n(M) = f^{n+1}(M)$ and $\ker(f^n) = \ker(f^{n+1})$. Then $f^t(H) = f^{t+1}(H)$ and $\ker(f^t) = \ker(f^{t+1})$ for every $t \geq n$, so that $f^n(H) = f^{2n}(H)$ and $\ker(f^n) = \ker(f^{2n})$. Let us show that $\ker(f^n)f^n(H) = H$. If $x \in H$, then $f^n(x) \in f^n(H) = f^{2n}(H)$, so that $f^n(x) = f^n(y)$ for some $y \in f^n(M)$. Therefore $z := xy^{-1} \in \ker(f^n)$, and $x = zy \in \ker(f^n)f^n(H)$. Next, if $x' \in \ker(f^n) \cap f^n(H)$, then $x' = f^n(y')$ for some $y' \in H$. Note that since $f^n(x') = f^{2n}(y') = 1$, we have $y' \in \ker(f^{2n}) = \ker(f^n)$, and so $x' = 1$. Therefore $\ker(f^n) \cap f^n(H) = 1$ and consequently $H = \ker(f^n) \times f^n(H)$.

(2) Let f be a G-endomorphism of H. Suppose f injective. Then, by (1), $\ker(f^n)f^n(H) = H$ for a suitable n. As f^n is injective, $\ker(f^n) = 1$, and therefore $f^n(H) = H$. In particular, f is an automorphism.

Now suppose f surjective. By (1), there exists an n for which $\ker(f^n) \cap f^n(H) = 1$. As f^n is surjective, $f^n(H) = H$, and therefore $\ker(f^n) = 1$. In particular, f is injective, and so f is an automorphism. $\square$

An endomorphism f of a group H is *nilpotent* if there exists an integer $n \geq 1$ such that $f^n = 0$.

LEMMA 5.3. *Let H be an indecomposable G-group in $\mathcal{C}_G$. If ${}_GH$ satisfies the ACC and the DCC on G-subgroups, then any G-endomorphism of H is either nilpotent or an automorphism.*

PROOF. Let $f \in \mathrm{End}_G(H)$. By Lemma 5.2, there is a positive integer n such that $H = \ker(f^n) \times f^n(H)$. Since H is G-indecomposable, either $f^n(H) = 1$ or $\ker(f^n) = 1$. Thus either $f^n = 0$ or f^n is an automorphism, hence f is either nilpotent or an automorphism. □

THEOREM 5.4. *Let H be a G-group in $\mathcal{C}_G$. If ${}_GH$ is an indecomposable G-group with the ACC and the DCC on G-subgroups, then ${}_GH$ is completely indecomposable.*

PROOF. In order to show that H is completely indecomposable, we must show that if the sum $f+g$ of two G-endomorphisms of H exists and is an automorphism, then either f or g is an automorphism. Replacing f and g with $f(f+g)^{-1}$ and $g(f+g)^{-1}$ respectively, we can suppose that the sum $f+g$ exists and is the identity endomorphism of G, and we must prove that either f or g is an automorphism. Suppose the contrary, that is, f and g are two non-invertible endomorphisms of H, but $f+g$ exists and is the identity of G. As we have just seen in the previous lemma, there exists a positive integer n such that $f^n = 0 = g^n$. Let us prove that $fg = gf$. From $f+g = 1$, we know that $f(x) = xg(x)^{-1}$ for every $x \in H$. Thus $fg(x) = g(x)g^2(x^{-1}) = g(xg(x^{-1})) = gf(x)$. Hence $fg = gf$. In view of the binomial theorem,

$$1 = (f+g)^{2n-1} = \sum_{i=0}^{2n-1} \binom{2n-1}{i} f^i g^{2n-i-1} = 0.$$

This implies that $H = 1$, which is a contradiction. □

COROLLARY 5.5. *If ${}_GH$ is a finite indecomposable G-group in $\mathcal{C}_G$, then ${}_GH$ is completely indecomposable.*

REMARK 5.6. What we have proved in this Section is very similar to what is presented in [**5**, pp. 81–84]. In that book, Ω-groups are considered. Here Ω is any set, and an *Ω-group* consists of a group H with a mapping $\alpha\colon \Omega \to \mathrm{End}(H)$ from Ω into the monoid of all group endomorphisms of the group H. Now every mapping $\alpha\colon \Omega \to \mathrm{End}(H)$ extends uniquely to a monoid morphism $\widetilde{\alpha}\colon W_\Omega \to \mathrm{End}(H)$ from the free monoid W_Ω of all words in the alphabet Ω. Thus, in the language of Remark 2.1, Robinson's Ω-groups are functors from the category $\mathcal{C}$ with a unique object $*$ whose endomorphism monoid is $\mathrm{End}_{\mathcal{C}}(*) = W_\Omega$ into the category **Grp** of groups and the natural transformations between two functors $\mathcal{C} \to$ **Grp** correspond to Ω-group morphisms. In the case of our G-groups, we have that $\mathcal{C}$ is the category with a unique object $*$ and with endomorphism monoid $\mathrm{End}_{\mathcal{C}}(*) := G$, the functors of this category $\mathcal{C}$ into the category **Grp** of groups correspond to G-groups, and the natural transformations between two functors $\mathcal{C} \to$ **Grp** correspond to G-group morphisms. Thus Robinson's Ω-groups and our G-groups cover different settings.

Notice that every free monoid W_Ω is antiisomorphic to itself via the involution $W_\Omega \to W_\Omega$, $\omega_1 \dots \omega_n \mapsto \omega_n \dots \omega_1$, so that the setting is left-right symmetric and there is no need in considering both right and left Ω-groups.

Acknowledgements

The authors are grateful to the referee for helpful comments and suggestions. This work was carried out while the first author was visiting the Università di Padova. She would like to express her gratitude to the members of the Department of Mathematics.

References

[1] M. J. A. Paniagua and A. Facchini, *Category of G-groups and its spectral category*, Comm. Algebra **45** (2017), no. 4, 1696–1710, DOI 10.1080/00927872.2016.1222409. MR3576687

[2] M. J. Arroyo Paniagua and A. Facchini, *G-groups and biuniform abelian normal subgroups*, Adv. Group Theory Appl. **2** (2016), 79–111. MR3593016

[3] N. Bourbaki, *Éléments de mathématique. Fasc. XXVI. Groupes et algèbres de Lie. Chapitre I: Algèbres de Lie* (French), Seconde édition. Actualités Scientifiques et Industrielles, No. 1285, Hermann, Paris, 1971. MR0271276

[4] A. Facchini and A. Lucchini, *The Krull-Schmidt theorem holds for finite direct products of biuniform groups*, Algebr. Represent. Theory **21** (2018), no. 2, 309–329, DOI 10.1007/s10468-017-9715-y. MR3780769

[5] D. J. S. Robinson, *A course in the theory of groups*, 2nd ed., Graduate Texts in Mathematics, vol. 80, Springer-Verlag, New York, 1996. MR1357169

Department of Mathematics, Hacettepe University, 06800 Beytepe, Ankara, Turkey
Email address: meltemaltun@hacettepe.edu.tr

Dipartimento di Matematica, Università di Padova, 35121 Padova, Italy
Email address: facchini@math.unipd.it

Contemporary Mathematics
Volume **727**, 2019
https://doi.org/10.1090/conm/727/14623

Good codes from metacyclic groups

Samir Assuena and César Polcino Milies

Abstract. In this paper, we consider semisimple group algebras $\mathbb{F}_q G$ of split metacyclic groups over finite fields. We construct left codes in $\mathbb{F}_q G$ in the case when the order G is $p^m \ell^n$, where p and ℓ are different primes such that $gcd(q,p,\ell) = 1$.

1. Introduction

A group G is **metacyclic** if G contains a cyclic normal subgroup H such that the factor group G/H is also cyclic. The dihedral groups and groups all whose Sylow subgroups are cyclic are examples of such groups, [**11**, 10.1.10].

Let G be a metacyclic group, $H = \langle a \rangle$ its cyclic normal subgroup, and set $G/H = \langle bH \rangle$. Then G has the following presentation

$$G = \left\langle a, b \mid a^m = 1, b^n = a^s, bab^{-1} = a^r \right\rangle$$

and the integers m,n,s, r satisfy the relations

$$s \mid m, \quad m \mid s(r-1) \quad , \quad r < m, \quad \gcd(r,m) = 1.$$

When $s = m$, we say G is *split*. In this case, $G = \langle a \rangle \rtimes \langle b \rangle$

A *group code* over a field $\mathbb{F}$ is any ideal I of the group algebra $\mathbb{F}G$ of a finite group G. A code is said to be metacyclic, abelian, or dihedral in case the given group G is of that kind. If I is two-sided, then it is called a *central* code. A *minimal central code* is an ideal I which is minimal in the set of all (two-sided) ideals of $\mathbb{F}G$.

The *weight* of an element $\alpha = \sum_{g \in G} \alpha_g g$ is

$$w(\alpha) = \mid \{g \mid \alpha_g \neq 0, \ g \in G\} \mid$$

that is, the number of elements of the support of α. The *weight* of an ideal I is

$$w(I) = \min\{w(\alpha) \mid \alpha \neq 0, \ \alpha \in I\}.$$

Given a code I, let A_k denote the number of elements in I of weight k. The set of numbers $A_0, A_1, \cdots, A_n$ is called the *weight distribution of I*.

In this paper we study left group codes and give a series of examples to show that they give a simple way to produce codes similar to the best know ones according to [**5**].

2010 *Mathematics Subject Classification.* Primary 20C05; Secondary 16S34.

Key words and phrases. Semisimple group algebras, finite groups, split metacyclic groups, primitive idempotents.

The second author was partially supported by CNPq., Proc. 300243/79-0(RN) and FAPESP, Proc 2015/09162-9.

2. Construction of non central idempotents

In this section we recall some known-facts and give useful facts about left group codes.

DEFINITION 2.1. Let G_1 and G_2 be finite groups of the same order and $\mathbb{F}$ be a field. Let $\mathbb{F}G_1$ and $\mathbb{F}G_2$ be the corresponding group algebras. A *combinatorial equivalence* is a linear isomorphism $\phi : \mathbb{F}G_1 \longrightarrow \mathbb{F}G_2$ induced by a bijection $\phi : G_1 \longrightarrow G_2$. Codes $C_1 \subset \mathbb{F}G_1$ and $C_2 \subset \mathbb{F}G_2$ are *combinatorially equivalent* if there exists a combinatorial equivalence $\phi : \mathbb{F}G_1 \longrightarrow \mathbb{F}G_2$ such that $\phi(C_1) = C_2$.

Clearly, combinatorially equivalent codes have the same weight distribuition and dimensions.

THEOREM 2.2. [**12**, Theorem 1] *Let G be a metacyclic group with presentation*

$$G = \langle a, b \mid a^m = 1 = b^n,\ \ bab^{-1} = a^r \rangle.$$

Let C_m and C_n be cyclic groups of order m and n, with generators $\tilde{a}$ and $\tilde{b}$ respectively and let $\phi : G \longrightarrow C_m \times C_n$ be the bijection given by $\phi(a^i b^j) = \tilde{a}^i \tilde{b}^j$. If C is a code which is generated by a central idempotent e in $\mathbb{F}G$, then $\phi(C)$ is the ideal of $\mathbb{F}[C_m \times C_n]$ generated by $\phi(e)$, an idempotent of $\mathbb{F}[C_m \times C_n]$. Consequently, every central metacyclic code is combinatorially equivalent to an abelian code.

For a subgroup S of a group $\mathcal{G}$ such that $gcd(q, \mid S \mid) = 1$, we set

$$\widehat{S} = \frac{1}{|S|} \sum_{x \in S} x.$$

Then $\widehat{S}$ is an idempotent of $\mathbb{F}_q\mathcal{G}$ and is central if and only if S is a normal subgroup. [**9**, Lemma 3.6.6].

The following result is proved in [**3**, Proposition 2.1] when $H \triangleleft G$ but, actually, the proof does not require normality.

LEMMA 2.3. *Let G be a finite group and let $\mathbb{F}$ a field. Let H, K be subgroups of G with $H \subset K$. Write $e = \widehat{H} - \widehat{K}$. Then:*

1. *$dim_{\mathbb{F}}[(\mathbb{F}G)e] = (G : H) - (G : K)$;*
2. *$w((\mathbb{F}G)e) = 2|H|$;*
3. *If $\mathcal{A}$ is a transversal of K in G and τ is a transversal of H in K containing 1, then the set*

$$\{r(1-t)\widehat{H} \mid r \in \mathcal{A}, t \in \tau \setminus \{1\}\}$$

is a basis of $(\mathbb{F}G)e$ over $\mathbb{F}$.

Let G be a split metacyclic group of order $p^m \ell^n$ with presentation

$$G = \langle a, b \mid a^{p^m} = 1 = b^{\ell^n}, bab^{-1} = a^r \rangle$$

and $\mathbb{F}_q$ be a finite field with q elements such that $gcd(q, p^m \ell^n) = 1$. In this case, the group algebra $\mathbb{F}_q G$ is semisimple and it can be decomposed as direct sum of simple rings. Futhermore, since $b^x a^j b^{-x} = a^{jr^x}$, the conjugacy class of the element $a^j \in G$ is

$$\mathcal{C}_{a^j} = \{a^j, a^{jr}, a^{jr^2}, \cdots, a^{jr^{\ell^n - 1}}\}.$$

Set $H = \langle a \rangle$ and let

$$H = H_0 \supseteq H_1 \supseteq \cdots \supseteq H_m = \{1\}$$

be the descending chain of all subgroups of H, i.e., $H_j = \langle a^{p^j}\rangle$, $0 \le j \le m$. Consider the idempotents

$$e_0 = \widehat{H} \quad \text{and} \quad e_j = \widehat{H_j} - \widehat{H_{j-1}},\ \ 1 \le j \le m,$$

which are central in $\mathbb{F}_q G$. Write $\widehat{H} = \widehat{a}$ and $\widehat{\langle b\rangle} = \widehat{b}$, so the elements $\widehat{b}e_j$, $1 \le j \le m$, are non central idempotents of $\mathbb{F}_q G$.

Note that, if $\widehat{b}e_j = 0$, for some j, then $e_j + \left(be_j + b^2 e_j + \cdots + b^{\ell^n - 1} e_j\right) = 0$. But $supp(e_j) \in H$ and $supp(be_j + b^2 e_j + \cdots + b^{\ell^n-1}e_j) \in bH$, so these elements have disjoint support. Consequently $\widehat{b}e_j \neq 0$ for all indices j, $0 \le j \le m$.

The next result will be useful.

PROPOSITION 2.4. [**10**, Proposition 12.1] *Let B be a simple associative algebra. The center $\mathcal{Z}(B)$ of B is a field.*

PROPOSITION 2.5. *Let e be a central idempotent. Then the left ideal $\mathbb{F}_q G \cdot \widehat{b}e$ is minimal if and only if the ideal $\mathbb{F}_q G \cdot e$ is minimal as a two-sided ideal.*

PROOF. Let e be a primitive central idempotent. Assume that $\mathbb{F}_q G \cdot \widehat{b}e$ is not minimal. Then, there exist orthogonal idempotents f_1, f_2 such that $\widehat{b}e = f_1 + f_2$ (see [**9**, Theorem 2.5.11].)

Since $0 = (1-\widehat{b})\widehat{b}e = (1-\widehat{b})f_1 + (1-\widehat{b})f_2$, multiplying on the right by f_1, we obtain $(1-\widehat{b})f_1 = 0$, and similarly, $(1-\widehat{b})f_2 = 0$. Hence $f_1 = \widehat{b}f_1$ and $f_2 = \widehat{b}f_2$.

Since $\widehat{b}e(1-\widehat{b}) = 0$, we also have that $f_1(1-\widehat{b}) + f_2(1-\widehat{b}) = 0$ and, as before, we get $f_1 = f_1\widehat{b}$ and $f_2 = f_2\widehat{b}$.

Then $f_1 = f_1\widehat{b} = \widehat{b}f_1$ so $f_1\widehat{b} = \widehat{b}f_1\widehat{b} = f_1$. Similarly, $\widehat{b}f_2\widehat{b} = f_2$.

Write $f_t = \sum_k \alpha_k a^k + \sum_{i,j} \beta_{ij} a^i b^j$. Since $bab^{-1} = a^r$, for each k, $1 \le k \le p^m - 1$, we have that $b^x a^k = a^{kr^x} b^x$ and

$$\begin{aligned}\widehat{b}a^k\widehat{b} &= (a^k + a^{kr}b + a^{kr^2}b^2 + a^{kr^3}b^3 + \cdots + a^{kr^{\ell^n-1}}b^{\ell^n-1})\widehat{b}\\ &= (a + a^{kr} + a^{kr^2} + a^{kr^3} + \cdots + a^{kr^{\ell^n-1}})\widehat{b}\\ &= \Gamma_{a^k}\widehat{b}\end{aligned}$$

where $\Gamma_{a^k} = \sum_{x \in \mathcal{C}_{a^k}} x \in \mathcal{Z}(\mathbb{F}_q G)$.

Hence, there exist elements $\lambda_1, \lambda_2 \in \mathcal{Z}(\mathbb{F}_q G) \cap \mathbb{F}_q\langle a\rangle$ such that $f_1 = \lambda_1\widehat{b}$ and $f_2 = \lambda_2\widehat{b}$ and as $f_t = f_t e$ we have $f_t = \lambda_t e\widehat{b}$ and we may assume that $\lambda_t \in \mathcal{Z}(\mathbb{F}_q G)\cdot e \cap \mathbb{F}_q\langle a\rangle$, $t = 1, 2$.

Finally, $f_1 f_2 = (\lambda_1\lambda_2)\widehat{b} = 0$, and as $\lambda_1\lambda_2 \in \mathbb{F}_q\langle a\rangle$, we have that $\lambda_1\lambda_2 = 0$. Since e is a primitive idempotent by, Proposition 2.4 $\lambda_1 = 0$ or $\lambda_2 = 0$, so $f_1 = 0$ or $f_2 = 0$ and the result follows.

Conversely, if $\mathbb{F}_q G \cdot e$ is not minimal, as before, there exist non zero orthogonal central idempotents e_1 and e_2 such that $e = e_1 + e_2$. Then, $\widehat{b}e = \widehat{b}e_1 + \widehat{b}e_2$, with $(\widehat{b}e_1)(\widehat{b}e_2) = 0$. So, the left ideal $\mathbb{F}_q G \cdot \widehat{b}e$ is not minimal. □

We shall show that the codes $\mathbb{F}_q G \cdot \widehat{b}e_j$ and $\mathbb{F}_q G \cdot (1-\widehat{b})e_j$ are combinatorially equivalent to cyclic codes. To see this, consider first a cyclic group C of order $p^m\ell^n$, which we write as $C = C_{p^m} \times C_{\ell^n}$, where $C_{p^m} = \langle u\rangle$ is the cyclic subgroup of C of order p^m and $C_{\ell^n} = \langle v\rangle$ of order ℓ^n. Denote by

$$C_{p^m} = K_0 \supseteq K_1 \supseteq \cdots \supseteq K_m = \{1\}$$

the descending chain of subgroups of C_{p^m} and, as before, write $f_j = \widehat{K_j} - \widehat{K_{j-1}}$.

PROPOSITION 2.6. *With the notations above, the map*

$$\gamma : G \longrightarrow C = C_{p^m} \times C_{\ell^n}$$
$$\gamma(a^i b^j) \longmapsto u^i v^j$$

extends linearly to $\mathbb{F}_q G$ *and is such that* $\gamma(\widehat{b}e_j) = \widehat{v} f_j$ *and* $\gamma((1-\widehat{b})e_j) = (1-\widehat{v})f_j$ *so*

$$\gamma(\mathbb{F}_q G \cdot \widehat{b}e_j) = \mathbb{F}_q C \cdot \widehat{v} f_j \quad \text{and} \quad \gamma(\mathbb{F}_q G \cdot (1-\widehat{b})e_j) = \mathbb{F}_q C \cdot (1-\widehat{v}) f_j.$$

Consequently, $\mathbb{F}_q G \cdot \widehat{b}e_j$ *and* $\mathbb{F}_q C \cdot (1-\widehat{b})e_j$ *are combinatorially equivalent to cyclic codes.*

PROOF. It follows directly, from the definition of γ, that $\gamma(\widehat{H_i}) = \widehat{K_i}$ and $\gamma(\widehat{bH_i}) = \widehat{v}\widehat{K_i}$. We claim that

$$\gamma(g\widehat{b}e_j) = \gamma(g)\widehat{v} f_j, \quad \text{for all } g \in G.$$

In fact, g is of the form $g = a^h b^s$, for some positives integers h, s.

Notice that $\widehat{b}\widehat{H_j} = \widehat{\langle b \rangle \cdot H_j}$ so $\gamma(a^h \widehat{b}\widehat{H_j}) = u^h \widehat{v}\widehat{K_j}$, thus

$$\gamma(g\widehat{b}e_j) = \gamma\left(a^h b^s \left(\widehat{b}e_j\right)\right) = \gamma\left[\left(a^h \widehat{b}\right)\left(\widehat{H_j} - \widehat{H_{j-1}}\right)\right] =$$
$$= \left(u^h \widehat{v}\right)\left(\widehat{K_j} - \widehat{K_{j-1}}\right) = (u^h v^s)(\widehat{v} f_j) = \gamma(g)\widehat{v} f_j.$$

Also, $\gamma(g(1-\widehat{b})) = \gamma(g)(1-\widehat{v})f_j$. Since γ is linear, the results follow. □

LEMMA 2.7. *For all* j, $0 \le j \le m$, *the elements* $e_j + \widehat{b}a(1-\widehat{b})e_j$ *are units inside the ideals* $\mathbb{F}_q G \cdot e_j$.

PROOF. Set $\alpha_j = e_j + \widehat{b}a(1-\widehat{b})e_j$ and let $\beta_j = e_j - \widehat{b}a(1-\widehat{b})e_j$. So

$$\alpha_j \beta_j = (e_j + \widehat{b}a(1-\widehat{b})e_j)(e_j - \widehat{b}a(1-\widehat{b})e_j)$$
$$= e_j - \widehat{b}a(1-\widehat{b})e_j + \widehat{b}a(1-\widehat{b})e_j = e_j.$$

Since e_j is the unity of $\mathbb{F}_q G \cdot e_j$ and $\beta_j \alpha_j = e_j$, the proof is completed. □

It is not difficult to see that $\alpha_j(\widehat{b}e_j)\alpha_j^{-1}$ and $\alpha_j^{-1}(\widehat{b}e_j)\alpha_j$ are equal to $(\widehat{b} \pm \widehat{b}a(1-\widehat{b}))e_j$, $1 \le j \le m$. Since conjugation is an automorphism, the elements of the form $(\widehat{b} \pm \widehat{b}a(1-\widehat{b}))e_j$, $1 \le j \le m$ are non central idempotents of $\mathbb{F}_q G \cdot e_j$.

By Lemma 2.3, the dimension of $\mathbb{F}_q G \cdot \widehat{b}e_j$ over $\mathbb{F}_q$ is $p^j - p^{j-1} = \varphi(p^j)$, where φ denotes Euler's totient finction. Hence the dimension of $\mathbb{F}_q G \cdot (\widehat{b} \pm \widehat{b}a(1-\widehat{b}))e_j$ over $\mathbb{F}_q$ is also $\varphi(p^j)$.

LEMMA 2.8. *The set*

$$\mathcal{B} = \{e_j, ae_j, a^2 e_j, \cdots, a^{\varphi(p^j)-1} e_j\}$$

is a basis of the ideal $\mathbb{F}_q G \cdot e_j$ *over* $\mathbb{F}_q$.

PROOF. We shall show that the elements $a^k e_j$ and $a^l e_j$ have disjoint supports if $k \neq l$, $0 \le k, l \le \varphi(p^j) - 1$. This readily shows that the set is linearly independent. As $|\mathcal{B}| = dim\, \mathbb{F}_q G \cdot e_j$, the result follows.

In fact, notice that, $supp(a^k e_j) \subset a^k \left\langle a^{p^{j-1}} \right\rangle$ and $supp(a^l e_j) \subset a^l \left\langle a^{p^{j-1}} \right\rangle$. Let $x \in a^k \left\langle a^{p^{j-1}} \right\rangle \cap a^l \left\langle a^{p^{j-1}} \right\rangle$, then we can write $x = a^k a^{tp^{j-1}} = a^l a^{sp^{j-1}}$ then $a^{k-l+p^{j-1}(t-s)} = 1$.

This implies that $p^m \mid (k-l+p^{j-1}(t-s))$. Since $p^{j-1} \mid p^m$, we have $p^{j-1} \mid k-l$, so $k-l=0$. □

PROPOSITION 2.9. *Write* $f = (\widehat{b} + \widehat{b}a(1-\widehat{b}))e_j$. *If* e_j *is a central primitive idempotent of* $\mathbb{F}_q\langle a\rangle$, *then the set*

$$\mathcal{B} = \{f, af, a^2f, \cdots, a^{\varphi(p^j)-1}f\}$$

is a basis of the left ideal $\mathbb{F}_qG\cdot f$ *over* $\mathbb{F}_q$.

PROOF. Let $\Gamma_a = \sum_{x\in\mathcal{C}_a} x$ and $f = (\widehat{b} + \widehat{b}a(1-\widehat{b}))e_j$. As in Proposition 2.5 we have $\widehat{b}a\widehat{b} = \Gamma_a\widehat{b}$. We can write f as

$$f = \left(\widehat{b} + \widehat{b}a - \Gamma_a\widehat{b}\right)e_j = \left(\widehat{b}(1+a-\Gamma_a)\right)e_j.$$

We shall show that $\mathcal{B}$ is linearly independent. Assume that

$$\alpha_0 f + \alpha_1 af + \ldots + \alpha_{\varphi(p^j)-1}a^{\varphi(p^j)-1}f = 0.$$

Then $\left(\sum_{i=0}^{\varphi(p^j)-1}\alpha_i a^i\right)f = \left(\sum_{i=0}^{\varphi(p^j)-1}\alpha_i a^i\right)\left(\left(\widehat{b}(1+a-\Gamma_a)\right)e_j\right) = 0$

As before, considering disjoint supports, we see that $\left(\sum_{i=0}^{\varphi(p^j)-1}\alpha_i a^i\right)(1+a-\Gamma_a)e_j = 0$. So $\left(\sum_{i=0}^{\varphi(p^j)-1}\alpha_i a^i\right)e_j(1+a-\Gamma_a)e_j = 0$

As e_j is a primitive idempotent in $\mathbb{F}_q\langle a\rangle$, the ideal $\mathbb{F}_q\langle a\rangle\cdot e_j$ is a field.

Computing the coefficient of 1 in $(1+a-\Gamma_a)e_j = ((1+a-\Gamma_a)(\widehat{a^{p^j}} - \widehat{a^{p^{j-1}}})$ it is easly seen that $(1+a-\Gamma_a)e_j \neq 0$; so $\sum_{i=0}^{\varphi(p^j)-1}(\alpha_i a^i)e_j = 0$, and by Lemma 2.8, $\alpha_i = 0$, for all i. Since the number of elements of $\mathcal{B}$ is $\varphi(p^j)$ and the dimension of $\mathbb{F}_qG\cdot f$ is also $\varphi(p^j)$, the result follows. □

Notice that all the elements of the basis found above have the same weight.

3. Examples

In the examples that follow, the weight distributions of the codes under consideration were obtained using a simple computer program written in Sage.

Example 1. Set $G = \langle a, b \mid a^7 = 1 = b^3, bab^{-1} = a^2\rangle$.

The central primitive idempotents of $\mathbb{F}_5G$ are $f_1 = \widehat{b}\widehat{a}, f_2 = (1-\widehat{b})\widehat{a}$ and $e_1 = 1-\widehat{a}$, and then by [**2**, Theorem 3]

$$\mathbb{F}_5G \cong \mathbb{F}_5 \oplus \mathbb{F}_{25} \oplus M_3(\mathbb{F}_{25}).$$

Compute

$f = (\widehat{b} + \widehat{b}a(1-\widehat{b}))e_1 = (\widehat{b} + \widehat{b}a(1-\widehat{b}))(1-\widehat{a}) = \widehat{b} + \widehat{b}a - \widehat{b}a\widehat{b} - \widehat{G}$

$= 2(1+b+b^2)+2(a+a^2b+a^4b)+a+ab+ab^2+a^2b+a^2b^2+a^2+a^4b^2+a^4+a^4b+4\widehat{G}.$

So its weight is $w(f) = 15$ and, by Proposition 2.5, this element is a primitive idempotent. Furthermore, the set

$$\mathcal{B} = \{f, af, a^2f, a^3f, a^4f, a^5f\}$$

is a basis of the left ideal $\mathbb{F}_5G \cdot f$ over $\mathbb{F}_5$ and the weight distribution of the ideal $\mathbb{F}_5G \cdot f$ is the following

weight	0	10	11	12	13	14	15	16	17	18	19	20	21
words	1	16	48	172	412	1192	1600	2892	3588	3008	1768	788	140

Consequently, the weight is 10. We have found a minimal [21,6,10] left code over $\mathbb{F}_5$ which improves the [21,6,8] minimal left code obtained [**8**, page 372], though it is not the best code known, according to [**5**].

Example 2. Set $G = \langle a, b \mid a^7 = 1 = b^3, bab^{-1} = a^2 \rangle$.

The central primitive idempotents of $\mathbb{F}_2G$ are $f_1 = \widehat{b}\widehat{a}$, $f_2 = (1 - \widehat{b})\widehat{a}$, $f_3 = \frac{1}{7}\left(3 + (\xi + \xi^2 + \xi^4)\Gamma_a + (\xi^3 + \xi^5 + \xi^6)\Gamma_{a^3}\right)$ and $f_4 = \frac{1}{7}\left(3 + (\xi^3 + \xi^5 + \xi^6)\Gamma_a + (\xi + \xi^2 + \xi^4)\Gamma_{a^3}\right)$, where ξ is a primitive 7th root of unity.

Then

$$\mathbb{F}_2G \cong \mathbb{F}_2 \oplus \mathbb{F}_4 \oplus M_3(\mathbb{F}_2) \oplus M_3(\mathbb{F}_2).$$

Take $e_1 = 1 + \widehat{a}$ which is not a central primitive idempotent and compute $f = (\widehat{b} + \widehat{b}a(1 + \widehat{b}))e_1 = (\widehat{b} + \widehat{b}a(1 + \widehat{b}))(1 + \widehat{a})$
$= 1 + b + b^2 + a + a^2b + a^4b + a + ab + ab^2 + a^2b + a^2b^2 + a^2 + a^4b^2 + a^4 + a^4b + \widehat{G}.$

So the weight of f is $w(f) = 12$ and, by Proposition 2.5, this idempotent is not primitive.

In this case, it is not difficult to see that the set

$$\mathcal{B} = \{f, af, a^2f, a^3f, a^4f, a^5f\}$$

is a basis of the left ideal $\mathbb{F}_2G \cdot f$ over $\mathbb{F}_2$. The weight distribution of this ideal is

weight	0	8	12
words	1	21	42

This is a [21,6,8]-code, which is not minimal and it has the same weight of the best known [21,6]-code (see [**5**]).

Example 3. Set $G = \langle a, b \mid a^{11} = 1 = b^5, bab^{-1} = a^3 \rangle$.

The central primitive idempotents of $\mathbb{F}_2G$ are $f_1 = \widehat{b}\widehat{a}$, $f_2 = (1 + \widehat{b})\widehat{a}$ and $e_1 = 1 + \widehat{a}$, then

$$\mathbb{F}_2G \cong \mathbb{F}_2 \oplus \mathbb{F}_{16} \oplus M_5(\mathbb{F}_4).$$

(see [**2**, Theorem 3]). Compute
$f = (\widehat{b} + \widehat{b}a(1 + \widehat{b}))e_1 = (\widehat{b} + \widehat{b}a(1 + \widehat{b}))(1 + \widehat{a})$

So its weight is $w(f) = 30$ and, by Proposition 2.5, this element is a primitive idempotent. Furthermore, the set

$$\mathcal{B} = \{f, af, a^2f, a^3f, a^4f, a^5f, a^6f, a^7f, a^8f, a^9f\}$$

is a basis of the left ideal $\mathbb{F}_2G \cdot f$ over $\mathbb{F}_2$. The weight distribution of this ideal is

weight	0	20	24	26	28	30	32	34
words	1	110	55	275	220	198	110	55

So, this is a [55,10,20] minimal left code which has the same weight as the minimal left code obtained in [**8**, Example 4] though, again, it is not the best possible (see[**5**]).

4. Left Dihedral Codes

Throughout this section, D will always denote the dihedral group of order $2p^m$ with presentation

$$D = \langle a, b \mid a^{p^m} = 1 = b^2,\ bab = a^{-1} \rangle.$$

As before, $\mathbb{F}_q$ will denote a finite field with q elements with $gcd(2p^m, q) = 1$. and we shall always assume that $\mathcal{U}(\mathbb{Z}_{p^m}) = \langle \overline{q} \rangle$; i.e. that the multiplicative order of q modulo p^n is $\varphi(p^n) = p^{n-1}(p-1)$.

Under these hypotheses, it has was shown in [**4**] that, for $A = \langle a \rangle$, the structure $\mathbb{F}A$ depends only on the subgroup structure of A. If

$$A = H_0 \supseteq H_1 \supseteq \cdots \supseteq H_m = \{1\}$$

is the descending chain of all subgroups of A, i.e., $H_j = \langle a^{p^j} \rangle$, $0 \leq j \leq m$, then the idempotents

$$e_{11} = \left(\tfrac{1+b}{2}\right)\widehat{A},\quad e_{22} = \left(\tfrac{1-b}{2}\right)\widehat{A},\quad e_j = \widehat{H_j} - \widehat{H_{j-1}},\ \ 1 \leq j \leq m$$

are the primitive central idempotents of $\mathbb{F}_q D$ ([**3**, Theorem 3.3]).

We shall construct some non-central ideals in a rather simple way and prove that these are not equivalent to abelian codes.

Let us denote by e, for short, one idempotent from the set $\{e_j,\ 1 \leq j \leq m\}$ which we fix for the rest of this section. It was shown in [**7**, Proposition 2.1] that the elements

$$e_{11} = \left(\tfrac{1+b}{2}\right)e, \qquad e_{12} = \left(\tfrac{1+b}{2}\right)a\left(\tfrac{1-b}{2}\right)e,$$
$$e_{21} = 4((a-a^{-1})e)^{-2}\left(\tfrac{1-b}{2}\right)a\left(\tfrac{1+b}{2}\right)e,\quad e_{22} = \left(\tfrac{1-b}{2}\right)e.$$

form a set of matrix units for $(\mathbb{F}D)e$; i.e., they verify the identities $e_{11} + e_{22} = e$ and $e_{ij}e_{hk} = \delta_{jh}e_{1k}$. Hence, by [**13**, Lemma VI.3.11], $(\mathbb{F}D)e$ is the full ring of 2×2 matrices over the centralizer of e_{12} in $(\mathbb{F}D)e$ and thus, e_{11} and e_{22} are primitive non-central idempotents.

Notice that, setting $H_j^* = \langle b \rangle . H_j$, $1 \leq j \leq m$, we have that $e_{11} = \widehat{H_j^*} - \widehat{H_{j-1}^*}$ so Lemma 2.3 readily gives

$$dim[(\mathbb{F}G) \cdot e_{11}] = \varphi(p^j) \quad \text{and} \quad w[(\mathbb{F}G) \cdot e_{11}] = 2|H_j| = 4p^{m-j}.$$

It is not difficult to show that $(\mathbb{F}G) \cdot e_{22}$ has the same dimension and weight.

It turns out that these particular idempotents generate codes $I_{11} = (\mathbb{F}D) \cdot e_{11}$ and $I_{22} = (\mathbb{F}D) \cdot e_{22}$ that are equivalent to cyclic codes. However, it will be easy to use them to produce better codes.

Example 4. Let D_9 be dihedral group of order 18, set $e = e_1 = \widehat{H_1} - \widehat{H_0}$, $f = e_{11} - e_{22}$ and let us consider the ideal $I = \mathbb{F}_q D_9 \cdot f$.

By Proposition 2.9, the set $\{f, af\}$ is a basis of this ideal. We have that

$$ae^{(1)} = (2a - a^2 + 1)e_1 + (2a + a^2 - 1)e_1 b$$

where

$$\begin{aligned}(2 - a + a^{-1})(\widehat{H_1} - \widehat{H_0}) &= 2\widehat{H_1} - a\widehat{H_1} + a^{-1}\widehat{H_1} - 2\widehat{H_0}\\ a(2 - a + a^{-1})(\widehat{H_1} - \widehat{H_0}) &= 2a\widehat{H_1} - a^2\widehat{H_1} + \widehat{H_1} - 2\widehat{H_0},\end{aligned}$$

consequently we can write an arbitrary element α of $\mathbb{F}_q D_9 f$ as

$$\begin{aligned}\alpha_0 f + \alpha_1 af \;=\; & (2\alpha_0 + \alpha_1)\widehat{H_1} + (-\alpha_0 + 2\alpha_1)a\widehat{H_1} + (\alpha_0 - \alpha_1)a^2\widehat{H_1} +\\ & (-2\alpha_0 - 2\alpha_1)\widehat{H_0} + (2\alpha_0 - \alpha_1)\widehat{H_1}b + (\alpha_0 + 2\alpha_1)a\widehat{H_1}b +\\ & (-\alpha_0 + \alpha_1)a^2\widehat{H_1}b + (-2\alpha_0 - 2\alpha_1)\widehat{H_0}b.\end{aligned}$$

A direct computation shows that, if the characteristic of $\mathbb{F}_q$ is different from 2,3,5 and 7, then at most one of the coefficients of α can be equal to zero. So, the weight of α is $w(\alpha) \geq 15$.

As it is easy to exibit elements of I of weight 15, we have $dim[I] = 2$ and $w(I) = 15$.

We remark that the weight of this code is the same as that of the best known code of same dimension (see [*www.codetables.de*]), for example in the case when the field is $\mathbb{F}_{11}$.

Finally, we shall show that, when the multiplicative order of q modulo 9 is $\varphi(9) = 6$, this code is not equivalent to any abelian code. Let A be an abelian group of order 18; then it can be written as $A = C_2 \times B$, where C_2 is a cyclic group of order 2 generated by an element t, and B is an abelian 3-group.

Suppose, by way of contradiction, that $I = \mathbb{F}_q D_9 f$ is combinatorially equivalent to a code J of $\mathbb{F}_q A$. Since $\mathcal{U}(\mathbb{Z}_{p^r}) = \langle \overline{q} \rangle$, by [**4**, Lemma 5 and Theorem 4.1], the primitive idempotents of $\mathbb{F}_q A$ are of the form $\left(\frac{1+t}{2}\right) e_H$, or $\left(\frac{1-t}{2}\right) e_H$ for H a subgroup of B such that $|B/H|$ is either 3 or 9. We have two possibilities:

(1) $J = I_{e_H}$.
Since the dimension of I is 2, there exists a subgroup H of B such that B/H is cyclic of order 3 and $J = (\mathbb{F}_q A)(\frac{1 \pm t}{2}) e_H$. By [**4**, Theorem 4.2], the weight of J is 12 , but the weight of $\mathbb{F}_q D_9 f$ is 15.

(2) $J = I_{e_{H_1}} \oplus I_{e_{H_2}}$.
Since the dimension of $\mathbb{F}_q D_9 f$ is 2, the dimension of $I_{e_{H_1}}$ and $I_{e_{H_2}}$ must be equal to 1. In this case, the order of B/H_1 and B/H_2 should be 2, which cannot happen since $|B| = 9$.

Example 5. Let $D_6 = S_3$ be the dihedral group of order 6 with the following presentation

$$D_6 = \langle a, b \mid a^3 = 1 = b^2, bab = a^2 \rangle.$$

and let $\mathbb{F}_q$ be a finite field with q elements such that $\mathcal{U}(\mathbb{Z}_3) = \langle \overline{q} \rangle$. By [**3**, Theorem 3.3]), the central primitive idempotents of $\mathbb{F}_q D_6$ are

$$\{e_{11}, e_{22}, e_1\}.$$

and we can write $f = e_{11} - e_{12}$ and set $I = \mathbb{F}_q D_6 \cdot f$. Since $dim[I] = 2$, the set $\{f, af\}$ is a basis over $\mathbb{F}_q$, and an element $\alpha \in \mathbb{F}_q D_6 \cdot f$ can be write as

$$\begin{aligned} \alpha = \alpha_0 f + \alpha_1 af = \tfrac{1}{12}[&(4\alpha_0 + \alpha_1)\, 1 + (-5\alpha_0 + 4\alpha_1)\, a + (\alpha_0 - 5\alpha_1)\, a^2 \\ &+ (4\alpha_0 - 5\alpha_1)\, b + (\alpha_0 + 4\alpha_1)\, ab + (-5\alpha_0 + \alpha_1)\, a^2 b]. \end{aligned}$$

If $q = 11$, a direct computation show us that $w(I) = 5$, the weight of the best known [6,2]-code. This is also the case for any field of characteristic different from 2,3,5 and 7.

References

[1] C. W. Curtis and I. Reiner, *Representation theory of finite groups and associative algebras*, Pure and Applied Mathematics, Vol. XI, Interscience Publishers, a division of John Wiley & Sons, New York-London, 1962. MR0144979

[2] G. K. Bakshi, S. Gupta, and I. B. S. Passi, *Semisimple metacyclic group algebras*, Proc. Indian Acad. Sci. Math. Sci. **121** (2011), no. 4, 379–396, DOI 10.1007/s12044-011-0045-4. MR2934345

[3] F. S. Dutra, R. A. Ferraz, and C. Polcino Milies, *Semisimple group codes and dihedral codes*, Algebra Discrete Math. **3** (2009), 28–48. MR2640386

[4] R. A. Ferraz and C. Polcino Milies, *Idempotents in group algebras and minimal abelian codes*, Finite Fields Appl. **13** (2007), no. 2, 382–393, DOI 10.1016/j.ffa.2005.09.007. MR2307135

[5] M. Grassl, *Bounds on the minimum distance of linear codes and quantum codes.* Online available at http://www.codetables.de.

[6] N. Jacobson, *Basic algebra. I*, 2nd ed., W. H. Freeman and Company, New York, 1985. MR780184

[7] E. Jespers, G. Leal, and C. Polcino Milies, *Units of integral group rings of some metacyclic groups*, Canad. Math. Bull. **37** (1994), no. 2, 228–237, DOI 10.4153/CMB-1994-034-0. MR1275709

[8] G. Olteanu and I. Van Gelder, *Construction of minimal non-abelian left group codes*, Des. Codes Cryptogr. **75** (2015), no. 3, 359–373, DOI 10.1007/s10623-014-9922-z. MR3336955

[9] C. Polcino Milies and S. K. Sehgal, *An introduction to group rings*, Algebra and Applications, vol. 1, Kluwer Academic Publishers, Dordrecht, 2002. MR1896125

[10] R. S. Pierce, *Associative algebras*, Graduate Texts in Mathematics, vol. 88, Springer-Verlag, New York-Berlin, 1982. Studies in the History of Modern Science, 9. MR674652

[11] D. J. S. Robinson, *A course in the theory of groups*, 2nd ed., Graduate Texts in Mathematics, vol. 80, Springer-Verlag, New York, 1996. MR1357169

[12] R. E. Sabin and S. J. Lomonaco, *Metacyclic error-correcting codes*, Appl. Algebra Engrg. Comm. Comput. **6** (1995), no. 3, 191–210, DOI 10.1007/BF01195337. MR1329366

[13] S. K. Sehgal, *Topics in group rings*, Monographs and Textbooks in Pure and Applied Math., vol. 50, Marcel Dekker, Inc., New York, 1978. MR508515

Centro Universitário da FEI, Av. Humberto de Alencar Castelo Branco 3972, São Bernardo do Campo-SP, CEP: 09850-901, Brazil

Email address: samir.assuena@fei.edu.br, sassuena@yahoo.com.br

Instituto de Matemática e Estatística, Universidade de São Paulo, Caixa Postal 66281, CEP-05315-970, São Paulo, SP, Brazil; and Universidade Federal do ABC, Av. dos Estados 5001, Santo André, CEP: 09210-580, SP, Brazil

Email address: polcino@ime.usp.br, polcino@ufabc.edu.br

Contemporary Mathematics
Volume **727**, 2019
https://doi.org/10.1090/conm/727/14624

The prime ideals and simple weight modules of the algebra $U(\mathfrak{b} \ltimes V_3)$

V. V. Bavula and T. Lu

ABSTRACT. The Euclidean algebra $\mathfrak{e}(3) = \mathfrak{sl}_2 \ltimes V_3$ is a Lie algebra which is a semidirect product of the Lie algebra $\mathfrak{sl}_2$ and its unique simple 3-dimensional module V_3. Let $\mathcal{A}$ be the universal enveloping algebra of the Lie subalgebra $\mathfrak{b} \ltimes V_3$ of $\mathfrak{e}(3)$ where $\mathfrak{b}$ is the Borel subalgebra of $\mathfrak{sl}_2$. The aim of this paper is to classify all the prime, maximal and primitive ideals of the algebra $\mathcal{A}$ and simple weight $\mathcal{A}$-modules. For each simple weight $\mathcal{A}$-module, an explicit finite set of generators is given for its annihilator. For each prime ideals of $\mathcal{A}$, an explicit set of generators is given. For prime ideals a containment criterion is given. As a consequence, a classification of prime, primitive and maximal ideals is given for the universal enveloping algebra of the Diamond Lie algebra.

1. Introduction

In the paper, $\mathbb{K}$ is a field of characteristic zero unless it is stated otherwise and module means a left module. The *Euclidean algebra* $\mathfrak{e}(3) = \mathfrak{sl}_2 \ltimes V_3$ is a Lie algebra which is a semidirect product of Lie algebras where $\mathfrak{sl}_2 = \mathbb{K}H \oplus \mathbb{K}E \oplus \mathbb{K}F$ and $V_3 = \mathbb{K}X \oplus \mathbb{K}Y \oplus \mathbb{K}Z$ is an abelian Lie algebra which is the unique three dimensional simple $\mathfrak{sl}_2$-module. The Euclidean algebra $\mathfrak{e}(3)$ is a 6-dimensional Lie algebra with basis H, E, F, X, Y, Z and the Lie structure is given by the rule:

$$\begin{array}{lllll} [H,E]=2E, & [H,F]=-2F, & [E,F]=H, & [H,X]=2X, & [H,Y]=-2Y, \\ [H,Z]=0, & [E,Y]=2Z, & [E,Z]=2X, & [E,X]=0, & [F,X]=Z, \\ [F,Z]=Y, & [F,Y]=0, & [X,Y]=[Y,Z]=[X,Z]=0. & & \end{array}$$

The universal enveloping algebra $\mathcal{U} := U(\mathfrak{e}(3))$ of $\mathfrak{e}(3)$ is a Noetherian domain of Gelfand–Kirillov dimension 6. A quantum analogue of $\mathcal{U}$, the, so-called, *quantum Euclidean algebra*, was defined and studied in [**8**].

Over the field $\mathbb{C}$ of complex numbers, the Euclidean algebra $\mathfrak{e}(3)$ is the complexification of the Lie algebra of the *Euclidean group* $E(3) \simeq SO(3) \ltimes \mathbb{R}^3$. Families of indecomposable representations of $\mathfrak{e}(3)$ have been constructed by embedding the Euclidean algebra $\mathfrak{e}(3)$ into the simple Lie algebra $\mathfrak{sl}(4,\mathbb{C})$ [**13**] or using the irreducible representations of $\mathfrak{sl}(2,\mathbb{C})$ [**14**].

2010 *Mathematics Subject Classification.* Primary 17B10, 16D25, 16D60, 16D70, 16P50.

Key words and phrases. Prime ideal, primitive ideal, weight module, Lie algebra, the universal enveloping algebra of a Lie algebra, simple module, centralizer.

The second author was supported by NSFC Grant 11701190.

Let $\mathfrak{g}$ be a finite dimensional Lie algebra and $U(\mathfrak{g})$ be its universal enveloping algebra. Every $\mathfrak{g}$-module is automatically a $U(\mathfrak{g})$-module, and vice versa. The problem of classification of *simple* $\mathfrak{g}$-modules is a very difficult one since there are plenty of simple $\mathfrak{g}$-modules. A slightly easier problem (but not much) is the problem of classification of prime and primitive ideals of the universal enveloping algebra $U(\mathfrak{g})$. An ideal is *primitive* if it is the annihilator of a simple module. Every primitive ideal is prime but not vice versa, in general.

Let $\mathcal{A}$ be the universal enveloping algebra of the Lie subalgebra $\mathfrak{a} := \mathfrak{b} \ltimes V_3$ of $\mathfrak{e}(3)$ where $\mathfrak{b} = \mathbb{K}H \oplus \mathbb{K}E$ is the *Borel* subalgebra of $\mathfrak{sl}_2$ and $\mathcal{D} := \mathcal{A}/(X)$. In Section 2, it is shown that the centre of the algebra $\mathcal{A}$ is a polynomial algebra $\mathbb{K}[C_1]$ where $C_1 = XY - \frac{1}{2}Z^2$ (Lemma 2.2) and the algebra $\mathcal{A}$ is a free $\mathbb{K}[C_1]$-module (Theorem 2.4).

The aim of the paper is to give an explicit description of prime, maximal and primitive ideals of the algebra $\mathcal{A}$ (Theorem 1.1, Corollary 1.1, Corollary 1.2). For each prime ideal of $\mathcal{A}$, an explicit finite set of generators is found (Theorem 1.1). For an algebra A, the set of prime and maximal ideals are denoted by $\operatorname{Spec}(A)$ and $\operatorname{Max}(A)$, respectively. Let $\vartheta := HZ + EY$.

THEOREM 1.1.

(1) $\operatorname{Spec}(\mathcal{A}) = \{(X, \mathfrak{p}) \mid \mathfrak{p} \in \operatorname{Spec}(\mathcal{D})\} \sqcup \{0, (\mathfrak{c}) \mid \mathfrak{c} \in \operatorname{Max}(\mathbb{K}[C_1])\}$.

(2) *The containments of prime ideals are given in the following diagram*

(1.1)

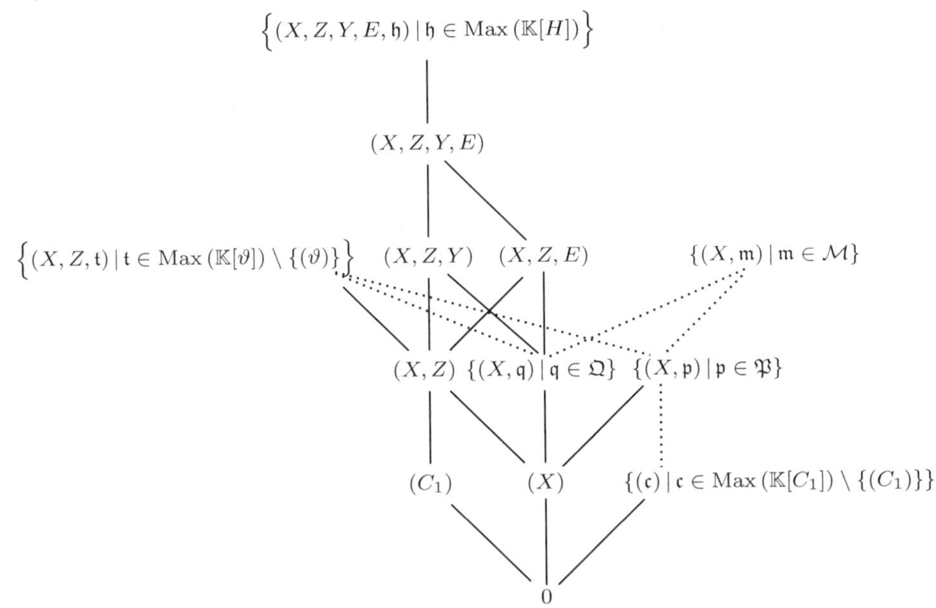

where the sets $\mathcal{M}$, $\mathfrak{P}$ *and* $\mathfrak{Q}$ *are defined in Proposition* 2.5. *Furthermore, let* $\mathfrak{c} \in \operatorname{Max}(\mathbb{K}[C_1]) \setminus \{(C_1)\}$, *i.e.,* $\mathfrak{c} = f(C_1)\mathbb{K}[C_1]$ *for some monic irreducible polynomial of* $\mathbb{K}[C_1]$ *such that* $f(0) \neq 0$, *and* $\mathfrak{p} \in \mathfrak{P}$. *Then* $(\mathfrak{c}) \subset (X, \mathfrak{p})$ *iff* $\mathfrak{p} = g(Z)\mathbb{K}[Z, \vartheta]$ *where* $g(Z) \in \mathbb{K}[Z]$ *is a monic irreducible polynomial of* $\mathbb{K}[Z]$ *which divides the polynomial* $f(-\frac{1}{2}Z^2)$.

Theorem 1.1.(2) gives a complete picture of containments of prime ideals: $P \subseteq Q$ iff there is a vertical path leading from P to Q. A dotted line between two sets of prime ideals means that each prime from the 'lower' set is contained in some prime ideals from the 'upper' set. These containments are made explicit in Proposition 2.5. The next corollary describes the set of maximal ideals of the algebra $\mathcal{A}$.

COROLLARY 1.1. $\mathrm{Max}\,(\mathcal{A}) = \{(X, Z, Y, E, \mathfrak{h}) \,|\, \mathfrak{h} \in \mathrm{Max}\,(\mathbb{K}[H])\} \sqcup \{(X, Z, \mathfrak{t}) \,|\, \mathfrak{t} \in \mathrm{Max}\,(\mathbb{K}[\vartheta]) \setminus \{(\vartheta)\}\} \sqcup \{(X, \mathfrak{m}) \,|\, \mathfrak{m} \in \mathcal{M}\}$ *where* $\mathcal{M} = \{\mathfrak{m} \in \mathrm{Max}\,(\mathbb{K}[Z, \vartheta]) \,|\, Z \notin \mathfrak{m}\}$.

PROOF. This follows from (1.1). □

The next corollary describes the set of primitive ideals $\mathrm{Prim}\,(\mathcal{A})$ of the algebra $\mathcal{A}$. Its proof is given in Section 2.

COROLLARY 1.2. $\mathrm{Prim}\,(\mathcal{A}) = \mathrm{Max}\,(\mathcal{A}) \sqcup \{(X, Z, Y), (X, Z, E), (C_1)\} \sqcup \{(\mathfrak{c}) \,|\, \mathfrak{c} \in \mathrm{Max}\,(\mathbb{K}[C_1]) \setminus \{(C_1)\}\}$.

In the literature, the Lie algebra $\mathfrak{d} := \mathfrak{a}/\mathbb{K}X$ is called the *Diamond Lie algebra*. So, $\mathcal{D} = U(\mathfrak{d})$. It is proved that the algebra $\mathcal{D}$ is a generalized Weyl algebra (Lemma 2.6). The prime, maximal and primitive ideals of the algebra $\mathcal{D}$ are described (Proposition 2.5, Corollary 2.6 and Corollary 2.7). In Section 3, simple weight $\mathcal{A}$-modules are classified up to isomorphism and, for each simple weight $\mathcal{A}$-module, an explicit finite set of generators is given for its annihilator (Theorem 3.11) where $\mathbb{K}$ is an algebraically closed field of characteristic zero. It is shown that for each simple weight $\mathcal{A}$-module all its weight spaces are either finite or infinite dimensional and they have the same dimension (Corollary 3.6). Classifications are given of simple weight $\mathcal{A}$-modules with finite and infinite dimensional weight spaces (Theorem 3.12 and Theorem 3.13).

An $\mathcal{A}$-module M is called a *weight module* if $M = \bigoplus_{\mu \in \mathbb{K}} M_\mu$ where $M_\mu = \{m \in M \,|\, Hm = \mu m\}$ is called a *weight subspace* of M provided $M_\mu \neq 0$ and such a μ is called a *weight* of M. The set of all weights of M is denoted by $\mathrm{Wt}\,(M)$. If M is a *simple* weight $\mathcal{A}$-module then every weight subspace M_μ of M is a *simple* module over the centralizer $C = \{a \in \mathcal{A} \,|\, aH = Ha\}$ of the element H in $\mathcal{A}$. So, the problem of classification of simple weight $\mathcal{A}$-modules is tightly connected with the problem of classification of *all* simple C-modules, which is the first step (the most difficult step). Another step is how to assemble some of the simple C-modules into a simple weight $\mathcal{A}$-module. In section 3, an explicit description of the algebra C and its centre is given (Proposition 3.1); simple C-modules are classified using [**6**] (Theorem 3.9). Prime, maximal and primitive ideals of the algebra C are classified (Theorem 3.14, Corollary 3.7, Corollary 3.8).

2. Prime ideals of the algebra $\mathcal{A}$

The aim of this section is to prove Theorem 1.1 and Corollary 1.2. One of the key steps in the proofs is to give a classification of prime, primitive and maximal ideals of the factor algebra $\mathcal{D} = \mathcal{A}/(X)$ of the algebra $\mathcal{A}$ (Proposition 2.5, Corollary 2.7 and Corollary 2.6). The algebra $\mathcal{D}$ turns out to be isomorphic to the universal enveloping algebra of the Diamond Lie algebra. Another key step in the proof of Theorem 1.1 is to use Proposition 2.1 in the case of the algebra $\mathcal{A}$ and the normal element $s = X$ of $\mathcal{A}$. The centre of $\mathcal{A}$ is found (Lemma 2.3) and it is proved that the algebra $\mathcal{A}$ is a free module over its centre (Theorem 2.4). Explicit generators are given for the semi-centre of the algebra $\mathcal{A}$ (Proposition 2.8).

The n-th *Weyl algebra* $A_n = A_n(\mathbb{K})$ is an associative algebra generated by elements $x_1, \dots, x_n, y_1, \dots, y_n$ subject to the defining relations: $[x_i, x_j] = 0$, $[y_i, y_j] = 0$ and $[y_i, x_j] = \delta_{ij}$ where $[a, b] := ab - ba$ and δ_{ij} is the Kronecker delta function. The Weyl algebra A_n is a central, simple, Noetherian domain of Gelfand–Kirillov dimension $2n$. For an algebra R, we denote by $Z(R)$ its centre. The algebra is called *central* if its centre is equal to the ground field. For a subset $S \subset R$, we denote by (S) the ideal of R generated by the elements of S. An element $a \in R$ is called a *normal element* if $aR = Ra$.

LEMMA 2.1. [**19**, Lemma 14.6.5] *Let B be a $\mathbb{K}$-algebra, $S = B \otimes A_n$ be the tensor product of the algebra B and the Weyl algebra A_n, δ be a $\mathbb{K}$-derivation of S and $T = S[t; \delta]$. Then there exists an element $s \in S$ such that the algebra $T = B[t'; \delta'] \otimes A_n$ is a tensor product of algebras where $t' = t + s$ and $\delta' = \delta + \mathrm{ad}_s$.*

The subalgebra $\mathcal{E}$. Let $\mathcal{E}$ be the subalgebra of $\mathcal{U}$ generated by the elements E, X, Y and Z. Then $\mathcal{E}$ is an iterated Ore extension $\mathcal{E} = \mathbb{K}[X, Z][E; \delta_1][Y; \delta_2]$ where $\delta_1(X) = 0$, $\delta_1(Z) = 2X$, $\delta_2(X) = \delta_2(Y) = 0$ and $\delta_2(E) = -2Z$. Clearly, X is a central element of $\mathcal{E}$ and the subalgebra $\mathbb{K}[X, Z][E; \delta_1]$ is isomorphic to the enveloping algebra of the three dimensional Heisenburg Lie algebra. Let $\mathcal{E}_X$ be the localization of $\mathcal{E}$ at the powers of the element X. Then

$$\mathcal{E}_X = \Big(\mathbb{K}[X^{\pm 1}] \otimes A_1\Big)[Y; \delta_2]$$

where $A_1 = \mathbb{K}\langle EX^{-1}, \frac{1}{2}Z\rangle$ is the first Weyl algebra since $[EX^{-1}, \frac{1}{2}Z] = 1$. Now, using Lemma 2.1 we can 'delete' the derivation δ_2. More precisely, the element $s = -\frac{1}{2}Z^2X^{-1}$ satisfies the conclusion of Lemma 2.1 since the element $Y' := Y + s = Y - \frac{1}{2}Z^2X^{-1}$ commutes with the elements of A_1. The element Y' also commutes with X, and so we have

$$\mathcal{E}_X = \mathbb{K}[X^{\pm 1}, Y'] \otimes A_1 = \mathbb{K}[X^{\pm 1}, C_1] \otimes A_1 \tag{2.1}$$

where $C_1 := Y'X = XY - \frac{1}{2}Z^2$. It can be easily verified that *the element C_1 belongs to the centre of the algebra $\mathcal{A}$*. Hence, $C_1 \in Z(\mathcal{E})$ and $C_1 \in Z(\mathcal{A})$.

LEMMA 2.2. $Z(\mathcal{E}) = \mathbb{K}[X, C_1]$.

PROOF. By (2.1), $Z(\mathcal{E}_X) = \mathbb{K}[X^{\pm 1}, C_1]$. Then $Z(\mathcal{E}) = \mathcal{E} \cap Z(\mathcal{E}_X) = \mathcal{E} \cap \mathbb{K}[X^{\pm 1}, C_1] = \mathbb{K}[X, C_1]$. □

The algebra $\mathcal{A}$. Since $\mathfrak{a}$ is a solvable Lie algebra, every prime ideal of $\mathcal{A}$ is completely prime [**19**, Corollary 14.5.5]. The algebra $\mathcal{A}$ is an Ore extension $\mathcal{A} = \mathcal{E}[H; \delta]$ where δ is a derivation of the algebra $\mathcal{E}$ defined by the rule $\delta(E) = 2E$, $\delta(X) = 2X$, $\delta(Y) = -2Y$ and $\delta(Z) = 0$. An element r of an algebra R is called a *normal* element if $Rr = rR$. The element X is a normal element of the algebra $\mathcal{A}$ since X is central in $\mathcal{E}$ and $XH = (H - 2)X$. Let $\mathcal{E}_X$ and $\mathcal{A}_X$ be the localizations of the algebras $\mathcal{E}$ and $\mathcal{A}$ at the powers of the element X. Then $\mathcal{A}_X = \mathcal{E}_X[H; \delta]$, by (2.1), $\mathcal{A}_X = \Big(\mathbb{K}[X^{\pm 1}, C_1] \otimes A_1\Big)[H; \delta]$. Since H commutes with the elements of A_1, the algebra $\mathcal{A}_X$ is a tensor product of three algebras

$$\mathcal{A}_X = \mathbb{K}[C_1] \otimes \mathbb{K}[X^{\pm 1}][H; \delta] \otimes A_1. \tag{2.2}$$

In particular, $\mathcal{A}_X$ is a Noetherian domain of Gelfand–Kirillov dimension 5. The algebra $\mathbb{K}[X^{\pm 1}][H; \delta]$ where $\delta(X) = 2X$ and the Weyl algebra A_1 are central simple algebras. Hence, $Z(\mathcal{A}_X) = \mathbb{K}[C_1]$.

LEMMA 2.3. $Z(\mathcal{A}) = Z(\mathcal{A}_X) = \mathbb{K}[C_1]$.

PROOF. Since $\mathbb{K}[C_1] \subseteq Z(\mathcal{A}) \subseteq \mathcal{A} \cap Z(\mathcal{A}_X) = \mathbb{K}[C_1]$, we have $Z(\mathcal{A}) = \mathbb{K}[C_1]$. □

The next theorem shows that the algebra $\mathcal{A}$ is a free module over its centre and presents an explicit free $Z(\mathcal{A})$-basis.

THEOREM 2.4. *The algebra $\mathcal{A}$ is a free module over its centre. Furthermore,*

$$\{H^i E^j X^k Z^m,\ H^i E^j Y^l Z^m \mid i,j,k \in \mathbb{N}, l \geqslant 1, m = 0,1\}$$

is a basis.

PROOF. The proof follows from the equality $C_1 = XY - \frac{1}{2}Z^2$. □

The factor algebra $\mathcal{D}$ of $\mathcal{A}$. In order to classify the prime ideals of the algebra $\mathcal{A}$, we classify first the prime ideals of the factor algebra $\mathcal{D} := \mathcal{A}/(X)$ of $\mathcal{A}$. Notice that the algebra $\mathcal{D}$ is isomorphic to the enveloping algebra $U(\mathfrak{d})$ of the Lie algebra $\mathfrak{d} := \mathfrak{a}/\mathbb{K}X$, i.e., the algebra $\mathcal{D}$ is generated by the elements H, E, Y, Z subject to the following relations:

$$[H,E] = 2E, \quad [H,Y] = -2Y, \quad [E,Y] = 2Z, \quad [Z,\cdot] = 0.$$

The four dimensional Lie algebra $\mathfrak{d}$ is called the *Diamond Lie algebra*. This algebra play an important role in conformal field theory, see for example [**20**]. Finite dimensional indecomposable representations of the Diamond Lie algebra were studied in [**11**]. The algebra $\mathcal{D}$ is an iterated Ore extension $\mathcal{D} = \mathbb{K}[Z,Y][E;\delta'][H;\delta'']$ where $\delta'(Z) = 0$, $\delta'(Y) = 2Z$, $\delta''(Z) = 0$, $\delta''(Y) = -2Y$ and $\delta''(E) = 2E$. Notice that Z is a central element of $\mathcal{D}$ and the subalgebra $\mathbb{K}[Z,Y][E;\delta']$ is isomorphic to the enveloping algebra of the three dimensional Heisenburg Lie algebra. Let $\mathcal{D}_Z$ be the localization of $\mathcal{D}$ at the powers of the element Z. Then

$$\mathcal{D}_Z = \left(\mathbb{K}[Z^{\pm 1}] \otimes A_1'\right)[H;\delta'']$$

where the algebra $A_1' := \mathbb{K}\langle E, \frac{1}{2}YZ^{-1}\rangle$ is isomorphic to the first Weyl algebra since $[E, \frac{1}{2}YZ^{-1}] = 1$. Now, using Lemma 2.1 we can 'delete' the derivation δ'' using the element $s = EYZ^{-1}$ which satisfies the conclusion of Lemma 2.1, and the element $H' := H + s = H + EYZ^{-1}$ commutes with the elements of A_1'. Hence,

$$\mathcal{D}_Z = \mathbb{K}[Z^{\pm 1}, H'] \otimes A_1' = \mathbb{K}[Z^{\pm 1}, \vartheta] \otimes A_1' \tag{2.3}$$

where $\vartheta := H'Z = HZ + EY$. Clearly, $Z(\mathcal{D}_Z) = \mathbb{K}[Z^{\pm 1}, \vartheta]$.

LEMMA 2.5. $Z(\mathcal{D}) = \mathbb{K}[Z,\vartheta]$ *and* $Z(\mathcal{D}_Z) = \mathbb{K}[Z^{\pm 1}, \vartheta]$.

PROOF. Since $\mathbb{K}[Z,\vartheta] \subseteq Z(\mathcal{D}) \subseteq \mathcal{D} \cap Z(\mathcal{D}_Z) = \mathcal{D} \cap \mathbb{K}[Z^{\pm 1}, \vartheta] = \mathbb{K}[Z,\vartheta]$, we have $Z(\mathcal{D}) = \mathbb{K}[Z,\vartheta]$. □

Generalized Weyl algebra. *Definition*, [**1–3**]. Let D be a ring, σ be an automorphism of D and a is an element of the centre of D. *The generalized Weyl algebra* $A := D(\sigma, a) := D[X,Y;\sigma,a]$ is a ring generated by D, X and Y subject to the defining relations:

$$X\alpha = \sigma(\alpha)X \text{ and } Y\alpha = \sigma^{-1}(\alpha)Y \text{ for all } \alpha \in D, \quad YX = a \text{ and } XY = \sigma(a).$$

The algebra $A = \oplus_{n\in\mathbb{Z}} A_n$ is $\mathbb{Z}$-graded where $A_n = Dv_n$, $v_n = X^n$ for $n > 0$, $v_n = Y^{-n}$ for $n < 0$ and $v_0 = 1$.

LEMMA 2.6. *The algebra $\mathcal{D}$ is the GWA $\mathbb{K}[\vartheta, Z, H][E, Y; \sigma, a = \vartheta - (H+2)Z]$ where $\sigma(\vartheta) = \vartheta$, $\sigma(Z) = Z$ and $\sigma(H) = H - 2$. In particular, the algebra $\mathcal{D}$ is a free module over its centre $Z(\mathcal{D}) = \mathbb{K}[\vartheta, Z]$ and the set $\{H^i, H^iE^j, H^iY^j \mid i \geqslant 0, j \geqslant 1\}$ is a free $Z(\mathcal{D})$-basis.*

PROOF. The algebra $\mathcal{D}$ satisfies the defining relations of the generalized Weyl algebra $\mathbb{K}[\vartheta, Z, H][E, Y; \sigma, a]$:

$$EY = \vartheta - HZ = \sigma(a), \qquad YE = EY + [Y, E] = \vartheta - HZ - 2Z = a,$$
$$Eu = \sigma(u)E \quad \text{and} \quad Yu = \sigma^{-1}(u)Y \quad \text{for all } u \in \mathbb{K}[\vartheta, Z, H].$$

Conversely, the GWA satisfies the defining relations of the algebra $\mathcal{D}$, i.e., both algebras coincide. Using the $\mathbb{Z}$-grading of the GWA $\mathcal{D}$ the statement about the centre is obvious. □

Prime ideals of the algebra $\mathcal{C}$. For an algebra R, let $\mathrm{Spec}\,(R)$ be the set of its prime ideals. The set $(\mathrm{Spec}\,(R), \subseteq)$ is a partially ordered set (poset) with respect to inclusion of prime ideals. Each element $r \in R$ determines two maps from R to R, $r\cdot : x \mapsto rx$ and $\cdot r : x \mapsto xr$ where $x \in R$.

A non-empty subset S of R is called a *multiplicative set* if $SS \subseteq S$ and $0 \notin S$. A multiplicative set S is called a *left Ore set* if $Sr \cap Rs \neq \emptyset$ for all $r \in R$ and $s \in S$ (*the left Ore condition*). A left Ore set S is called a *left denominator set* if $rs = 0$ for some $r \in R$ and $s \in S$ then $s'r = 0$ for some $s' \in S$. If S is a left denominator set of R then the ring of left fraction $S^{-1}R = \{s^{-1}r \mid s \in S, r \in R\}$ is called the *left localization of R at S*.

PROPOSITION 2.1. *(**[7]**.) Let R be a Noetherian ring and s be an element of R such that $\mathcal{S}_s := \{s^i \mid i \in \mathbb{N}\}$ is a left denominator set of the ring R and $(s^i) = (s)^i$ for all $i \geqslant 1$ (e.g., s is a normal element such that $\ker(\cdot s_R) \subseteq \ker(s_R\cdot)$). Then $\mathrm{Spec}\,(R) = \mathrm{Spec}(R, s) \sqcup \mathrm{Spec}_s(R)$ where $\mathrm{Spec}(R, s) := \{\mathfrak{p} \in \mathrm{Spec}\,(R) \mid s \in \mathfrak{p}\}$, $\mathrm{Spec}_s(R) = \{\mathfrak{q} \in \mathrm{Spec}\,(R) \mid s \notin \mathfrak{q}\}$ and*

(a) *the map $\mathrm{Spec}\,(R, s) \to \mathrm{Spec}\,(R/(s))$, $\mathfrak{p} \mapsto \mathfrak{p}/(s)$, is a bijection with the inverse $\mathfrak{q} \mapsto \pi^{-1}(\mathfrak{q})$ where $\pi : R \to R/(s), r \mapsto r + (s)$,*
(b) *the map $\mathrm{Spec}_s(R) \to \mathrm{Spec}\,(R_s)$, $\mathfrak{p} \mapsto \mathcal{S}_s^{-1}\mathfrak{p}$, is a bijection with the inverse $\mathfrak{q} \mapsto \sigma^{-1}(\mathfrak{q})$ where $\sigma : R \to R_s := \mathcal{S}_s^{-1}R$, $r \mapsto \frac{r}{1}$.*
(c) *For all $\mathfrak{p} \in \mathrm{Spec}\,(R, s)$ and $\mathfrak{q} \in \mathrm{Spec}_s(R), \mathfrak{p} \not\subseteq \mathfrak{q}$.*

In this paper, we *identify* the sets in the statements (a) and (b) via the bijections given there.

Let $\mathcal{C} := \mathcal{D}/(Z)$ be the factor algebra of $\mathcal{D}$ where (Z) is the ideal of $\mathcal{D}$ generated by the central element $Z \in \mathcal{D}$. Then the algebra $\mathcal{C}$ is isomorphic to the enveloping algebra $U(\mathfrak{e}(2))$ of the *three dimensional Euclidean algebra* $\mathfrak{e}(2) := \mathfrak{d}/\mathbb{K}Z$. So, the algebra $\mathcal{C}$ is generated by the elements H, E and Y subject to the defining relations:

$$[H, E] = 2E, \quad [H, Y] = -2Y, \quad [E, Y] = 0.$$

Notice that $\mathcal{C} \simeq \mathcal{A}/(X, Z)$. Therefore, $\mathrm{Spec}\,(\mathcal{C}) \subseteq \mathrm{Spec}\,(\mathcal{D}) \subseteq \mathrm{Spec}\,(\mathcal{A})$. The finite dimensional indecomposable representations of $\mathfrak{e}(2)$ are studied in **[15, 16, 22]**. Clearly, the element $\phi := EY$ belongs to the centre of $\mathcal{C}$. Let $\mathcal{C}_Y$ be the localization of $\mathcal{C}$ at the powers of the element Y. Then $\mathcal{C}_Y = \mathbb{K}[\phi] \otimes \mathbb{K}[H][Y^{\pm 1}; \sigma]$ where σ is the automorphism of $\mathbb{K}[H]$ such that $\sigma(H) = H + 2$. Since $\mathbb{K}[H][Y^{\pm 1}; \sigma]$ is a central simple algebra, $Z(\mathcal{C}_Y) = \mathbb{K}[\phi]$. Then $Z(\mathcal{C}) = \mathbb{K}[\phi]$.

PROPOSITION 2.2. $\mathrm{Spec}\,(\mathcal{C}) = \{0, (E)\} \sqcup \{(\mathfrak{t}) \mid \mathfrak{t} \in \mathrm{Max}\,(\mathbb{K}[\phi]) \setminus \{(\phi)\}\} \sqcup \{(Y), (Y, E), (Y, E, \mathfrak{h}) \mid \mathfrak{h} \in \mathrm{Max}\,(\mathbb{K}[H])\}$. *The containments of prime ideals are shown in the following diagram*

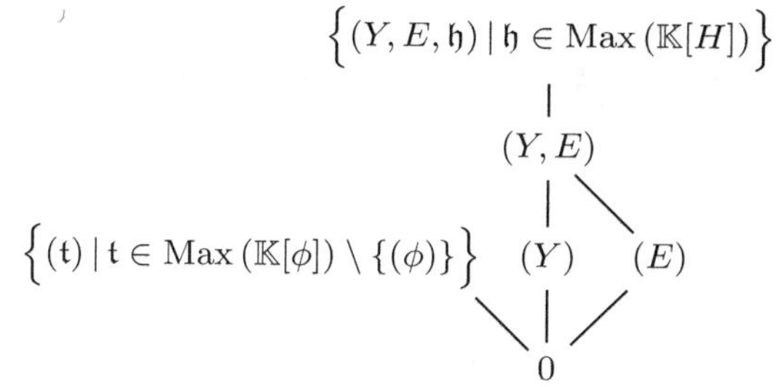

(2.4)

PROOF. Notice that Y is a normal element of the algebra $\mathcal{C}$. Then, by Proposition 2.1, $\mathrm{Spec}\,(\mathcal{C}) = \mathrm{Spec}\,(\mathcal{C}, Y) \sqcup \mathrm{Spec}_Y(\mathcal{C})$. The factor algebra $\mathcal{C}/(Y)$ is isomorphic to the enveloping algebra $U(\mathfrak{b})$ where $\mathfrak{b} := \mathbb{K}H \oplus \mathbb{K}E$ is the Borel subalgebra of $\mathfrak{sl}_2$. It is easy to see that $\mathrm{Spec}\,(U(\mathfrak{b})) = \{0,\ (E),\ (E, \mathfrak{h}) \mid \mathfrak{h} \in \mathrm{Max}\,(\mathbb{K}[H])\}$. Then, by Proposition 2.1.(a), $\mathrm{Spec}\,(\mathcal{C}, Y) = \{(Y),\ (Y, E),\ \ (Y, E, \mathfrak{h}) \mid \mathfrak{h} \in \mathrm{Max}\,(\mathbb{K}[H])\}$. Since $\mathcal{C}_Y = \mathbb{K}[\phi] \otimes \mathbb{K}[H][Y^{\pm 1}; \sigma]$ and the algebra $\mathbb{K}[H][Y^{\pm 1}; \sigma]$ is central and simple, $\mathrm{Spec}\,(\mathcal{C}_Y) = \{0, (\mathfrak{t})_Y \mid \mathfrak{t} \in \mathrm{Max}\,(\mathbb{K}[\phi])\}$. Then, by Proposition 2.1.(b), $\mathrm{Spec}_Y(\mathcal{C}) = \{0, (\mathfrak{t})_Y \cap \mathcal{C} \mid \mathfrak{t} \in \mathrm{Max}\,(\mathbb{K}[\phi])\}$. It remains to show that $(\phi)_Y \cap \mathcal{C} = (E)$ and $(\mathfrak{t})_Y \cap \mathcal{C} = (\mathfrak{t})$ for $\mathfrak{t} \in \mathrm{Max}\,(\mathbb{K}[\phi]) \setminus \{(\phi)\}$. Recall that $\phi = EY$, then $(\phi)_Y \cap \mathcal{C} = (E)_Y \cap \mathcal{C}$. Notice that (E) is a completely prime ideal of $\mathcal{C}$ and $Y \notin (E)$, we have $(E)_Y \cap \mathcal{C} = (E)$, and so, $(\phi)_Y \cap \mathcal{C} = (E)$. Now, we prove that $(\mathfrak{t}) = \mathcal{C} \cap \mathfrak{t}\mathcal{C}_Y$ for all $\mathfrak{t} \in \mathrm{Max}\,(\mathbb{K}[\phi]) \setminus \{(\phi)\}$. The element $\phi = EY$ is a unit of the field $L_{\mathfrak{t}} := \mathbb{K}[\phi]/\mathfrak{t}$. Therefore, the commuting elements E and Y are units in the factor algebra $\mathcal{C}/(\mathfrak{t})$. Then $\mathcal{C}/(\mathfrak{t}) \simeq \mathcal{C}_Y/\mathfrak{t}\mathcal{C}_Y$, and so $(\mathfrak{t}) = \mathcal{C} \cap \mathfrak{t}\mathcal{C}_Y$.

The containments of primes in the diagram (2.4) are obvious. By Proposition 2.1, for all $\mathfrak{p} \in \mathrm{Spec}\,(\mathcal{C}, Y)$ and $\mathfrak{q} \in \mathrm{Spec}_Y(\mathcal{C})$, $\mathfrak{p} \not\subseteq \mathfrak{q}$. Now, since all nonzero ideals in $\mathrm{Spec}\,(\mathcal{C}_Y)$ are maximal ideals of the algebra $\mathcal{C}_Y$ (hence are pairwise co-prime, i.e., $\mathfrak{q} + \mathfrak{q}' = \mathcal{C}_Y$), there are no additional containments in the diagram (2.4). $\square$

COROLLARY 2.3. $\mathrm{Max}\,(\mathcal{C}) = \Big\{(Y, E, \mathfrak{h}) \mid \mathfrak{h} \in \mathrm{Max}\,(\mathbb{K}[H])\Big\} \sqcup \Big\{(\mathfrak{t}) \mid \mathfrak{t} \in \mathrm{Max}\,(\mathbb{K}[\phi]) \setminus \{(\phi)\}\Big\}$.

PROOF. The statement follows from (2.4). $\square$

PROPOSITION 2.4. $\mathrm{Prim}\,(\mathcal{C}) = \mathrm{Max}\,(\mathcal{C}) \sqcup \{(Y),\ (E)\}$.

PROOF. Maximal ideals are primitive ideals. The zero ideal is not a primitive ideal since $Z(\mathcal{C}) = \mathbb{K}[\phi]$. The ideal (Y, E) is not a primitive ideal since the factor algebra $\mathcal{C}/(Y, E) \simeq \mathbb{K}[H]$ is a polynomial algebra. It remains to show that (Y) and (E) are primitive ideals. Notice that the algebra $\mathcal{C}$ is a Jacobson algebra, hence every prime ideal is an intersection of primitive ideal [**19**, Corollary 9.1.8]. Then it follows from the diagram (2.4) that (Y) and (E) must be primitive ideals. $\square$

Prime ideals of the algebra $\mathcal{D}$. For a prime ideal $\mathfrak{p}$ of a ring, $\mathrm{ht}\,(\mathfrak{p})$ is its height. The following proposition describes the prime ideals of the algebra $\mathcal{D}$, together with their containments.

PROPOSITION 2.5. *Let* $\mathcal{P} := \{\mathfrak{p} \in \operatorname{Spec}(\mathbb{K}[Z,\vartheta]) \mid \operatorname{ht}(\mathfrak{p}) = 1,\ \mathfrak{p} \neq (Z)\}$ *and* $\mathcal{M} := \{\mathfrak{m} \in \operatorname{Max}(\mathbb{K}[Z,\vartheta]) \mid Z \notin \mathfrak{m}\}$. *Then*

(1) $\operatorname{Spec}(\mathcal{D}) = \{(Z,\mathfrak{q}) \mid \mathfrak{q} \in \operatorname{Spec}(\mathcal{C})\} \sqcup \{0, (\mathfrak{p}), (\mathfrak{m}) \mid \mathfrak{p} \in \mathcal{P},\ \mathfrak{m} \in \mathcal{M}\}$.

(2) *The containments of prime ideals are shown in the following diagram*

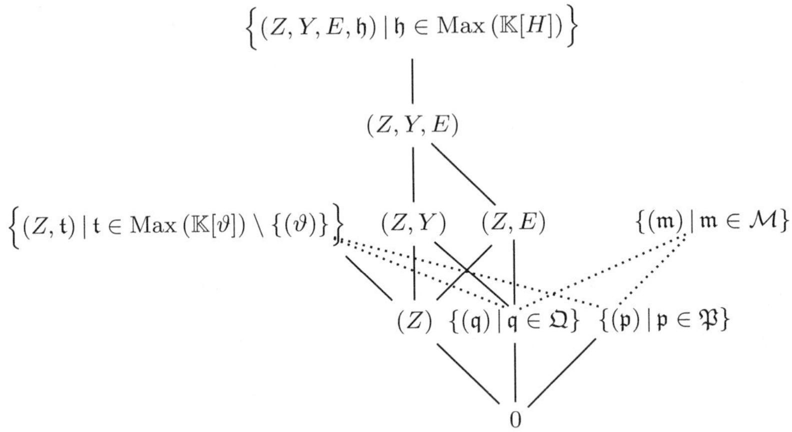

(2.5)

where $\mathfrak{P} = \{\mathfrak{p} \in \operatorname{Spec}(\mathbb{K}[Z,\vartheta]) \mid \operatorname{ht}(\mathfrak{p}) = 1,\ \mathfrak{p} \neq (Z),\ \mathfrak{p} \nsubseteq (Z,\vartheta)\}$ *and* $\mathfrak{Q} = \{\mathfrak{p} \in \operatorname{Spec}(\mathbb{K}[Z,\vartheta]) \mid \operatorname{ht}(\mathfrak{p}) = 1,\ \mathfrak{p} \neq (Z),\ \mathfrak{p} \subseteq (Z,\vartheta)\}$. *Moreover, for* $\mathfrak{q} \in \mathfrak{Q}$, $\mathfrak{p} \in \mathfrak{P}$ *and* $\mathfrak{t} \in \operatorname{Max}(\mathbb{K}[\vartheta]) \setminus \{(\vartheta)\}$ *we have* $(\mathfrak{q}) \subset (\mathfrak{m})$ *iff* $\mathfrak{q} \subset \mathfrak{m}$; $(\mathfrak{p}) \subset (\mathfrak{m})$ *iff* $\mathfrak{p} \subset \mathfrak{m}$; $(\mathfrak{q}) \subset (Z,\mathfrak{t})$ *iff* $\mathfrak{q} \subset Z\mathbb{K}[Z,\vartheta] + \mathfrak{t}\mathbb{K}[Z,\vartheta]$, *and* $(\mathfrak{p}) \subset (Z,\mathfrak{t})$ *iff* $\mathfrak{p} \subset Z\mathbb{K}[Z,\vartheta] + \mathfrak{t}\mathbb{K}[Z,\vartheta]$. *The dotted lines represent the containments described in the previous sentence.*

PROOF. 1. Since Z is a central element of $\mathcal{D}$, by Proposition 2.1, $\operatorname{Spec}(\mathcal{D}) = \operatorname{Spec}(\mathcal{D},Z) \sqcup \operatorname{Spec}_Z(\mathcal{D})$. Since $\mathcal{D}/(Z) = \mathcal{C}$, by Proposition 2.1.(a), $\operatorname{Spec}(\mathcal{D},Z) = \{(Z,\mathfrak{q}) \mid \mathfrak{q} \in \operatorname{Spec}(\mathcal{C})\}$. Notice that $\vartheta \equiv \phi \mod (Z)$, then $\{(Z,\mathfrak{t}) \mid \mathfrak{t} \in \operatorname{Max}(\mathbb{K}[\phi]) \setminus \{(\phi)\}\} = \{(Z,\mathfrak{t}) \mid \mathfrak{t} \in \operatorname{Max}(\mathbb{K}[\vartheta]) \setminus \{(\vartheta)\}\}$.

Let us describe the set $\operatorname{Spec}_Z(\mathcal{D})$. By (2.3), $\mathcal{D}_Z = \mathbb{K}[Z^{\pm 1},\vartheta] \otimes A_1'$ is a tensor product of algebras where A_1' is central simple. Hence, extension and contraction provides mutually inverse bijections between the sets $\operatorname{Spec}(\mathcal{D}_Z)$ and $\operatorname{Spec}(\mathbb{K}[Z^{\pm 1},\vartheta])$. So, $\operatorname{Spec}(\mathcal{D}_Z) = \{0,\ (\mathfrak{p})_Z,\ (\mathfrak{m})_Z \mid \mathfrak{p} \in \mathcal{P}, \mathfrak{m} \in \mathcal{M}\}$. By Proposition 2.1.(b), $\operatorname{Spec}_Z(\mathcal{D}) = \{0,\ (\mathfrak{p})_Z \cap \mathcal{D},\ (\mathfrak{m})_Z \cap \mathcal{D} \mid \mathfrak{p} \in \mathcal{P}, \mathfrak{m} \in \mathcal{M}\}$. For every $\mathfrak{m} \in \mathcal{M}$, $(\mathfrak{m})$ is a maximal ideal of $\mathcal{D}$: The element Z is a unit in the field $F_\mathfrak{m} := \mathbb{K}[Z,\vartheta]/\mathfrak{m}$, then $\mathcal{D}/\mathfrak{m}\mathcal{D} \simeq \mathcal{D}_Z/\mathfrak{m}\mathcal{D}_Z \simeq F_\mathfrak{m} \otimes A_1'$ is a simple algebra. Hence, $(\mathfrak{m})_Z \cap \mathcal{D} = (\mathfrak{m})$.

Let us show that $(\mathfrak{p})_Z \cap \mathcal{D} = (\mathfrak{p})$ for all $\mathfrak{p} \in \mathcal{P}$. The inclusion $(\mathfrak{p}) \subseteq (\mathfrak{p})_Z \cap \mathcal{D}$ is obvious. Now, let $u \in (\mathfrak{p})_Z \cap \mathcal{D}$. Then $Z^i u \in (\mathfrak{p})$ for some $i \in \mathbb{N}$. Choose i to be minimal. Notice that $(\mathfrak{p}) = (f)$ for some irreducible polynomial $f \in \mathbb{K}[Z,\vartheta]$ and $f \neq Z$. Then $Z^i u = fd$ for some $d \in \mathcal{D}$ since f is a central element of $\mathcal{D}$. If $i \geqslant 1$ then $fd \in (Z)$. But notice that $f \notin (Z)$, we have $d \in (Z)$, i.e., $d = Zd'$ for some $d' \in \mathcal{D}$. Then the equality $Z^i u = fd = Zfd'$ yields the equality $Z^{i-1}u = fd' \in (f) = (\mathfrak{p})$, which contradicts the minimality of i. Therefore, $i = 0$ and $(\mathfrak{p})_Z \cap \mathcal{D} = (\mathfrak{p})$.

2. By statement 1, all the prime ideals are present in the diagram (2.5). By Proposition 2.2, we have all the containments between primes above the prime ideal (Z). By Proposition 2.1, for all $P \in \operatorname{Spec}(\mathcal{D},Z)$ and $Q \in \operatorname{Spec}_Z(\mathcal{D})$, $P \nsubseteq Q$. Notice

that $(Z, Y) = (Z, Y, \vartheta)$ and $(Z, E) = (Z, E, \vartheta)$ then $(\mathfrak{q}) \subset (Z, Y)$ and $(\mathfrak{q}) \subseteq (Z, E)$ for all $\mathfrak{q} \in \mathfrak{Q}$, and, clearly, $(\mathfrak{p}) \not\subset (Z, Y)$ and $(\mathfrak{p}) \not\subset (Z, E)$ for all $\mathfrak{p} \in \mathfrak{P}$. The rest is obvious. $\square$

As a corollary of Proposition 2.4, we have a description of maximal ideals of the algebra $\mathcal{D}$.

COROLLARY 2.6. $\mathrm{Max}\,(\mathcal{D}) = \Big\{(Z, Y, E, \mathfrak{h}) \,|\, \mathfrak{h} \in \mathrm{Max}\,(\mathbb{K}[H])\Big\} \sqcup \Big\{(Z, \mathfrak{t}) \,|\, \mathfrak{t} \in \mathrm{Max}\,(\mathbb{K}[\vartheta]) \setminus \{(\vartheta)\}\Big\} \sqcup \{(\mathfrak{m}) \,|\, \mathfrak{m} \in \mathcal{M}\}$ *where* $\mathcal{M} = \{\mathfrak{m} \in \mathrm{Max}\,(\mathbb{K}[Z, \vartheta]) \,|\, Z \notin \mathfrak{m}\}$.

PROOF. This follows from (2.5). $\square$

Proof of Theorem 1.1. Recall that X is a normal element of the algebra $\mathcal{A}$. By Proposition 2.1, $\mathrm{Spec}\,(\mathcal{A}) = \mathrm{Spec}\,(\mathcal{A}, X) \sqcup \mathrm{Spec}_X(\mathcal{A})$. Since $\mathcal{A}/(X) = \mathcal{D}$, by Proposition 2.1.(a), $\mathrm{Spec}\,(\mathcal{A}, X) = \{(X, \mathfrak{p}) \,|\, \mathfrak{p} \in \mathrm{Spec}\,(\mathcal{D})\}$. It remains to describe the set $\mathrm{Spec}_X(\mathcal{A})$. By (2.2), extension and contraction provides mutually inverse bijections between the sets $\mathrm{Spec}\,(\mathcal{A}_X)$ and $\mathrm{Spec}\,(\mathbb{K}[C_1])$, we have $\mathrm{Spec}\,(\mathcal{A}_X) = \{0, (\mathfrak{c})_X \,|\, \mathfrak{c} \in \mathrm{Max}\,(\mathbb{K}[C_1])\}$. Then by Proposition 2.1.(b), $\mathrm{Spec}_X(\mathcal{A}) = \{0, (\mathfrak{c})_X \cap \mathcal{A} \,|\, \mathfrak{c} \in \mathrm{Max}\,(\mathbb{K}[C_1])\}$. We have to show that $(\mathfrak{c})_X \cap \mathcal{A} = (\mathfrak{c})$ where $\mathfrak{c} \in \mathrm{Max}\,(\mathbb{K}[C_1])$. The inclusion $(\mathfrak{c}) \subseteq (\mathfrak{c})_X \cap \mathcal{A}$ is obvious. Now, let $u \in (\mathfrak{c})_X \cap \mathcal{A}$. Then $X^i u \in (\mathfrak{c})$ for some $i \in \mathbb{N}$. Choose i to be minimal. Notice that $\mathfrak{c} = (f(C_1))$ for some irreducible polynomial $f(C_1) \in \mathbb{K}[C_1]$. Then $X^i u = f(C_1)a$ for some $a \in \mathcal{A}$ since $f(C_1)$ is a central element of $\mathcal{A}$. If $i \geqslant 1$, then the equality $X^i u = f(C_1)a$ yields that $f(-\frac{1}{2}Z^2)a \equiv 0 \mod (X)$ since $C_1 = XY - \frac{1}{2}Z^2$. Notice that $f(-\frac{1}{2}Z^2) \notin (X)$, we have $a \in (X)$. Then $a = Xb$ for some $b \in \mathcal{A}$ (by the normality of X). Then the equality $X^i u = f(C_1)a = Xf(C_1)b$ yields the inclusion $X^{i-1}u = f(C_1)b \in (\mathfrak{c})$, which contradicts the minimality of i. Hence $i = 0$ and $(\mathfrak{c})_X \cap \mathcal{A} = (\mathfrak{c})$.

Now, it is obvious that all the prime ideals of the algebra $\mathcal{A}$ are present in the diagram (1.1). By Proposition 2.5, we have all the containments between primes above the prime ideal (X). The rest of the containments in the diagram (1.1) are obvious. It remains to show that these are all the containments between primes. By Proposition 2.1, for all primes $P \in \mathrm{Spec}\,(\mathcal{A}, X)$ and $Q \in \mathrm{Spec}_X(\mathcal{A})$, $P \not\subseteq Q$. The ideals from the set $\{(C_1), (\mathfrak{c}) \,|\, \mathfrak{c} \in \mathrm{Max}\,(\mathbb{K}[C_1]) \setminus \{(C_1)\}\}$ are pairwise co-prime. It follows from this fact and the diagram (1.1) that all the containments in (1.1) are present (otherwise, we would have a contradiction).

Clearly, $(C_1) \subseteq (Z, Y)$ since $C_1 = XY - \frac{1}{2}Z^2$. Then $(\mathfrak{c}) \not\subseteq (Z, Y)$ for all $\mathfrak{c} \in \mathrm{Max}\,(\mathbb{K}[C_1]) \setminus \{(C_1)\}$ since $(\mathfrak{c}) + (C_1) = (1)$. The maximal ideal $\mathfrak{c}$ of $\mathbb{K}[C_1]$ such that $\mathfrak{c} \neq (C_1)$ is generated by a monic, irreducible polynomial $f(C_1)$ such that $f(0) \neq 0$. Then $(\mathfrak{c}) \not\subseteq (X, \mathfrak{q})$ for all $\mathfrak{q} \in \mathfrak{Q}$ since otherwise $f(C_1) \in (X, \mathfrak{q}) \subseteq (X, Z, \vartheta)$, and so $\mathbb{K}^* \ni f(0) \in (X, Z, \vartheta)$, a contradiction.

Let us show that each ideal $(\mathfrak{c})$ is contained in a prime ideal $(X, \mathfrak{p})$ for some $\mathfrak{p} \in \mathfrak{P}$. More precisely, $(\mathfrak{c}) \subseteq (X, \mathfrak{p})$ iff $f(-\frac{1}{2}Z^2) \in \mathfrak{p}$ (where $f(C_1)$ is as above). Indeed, $(\mathfrak{c}) \subseteq (X, \mathfrak{p})$ iff $(f(C_1)) \subseteq (X, \mathfrak{p})$ iff $(f(C_1)) + (X) = (f(-\frac{1}{2}Z^2)) + (X) \subseteq (\mathfrak{p})$ in $\mathcal{D} = \mathcal{A}/(X)$ iff $f(-\frac{1}{2}Z^2) \in \mathfrak{p}$, by Lemma 2.6 (by using the $\mathbb{Z}$-grading of the GWA $\mathcal{D}$) iff $\mathfrak{p} = g(Z)\mathbb{K}[Z, \vartheta]$ where $g(Z) \in \mathbb{K}[Z]$ is a monic, irreducible polynomial of $\mathbb{K}[Z]$ which is a factor of $f(-\frac{1}{2}Z^2)$. $\square$

A prime ideal P of a ring R is said to be *locally closed* if the set $\{P\}$ is locally closed in the topological space $\mathrm{Spec}\,(R)$ where $\mathrm{Spec}\,(R)$ is equipped with Zariski topology [**10**, II.1.1]. A prime ideal P of a Noetherian $\mathbb{K}$-algebra R is said to be *rational* if the field $Z\Big(\operatorname{Frac}(R/P)\Big)$ is algebraic over $\mathbb{K}$ where $\operatorname{Frac}(R/P)$ is the left (right) quotient ring of the Noetherian prime algebra R/P. We say that the *Dixmier-Moeglin equivalence* holds for a Noetherian $\mathbb{K}$-algebra A if for each prime ideal P of A we have the following equivalences:

$$P \text{ is locally closed} \iff P \text{ is primitive} \iff P \text{ is rational.}$$

Proof of Corollary 1.2. Since $\mathcal{A}$ is a universal enveloping algebra of a finite dimensional Lie algebra it satisfies the Dixmier-Moeglin equivalence. By [**10**, Lemma II.7.7], a prime ideal P in a ring R is locally closed iff the intersection of all prime ideals properly containing P is also an ideal properly containing P. Clearly, all the maximal ideals are primitive ideals. By (1.1), the set of locally closed prime ideals is $\operatorname{Max}(\mathcal{A}) \sqcup \{(X,Z,Y),(X,Z,E),(C_1)\} \sqcup \{(\mathfrak{c}) \,|\, \mathfrak{c} \in \operatorname{Max}(\mathbb{K}[C_1]) \setminus \{(C_1)\}\}$. Then the corollary follows from the Dixmier-Moeglin equivalence for $\mathcal{A}$. □

The next corollary describes the set of primitive ideals of the algebra $\mathcal{D}$.

COROLLARY 2.7. $\operatorname{Prim}(\mathcal{D}) = \operatorname{Max}(\mathcal{D}) \sqcup \{(Z,Y),\ (Z,E)\}$.

PROOF. The statement follows from Proposition 2.5.(2) and Corollary 1.2. □

Semi-centre of $\mathcal{A}$. Recall that $\mathcal{A} = U(\mathfrak{a})$. For each $\lambda \in \mathfrak{a}^* = \operatorname{Hom}_{\mathbb{K}}(\mathfrak{a},\mathbb{K})$, let $\mathcal{A}_\lambda$ be the set of all $a \in \mathcal{A}$ such that $\operatorname{ad}_x(a) := [x,a] = \lambda(x)a$ for all $x \in \mathfrak{a}$. Any nonzero element $a \in \mathcal{A}_\lambda$ is called the *semi-invariant* with weight λ. The sum of the vector spaces $\mathcal{A}_\lambda$ is a direct one, $Sz(\mathcal{A}) := \oplus_{\lambda\in\mathfrak{a}^*}\mathcal{A}_\lambda$, and is called the *seimi-centre* of $\mathcal{A}$. Let $\mathfrak{g}$ be a Lie algebra over $\mathbb{K}$ with basis $x_1,\dots,x_n$. For $f \in \mathfrak{g}^*$, consider the *skew symmetric bilinear form* B_f on $\mathfrak{g}$ given by the rule $B_f(x,y) = f([x,y])$ for $x,y \in \mathfrak{g}$. The linear subspace $\mathfrak{g}_f := \{x \in \mathfrak{g} \,|\, f([x,y]) = 0 \text{ for all } y \in \mathfrak{g}\}$ is called the *kernel* of B_f and $\operatorname{ind}(\mathfrak{g}) = \min_{f\in\mathfrak{g}^*} \dim \mathfrak{g}_f$ is called the *index* of $\mathfrak{g}$. By [**12**, 1.14.13],

$$\operatorname{ind}(\mathfrak{g}) = \dim \mathfrak{g} - \operatorname{rank}_{R(\mathfrak{g})}([x_i,x_j]) \tag{2.6}$$

where $R(\mathfrak{g})$ is the quotient field of the symmetric algebra $S(\mathfrak{g})$. A Lie algebra $\mathfrak{g}$ is called *Frobenius* if $\operatorname{ind}(\mathfrak{g}) = 0$. We apply [**21**, Theorem 3.5] to determine the semi-centre of $\mathcal{A}$. Using (2.6), we see that $\operatorname{ind}(\mathfrak{a}) = 1$. Therefore, we need to find a 1-dimensional split torus $T \subset \operatorname{Der}(\mathfrak{a})$ (i.e., an abelian Lie algebra consisting of diagonalizable derivations of $\mathfrak{a}$) such that the semi-direct product $\mathfrak{g} := T \ltimes \mathfrak{a}$ is Frobenius. We can take $T = \langle t \rangle$ with $[t,H] = 0$, $[t,E] = E$, $[t,X] = 2X$, $[t,Y] = 0$ and $[t,Z] = Z$. The matrix of the Lie brackets of $\mathfrak{g}$ is

	t	H	E	X	Y	Z
t	0	0	E	$2X$	0	Z
H	0	0	$2E$	$2X$	$-2Y$	0
E	$-E$	$-2E$	0	0	$2Z$	$2X$
X	$-2X$	$-2X$	0	0	0	0
Y	0	$2Y$	$-2Z$	0	0	0
Z	$-Z$	0	$-2X$	0	0	0

PROPOSITION 2.8. $Sz(\mathcal{A}) = \mathbb{K}[X, C_1]$.

PROOF. The determinant of the matrix of the Lie brackets of $\mathfrak{g}$ is $\det(\mathfrak{g}) = 64X^2(XY - \frac{1}{2}Z^2)^2$ which is nonzero. So, the Lie algebra $\mathfrak{g}$ is Frobenius. Clearly, X

and $(XY - \frac{1}{2}Z^2)$ are the irreducible factors of det $(\mathfrak{g})$. Hence, by [**21**, Theorem 3.5], the semi-centre $Sz(\mathcal{A})$ of $\mathcal{A}$ is a polynomial algebra and $Sz(\mathcal{A}) = \mathbb{K}[X, C_1]$. $\square$

3. A classification of simple weight $\mathcal{A}$-modules

In this section, the field $\mathbb{K}$ is an algebraically closed field of characteristic zero. The aim of the section is to give a classification of simple, weight $\mathcal{A}$-modules (Theorem 3.11). As a consequence, classifications are given of simple weight $\mathcal{A}$-modules with finite and infinite dimensional weight spaces (Theorem 3.12, Theorem 3.13). The centralizer C of the element H in $\mathcal{A}$ is studied (Proposition 3.1). All simple C-modules are described (Theorem 3.9). Prime, maximal and primitive ideals of the algebra C are classified (Theorem 3.14, Corollary 3.7, Corollary 3.8).

The element $X \in \mathcal{A}$ is normal and a non-zero-divisor. It determines a $\mathbb{K}$-automorphism ω_X of the algebra $\mathcal{A}$ by the rule $Xa = \omega_X(a)X$ for all $a \in \mathcal{A}$. More precisely,

$$\omega_X : \mathcal{A} \to \mathcal{A}, \;\; E \mapsto E, \; H \mapsto H - 2, \; X \mapsto X, \; Y \mapsto Y, \; Z \mapsto Z. \tag{3.1}$$

The centralizer $C_{\mathcal{A}}(H)$ of H in $\mathcal{A}$. Let R be an algebra. For each element $r \in R$, the set $C_R(r) := \{s \in R \,|\, rs = sr\}$ is called the *centralizer* of the element $r \in R$ in R. It is a subalgebra of R. We aim to determine generators and defining relations of the centralizer $C := C_{\mathcal{A}}(H)$ of the element H in $\mathcal{A}$ (Proposition 3.1). But first we do it for the centralizer $C_{\mathcal{A}_X}(H)$.

LEMMA 3.1. (1) *$C_{\mathcal{A}_X}(H) = \mathbb{K}[C_1, H] \otimes A_1$.*

(2) *$\mathcal{A}_X = C_{\mathcal{A}_X}(H)[X^{\pm 1}; \sigma]$ where $\sigma(C_1) = C_1, \sigma(H) = H - 2, \sigma(EX^{-1}) = EX^{-1}$ and $\sigma(Z) = Z$.*

PROOF. 1. By (2.2), $C_{\mathcal{A}_X}(H) = \mathbb{K}[C_1] \otimes \mathbb{K}[H] \otimes A_1 = \mathbb{K}[C_1, H] \otimes A_1$.
2. Statement 2 follows from statement 1 and (2.2). $\square$

PROPOSITION 3.1.

(1) *The algebra $C = C_{\mathcal{A}}(H)$ is generated by C_1, H, Z and $\phi = EY$ subject to the following defining relations*

$$[\phi, Z] = Z^2 + 2C_1, \quad [C_1, \cdot] = [H, \cdot] = 0. \tag{3.2}$$

(2) *$C = \mathbb{K}[C_1, H, Z][\phi; \delta]$ is a Noetherian domain of GK dimension 4 where $\delta(C_1) = \delta(H) = 0$ and $\delta(Z) = Z^2 + 2C_1$.*

(3) *$Z(C) = \mathbb{K}[C_1, H]$.*

(4) *$\omega_X(C) = C$.*

PROOF. 1 and 2. The polynomial algebra $\mathbb{K}[X, Y, Z] = \oplus_{i \in \mathbb{Z}} \mathbb{K}[Z, C_1]w_i$ is a $\mathbb{Z}$-graded algebra where $w_0 = 1$ and for $i \geqslant 1$, $w_i = X^i$ and $w_{-i} = Y^i$. Now, using $U(\mathfrak{b}) = \oplus_{j \in \mathbb{N}} \mathbb{K}[H]E^j$ and $\mathcal{A} = \mathbb{K}[X, Y, Z] \otimes U(\mathfrak{b})$ (a tensor product of vector spaces), we have that

$$\mathcal{A} = \bigoplus_{i \in \mathbb{Z}} \mathbb{K}[Z, C_1]w_i \otimes \bigoplus_{j \in \mathbb{N}} \mathbb{K}[H]E^j = \bigoplus_{i \in \mathbb{Z}, j \in \mathbb{N}} \mathbb{K}[Z, C_1, H]w_iE^j.$$

Now, it is clear that $C = \oplus_{i \in \mathbb{N}} \mathbb{K}[Z, C_1, H]Y^iE^i = \oplus_{i \in \mathbb{N}} \mathbb{K}[Z, C_1, H](EY)^i$. Then statements 1 and 2 follow.

3. By Lemma 3.1.(1), $Z(C_{\mathcal{A}_X}(H)) = \mathbb{K}[C_1, H]$. Hence, $Z(C) = \mathbb{K}[C_1, H]$.
4. Statement 4 follows from the equality $XH = (H - 2)X$. $\square$

Choose the following generators for the Weyl algebra A_1 in the equality (2.1) and Lemma 3.1: $\partial := \frac{1}{2}EX^{-1}$ and Z ($[\partial, Z] = 1$). The Weyl algebra $A_1 = \mathbb{K}\langle Z, \partial \rangle$ is the GWA $A_1 = \mathbb{K}[h][Z, \partial; \sigma, a = h]$ where $\sigma(h) = h - 1$ and $h := \partial Z$. So, the Weyl algebra $A_1 = \bigoplus_{i\in\mathbb{Z}} A_{1,i}$ is a $\mathbb{Z}$-graded algebra where $A_{1,0} := \mathbb{K}[h]$ and $A_{1,\pm i} = \mathbb{K}[h]v_{\pm i}$ ($i \geqslant 1$) where $v_i = Z^i$ and $v_{-i} = \partial^i$. Let $v_0 := 1$.

The algebra $\mathcal{C}$ is a subalgebra of $C_{\mathcal{A}_X}(H) = \mathbb{K}[C_1, H] \otimes A_1 = \mathbb{K}[C_1, H] \otimes \mathbb{K}\langle Z, \partial \rangle$ where

$$\phi = \partial(2C_1 + Z^2) = 2C_1\partial + hZ \tag{3.3}$$

since $\phi = EY = EX^{-1} \cdot XY = 2\partial(C_1 + \frac{1}{2}Z^2) = \partial(2C_1 + Z^2)$.

An $\mathcal{A}$-module M is called a *weight* $\mathcal{A}$-module (or an H-*weight* $\mathcal{A}$-module) if it is a (direct) sum of eigenspaces, i.e.,

$$M = \bigoplus_{\lambda\in\mathbb{K}} M_\lambda \quad \text{where } M_\lambda = \ker_M(H - \lambda) = \{m \in M \mid Hm = \lambda m\}.$$

The set of eigenvalues of the linear map $H_M : M \to M$, $m \mapsto Hm$, is denoted by $\mathrm{Wt}\,(M)$ and the elements of $\mathrm{Wt}\,(M)$ are called *weights* of M. So, $M = \bigoplus_{\lambda\in \mathrm{Wt}\,(M)} M_\lambda$ and $M_\lambda \neq 0$ for all $\lambda \in \mathrm{Wt}\,(M)$.

By Corollary 2.3, for each maximal ideal $\mathfrak{t} \in \mathrm{Max}\,(\mathbb{K}[\phi]) \setminus \{(\phi)\}$, $\mathbb{K}[\phi]/\mathfrak{t} \simeq \mathbb{K}$ (since $\mathbb{K}$ is an algebraically closed field) and the ideal $(\mathfrak{t})$ is a maximal ideal of the algebra $\mathcal{C}$ and $\mathcal{C}/(\mathfrak{t}) \simeq \mathbb{K}[H][Y^{\pm 1}; \sigma]$ where σ is a $\mathbb{K}$-automorphism of the polynomial algebra $\mathbb{K}[H]$ such that $\sigma(H) = H + 2$.

The group $\langle\sigma\rangle := \{\sigma^i \mid i \in \mathbb{Z}\}$ acts on the set $\mathrm{Max}\,(\mathbb{K}[H])$ of maximal ideals of $\mathbb{K}[H]$ in the obvious way. The field $\mathbb{K}$ has characteristic zero, so, for each $\mathfrak{m} \in \mathrm{Max}\,(\mathbb{K}[H])$, its $\langle\sigma\rangle$-orbit $\mathcal{O}(\mathfrak{m}) := \{\sigma^i(\mathfrak{m}) \mid i \in \mathbb{Z}\}$ is infinite. We denote by $\mathrm{Max}\,(\mathbb{K}[H])/\langle\sigma\rangle$ the set of $\langle\sigma\rangle$-orbits and for each $\langle\sigma\rangle$-orbit $\mathcal{O}$ we fix a representative $\mathfrak{m}_\mathcal{O} = (H - \lambda_\mathcal{O})$ where $\lambda_\mathcal{O} \in \mathbb{K}$. The set $\mathrm{Max}\,(\mathbb{K}[H])$ can be identified with the field $\mathbb{K}$ via the bijection $(H - \lambda) \mapsto \lambda$. Then each $\langle\sigma\rangle$-orbit $\mathcal{O}(\mathfrak{m} = (H - \lambda))$ can be identified with the set $\lambda + 2\mathbb{Z}$ and the set of all $\langle\sigma\rangle$-orbits can be identified with the factor group $\mathbb{K}/2\mathbb{Z}$.

Simple weight $\mathcal{C}$-modules. For an algebra A, we denote by $\widehat{A}$ the set of isomorphism classes of simple A-modules. For a simple A-module M, we denote by $[M]$ its isomorphism class. We often omit the brackets '[ ]'. Let $\widehat{\mathcal{C}}\,(\text{weight})$ be the set of isomorphism classes of simple weight $\mathcal{C}$-modules. Recall that Proposition 2.4 is an explicit description of the set of primitive ideals of the algebra $\mathcal{C}$. For each $\mathfrak{p} \in \mathrm{Prim}\,(\mathcal{C})$, let $\widehat{\mathcal{C}}\,(\text{weight}, \mathfrak{p}) := \{M \in \widehat{\mathcal{C}}\,(\text{weight}) \mid \mathrm{ann}_\mathcal{C}(M) = \mathfrak{p}\}$. Clearly,

$$\widehat{\mathcal{C}}\,(\text{weight}) = \bigsqcup_{\mathfrak{p}\in\mathrm{Prim}\,(\mathcal{C})} \widehat{\mathcal{C}}\,(\text{weight}, \mathfrak{p}). \tag{3.4}$$

For each non-empty set $S \subseteq \mathrm{Prim}\,(\mathcal{C})$, let $\widehat{\mathcal{C}}\,(\text{weight}, S) := \bigsqcup_{\mathfrak{p}\in S} \widehat{\mathcal{C}}\,(\text{weight}, \mathfrak{p})$. The following proposition describes the set $\widehat{\mathcal{C}}\,(\text{weight}, \mathfrak{p})$ for each $\mathfrak{p} \in \mathrm{Prim}\,(\mathcal{C})$. Hence, it gives classification of simple weight $\mathcal{C}$-modules, by (3.4).

PROPOSITION 3.2. *Let* $S_1 = \Big\{(Y, E, \mathfrak{h}) \mid \mathfrak{h} \in \mathrm{Max}\,(\mathbb{K}[H])\Big\}$ *and* $S_2 = \Big\{(\mathfrak{t}) \mid \mathfrak{t} \in \mathrm{Max}\,(\mathbb{K}[\phi]) \setminus \{(\phi)\}\Big\}$. *Then* $\mathrm{Prim}\,(\mathcal{C}) = S_1 \sqcup S_2 \sqcup \{(Y), (E)\}$ *(Corollary* 2.3 *and Proposition* 2.4*) and*

(1) $\widehat{\mathcal{C}}\,(\text{weight}, (Y)) = \emptyset$.

(2) $\widehat{\mathcal{C}}\,(\text{weight}, (E)) = \emptyset$.

(3) $\widehat{\mathcal{C}}\,(\text{weight}, S_1) = \{\mathcal{C}/(Y, E, \mathfrak{h}) \simeq \mathbb{K}[H]/\mathfrak{h} \mid \mathfrak{h} \in \text{Max}\,(\mathbb{K}[H])\}$. *Let* $M = \mathcal{C}/(Y, E, \mathfrak{h})$ *and* $\mathfrak{h} = (H - \mu)$. *Then* $\dim_{\mathbb{K}}(M) = 1$, $\text{Wt}\,(M) = \{\mu\}$ *and* $\text{ann}_{\mathcal{A}}(M) = (X, Y, Z, E, H - \mu)$.

(4) $\widehat{\mathcal{C}}(\text{weight}, S_2) = \bigsqcup_{(\mathfrak{t}) \in S_2} \widehat{\mathcal{C}}\,(\text{weight}, (\mathfrak{t}))$ *and* $\widehat{\mathcal{C}}\,(\text{weight}, (\mathfrak{t})) = \widehat{\mathcal{C}/(\mathfrak{t})}\,(\text{weight})$. *The map*

$$\text{Max}(\mathbb{K}[H])/\langle\sigma\rangle \to \widehat{\mathcal{C}/(\mathfrak{t})}(\text{weight}), \quad \mathcal{O}(\mathfrak{m}) \mapsto \left[\frac{\mathcal{C}/(\mathfrak{t})}{\mathcal{C}/(\mathfrak{t}) \cdot \mathfrak{m}} = \bigoplus_{i \in \mathbb{Z}} Y^i \otimes_{\mathbb{K}[H]} (\mathbb{K}[H]/\mathfrak{m})\right],$$

is a bijection with the inverse $M \mapsto \text{Wt}\,(M)$. *Let* $N = \frac{\mathcal{C}/(\mathfrak{t})}{\mathcal{C}/(\mathfrak{t}) \cdot \mathfrak{m}}$ *and* $\mathfrak{m} = (H - \mu)$. *Then* $\dim_{\mathbb{K}}(N) = \infty$, $\text{Wt}(N) = \mu + 2\mathbb{Z}$, $\dim_{\mathbb{K}}(N_\gamma) = 1$ *for all* $\gamma \in \text{Wt}\,(N)$ *and* $\text{ann}_{\mathcal{A}}(N) = (X, Z, \mathfrak{t})$.

PROOF. 1. The algebra $\mathcal{C}/(Y)$ is isomorphic to the algebra $U^+ := \mathbb{K}[H][E; \sigma]$ where $\sigma(H) = H - 2$. So, $\widehat{\mathcal{C}}\,(\text{weight}, (Y)) = \widehat{U^+}\,(\text{weight}, 0)$. For each $\mathfrak{p} \in \text{Max}\,(\mathbb{K}[H])$, the U^+-module $U^+(\mathfrak{p}) := U^+/U^+\mathfrak{p} \simeq \bigoplus_{i \geqslant 0} E^i \otimes \mathbb{K}[H]/\mathfrak{p}$ contains the largest submodule $U^+(\mathfrak{p})' = \bigoplus_{i \geqslant 1} E^i \otimes \mathbb{K}[H]/\mathfrak{p}$ and the factor module $U^+\mathfrak{p}/U^+(\mathfrak{p})' \simeq \mathbb{K}[H]/\mathfrak{p}$ is simple with nonzero annihilator, and the statement follows.

2. The algebra $\mathcal{C}/(E)$ is isomorphic to the algebra $U^- := \mathbb{K}[H][Y; \sigma^{-1}]$ where $\sigma^{-1}(H) = H + 2$. Now, we can repeat the proof of statement 1 by replacing U^+ with U^- to get the result.

3. Evident.

4. For each $(\mathfrak{t}) \in S_2$, we have $\widehat{\mathcal{C}}\,(\text{weight}, (\mathfrak{t})) = \widehat{\mathcal{C}/(\mathfrak{t})}\,(\text{weight})$. Since the element EY is a unit in the field $\mathbb{K}[\phi]/(\mathfrak{t})$, $\mathcal{C}/(\mathfrak{t}) \simeq \mathcal{C}_Y/\mathcal{C}_Y\mathfrak{t} \simeq \mathbb{K}[H][Y^{\pm 1}; \sigma]$. The rest is obvious, see also [**3**]. □

Simple weight $\mathcal{D}$-modules. For $\lambda \in \mathbb{K}$, let $\mathcal{D}(\lambda) := \mathcal{D}/(Z - \lambda)$. Then $\mathcal{D}(0) = \mathcal{C}$ and the set $\widehat{\mathcal{D}}\,(\text{weight})$ is a disjoint union

$$\widehat{\mathcal{D}}\,(\text{weight}) = \bigsqcup_{\lambda \in \mathbb{K}} \widehat{\mathcal{D}(\lambda)}\,(\text{weight}) = \widehat{\mathcal{C}}\,(\text{weight}) \sqcup \bigsqcup_{\lambda \in \mathbb{K}^*} \widehat{\mathcal{D}(\lambda)}\,(\text{weight}). \tag{3.5}$$

In view of Proposition 3.2, in order to give a classification of simple weight $\mathcal{D}$-modules it remains to describe the set $\widehat{\mathcal{D}(\lambda)}\,(\text{weight})$ for each $\lambda \in \mathbb{K}^*$. For $\lambda, \mu \in \mathbb{K}$, let $\mathcal{D}(\lambda, \mu) := \mathcal{D}/(Z - \lambda, \vartheta - \mu)$. For a given $\lambda \in \mathbb{K}^*$,

$$\widehat{\mathcal{D}(\lambda)}\,(\text{weight}) = \bigsqcup_{\mu \in \mathbb{K}} \widehat{\mathcal{D}(\lambda, \mu)}\,(\text{weight}). \tag{3.6}$$

By Lemma 2.6, the algebra $\mathcal{D}(\lambda, \mu) = \mathbb{K}[H][E, Y; \sigma, a(\lambda, \mu) := \mu + (H + 2)\lambda]$ is a GWA where $\sigma(H) = H - 2$. All the simple $\mathcal{D}(\lambda, \mu)$-modules (including weight ones) were classified in [**3**].

The additive subgroup $2\mathbb{Z}$ of $(\mathbb{K}, +)$ acts on the field $\mathbb{K}$ by the addition. For each element $\gamma \in \mathbb{K}$, $\mathcal{O}(\gamma) := \gamma + 2\mathbb{Z}$ is its $2\mathbb{Z}$-orbit. The orbit $\mathcal{O}(\lambda^{-1}\mu) = \lambda^{-1}\mu + 2\mathbb{Z}$ is called a *degenerate* orbit (this is a unique orbit that contains the root $\lambda^{-1}\mu - 2$ of the polynomial $a(\lambda, \mu)$). Two elements of the field $\mathbb{K}$, γ and γ', are called *equivalent*, $\gamma \sim \gamma'$, if either they both belong to the same non-degenerate orbit or to one of the sets $\Gamma^+ := \lambda^{-1}\mu + 2\mathbb{N}$, $\Gamma^- := \lambda^{-1}\mu - 2 - 2\mathbb{N}$. Notice that $\mathcal{O}(\lambda^{-1}\mu) = \Gamma^- \sqcup \Gamma^+$. We denote by $\mathbb{K}/\sim$ the set of equivalent classes.

THEOREM 3.2. *Let $\lambda \in \mathbb{K}^*$ and $\mu \in \mathbb{K}$. The map* $\mathbb{K}/\sim \longrightarrow \widehat{\mathcal{D}(\lambda,\mu)}\,(\text{weight})$, $\Gamma \mapsto [L(\Gamma)]$, *is a bijection with the inverse* $[M] \mapsto \mathrm{Wt}\,(M)$ *where*

(1) *If* $\Gamma = \gamma + 2\mathbb{Z}$ *is a non-degenerate orbit (i.e.,* $\Gamma \neq \lambda^{-1}\mu + 2\mathbb{Z}$*) then* $L(\Gamma) = \mathcal{D}(\lambda,\mu)/\mathcal{D}(\lambda,\mu)(H-\gamma)$.

(2) *If* $\Gamma = \Gamma^-$ *then* $L(\Gamma^-) = \mathcal{D}(\lambda,\mu)/\mathcal{D}(\lambda,\mu)(H-\lambda^{-1}\mu+2, X)$.

(3) *If* $\Gamma = \Gamma^+$ *then* $L(\Gamma^+) = \mathcal{D}(\lambda,\mu)/\mathcal{D}(\lambda,\mu)(H-\lambda^{-1}\mu, Y)$.

All the simple, weight $D(\lambda,\mu)$-modules are faithful, i.e., the ideal $(X, Z-\lambda, \vartheta-\mu)$ *of $\mathcal{A}$ is their annihilator in $\mathcal{A}$, with 1-dimensional weight spaces.*

PROOF. The result follows from [**3**]. □

REMARK 3.3. (3.5), (3.6), Proposition 3.2 and Theorem 3.2 give a classification of simple weight $\mathcal{D}$-modules.

Simple weight $\mathcal{A}$-modules. Recall that X is a normal element of $\mathcal{A}$. Then

$$\widehat{\mathcal{A}}\,(\text{weight}) = \widehat{\mathcal{A}}\,(\text{weight}, X\text{-torsion}) \sqcup \widehat{\mathcal{A}}\,(\text{weight}, X\text{-torsionfree}). \tag{3.7}$$

Notice that for any $M \in \widehat{\mathcal{A}}\,(\text{weight}, X\text{-torsion})$, $XM = 0$ and $\mathcal{A}/(X) = \mathcal{D}$. We have

$$\widehat{\mathcal{A}}\,(\text{weight}, X\text{-torsion}) = \widehat{\mathcal{D}}\,(\text{weight}). \tag{3.8}$$

Let $M \in \widehat{\mathcal{A}}\,(\text{weight}, X\text{-torsionfree})$. Since X is a normal element of $\mathcal{A}$, it acts on M *bijectively*. Hence, the map

(3.9)
$$\begin{aligned}\widehat{\mathcal{A}}\,(\text{weight}, X\text{-torsionfree}) \to \widehat{\mathcal{A}_X}\,(\text{weight}, \mathrm{soc}_{\mathcal{A}} \neq 0)&\\ := \{[N] \in \widehat{\mathcal{A}_X}\,(\text{weight}) \mid \mathrm{soc}_{\mathcal{A}}(N) \neq 0\}&,\\ [M] \mapsto [M_X] = [M]&,\end{aligned}$$

is a *bijection* with the inverse $[N] \mapsto [\mathrm{soc}_{\mathcal{A}}(N)] = [{}_{\mathcal{A}}N]$. So, we can write

$$\widehat{\mathcal{A}}\,(\text{weight}, X\text{-torsionfree}) = \widehat{\mathcal{A}_X}\,(\text{weight}, \mathrm{soc}_{\mathcal{A}} \neq 0). \tag{3.10}$$

The algebra $C^{\lambda,\mu}$. Recall that the centre of the algebra C is $\mathbb{K}[C_1, H]$ (Proposition 3.1.(3)). For each pair $\lambda, \mu \in \mathbb{K}$, let

$$C^{\lambda,\mu} := C/C(C_1-\lambda, H-\mu).$$

Similarly, let $C^{\lambda,\mu}_{\mathcal{A}_X} := C_{\mathcal{A}_X}(H)/(C_1-\lambda, H-\mu)$. By Lemma 3.1.(1), $C^{\lambda,\mu}_{\mathcal{A}_X} \simeq A_1 = \mathbb{K}\langle \partial, Z\rangle$. We identify these two algebras via the isomorphism.

LEMMA 3.4. *Let $\lambda, \mu \in \mathbb{K}$. Then*

(1) *As an abstract algebra, the algebra $C^{\lambda,\mu}$ is generated by the elements Z and ϕ subject to the defining relation $[\phi, Z] = Z^2 + 2\lambda$. The algebra $C^{\lambda,\mu}$ is the skew polynomial algebra $\mathbb{K}[Z][\phi; \delta = (Z^2+2\lambda)\partial]$ where $\partial = \frac{\partial}{\partial Z}$ (δ is a $\mathbb{K}$-derivation of the polynomial algebra $\mathbb{K}[Z]$). The algebra $C^{\lambda,\mu}$ is a central Noetherian domain of Gelfand–Kirillov dimension 2.*

(2) *The algebra homomorphism $C^{\lambda,\mu} \to C^{\lambda,\mu}_{\mathcal{A}_X} = A_1$, $Z \mapsto Z$, $\phi \mapsto \partial(Z^2 + 2\lambda) = 2\lambda\partial + hZ$, is a monomorphism.*

PROOF. By Lemma 3.1.(1), $C^{\lambda,\mu} \subseteq C^{\lambda,\mu}_{\mathcal{A}_X} = A_1$. Now, the statements follow (see (3.3)). □

Notice that $\phi = \partial s$ where $s := s_\lambda := Z^2 + 2\lambda \in C^{\lambda,\mu}$, and

$$C^{\lambda,\mu} \subset C^{\lambda,\mu}_{A_X} = A_1 \subset C^{\lambda,\mu}_s = A_{1,s} \tag{3.11}$$

where the subscript 's' means the localization of an algebra at the powers of the element s. Clearly, $s = (Z - \nu)(Z + \nu)$ where $\nu = \sqrt{-2\lambda}$. The elements s, $Z - \nu$ and $Z + \nu$ are normal elements of the algebra $C^{\lambda,\mu}$ since $Zs = sZ$ and $\phi s = s\phi + [\phi, s] = s\phi + [\partial s, s] = s\phi + [\partial, Z^2 + 2\lambda]s = s\phi + 2Zs = s(\phi + 2Z)$.

The $C^{\lambda,\mu}$-module $V(\alpha)$ where $\alpha \in \mathbb{K}$. For each $\alpha \in \mathbb{K}$, consider the A_1-module,

$$V(\alpha) := A_1/A_1(Z - \alpha) = \bigoplus_{i \geqslant 0} \mathbb{K}\partial^i \bar{1} \quad \text{where } \bar{1} = 1 + A_1(Z - \alpha).$$

Since $C^{\lambda,\mu} \subseteq A_1$, $V(\alpha)$ is also a $C^{\lambda,\mu}$-module. The $C^{\lambda,\mu}$-modules $V(\alpha)$ play a key role in description of the primitive spectrum of the algebra $C^{\lambda,\mu}$ (Corollary 3.3). Since

$$(Z - \alpha)\bar{1} = 0 \text{ and } (Z - \alpha)\partial^i\bar{1} = i\partial^{i-1}\bar{1} \text{ for all } i \geqslant 1, \tag{3.12}$$

the element $Z - \alpha$ acts locally nilpotently on $V(\alpha)$, i.e., $V(\alpha) = \bigcup_{i \geqslant 1} \ker_{V(\alpha)}(Z - \alpha)^i$, and $\ker_{V(\alpha)}(Z - \alpha)^i = \bigoplus_{j=0}^{i-1} \mathbb{K}\partial^j\bar{1}$. Furthermore, by (3.12), the $\mathbb{K}[Z]$-module $V(\alpha)$ is a uniserial module and the chain

$$0 \subset \ker_{V(\alpha)}(Z - \alpha) \subset \cdots \subset \ker_{V(\alpha)}(Z - \alpha)^i \subset \cdots \subset V(\alpha) \tag{3.13}$$

contains all the $\mathbb{K}[Z]$-submodules of $V(\alpha)$. The chain of $\mathbb{K}[Z]$-submodules (3.13) determines the ascending, finite dimensional filtration by the degree of the element ∂ as $\ker_{V(\alpha)}(Z - \alpha)^i = \bigoplus_{j=0}^{i-1} \mathbb{K}\partial^j\bar{1}$. The following lemma follows at once from the equality

$$\phi = \partial(2\lambda + Z^2) = (2\lambda + \alpha^2)\partial + 2\partial(Z - \alpha) + \partial(Z - \alpha)^2$$

and the fact that the element $Z - \alpha$ acts locally nilpotently on $V(\alpha)$.

LEMMA 3.5.

(1) *The $C^{\lambda,\mu}$-module $V(\alpha)$ is simple iff $\alpha^2 \neq -2\lambda$. In this case, $V(\alpha) \simeq C^{\lambda,\mu}/C^{\lambda,\mu}(Z - \alpha)$.*

(2) *If $\alpha^2 \neq -2\lambda$ (i.e., $\alpha = 0$ if $\lambda = 0$ and $\alpha = \pm\nu$ if $\lambda \neq 0$ where $\nu = \sqrt{-2\lambda}$) then the $C^{\lambda,\mu}$-module $V(\alpha)$ is a uniserial $C^{\lambda,\mu}$-module and the chain (3.13) contains precisely all the $C^{\lambda,\mu}$-submodules of $V(\alpha)$. In particular, $\operatorname{soc}_{C^{\lambda,\mu}}(V(\alpha)) = \mathbb{K}\bar{1} \simeq C^{\lambda,\mu}/C^{\lambda,\mu}(Z - \alpha, \phi)$ is a simple, 1-dimensional $C^{\lambda,\mu}$-module.*

The prime and primitive spectra of the algebra $C^{\lambda,\mu}$. The next lemma describes the prime spectrum of the algebra $C^{\lambda,\mu}$.

LEMMA 3.6.

(1) *If $\lambda = 0$ then* $\operatorname{Spec}(C^{0,\mu}) = \{0, (Z), (Z, \phi - \gamma) \mid \gamma \in \mathbb{K}\}$, $C^{0,\mu}/(Z) \simeq \mathbb{K}[\phi]$ *and* $C^{0,\mu}/(Z, \phi - \gamma) \simeq \mathbb{K}$.

(2) *If $\lambda \neq 0$ then* $\operatorname{Spec}(C^{\lambda,\mu}) = \{0, (Z \pm \nu), (Z \pm \nu, \phi - \gamma) \mid \gamma \in \mathbb{K}\}$, $C^{\lambda,\mu}/(Z \pm \nu) \simeq \mathbb{K}[\phi]$ *and* $C^{\lambda,\mu}/(Z \pm \nu, \phi - \nu) \simeq \mathbb{K}$ *(where $\nu = \sqrt{-2\lambda}$).*

PROOF. 1. If $\lambda = 0$ then the element Z is a normal element of the algebra $C^{0,\mu}$ such that $C^{0,\mu}_Z = C^{0,\mu}_{Z^2} = C^{0,\mu}_s = A_{1,s}$ is a simple algebra (see (3.11)). By

Proposition 2.1, $\mathrm{Spec}\,(C^{0,\mu}) = \mathrm{Spec}\,(C^{0,\mu}/(Z)) \sqcup \{0\}$. Since $C^{0,\mu}/(Z) \simeq \mathbb{K}[\phi]$ (Lemma 3.4.(1)) and the field $\mathbb{K}$ is algebraically closed, statement 1 follows.

2. If $\lambda \neq 0$ then the element $s = (Z-\nu)(Z+\nu)$ is a normal element of the algebra $C^{\lambda,\mu}$ such that $C_s^{\lambda,\mu} = A_{1,s}$ is a simple algebra (see (3.11)). By Proposition 2.1, $\mathrm{Spec}\,(C^{\lambda,\mu}) = \mathrm{Spec}\,(C^{\lambda,\mu}/(s)) \sqcup \{0\}$. The ideal (s) is a product of two ideals $(Z-\nu)(Z+\nu)$ generated by the normal elements $Z-\nu$ and $Z+\nu$ of the algebra $C^{\lambda,\mu}$, respectively. A prime ideal of the algebra $C^{\lambda,\mu}$ contains the element s iff it contains *precisely* one of the elements $Z-\nu$ or $Z+\nu$ (since $\nu \neq 0$ and $1 = (2\nu)^{-1}(Z+\nu-(Z-\nu))$). Now, statement 2 follows from $C^{\lambda,\mu}/(Z\pm\nu) \simeq \mathbb{K}[\phi]$ and the fact that $\mathbb{K}$ is an algebraically closed field. □

As a corollary of Lemma 3.6, we have a description of maximal and primitive ideals of the algebra $C^{\lambda,\mu}$.

COROLLARY 3.3.
(1) *If* $\lambda = 0$ *then* $\mathrm{Max}\,(C^{0,\mu}) = \{(Z, \phi-\gamma)\,|\,\gamma \in \mathbb{K}\}$ *and* $\mathrm{Prim}\,(C^{0,\mu}) = \mathrm{Max}\,(C^{0,\mu}) \sqcup \{0\}$.
(2) *If* $\lambda \neq 0$ *then* $\mathrm{Max}\,(C^{\lambda,\mu}) = \{(Z\pm\nu, \phi-\gamma)\,|\,\gamma \in \mathbb{K}\}$ *and* $\mathrm{Prim}\,(C^{\lambda,\mu}) = \mathrm{Max}\,(C^{\lambda,\mu}) \sqcup \{0\}$.

PROOF. In both cases, $\mathrm{Max}\,(C^{\lambda,\mu}) \subseteq \mathrm{Prim}\,(C^{\lambda,\mu})$. If $\lambda = 0$ then the prime ideal (Z) of $C^{0,\mu}$ is not a primitive ideal since $C^{0,\mu}/(Z) \simeq \mathbb{K}[\phi]$ (Lemma 3.6.(1)). Similarly, if $\lambda \neq 0$ the prime ideals $(Z\pm\nu)$ of $C^{\lambda,\mu}$ are not primitive ideals since $C^{\lambda,\mu}/(Z\pm\nu) \simeq \mathbb{K}[\phi]$ (Lemma 3.6.(2)). In view of Lemma 3.6, to finish the proof of the lemma we have to show that the zero ideal is primitive in both cases.

Let $\alpha \in \mathbb{K}$ be such that $\alpha^2 \neq 2\lambda$. By Lemma 3.5.(1), $V(\alpha)$ is a simple, infinite dimensional $C^{\lambda,\mu}$-module. Then necessarily its annihilator is zero (by Lemma 3.6) since the ideals $(Z\pm\nu)$ if $\lambda\neq 0$ and (Z) if $\lambda = 0$ are not primitive, every maximal ideal of $C^{\lambda,\mu}$ has codimension 1 and $\dim_{\mathbb{K}}(V(\alpha)) = \infty$. □

Classification of simple $C^{\lambda,\mu}$-modules. Recall that $s = Z^2 + 2\lambda$ and the set $S_s = \{s^i\,|\,i \in \mathbb{N}\}$ is an Ore set of the algebra $C^{\lambda,\mu}$ such that $C_s^{\lambda,\mu} = A_{1,s}$, see (3.11). Then

$$\widehat{C^{\lambda,\mu}} = \widehat{C^{\lambda,\mu}}\,(S_s\text{-torsion}) \sqcup \widehat{C^{\lambda,\mu}}\,(S_s\text{-torsionfree}). \tag{3.14}$$

By Corollary 3.3, $\mathrm{Max}\,(C^{\lambda,\mu}) = \{(Z-\alpha, \phi-\gamma)\,|\,\alpha^2 = -2\lambda, \gamma \in \mathbb{K}\}$. For each maximal ideal $(Z-\alpha, \phi-\gamma)$ of the algebra $C^{\lambda,\mu}$, the 1-dimensional factor algebra $C^{\lambda,\mu}(\alpha,\gamma) := C^{\lambda,\mu}/(Z-\alpha, \phi-\gamma) \simeq \mathbb{K}$ is a simple, S_s-torsion $C^{\lambda,\mu}$-module. The next proposition shows that these are all simple, S_s-torsion $C^{\lambda,\mu}$-modules.

PROPOSITION 3.4.

$$\widehat{C^{\lambda,\mu}}\,(S_s\text{-torsion}) = \{C^{\lambda,\mu}/\mathfrak{m}\,|\,\mathfrak{m} \in \mathrm{Max}\,(C^{\lambda,\mu})\} = \{C^{\lambda,\mu}(\alpha,\gamma)\,|\,\alpha^2 = -2\lambda, \gamma \in \mathbb{K}\}$$

and $\dim_{\mathbb{K}}(C^{\lambda,\mu}(\alpha,\gamma)) = 1$ *for all* α, γ.

PROOF. If M is a simple, S_s-torsion $C^{\lambda,\mu}$-module then $sM = 0$ since s is a normal element of $C^{\lambda,\mu}$. By Corollary 3.3, $\mathfrak{m} := \mathrm{ann}_{C^{\lambda,\mu}}(M) \in \mathrm{Max}\,(C^{\lambda,\mu})$, i.e., M is a simple module over the field $C^{\lambda,\mu}/\mathfrak{m} \simeq \mathbb{K}$. Therefore, $M \simeq C^{\lambda,\mu}(\alpha,\gamma)$ where $\mathfrak{m} = (Z-\alpha, \phi-\gamma)$ for some $\alpha, \gamma \in \mathbb{K}$ such that $\alpha^2 = -2\lambda$. □

Let D be a Dedekind domain, σ be its automorphism and δ be its σ-derivation ($\delta(ab) = \delta(a)b + \sigma(a)\delta(b)$ for all $a, b \in D$). In [**6**], a classification of simple modules is given for the *Ore extension* $A = D[X; \sigma, \delta]$ ($Xd = \sigma(d)X + \delta(d)$ for all $d \in D$). The algebra $C^{\lambda,\mu} = \mathbb{K}[Z][\phi; \delta]$ is a very special example of the Ore extension A where $D = \mathbb{K}[Z]$, $\sigma = \mathrm{id}_D$ and $\delta = s\partial$. "Very special" in the sense that general situation is much more complex in comparison with the case where $A = C^{\lambda,\mu}$.

Two extreme cases, the ring of differential operators $D[X; \delta]$ and the skew polynomial ring $D[X; \sigma]$ were considered in [**9**] and [**4, 5**]. In [**9**], a classification is given under some (rather strong) restriction–the condition (3.1.1) in [**9**].

Let us apply the results of the paper [**6**] to the algebra $C^{\lambda,\mu}$ in order to obtain a classification of its simple modules. The set $S = \mathbb{K}[Z] \setminus \{0\}$ is an Ore set of the Noetherian domain $C^{\lambda,\mu}$. The localization of the algebra $C^{\lambda,\mu}$ at S is a (left and right) *Euclidean ring* $B := S^{-1}C^{\lambda,\mu} = \mathbb{K}(Z)[\phi; \delta = s\partial]$ with coefficient from the field $\mathbb{K}(Z)$ of the rational functions in Z. So, a B-module is simple if it is isomorphic to a B-module B/Bb where b is an irreducible element of B; and $B/Bb \simeq B/Bb'$ iff the elements b and b' are *similar* (see [**18**] for details). Then

$$\begin{aligned}\widehat{C^{\lambda,\mu}} &= \widehat{C^{\lambda,\mu}}\,(S\text{-torsion}) \sqcup \widehat{C^{\lambda,\mu}}\,(S\text{-torsionfree}) \\ &= \widehat{C^{\lambda,\mu}}\,(S_s\text{-torsion}) \sqcup \widehat{C^{\lambda,\mu}}\,(S\text{-torsion}, S_s\text{-torsionfree}) \sqcup \widehat{C^{\lambda,\mu}}\,(S\text{-torsionfree}).\end{aligned} \tag{3.15}$$

THEOREM 3.7. $\widehat{C^{\lambda,\mu}}\,(S\text{-torsion}, S_s\text{-torsionfree}) = \{V(\alpha) = C^{\lambda,\mu}/C^{\lambda,\mu}(Z - \alpha) \mid \alpha \in \mathbb{K}, \alpha^2 \neq -2\lambda\}$*; the* $C^{\lambda,\mu}$*-modules* $V(\alpha)$ *and* $V(\alpha')$ *are isomorphic iff* $\alpha = \alpha'$*; all* $C^{\lambda,\mu}$*-modules* $V(\alpha)$ *(where* $\alpha^2 \neq -2\lambda$*) are faithful and infinite dimensional.*

PROOF. Let M be a simple, S-torsion, S_s-torsionfree $C^{\lambda,\mu}$-module. Then M is an epimorphic image of the $C^{\lambda,\mu}$-module $C^{\lambda,\mu}/C^{\lambda,\mu}(Z - \alpha)$ for some $\alpha \in \mathbb{K}$. By Proposition 3.4, $\alpha^2 \neq -2\lambda$. Then, by Lemma 3.5, the $C^{\lambda,\mu}$-module $C^{\lambda,\mu}/C^{\lambda,\mu}(Z - \alpha)$ is simple and equal to $V(\alpha)$. □

In [**6**], for the algebra $A = D[X; \sigma, \delta]$, Cyc_1 denotes all the σ-invariant maximal ideals of the ring D. If $A = C^{\lambda,\mu}$ then $\mathrm{Cyc}_1 = \mathrm{Max}\,(\mathbb{K}[Z])$ and

$$\mathrm{Cyc}_1 = \mathrm{Cyc}_1(\mathrm{inv}) \sqcup \mathrm{Cyc}_1(\infty)$$

where $\mathrm{Cyc}_1(\mathrm{inv}) = \{(Z - \alpha)\mathbb{K}[Z] \mid \alpha^2 = -2\lambda\}$ (resp. $\mathrm{Cyc}_1(\infty)$) is the set of δ-invariant (resp. not δ-invariant) maximal ideals of $\mathbb{K}[Z]$.

In [**6**], for each maximal $\mathfrak{p} = (Z - \alpha)\mathbb{K}[Z] \in \mathrm{Cyc}_1(\infty)$ (i.e., $\alpha^2 \neq -2\lambda$), the valuation $v_\mathfrak{p}$ on the field $\mathbb{K}(Z)$ that determined by $\mathfrak{p}$ is extended to a valuation of the algebra $B = \mathbb{K}(Z)[\phi; \delta]$ [**6**, Corollary 6.4] by the rule: given $b = \sum_{i \geqslant 0} b_i \phi^i$ where $b_i \in \mathbb{K}(Z)$, $v_\mathfrak{p}(b) = \min\{v_\mathfrak{p}(b_i) - i \mid i \geqslant 0\}$.

In [**6**], for each element $b = \sum b_i \phi^i \in B$, the, so-called, *principle polynomial* $Q_{\mathfrak{p},b} \in \mathbb{K}[H]$ *of* b *relative to* $\mathfrak{p}$ is defined (see the Equality (6.12) in [**6**]):

$$Q_{\mathfrak{p},b} := \sum_{1 \leqslant i : v_\mathfrak{p}(b_i) - i = v_\mathfrak{p}(b)} \eta_\mathfrak{p}\left(b_i p^{-v_\mathfrak{p}(b_i)}\right) H(H-1)\cdots(H-i+1) + \eta_\mathfrak{p}\left(b_0 p^{-v_\mathfrak{p}(b)}\right), \tag{3.16}$$

where $\eta_\mathfrak{p}$ is the ring epimorphism $\mathbb{K}[Z]_\mathfrak{p} \to \mathbb{K}[Z]_\mathfrak{p}/\mathfrak{p}\mathbb{K}[Z]_\mathfrak{p} = \mathbb{K}$ (where $\mathbb{K}[Z]_\mathfrak{p}$ is the localization of $\mathbb{K}[Z]$ at the maximal ideal $\mathfrak{p}$) and $p \in \mathfrak{p} \backslash \mathfrak{p}^2$ satisfies $\delta(P) \equiv 1 \pmod{\mathfrak{p}}$ (for example, $p = (\alpha^2 + 2\lambda)^{-1}Z$ if $\mathfrak{p} = (Z - \alpha)$ where $\alpha^2 \neq 2\lambda$).

In general situation, the formula for the polynomial $Q_{\mathfrak{p},b}$ is more complicated. The formula given here is more simple due to a special ("simple") nature of the algebra $C^{\lambda,\mu}$. The expression (3.16) appeared in [**9**]. We say that an element $b \in B$ is *preserving relative to* $\mathfrak{p} \in \mathrm{Cyc}_1(\infty)$ if there is no root of $Q_{\mathfrak{p},b}$ that belongs to the set $\{-1,-2,\dots\}$. We say that the element $b \in B$ is *preserving* if it is preserving relative to all $\mathfrak{p} \in \mathrm{Cyc}_1(\infty)$. By [**6**, Theorem 1.2], the set $\widehat{C^{\lambda,\mu}}$ (S-torsionfree) is a *non-empty set* (since the set $\bigcup_{n\geqslant 2}\mathrm{Cyc}_n \cup \mathrm{Cyc}_1(\mathrm{inv}) \cup \mathrm{Cyc}_1(\mathrm{finite})$, see [**6**] for details, is a *finite* set for the algebra $C^{\lambda,\mu}$ which is equal to $\mathrm{Cyc}_1(\mathrm{inv})$).

The next theorem is a description of the set $\widehat{C^{\lambda,\mu}}$ (S-torsionfree).

THEOREM 3.8. [**6**, Theorem 1.3] *Let $b=\sum_{i\geqslant 0} b_i\phi^i$ be an irreducible preserving element. Then the S-torsionfree $C^{\lambda,\mu}$-module $M_b := C^{\lambda,\mu}/C^{\lambda,\mu}\cap Bb$ is simple iff*

(CO1) all $b_ib_0^{-1} \in (Z-\alpha)\mathbb{K}[Z]_{(Z-\alpha)}$ for all $\alpha \in \mathbb{K}$ such that $\alpha^2 \neq -2\lambda$.

Up to isomorphism, every simple $\mathbb{K}[h]$-torsionfree $C^{\lambda,\mu}$-module arises in this way and two such $C^{\lambda,\mu}$-modules are isomorphic, $M_b \simeq M_{b'}$, iff the B-modules B/Bb and B/Bb' are isomorphic, i.e., the elements b and b' are similar.

So, the classification of simple $C^{\lambda,\mu}$-modules is complete.

THEOREM 3.9. *A classification of simple $C^{\lambda,\mu}$-modules is given in Proposition* 3.4, *Theorem* 3.7 *and Theorem* 3.8.

PROOF. The statement follows from (3.15). □

The set $\widehat{\mathcal{A}}$**(weight, X-torsionfree).** The next theorem is a classification of simple X-torsionfree $\mathcal{A}$-modules.

THEOREM 3.10. *For each orbit $\mathcal{O} \in \mathbb{K}/2\mathbb{Z}$, we fix a representative $\mu_{\mathcal{O}}$ ($\mathcal{O} = \mu_{\mathcal{O}} + 2\mathbb{Z}$). Then*

$$\widehat{\mathcal{A}}\,(\text{weight}, X\text{-torsionfree}) = \bigsqcup_{\lambda\in\mathbb{K},\mathcal{O}\in\mathbb{K}/2\mathbb{Z}} \mathcal{W}(\lambda,\mathcal{O})$$

where $\mathcal{W}(\lambda,\mathcal{O}) = \big\{[\mathcal{M}] \in \widehat{\mathcal{A}}\,(\text{weight}, X\text{-torsionfree}) \mid \mathrm{Wt}(\mathcal{M}) = \mathcal{O},\ (C_1-\lambda)M = 0\big\}$. *Furthermore,*

$$\mathcal{W}(\lambda,\mathcal{O}) = \big\{[\widetilde{M} = \mathcal{A}_X \otimes_{C_{\mathcal{A}_X}(H)} M] \mid M \in \widehat{C^{\lambda,\mu_{\mathcal{O}}}}\,(S_{s_\lambda}\text{-torsionfree})\big\}$$

where $s_\lambda = Z^2+2\lambda$; $\mathrm{Wt}\,(\widetilde{M}) = \mathcal{O}$ and $\dim(\widetilde{M}_\gamma) = \infty$ for all $\gamma \in \mathcal{O}$; $\mathrm{ann}_{\mathcal{A}}(\widetilde{M}) = (C_1-\lambda)$; the $\mathcal{A}$-modules $\widetilde{M}$ and $\widetilde{M}'$ are isomorphic iff $\lambda=\lambda'$ and $\mathcal{O}=\mathcal{O}'$ (where $M' \in \widehat{C^{\lambda',\mu_{\mathcal{O}'}}}\,(S_{s_{\lambda'}}\text{-torsionfree})$); the elements X and Y act bijectively on each module $\widetilde{M}$.

PROOF. Let $\mathcal{M} \in \widehat{\mathcal{A}}\,(\text{weight}, X\text{-torsionfree})$. Then $\mathcal{M} = \mathcal{M}_X$ is a simple, weight $\mathcal{A}_X$-module. In particular, the element X acts bijectively on $\mathcal{M}$. Therefore, $\mathrm{Wt}\,(\mathcal{M}) = \mathcal{O}$ for some $\mathcal{O} \in \mathbb{K}/2\mathbb{Z}$. So, the first equality of the theorem holds. In particular, $\mathcal{M} \in \mathcal{W}(\lambda,\mathcal{O})$ for some $\lambda \in \mathbb{K}$. Let $\mu = \mu_{\mathcal{O}}$. Then $M := \mathcal{M}_\mu$ is a simple $C^{\lambda,\mu}$-module since the weight $\mathcal{A}$-module $\mathcal{M}$ is simple. Since $\mathcal{M} = \mathcal{M}_X$, the vector space M is also a simple $C^{\lambda,\mu}_{\mathcal{A}_X}$-module. Similarly, for all $i \in \mathbb{Z}$, the weight space $\mathcal{M}_{\mu+2i}$ is a simple $C^{\lambda,\mu+2i}$-module and $C^{\lambda,\mu+2i}_{\mathcal{A}_X}$-module. Since $\mathcal{A}_X = C_{\mathcal{A}_X}(H)[X^{\pm 1};\sigma]$, we have

$$\mathcal{M} = \mathcal{M}_X = \mathcal{A}_X\mathcal{M}_\mu = \bigoplus_{i\in\mathbb{Z}} X^iM \simeq \mathcal{A}_X \otimes_{C_{\mathcal{A}_X}(H)} M = \widetilde{M}$$

and $\mathcal{M}_{\mu+2i} = X^i M$ for all $i \in \mathbb{Z}$.

(i) $M = \mathcal{M}_\mu \in \widehat{C^{\lambda,\mu}}\,(S_{s_\lambda}\text{-torsionfree})$: The element s_λ is a normal element in $C^{\lambda,\mu}$ and the $C^{\lambda,\mu}$-module M is simple. So, either $s_\lambda M = 0$ or, otherwise, the elements s_λ acts bijectively on M. Suppose that $s_\lambda M = 0$. Then $YM = 0$ since $s_\lambda = 2XY$ in $C^{\lambda,\mu}$ and the map $X_{\mathcal{M}}$ is a bijection. Then $\mathcal{M} = \mathcal{A}M \subseteq \bigoplus_{i\geqslant 0} X^i M$, a contradiction. Therefore, $M \in \widehat{C^{\lambda,\mu}}\,(S_{s_\lambda}\text{-torsionfree})$.

(ii) $\mathcal{M}_{\mu+2i} \in \widehat{C^{\lambda,\mu+2i}}\,(S_{s_\lambda}\text{-torsionfree})$ *for all* $i \in \mathbb{Z}$: Repeat the arguments in the proof of the statement (i).

(iii) *The element Y acts bijectively on $\mathcal{M}$*: We have seen that, for all $i \in \mathbb{Z}$, the element $s_\lambda = 2XY$ is a unit in the algebra $C^{\lambda,\mu+2i}$. In particular, the element s_λ acts bijectively on every weight subspace $\mathcal{M}_{\mu+2i}$. Hence, the element s_λ acts bijectively on $\mathcal{M}$, and so does the element Y (since X acts bijectively on $\mathcal{M}$).

(iv) $\dim(\mathcal{M}_{\mu+2i}) = \infty$ *for all* $i \in \mathbb{Z}$: By the statement (ii), the module $\mathcal{M}_{\mu+2i} \in \widehat{C^{\lambda,\mu+2i}}\,(S_{s_\lambda}\text{-torsionfree})$ hence $\dim(\mathcal{M}_{\mu+2i}) = \infty$.

(v) *For each* $M \in \widehat{C^{\lambda,\mu}}\,(S_{s_\lambda}\text{-torsionfree})$, *the $\mathcal{A}$-module* $\widetilde{M} = \mathcal{A}_X \otimes_{C_{\mathcal{A}_X}(H)} M \in \mathcal{W}(\lambda, \mathcal{O} = \mu + 2\mathbb{Z})$: Since the module M is a simple, s_λ-torsionfree $C^{\lambda,\mu}$-module, it is also a simple $C^{\lambda,\mu}_{\mathcal{A}_X}$-module (by Lemma 3.1.(1)) since $EX^{-1} = EY(XY)^{-1} = EY2s_\lambda^{-1}$ (recall that the element s_λ acts bijectively on $\mathcal{M}$). So, the expression $\widetilde{M}$ makes sense. Since $\widetilde{M} = \bigoplus_{i\in\mathbb{Z}} X^i M$, $\widetilde{M}_{\mu+2i} = X^i M$ for all $i \in \mathbb{Z}$ and $\mathrm{Wt}\,(\widetilde{M}) = \mathcal{O}$. For each $i \in \mathbb{Z}$, the C-module $\widetilde{M}_{\mu+2i} = X^i M$ is isomorphic to the twisted C-module ${}^{\omega_X^{-i}}M$ by the automorphism ω_X^{-i} of the algebra C. In particular, all C-modules $\widetilde{M}_{\mu+2i}$ are simple. Therefore, the weight $\mathcal{A}$-module $\widetilde{M}$ is simple.

(vi) *Let* $M \in \widehat{C^{\lambda,\mu}}\,(S_{s_\lambda}\text{-torsionfree})$ *and* $M' \in \widehat{C^{\lambda',\mu'}}\,(S_{s_{\lambda'}}\text{-torsionfree})$ *then the $\mathcal{A}$-modules $\widetilde{M}$ and $\widetilde{M}'$ are isomorphic iff $\lambda = \lambda'$ and $\mathcal{O}(\mu) = \mathcal{O}(\mu')$*: The statement follows from the fact that $(C - \lambda)\widetilde{M} = 0$ and $\mathrm{Wt}\,(\widetilde{M}) = \mathcal{O}(M)$.

(vii) *Let* $\mathcal{M} \in \mathcal{W}(\lambda, \mathcal{O})$ *then* $\mathrm{ann}_{\mathcal{A}}(\mathcal{M}) = (C_1 - \lambda)$: This follows from Theorem 3.8 (since B is a simple Noetherian algebra). □

The next theorem is a classification of simple weight $\mathcal{A}$-modules.

THEOREM 3.11. *Proposition 3.2, Theorem 3.2 and Theorem 3.10 give a classification of simple weight $\mathcal{A}$-modules, and for each simple weight $\mathcal{A}$-module an explicit finite set of generators is given for its annihilator.*

PROOF. By (3.7) and (3.8), $\widehat{\mathcal{A}}(\text{weight}) = \widehat{\mathcal{D}}(\text{weight}) \sqcup \widehat{\mathcal{A}}(\text{weight}, X\text{-torsionfree})$. The set $\widehat{\mathcal{D}}\,(\text{weight})$ is described by Proposition 3.2 and Theorem 3.2 (see also (3.5) and (3.6)). The set $\widehat{\mathcal{A}}\,(\text{weight}, X\text{-torsionfree})$ is described by Theorem 3.10. □

Simple finite dimensional $\mathcal{A}$-modules. The next proposition is a classification of simple finite dimensional $\mathcal{A}$-modules. For a Lie algebra $\mathfrak{g}$, $[\mathfrak{g}, \mathfrak{g}]$ is the linear span in $\mathfrak{g}$ of the set $\{[a, b] \mid a, b \in \mathfrak{g}\}$. The vector space $[\mathfrak{g}, \mathfrak{g}]$ is the smallest ideal $\mathfrak{g}'$ of $\mathfrak{g}$ such that the Lie factor algebra $\mathfrak{g}/\mathfrak{g}'$ is abelian.

PROPOSITION 3.5. $\widehat{\mathcal{A}}\,(\text{f. dim.}) = \widehat{U(\mathfrak{a}/[\mathfrak{a}, \mathfrak{a}])} = \widehat{\mathbb{K}[H]}$ *since* $\mathfrak{a}/[\mathfrak{a}, \mathfrak{a}] = \mathbb{K}H$ *and* $\widehat{\mathcal{A}}\,(\text{f. dim.}) = \widehat{\mathcal{C}}\,(\text{weight},\ S_1)$.

PROOF. The Lie algebra $\mathfrak{a}$ is solvable and the field $\mathbb{K}$ is an algebraically closed field of characteristic zero. So, every simple $\mathfrak{a}$-module is 1-dimensional. Hence, the first equality in statement 2 holds. It follows from the commutation relations $[H,E]=2E, [H,X]=X, [H,Y]=-Y$ and $[E,Y]=2Z$ that $[\mathfrak{a},\mathfrak{a}]\supseteq V=\mathbb{K}E\oplus\mathbb{K}X\oplus\mathbb{K}Z\oplus\mathbb{K}Y$. Since V is an ideal of the Lie algebra $\mathfrak{a}$ such that $\mathfrak{a}/V=\mathbb{K}H$ is abelian, we have $V\supseteq[\mathfrak{a},\mathfrak{a}]$, and so $V=[\mathfrak{a},\mathfrak{a}]$ and $\mathfrak{a}/[\mathfrak{a},\mathfrak{a}]=\mathbb{K}H$. Now, the second equality in statement 2 follows. □

Simple weight $\mathcal{A}$-modules with finite dimensional weight spaces. The next theorem is a classification of simple weight $\mathcal{A}$-modules with finite dimensional weight spaces.

THEOREM 3.12. $\widehat{\mathcal{A}}\,(\text{w., f. dim. w. spaces})=\widehat{\mathcal{D}}\,(\text{weight})$. *In particular, a simple weight $\mathcal{A}$-module has finite dimensional weight spaces iff it is annihilated by X.*

PROOF. The theorem follows from the classification of simple weight $\mathcal{A}$-modules (Theorem 3.11). □

The next theorem is a classification of simple weight $\mathcal{A}$-modules with infinite dimensional weight spaces.

THEOREM 3.13. $\widehat{\mathcal{A}}\,(\text{w., }\infty\text{-dim. w. spaces})=\widehat{\mathcal{A}}\,(\text{weight}, X\text{-torsionfree})$. *In particular, a simple weight $\mathcal{A}$-module has infinite dimensional weight spaces iff it is not annihilated by X.*

PROOF. By (3.7) and (3.8), $\widehat{\mathcal{A}}(\text{weight})=\widehat{\mathcal{D}}(\text{weight})\sqcup\widehat{\mathcal{A}}(\text{weight}, X\text{-torsionfree})$. Every module in $\widehat{\mathcal{A}}\,(\text{weight}, X\text{-torsionfree})$ (resp. $\widehat{\mathcal{D}}\,(\text{weight})$) has infinite (resp. finite) dimensional weight spaces, and the result follows. □

The next result shows that a *finite-infinite dimensional weight spaces dichotomy* holds for simple weight $\mathcal{A}$-modules as well as the *equidimensionality* of all weight spaces does.

COROLLARY 3.6. *All weight spaces of a simple weight $\mathcal{A}$-module are either finite dimensional or infinite dimensional. Furthermore, all of them have the same dimension. More precisely, all of them are either 1-dimensional or infinite dimensional.*

PROOF. See the proof of Theorem 3.13. □

Spec (C). The element $s:=Z^2+2C_1$ is a normal element in C. By Proposition 2.1,

$$\operatorname{Spec}(C)=\operatorname{Spec}(C/(s))\ \sqcup\ \operatorname{Spec}_s(C).$$

By Proposition 3.1.(2), the factor algebra $\overline{C}:=C/(s)\simeq\mathbb{K}[H,Z,\phi]$ is a polynomial algebra and $C_s\simeq(\mathbb{K}[C_1,H]\otimes A_1)_s$ where $A_1=\mathbb{K}\langle Z,s^{-1}\phi\rangle$ is the Weyl algebra since $[s^{-1}\phi,Z]=1$. Hence,

$$\operatorname{Spec}(C_s)=\{(\mathfrak{p}\otimes A_1)_s\,|\,\mathfrak{p}\in\operatorname{Spec}(\mathbb{K}[C_1,H])\}\simeq\operatorname{Spec}(\mathbb{K}[C_1,H])$$

via $(\mathfrak{p}\otimes A_1)_s\mapsto\mathfrak{p}$ (since $(\mathfrak{p}\otimes A_1)_s\cap\mathbb{K}[C_1,H]=\mathfrak{p}$). By Proposition 3.1.(2), the algebra C is a free $\mathbb{K}[C_1,H,Z]$-module, hence $C\cap(\mathfrak{p}\otimes A_1)_s=\mathfrak{p}C$ for all $\mathfrak{p}\in\operatorname{Spec}(\mathbb{K}[C_1,H])$ since $s\in\mathbb{K}[C_1,H,Z]$. Therefore, $\operatorname{Spec}_s(C)=\{\mathfrak{p}C\,|\,\mathfrak{p}\in\operatorname{Spec}(\mathbb{K}[C_1,H])\}\simeq\operatorname{Spec}(\mathbb{K}[C_1,H])$ via $\mathfrak{p}C\mapsto\mathfrak{p}$ (since $\mathfrak{p}C\cap\mathbb{K}[C_1,H]=\mathfrak{p}$).

THEOREM 3.14. (1) $\mathrm{Spec}(C) = \mathrm{Spec}(\overline{C}) \sqcup \mathrm{Spec}_s(C)$ *where* $\overline{C} = \mathbb{K}[H, Z, \phi]$ *is a polynomial algebra and* $\mathrm{Spec}_s(C) = \{\mathfrak{p}C \mid \mathfrak{p} \in \mathrm{Spec}(\mathbb{K}[C_1, H])\} \simeq \mathrm{Spec}(\mathbb{K}[C_1, H])$.

(2) *Let* $P \in \mathrm{Spec}(\overline{C})$ *and* $\mathfrak{p}C \in \mathrm{Spec}_s(C) \setminus \{0\}$. *Then* $P \not\subseteq \mathfrak{p}C$; *and* $\mathfrak{p}C \subseteq P$ *iff* $f(\mathfrak{p}) \subseteq P$ *where* $f : \mathbb{K}[C_1, H] \mapsto \overline{C}$, $C_1 \mapsto -\frac{1}{2}Z^2$, $H \mapsto H$.

PROOF. 1. Statement 1 has already been proven above.

2. By Proposition 2.1.(c), $P \not\subseteq \mathfrak{p}C$. Notice that the algebra homomorphism f is the restriction to the subalgebra $\mathbb{K}[C_1, H]$ of C of the algebra epimorphism $C \to \overline{C}$, $c \mapsto c + (s)$. Then $\mathfrak{p}C \subseteq P$ iff $f(\mathfrak{p}) \subseteq P$. □

COROLLARY 3.7. $\mathrm{Max}(C) = \mathrm{Max}(\overline{C})$.

PROOF. The statement follows from Corollary 3.3 and Theorem 3.14. □

COROLLARY 3.8. $\mathrm{Prim}(C) = \mathrm{Max}(C) \sqcup \{\mathfrak{m}C \mid \mathfrak{m} \in \mathrm{Max}(\mathbb{K}[C_1, H])\}$.

PROOF. The corollary follows from Corollary 3.3 and Theorem 3.14. □

References

[1] V. V. Bavula, *Finite-dimensionality of* Ext^n *and* Tor_n *of simple modules over a class of algebras* (Russian), Funktsional. Anal. i Prilozhen. **25** (1991), no. 3, 80–82, DOI 10.1007/BF01085496; English transl., Funct. Anal. Appl. **25** (1991), no. 3, 229–230 (1992). MR1139880

[2] V. V. Bavula, *Simple* $D[X, Y; \sigma, a]$*-modules* (Russian, with Russian and Ukrainian summaries), Ukraïn. Mat. Zh. **44** (1992), no. 12, 1628–1644, DOI 10.1007/BF01061275; English transl., Ukrainian Math. J. **44** (1992), no. 12, 1500–1511 (1993). MR1215036

[3] V. V. Bavula, *Generalized Weyl algebras and their representations* (Russian), Algebra i Analiz **4** (1992), no. 1, 75–97; English transl., St. Petersburg Math. J. **4** (1993), no. 1, 71–92. MR1171955

[4] V. Bavula, *Classification of the simple modules of the quantum Weyl algebra and the quantum plane*, Quantum groups and quantum spaces (Warsaw, 1995), Banach Center Publ., vol. 40, Polish Acad. Sci. Inst. Math., Warsaw, 1997, pp. 193–201. MR1481744

[5] V. Bavula and F. van Oystaeyen, *The simple modules of certain generalized crossed products*, J. Algebra **194** (1997), no. 2, 521–566, DOI 10.1006/jabr.1997.7038. MR1467166

[6] V. Bavula, *The simple modules of the Ore extensions with coefficients from a Dedekind ring*, Comm. Algebra **27** (1999), no. 6, 2665–2699, DOI 10.1080/00927879908826587. MR1687317

[7] V. V. Bavula and T. Lu, *The prime spectrum and simple modules over the quantum spatial ageing algebra*, Algebr. Represent. Theory **19** (2016), no. 5, 1109–1133, DOI 10.1007/s10468-016-9613-8. MR3550404

[8] V. V. Bavula and T. Lu, *The quantum Euclidean algebra and its prime spectrum*, Israel J. Math. **219** (2017), no. 2, 929–958, DOI 10.1007/s11856-017-1503-1. MR3649612

[9] R. E. Block, *The irreducible representations of the Lie algebra* $\mathfrak{sl}(2)$ *and of the Weyl algebra*, Adv. in Math. **39** (1981), no. 1, 69–110, DOI 10.1016/0001-8708(81)90058-X. MR605353

[10] K. A. Brown and K. R. Goodearl, *Lectures on algebraic quantum groups*, Advanced Courses in Mathematics. CRM Barcelona, Birkhäuser Verlag, Basel, 2002. MR1898492

[11] P. Casati, S. Minniti, and V. Salari, *Indecomposable representations of the Diamond Lie algebra*, J. Math. Phys. **51** (2010), no. 3, 033515, 20, DOI 10.1063/1.3316063. MR2647894

[12] J. Dixmier, *Enveloping algebras*, Graduate Studies in Mathematics, vol. 11, American Mathematical Society, Providence, RI, 1996. Revised reprint of the 1977 translation. MR1393197

[13] A. Douglas and J. Repka, *Embeddings of the Euclidean algebra* $e(3)$ *into* $\mathfrak{sl}(4, \mathbb{C})$ *and restrictions of irreducible representations of* $\mathfrak{sl}(4, \mathbb{C})$, J. Math. Phys. **52** (2011), no. 1, 013504, 9, DOI 10.1063/1.3531984. MR2791126

[14] A. Douglas and H. de Guise, *Some nonunitary, indecomposable representations of the Euclidean algebra* $\mathfrak{e}(3)$, J. Phys. A **43** (2010), no. 8, 085204, 13, DOI 10.1088/1751-8113/43/8/085204. MR2592347

[15] A. Douglas, *Finite dimensional representations of the Euclidean algebra* $\mathfrak{e}(2)$ *having two generators*, J. Math. Phys. **47** (2006), no. 5, 053506, 14, DOI 10.1063/1.2197688. MR2239367

[16] A. Douglas, J. Repka, and W. Joseph, *The Euclidean algebra in rank 2 classical Lie algebras*, J. Math. Phys. **55** (2014), no. 6, 061701, 11, DOI 10.1063/1.4880195. MR3390644

[17] B. Dubsky, R. Lü, V. Mazorchuk, and K. Zhao, *Category* $\mathcal{O}$ *for the Schrödinger algebra*, Linear Algebra Appl. **460** (2014), 17–50, DOI 10.1016/j.laa.2014.07.030. MR3250530

[18] N. Jacobson, *The Theory of Rings*, American Mathematical Society Mathematical Surveys, vol. II, American Mathematical Society, New York, 1943. MR0008601

[19] J. C. McConnell and J. C. Robson, *Noncommutative Noetherian rings*, Revised edition, Graduate Studies in Mathematics, vol. 30, American Mathematical Society, Providence, RI, 2001. With the cooperation of L. W. Small. MR1811901

[20] C. R. Nappi and E. Witten, *Wess-Zumino-Witten model based on a nonsemisimple group*, Phys. Rev. Lett. **71** (1993), no. 23, 3751–3753, DOI 10.1103/PhysRevLett.71.3751. MR1247150

[21] A. I. Ooms, *Computing invariants and semi-invariants by means of Frobenius Lie algebras*, J. Algebra **321** (2009), no. 4, 1293–1312, DOI 10.1016/j.jalgebra.2008.10.026. MR2489901

[22] J. Repka and H. de Guise, *Some finite-dimensional indecomposable representations of* E(2), J. Math. Phys. **40** (1999), no. 11, 6087–6109, DOI 10.1063/1.533072. MR1722355

Department of Pure Mathematics, University of Sheffield, Hicks Building, Sheffield S3 7RH, United Kingdom

Email address: v.bavula@sheffield.ac.uk

School of Mathematical Sciences, Huaqiao University, Quanzhou, Fujian 362021, China

Email address: lutao.sdu@gmail.com

Contemporary Mathematics
Volume **727**, 2019
https://doi.org/10.1090/conm/727/14625

n-torsion clean rings

Peter Danchev and Jerzy Matczuk

ABSTRACT. Let n be an arbitrary natural number. The class of (strongly) n-torsion clean rings is introduced and investigated. Abelian n-torsion clean rings are somewhat characterized and a complete characterization of strongly n-torsion clean rings is given in the case when n is odd. Some open questions are posed at the end.

1. Introduction

Everywhere in the text, all rings are assumed to be associative with unity. Our notations and notions are in agreement with those from [**L**]. For instance, for such a ring R, $U(R)$ denotes the group of units, $\mathrm{Id}(R)$ the set of idempotents and $J(R)$ the Jacobson radical of R, respectively. Besides, the finite field with m elements will be denoted by $\mathbb{F}_m$, and $\mathbb{M}_k(R)$ will stand for the $k \times k$ matrix ring over R; $k \in \mathbb{N}$. For an element u of a group G, $o(u)$ will denote the order of u. The symbol $LCM(n_1, \ldots, n_k)$ will be reserved for the least common multiple of $n_1, \ldots, n_k \in \mathbb{N}$.

We will say a nil ideal I of R is nil of index k if, for any $r \in I$, we have $r^k = 0$ and k is the minimal natural number with this property. Likewise, we will say that I is nil of bounded index if it is nil of index k, for some fixed k.

Let us recall that a ring R is said to be *clean* if, for every $r \in R$, there are $u \in U(R)$ and $e \in \mathrm{Id}(R)$ with $r = e + u$. If, in addition, the commutativity condition $ue = eu$ is satisfied, the clean ring R is called *strongly clean*. These rings were introduced by Nicholson in [**N1**] and [**N2**]. Both clean rings and their various specializations or generalizations are intensively studied since then (see, for example, [**BDZ**], [**C**], [**D**], [**DM**], [**Di**], [**M**] and references within).

A decomposition $r = e + u$ of an element r in a ring R will be called *n-torsion clean decomposition* of r if $e \in \mathrm{Id}(R)$ and $u \in U(R)$ is n-torsion, i.e. $u^n = 1$. We will say that such a decomposition of r is *strongly n-torsion clean*, if additionally e and u commute.

The aim of this article is to investigate in detail the following proper subclasses of (strongly) clean rings:

DEFINITION 1.1. A ring R is said to be (*strongly*) *n-torsion clean* if there is $n \in \mathbb{N}$ such that every element of R has a (strongly) n-torsion clean decomposition and n is the smallest possible natural number with the above property.

2010 *Mathematics Subject Classification.* Primary 16D60, 16S34, 16U60.

Key words and phrases. Clean rings, strongly clean rings, n-torsion clean rings, strongly n-torsion clean rings, idempotents, nilpotents, units.

It is easy to see that boolean rings are precisely the rings which are (strongly) 1-torsion clean. Thus the classes introduced above can be treated as natural generalizations of boolean rings.

Let us notice that in [**D**] the class of *(strongly) invo-clean* rings was investigated. In our terminology, (strongly) invo-clean rings are precisely rings which are either (strongly) 1-torsion clean or (strongly) 2-torsion clean.

It is clear that every clean ring having the unit group of bounded exponent s is n-torsion clean for some n with $1 \leq n \leq s$. We will see below that n has to divide s, but does not have to be equal to s. Let us also observe that a homomorphic image of an n-torsion clean ring is always m-torsion clean, for some $m \leq n$. Hoverer, it is not clear whether n is a multiple of m. Notice that finite rings are always clean, so they are n-torsion clean for suitable n and it would be of interest to compute n for some classes of finite rings; for instance, for matrix rings over finite fields.

In the present paper we mainly concentrate on the case of strongly n-torsion clean rings. Our work is organized as follows: The first short section is of introductory character and it contains some basic observations and examples. Strongly n-torsion clean rings are investigated in Section 2. In particular, it is shown in Theorem 3.4 that such rings have to satisfy a polynomial identity of degree $2n$ and that their Jacobson radical is nil of bounded index. Theorem 3.13 offers a description of such rings which are abelian. Surprisingly, when n is odd, strongly n-torsion clean rings have to be commutative. Their precise description is given in the subsequent Theorem 3.15. We close the paper with some open questions.

2. Preliminaries and Examples

We begin with the following simple but useful observation. Its proof is provided for the sake of completeness.

LEMMA 2.1. *Let R be a (strongly) n-torsion clean ring. Then there exists a finite number of elements $r_1, \dots, r_k \in R$ with (strongly) clean decompositions $r_i = e_i + u_i$, $1 \leq i \leq k$, such that $n = LCM(o(u_1), \dots, o(u_k))$. In particular:*

1. *When the group $U(R)$ has finite exponent s, then n divides s.*
2. *When R is commutative, then $U(R)$ contains an element of order n.*

PROOF. For $r \in R$, let us set

$$r_{min} = \min\{o(u) \mid r = e + u \text{ is a (strongly) } n\text{-torsion clean decomposition of } r \text{ and } o(u) \text{ divides } n\}.$$

Then each r_{min} divides n. Thus $LCM(r_{min} \mid r \in R)$ exists and also divides n. Moreover, we can pick elements $r_1, \dots, r_k \in R$ such that $LCM(r_{min} \mid r \in R) = LCM(r_{1min}, \dots, r_{kmin})$. The minimality of n gives $LCM(r_{min} \mid r \in R) = n$. This completes the proof of the main statement. Subsequently, (1) and (2) follow. □

It is well known that $1 + J(R) \subseteq U(R)$. In the class of rings for which the equality holds, the notation of n-torsion clean rings boils down to rings R for which the unit group $U(R)$ is of finite exponent n. Indeed, we have:

PROPOSITION 2.2. *Let R be a ring and $n \in \mathbb{N}$. Then:*

1. *If $r \in J(R)$, then the unit $1 + r$ has exactly one clean decomposition.*

(2) *Suppose $U(R) = 1 + J(R)$. Then the following two conditions are equivalent:*

(a) *R is (strongly) n-torsion clean.*

(b) *R is (strongly) clean and the group $U(R)$ is of finite exponent n.*

Moreover, if one of the equivalent conditions holds, then $R/J(R)$ is a boolean ring.

PROOF. (1) Let $r \in J(R)$. Observe that if $1+r = e+u$ is a clean decomposition of $1+r$, then $1-e = u-r \in \mathrm{Id}(R) \cap U(R) = \{1\}$, that is, $e = 0$. This implies that $1+r$ has the unique clean decomposition $r+1 = 0+(1+r)$.

(2) Suppose R is (strongly) n-torsion clean and $u \in U(R) = 1 + J(R)$. Then, by (1), $u = 0+u$ is the only clean decomposition of u and $u^n = 1$ follows, i.e. $U(R)$ is of finite exponent $s \leq n$.

Conversely suppose that R is (strongly) clean and $U(R)$ is a group of finite exponent s. Then it is clear that R is n-torsion clean ring, for some $n \leq s$. This yields the equivalence $(a) \Leftrightarrow (b)$.

Since units always lift modulo the Jacobson radical, we have $U(R/J(R)) = \{1\}$. If R is strongly n-torsion clean, then $R/J(R)$ is m-torsion clean for some $m \leq n$. The above yields that $m = 1$, i.e. $R/J(R)$ is a boolean ring. □

Notice that the ring $T_m(\mathbb{F}_2)$ of all upper triangular $m \times m$ matrices over the field $\mathbb{F}_2$ is clean, its Jacobson radical J consists of all strictly upper triangular matrices and $U(T_m(\mathbb{F}_2)) = 1 + J$. Thus, with Proposition 2.2 at hand, we deduce:

EXAMPLE 2.3. Let $m \in \mathbb{N}$ and let k be the smallest nonnegative integer such that $m \leq 2^k$. Then the ring $T_m(\mathbb{F}_2)$ is (strongly) 2^k-torsion clean.

Recall that a ring R is uniquely clean if every element of R has a unique clean presentation. Such rings were characterized in [**NZ**] as those abelian rings R such that $R/J(R)$ is boolean (whence $U(R) = 1 + J(R)$) and idempotents lift modulo $J(R)$. Notice that, as idempotents always lift modulo nil ideals, every ring R such that $R/J(R)$ is boolean and $J(R)$ is nil must be clean. Therefore, the above proposition also gives the following corollary. Its second statement generalizes Example 2.3.

COROLLARY 2.4. (1) *Let R be a uniquely clean ring. Then R is n-torsion clean if and only if $U(R)$ is of exponent n;*

(2) *Let R be a ring such that $R/J(R)$ is boolean and $J(R)$ is nil of bounded index. Then R is n-torsion clean, where n is the exponent of $U(R)$. Moreover, n is a power of 2.*

PROOF. (1) being an immediate consequence of the preceding discussion, let the ring R be as in (2). Then R is a UU ring (i.e. all units are unipotent) and so [**DL**, Theorem 3.4 (2)] implies that $U(R)$ is a 2-group. Now the thesis is a simple consequence of Proposition 2.2. □

Proposition 2.2 demonstrates that, from the point of view of n-torsion clean property, rings with $U(R) = 1 + J(R)$ are, in some sense, not too interesting. The situation when the ring is Jacobson semisimple and has non-trivial group of units is much more interesting. The next example is of such nature and it shows that a ring can be n-torsion clean with n strictly smaller than the exponent of the group $U(R)$.

EXAMPLE 2.5. Let $R = \mathbb{M}_2(\mathbb{F}_2)$. Then R is 2-torsion clean and strongly 6-torsion clean.

PROOF. The ring R is nil clean by virtue of [**BCDM**]. Since the index of nilpotence of elements of R is at most 2, [**D**, Corollary 2.11] implies that R invo-clean which is not boolean, so that it is 2-torsion clean. The above can be also checked by direct computations. For instance, the unit $r = \begin{pmatrix} 1 & 1 \\ 1 & 0 \end{pmatrix}$ of order 3 has a 2-torsion clean decomposition $r = \begin{pmatrix} 1 & 0 \\ 0 & 0 \end{pmatrix} + \begin{pmatrix} 0 & 1 \\ 1 & 0 \end{pmatrix}$ (but it does not have strongly 2-torsion clean decomposition). It is also easy to make elementary computations showing that R is strongly 6-torsion clean. □

Let us notice that the unit group of $\mathbb{M}_2(\mathbb{F}_2)$ is isomorphic to the symmetric group $\mathcal{S}_3$.

PROPOSITION 2.6. *Let $m, k \in \mathbb{N}$ be such that $m \leq 2^k$. Then $\mathbb{M}_m(\mathbb{F}_2)$ is n-torsion clean for some natural $n \leq 2^k$.*

PROOF. Set $R = \mathbb{M}_m(\mathbb{F}_2)$. It is known (cf. [**BCDM**, Theorem 3]) that R is a nil-clean ring. Thus every element x of R can be written as $x = e + z$ with $e = e^2$ and $z^m = 0$, as the index of nilpotence of elements in R is bounded by m. Now we can write $x = (1 + e) + (1 + z)$, where $1 + e$ is an idempotent and $(1+z)^{2^k} = 1 + z^{2^k} = 1$, as $m \leq 2^k$. This enables us to conclude that every element of R has 2^k-torsion clean decomposition. In particular, R is n-torsion clean for some $n \leq 2^k$. □

With the help of this proposition, we derive:

EXAMPLE 2.7. The rings $\mathbb{M}_3(\mathbb{F}_2)$ and $\mathbb{M}_4(\mathbb{F}_2)$, in view of Proposition 2.6, are n-torsion clean for some $n \leq 2^2$. The rings are, however, not invo-clean by virtue of [**D**, Corollary 2.11], and so $n \in \{3, 4\}$.

The linear group $GL(3, \mathbb{F}_2)$ is the unit group of $\mathbb{M}_3(\mathbb{F}_2)$. The group is known to be simple of order 168 and exponent 84.

3. Strongly n-torsion clean rings

The following technical lemma is crucial for our further considerations.

LEMMA 3.1. *Suppose that R is a ring and the element $a \in R$ possesses strongly n-torsion clean decomposition. Then the equality $(a^n - 1)((a - 1)^n - 1) = 0$ holds.*

PROOF. Let $a = e + v$ be a strongly n-torsion clean decomposition of a. Since e, v commute and $v^n = 1$, we deduce:

$$a^n - 1 = (e + v)^n - 1 = \sum_{i=0}^{n} \binom{n}{i} e^i v^{n-i} - 1 = \sum_{i=1}^{n} \binom{n}{i} e^i v^{n-i} \in Re.$$

This implies that

$$(a^n - 1)(1 - e) = 0 \text{ and, consequently, } a^n - 1 = (a^n - 1)e. \tag{3.1}$$

Using $ve = (a - 1)e$, we also have $1 = v^n = (a - e)^n = (a - 1)^n e - a^n e + a^n$. This yields

$$a^n - 1 = (a^n - (a - 1)^n)e. \tag{3.2}$$

Applying (3.2) and the second equation of (3.1), we get $((a-1)^n-1)e = 0$. This equality and the first equation of (3.1) now give together that $(a^n-1)((a-1)^n-1) = (a^n-1)((a-1)^n-1)(e+(1-e)) = 0$, as desired. □

The following assertion is central.

LEMMA 3.2. *Let $n \in \mathbb{N}$ and let R be a ring satisfying the identity $(x^n-1)((x-1)^n-1) = 0$. Then:*

(1) $\mathrm{char}(R) := |1 \cdot \mathbb{Z}|$ *is finite and $J(R)$ is a nil ideal;*
(2) *If n is odd, then R is a reduced ring of characteristic 2 and $J(R) = 0$;*
(3) *If R is an algebra over a field F, then either R is abelian (i.e. all idempotents of R are central) or* $\mathrm{char}(F)$ *divides n.*

PROOF. Set $\phi(x) = (x^n-1)((x-1)^n-1) \in \mathbb{Z}[x]$.

(1) Substituting $x = 3 \cdot 1$ in the identity $\phi(x) = 0$, we see that there exists $0 \neq m \in \mathbb{Z}$ such that $1 \cdot m = 0$ in R. This shows that characteristic $\mathrm{char}(R)$ of R is finite.

Let $r \in J(R)$. Assume that r is not nilpotent. Then the multiplicatively closed set $S = \{r^k \mid k \in \mathbb{N}\}$ does not contain 0. Let P be a maximal ideal of R in the class of all ideals having empty intersection with S. So, P is a prime ideal of R, the ring $\bar{R} = R/P$ satisfies the same identity as R does and $\bar{r} = r + P \in J(\bar{R})$. Moreover $\bar{r}$ is not nilpotent, as $S \cap P = \emptyset$. Thus, eventually replacing R by $\bar{R}$, we may additionally assume that the ring R is prime. Then, its subring $1 \cdot \mathbb{Z} = F$ of R is a domain. By the first part of the proof, $1 \cdot \mathbb{Z}$ is finite, so it is a field. This means that the element $r \in J(R)$ is algebraic over the field F and, as such, has to be nilpotent (cf. [**L**, Proposition 4.18]). This contradicts the choice of r and shows that every element of $J(R)$ is nilpotent.

(2) Suppose n is odd. Then, substituting $x = 0$ in the identity $\phi(x) = 0$, we obtain $2 = 0$, i.e. $\mathrm{char}(R) = 2$.

Let $r \in R$ be such that $r^2 = 0$. If $n = 1$, then the identity $\phi(x) = 0$ shows that $(r-1)r = 0$ and $r = 0$ follows immediately, as $r - 1$ is invertible.

Suppose now that $n \geq 3$. Notice that $(r-1)^2 = 1$. Thus, as n is odd, $(r-1)^n = (r-1)$. Therefore, $r = \phi(r) = 0$. This shows that R has no nonzero nilpotent elements, i.e. R is reduced. Then also $J(R) = 0$ as, by (1), $J(R)$ is a nil ideal.

(3) Suppose R is an algebra over a field F. By (1), $\mathrm{char}(F) = p \neq 0$. If n is odd then, using (2), R is a reduced ring, so it is abelian. Suppose now that n is even and p does not divide n. Thus $1 \cdot n$ is invertible in R. Let $e = e^2, r \in R$. Substituting $x := er(1-e)$ in the identity $\phi(x) = 0$ and using the fact that n is even, we obtain $0 = ((er(1-e))^n - 1)((er(1-e)-1)^n - 1) = ner(1-e)$ and thus the equality $er(1-e) = 0$ follows. Similarly $(1-e)re = 0$. The above shows that every idempotent e of R is central, provided that $\mathrm{char}(R)$ does not divide n. This completes the proof of the lemma. □

REMARK 3.3. In regard to point (2) stated above, a routine argument demonstrates that when R is an n-torsion clean ring and n is odd, then $J(R) = 0$ and $\mathrm{char}(R) = 2$. Indeed, let $0 = f + v$ be an n-torsion clean decomposition of 0. Then $-f = (-f)^n = v^n = 1$ and $\mathrm{char}(R) = 2$ follows. Now, Proposition 2.2(1) yields that $1 = (r-1)^n = \sum_{i=0}^{n} \binom{n}{i}(-1)^{n-i} r^i$, for any $r \in J(R)$. As n is odd, this equation gives $0 = rw$, where $w = \sum_{i=1}^{n} \binom{n}{i}(-1)^{n-i}r^{i-1} \in 1 + J(R)$ is invertible in R, i.e. $r = 0$, as required.

Now we are ready to establish the following theorem.

THEOREM 3.4. *Let $n \in \mathbb{N}$. Suppose R is a strongly n-torsion clean ring. Then:*

1. *R is a PI-ring satisfying the polynomial identity $(x^n - 1)((x-1)^n - 1) = 0$;*
2. *R has finite characteristic $\mathrm{char}(R) = |1 \cdot \mathbb{Z}|$;*
3. *$J(R)$ is a nil ideal of index smaller than $(\mathrm{char}(R))^n$;*
4. *When n is odd, then R is a reduced ring of characteristic 2 and $J(R) = 0$;*
5. *If R is an algebra over a field F, then:*
   (i) $J(R)$ is a nil ideal of index bounded by n;
   (ii) either R is abelian (i.e. all idempotents of R are central) or $\mathrm{char}(F)$ divides n.

PROOF. The first statement is a direct consequence of Lemma 3.1. Notice that, in virtue of Lemma 3.2, for completing the proof it remains only to show that $J(R)$ is nil of index bounded as indicated in the theorem.

Let $r \in J(R)$. We claim that $r^{(\mathrm{char}(R))^n} = 0$. By Lemma 3.2(1), r is nilpotent. Furthermore, Proposition 2.2 (1) shows that the unit $1 + r$ has exactly one clean presentation. Thus $(1+r)^n = 1$ follows, as R is n-torsion clean. Therefore $r^n \in S = (1 \cdot \mathbb{Z})[r] = (1 \cdot \mathbb{Z})r^{n-1} + \ldots + (1 \cdot \mathbb{Z})$. By (1), $1 \cdot \mathbb{Z}$ is a finite ring with $c := \mathrm{char}(R)$ elements. Hence the ring S is finite and has at most c^n elements. As $r \in S$ is nilpotent, its index of nilpotence has to be smaller than $|S| \leq c^n$ (to argue this, just consider the set $A \subseteq S$ of all powers of the element r and show that $|A|$ is the nilpotence index of r). This gives (3).

Suppose now that R is an algebra over a field F. Then S defined as above is, in this case, a finite dimensional algebra over $\mathbb{F}_p = 1 \cdot \mathbb{Z} \subseteq F$ of dimension not bigger than n. The dimension argument applied to the sequence of subspaces $S \supseteq Sr \supseteq Sr^2 \supseteq \ldots$ shows that $r^n = 0$, when $r \in S$ is nilpotent. This yields (5)(i) and completes the proof of the theorem. □

It is an important open question (see [**N2**, Question 2] and [**CDN**]) whether strongly clean rings are Dedekind finite. Since PI rings are Dedekind finite, the above theorem gives immediately the following corollary:

COROLLARY 3.5. *Strongly n-torsion clean rings are Dedekind finite.*

COROLLARY 3.6. *Let R be a strongly n-torsion clean ring. If R is a finitely generated algebra over a central noetherian subring, then $J(R)$ is nilpotent.*

PROOF. By Theorem 3.4(1), R satisfies a monic polynomial identity and $J(R)$ is a nil ideal. Now the thesis is a direct consequence of [**B**, Theorem 2.5]. □

The following example shows that, in general, Jacobson radical of strongly n-torsion clean rings does not have to be nilpotent.

EXAMPLE 3.7. Let $F = \mathbb{F}_{p^k}$ and $F[X]$ be the polynomial ring in infinitely many commuting indeterminates from the set X. Set $R = F[X]/I$, where I is the ideal of $F[X]$ generated by all elements x^p, $x \in X$. Then R is a local ring and its Jacobson radical is not nilpotent. Making use of Propositions 2.2(1) and 3.8(1), one can easily check that R is $p(p^k - 1)$-torsion clean.

In Theorem 3.15, stated in the sequel, we will present a complete characterization of strongly n-torsion clean rings in the case when n is odd. For doing so, the following proposition, which gives a characterization of strongly n-torsion clean rings which are subdirect products of fields, is needed.

PROPOSITION 3.8. (1) *Let F be a field. Then F is n-torsion clean if and only if F is finite and $n = |F| - 1$.*

(2) *A product of fields $\mathbb{F}_{p_1^{k_1}} \times \ldots \times \mathbb{F}_{p_t^{k_t}}$ is n-torsion clean, where n is equal to $LCM(p_1^{k_1} - 1, \ldots, p_t^{k_t} - 1)$;*

(3) *A product $\prod_{i \in I} F_i$ of fields is n-torsion clean if and only if all fields F_i, $i \in I$, are finite, $LCM(|F_i| - 1 \mid i \in I)$ exists and is equal to n;*

(4) *Let R be a subdirect product of fields F_i, $i \in I$. Then R is n-torsion clean if and only if $\prod_{i \in I} F_i$ is n-torsion clean.*

PROOF. (1) Notice that any finite field F is n-torsion clean for some divisor n of $|F| - 1$. On the other hand, if F is any field which is n-torsion clean then, by Theorem 3.4, every element of F is a root of the polynomial $(x^n - 1)((x-1)^n - 1) \in F[x]$, so $|F|$ is finite and $|F| \leq 2n$. Suppose that F is a finite n-torsion clean field and let $|F| - 1 = l \cdot n$. By what we have just shown it follows that $l = \frac{|F|-1}{n} \leq 2 - \frac{1}{n} < 2$ and so $l = 1$ holds, i.e. $s = |F| - 1$, as required.

(2) Let $T = \mathbb{F}_{p_1^{k_1}} \times \ldots \times \mathbb{F}_{p_t^{k_t}}$ and let n be as defined in (2). Notice that $n = \max\{o(u) \mid u \in U(T)\}$ and the order of any $u \in U(T)$ divides n. Therefore, T is m-torsion clean for some $m \leq n$.

For showing that $n = m$, it suffices to show that $n_i = p_i^{k_i} - 1$ divides n, for any $1 \leq i \leq t$. Note that, by (1), $F_i = \mathbb{F}_{p_i^{k_i}}$ is n_i-torsion clean. Furthermore, using Lemma 2.1, we can pick elements $r_1, \ldots, r_s \in T$ and their clean decompositions $r_j = e_j + u_j$, $1 \leq j \leq s$, such that $m = LCM(u_1, \ldots, u_s)$. For a fixed $1 \leq i \leq t$ consider the set $\{\pi_i(r_1), \ldots, \pi_i(r_s)\} \subseteq F_i$, where π_i denotes the canonical projection of R onto F_i. Then, for every $a \in F_i$, a can be presented as $e + u$ with $u^{z_i} = 1$, where $z_i = LCM(o(\pi_i(u_1)), \ldots, o(\pi_i(u_s)))$. Thus $n_i \leq z_i$. As z_i is a LCM of orders of elements in a cyclic group $U(F_i)$ of order n_i, we also deduce that $z_i \leq n_i$, i.e. $z_i = n_i$. This implies that n_i divides m, for any $1 \leq i \leq t$, as desired.

(3) Suppose the product $\prod_{i \in I} F_i$ is n-torsion clean. Then every field F_i is a homomorphic image of $\prod_{i \in I} F_i$. Thus, owing to (1), each F_i is a finite field. If $LCM(|F_i| - 1 \mid i \in I)$ would not exist, then there would exist indexes $i_1, \ldots, i_k \in I$ such that $m = LCM(|F_{i_1}| - 1, \ldots, |F_{i_k}| - 1) > n$. However, in virtue of (2), $T = F_{i_1} \times \cdots \times F_{i_k}$ is m-torsion clean and $m \leq n$, as T is a homomorphic image of R. Thus $LCM(|F_i| - 1 \mid i \in I)$ do exist and we can assume that $LCM(|F_i| - 1 \mid i \in I) = LCM(|F_{i_1}| - 1, \ldots, |F_{i_k}| - 1) = m$. Then it is clear that $n \leq m$. Notice also that $m \leq n$, as T is a homomorphic image of R, i.e. $n = m$. This gives (3).

(4) Let R be a subdirect product of fields F_i, $i \in I$. Suppose R is m-torsion clean. For every $i \in I$, F_i is a homomorphic image of R so, with the aid of (1), the field F_i is n_i-torsion clean, where $n_i = |F_i| - 1$. We also have $n_i \leq m$. Therefore, $LCM(|F_i| - 1 \mid i \in I)$ exists and the statement (3) shows that $\prod_{i \in I} F_i$ is n-torsion clean, where $n = LCM(|F_i| - 1 \mid i \in I)$. In particular, the order of any unit of R divides n and thus $m \leq n$ follows.

Let us fix $i \in I$ and let $F = F_i$ with $s = n_i$. Then, any $a \in F$ can be presented as $a = e + u$ with $u^m = 1$. Let $k_1, \ldots k_s$ be orders of units in such presentations of all elements of F. Then, by construction, $k = LCM(k_1, \ldots, k_s)$ divides m and also divides $s = |F| - 1$ (as s is equal to the order of the group $U(F)$). In particular $k \leq s$. On the other hand, appealing to (1), F is s-torsion clean and this forces that $s \leq k$. However, this shows that $k = s = n_i$ divides m. This means that, for any $i \in I$, $n_i = s$ divides n. Consequently, $n = LCN(n_i \mid i \in I)$ divides m. By the first part of the proof $m \leq n$, so $n = m$ really follows.

Suppose now that $\prod_{i \in I} F_i$ is n-torsion clean and R is a subdirect products of fields F_i, $i \in I$. To complete the proof, it is enough to show R is m-torsion clean for some m. The statement (3) implies that the group $U(R)$ is of finite exponent, say k is the exponent. Then, for any $r \in R$, $e = r^k$ is an idempotent, and $r = (1-e) + ((e-1)+r)$ is a clean decomposition of a with $((e-1)+r)^k = 1$. This allows us to conclude that R is m-torsion clean, for some $m \leq k$, as required. $\square$

The following result, which is needed later in the text, is also of some independent interest. Before stating it, let us recall that idempotents lift modulo an ideal J of R if, for any $a \in R$ such that $a^2 - a \in J$, there exists an idempotent $e \in R$ such that $e - a \in J$. If the idempotent e is uniquely determined by the element a, then we say that idempotents lift uniquely modulo I.

It is known that idempotents lift modulo nil ideals, thus the following lemma applies when J is a nil ideal of a ring R.

LEMMA 3.9. *Let $J \subseteq J(R)$ be an ideal of R. Suppose that idempotents lift modulo J. Then the following conditions are equivalent:*

(1) *R is an abelian ring;*
(2) *R/J is an abelian ring and idempotents lift uniquely modulo J.*

PROOF. Let $\pi \colon R \to R/J$ denotes the canonical homomorphism.

(1) $\Rightarrow$ (2). Suppose the ring R is abelian. Since idempotents lift modulo J, $\mathrm{Id}(R/J) = \pi(\mathrm{Id}(R))$. Thus the ring R/J is abelian, as R is such. Let $e, f \in \mathrm{Id}(R)$ be such that $e - f \in J \subseteq J(R)$. Then, by [**KLM**, Corollary 11], e and f are conjugate in R, i.e. there exists $u \in U(R)$ such that $e = ufu^{-1}$. However, all idempotents of R are central, so $e = f$. This, together with the assumption that idempotents lift modulo J yield that idempotents lift uniquely modulo J.

(2) $\Rightarrow$ (1). The commutator of elements $a, b \in R$ will be denoted by $[a, b] := ab - ba$. Suppose (2) holds and let $e \in \mathrm{Id}(R)$, $r \in R$. Then $f = e + er(1-e)$ is also an idempotent and $[f, e] = er(1-e)$. By assumption R/J is abelian, so $\pi([f, e]) = 0$. This shows that $er(1-e) \in J$. Since $\pi(e) = \pi(f)$ and, by assumption, idempotents lift uniquely modulo J, we obtain $e = f$, i.e. $er(1-e) = 0$. Now, replacing e by $1-e$, we also have $(1-e)re = 0$, for any $r \in R$. This means that every idempotent e of R is central, i.e. R is abelian, as required. $\square$

We will need in the sequel the following direct application of [**DL**, Theorem 3.2.].

LEMMA 3.10. *Let R be a ring and $u \in R$. Suppose that $m := \mathrm{char}(R)$ is finite and $J(R)$ is a nil ideal of index $s+1$, where $s \geq 0$. If $u^t - 1 \in J(R)$, then $u^{tm^s} = 1$.*

PROOF. [**DL**, Theorem 3.2.] states that if R is a ring satisfying assumptions of the lemma, then $(1-r)^{m^s} = 1$, for any $r \in J(R)$. Now, if $u^t - 1 \in J(R)$, then there exists $r \in J(R)$ such that $u^t = 1 - r$ and $u^{tm^s} = 1$ follows. $\square$

The above lemma gives immediately the following corollary:

COROLLARY 3.11. *Let R be a ring of such that* $\operatorname{char}(R)$ *is finite and $J(R)$ is nil of bounded index. If the group $U(R/J(R))$ is of finite exponent, then so is $U(R)$. If additionally $R/J(R)$ is clean (so R is also clean, as units and idempotents lift modulo nil ideals), then R is n-torsion clean, for some $n \in \mathbb{N}$.*

COROLLARY 3.12. *Let R be a ring of finite characteristic and J a nil ideal of R of bounded index. Then the following conditions are equivalent:*

(1) *R is an n-torsion clean ring, for some $n \in \mathbb{N}$.*
(2) *R/J is an t-torsion clean ring, for some $t \in \mathbb{N}$.*

PROOF. Suppose R/J is an m-torsion clean ring, for some $m \in \mathbb{N}$. Let $r \in R$. Since units and idempotents lift modulo J we can find $e \in \operatorname{Id}(R)$ and $u \in U(R)$ such that $\bar{r} = \bar{e} + \bar{u}$ is an t-torsion clean decomposition of $\bar{r}$ in R/J, where $\bar{r}$ denotes the natural image of r in R/J. By Lemma 3.10, $u = 1$, where $m = \operatorname{char}(R)$ and $s+1$ is the nil index of the ideal J. This implies that R is n-torsion clean, for some $n \leq tm^s$.

The reverse implication is clear. □

The following theorem offers a characterization of strongly n-torsion clean abelian rings (compare with Theorem 3.4).

THEOREM 3.13. *For a ring R, the following conditions are equivalent:*

(1) *There exists $n \in \mathbb{N}$ such that R is an n-torsion clean abelian ring.*
(2) (a) $\operatorname{char}(R)$ *is finite;*
    (b) *$J(R)$ is nil of bounded index;*
    (c) *Idempotents lift uniquely modulo $J(R)$;*
    (d) *$R/J(R)$ is a subdirect product of finite fields F_i, where i ranges over some index set I, such that $LCM(|F_i| - 1 \mid i \in I)$ exists.*
(3) *R is an abelian clean ring such that $U(R)$ is of finite exponent.*

PROOF. (1) $\Rightarrow$ (2). Suppose R is an n-torsion clean abelian ring. Then, Theorem 3.4 guarantees that $\operatorname{char}(R)$ is finite and $J(R)$ is nil of finite index. In particular, R has properties (a) and (b). Since $J(R)$ is a nil ideal, idempotents lift modulo $J(R)$ and, by Lemma 3.9, they lift uniquely, so (c) holds.

Finally, by Theorem 3.4(1), $R/J(R)$ satisfies the polynomial identity $\phi(x) = 0$, where $\phi(x) = (x^n - 1)((x-1)^n - 1) \in \mathbb{Z}[x]$. Therefore, $R/J(R)$ is a subdirect product of primitive PI-rings, say $R/J(R)$ is a subdirect product of primitive rings $\{R_i\}_{i \in I}$, for some index set I. Let us fix $i \in I$. Then R_i, as a homomorphic image of R, also satisfies the identity $\phi(x) = 0$. Consequently, by the classical Kaplansky's theorem (cf. [**R**]), each R_i has to be a central simple algebra, finite dimensional over its center C. Notice that, as $\operatorname{char}(R)$ is finite, C is a field of nonzero characteristic, say $\mathbb{F}_p \subseteq C$. Observe also that, by Lemma 3.9, R_i is an abelian ring. This implies that R_i has to be a division algebra over $\mathbb{F}_p$. It is known (cf. [**F**, Corollary from page 48]) that every division algebra which is algebraic over a finite field is necessarily commutative. In particular, R_i has to be a field. In fact, it is a finite field, as R_i is contained in the spitting field of $\phi(x) \in \mathbb{F}_p[x]$. The above shows that $R/J(R)$ is a subdirect product of finite fields. Moreover, $R/J(R)$ is also, as a homomorphic image of R, strongly n'-torsion clean, for some $n' \leq n$. Therefore, making use of Proposition 3.8, we see that R satisfies the property (d). This completes the proof of the implication.

(2) $\Rightarrow$ (3). Suppose (2) holds. We know, by (d) and Proposition 3.8, that $R/J(R)$ is a clean ring with the unit group $U(R)$ of finite exponent. The property (b) guarantee that $J(R)$ is a nil ideal and Corollary 3.11 yields that R is a clean ring with the unit group $U(R)$ is of finite exponent. Finally, properties (d), (c) together with Lemma 3.9 imply that R is an abelian ring.

The implication (3) $\Rightarrow$ (1) is obvious. □

In parallel to Theorem 3.13, one can state the following:

THEOREM 3.14. *For a ring R, the following conditions are equivalent:*

(1) *R is strongly n-torsion clean, for some $n \in \mathbb{N}$.*
(2) *R is strongly clean and $U(R)$ is of finite exponent.*

PROOF. (1) $\Rightarrow$ (2). Suppose R is strongly n-torsion. Then clearly R is strongly clean. Next, observe that Theorem 3.4 implies that R is a PI-ring satisfying an identity of degree $2n$ and $J(R)$ is a nil ideal of bounded index. Using similar arguments as in the proof of Theorem 3.13, one can see that the quotient $R/J(R)$ is a subdirect product of a matrix rings, say $R_i = M_{m_i}(F_i)$, over finite fields F_i. Notice that, as char(R) is finite, the set of characteristics of fields from the set $\mathcal{F} = \{F_i \mid i \in I\}$ is finite and also the number of fields of a given characteristic p is finite, as every such field is contained in the splitting field of a given polynomial of degree $2n$. Thus there are only finitely many classes of isomorphic fields in the set $\mathcal{F}$. Moreover, by the classical Amitsur-Levitzki's theorem (cf. [**R**]), each m_i is not grater than n, as every R_i satisfies a polynomial identity of degree $2n$. Therefore, the unit group of the product $\prod_{i\in I} R_i$ is a group of finite exponent. By Theorem 3.4, char(R) is finite and $J(R)$ is a nil ideal of bounded index. Now, we can apply Lemma 3.10 to obtain that the group $U(R)$ is of finite exponent.

The implication (2) $\Rightarrow$ (1) is clear. □

We now have at our disposal all the necessary information to present a satisfactory structural characterization of strongly n-torsion clean rings, for all odd n.

THEOREM 3.15. *Suppose $n \in \mathbb{N}$ is odd. For a ring R, the following conditions are equivalent:*

(1) *R is a strongly n-torsion clean ring;*
(2) *There exist integers $k_1, \dots, k_t \geq 1$ such that $n = LCM(2^{k_1}-1, \dots, 2^{k_t}-1)$ and R is a subdirect product of copies of fields $\mathbb{F}_{2^{k_i}}$, $1 \leq i \leq t$;*
(3) *R is a clean ring in which orders of all units are odd, bounded by n and there exists a unit of order n.*

PROOF. (1) $\Rightarrow$ (2). Suppose R is a strongly n-torsion clean ring. Then, by Theorem 3.4 (4), R is a reduced ring of characteristic 2 and $J(R) = 0$. Thus, as every reduced ring is abelian, we can apply Theorem 3.13 to obtain that R is a subdirect product of finite fields F_i of characteristic 2, where $i \in I$, for some index set I. Now, Proposition 3.8 completes the proof of the implication.

The reverse implication (2) $\Rightarrow$ (1) is a direct consequence of Proposition 3.8. The implication (2) $\Rightarrow$ (3) is a tautology.

(3) $\Rightarrow$ (1). Let R be as in (3). Then, as $(-1)^2 = 1$ and R has no units of even order, $-1 = 1$, i.e., char$(R) = 2$. Let us observe that R has to be reduced. Indeed, if $r^2 = 0$ for some $r \in R$, then $(1+r)^2 = 1 + r^2 = 1$. Using again the fact that R has no units of even order, we get $r = 0$. It is known that in a reduced ring

all idempotents are central. Moreover, by assumption, R is a clean ring and, as every unit of R is of finite order bounded by n, the ring must be strongly m-torsion clean, for some $m \leq n!$. Now, because orders of units are odd, m has to be odd (as $u^{2k} = 1$ yields $u^k = 1$, when $o(u)$ is odd). Furthermore, bearing in mind the equivalence of statements (1) and (2), we conclude that $n = m$, as required. □

It is worth to mention certain slightly unexpected, non-trivial consequences of the above theorem. Namely, not every odd natural number n can serve as torsion degree of strongly n-torsion clean rings and, for odd, n-torsion clean rings are always commutative.

Notice that, because every finite ring is clean, Theorem 3.15 forces the following:

COROLLARY 3.16. *For a finite ring R the following conditions are equivalent:*

(1) *R is strongly n-torsion clean for some odd n;*
(2) *R has no units of even order;*
(3) *R is isomorphic to a finite direct product of fields of characteristic* 2.

PROOF. By the Chinese Remainder Theorem, any subdirect product of finite number of fields is isomorphic to a direct product of fields. Now, the corollary is a straightforward consequence Theorem 3.15. □

We close the paper with some problems of interest.

QUESTION 1. The matrix ring $\mathbb{M}_n(\mathbb{F}_{2^k})$ is always m-torsion clean for some m. Compute m in terms of n and k; is $m = n$ if $k = 1$?

Recall that some basic observations related to the above problem can be found in Proposition 2.6 and Examples 2.5 and 2.7. In particular $M_2(\mathbb{F}_2)$ is 2-torsion clean and, when $n \in \{3, 4\}$ then $M_n(\mathbb{F}_2)$ is m-torsion clean, where $2 < m \leq 4$. Christian Lomp checked for us, with the help of SageMath, that $n = m$ in the above cases.

We have seen in Theorem 3.14 that strongly n-torsion clean rings have unit groups $U(R)$ of finite exponent. For odd n, by Theorem 3.15, $n = \exp(U(R))$. Example 2.5 shows also such equality in the case of the ring $M_2(\mathbb{F}_2)$. We were kindly informed by Pace Nielsen, that such equality also holds for $M_3(\mathbb{F}_2)$, i.e. $M_3(\mathbb{F}_2)$ is strongly 84-torsion clean. Notice also that Example 3.7 offers yet another instance of equality $n = \exp(U(R))$.

Thus we pose the following two questions.

QUESTION 2. Let R be a strongly n-torsion clean ring. Is it true that $n = \exp(U(R))$?

QUESTION 3. Let R be an n-torsion clean ring. Is then necessary $U(R)$ of finite exponent?

In regard to both questions, are clean rings with cyclic units also strongly clean? For odd n, strongly n-torsion clean rings were characterized in Theorem 3.15. Besides, Theorem 3.13 offers a description of strongly n-torsion clean rings with extra assumption that the considered rings are abelian. So, we come to

QUESTION 4. Characterize strongly n-torsion clean rings, for even $n \in \mathbb{N}$.

If R is not abelian, then Theorem 3.4 (5) and arguments used in the proof of Theorem 3.14 show that, modulo the Jacobson radical (which is nil of bounded index), Question 4 essentially reduces to the investigation of matrix rings over finite fields of characteristic dividing n.

It is also worthwhile noticing that (strongly) 2-torsion clean rings were classified in [**D**] under the name (strongly) invo-clean rings by using another approach. In fact, R is strongly invo-clean if and only if $R \cong R_1 \times R_2$, where R_1 is a ring for which $R_1/J(R_1)$ is boolean with $z^2 = 2z$ for every $z \in J(R_1)$, and R_2 is a ring which can be embedded in a direct product of copies of the field $\mathbb{F}_3$.

Acknowledgments. We would like to thank Christian Lomp and Pace Nielsen for helpful discussion and support in computations.

References

[B] A. Braun, *The nilpotency of the radical in a finitely generated PI ring*, J. Algebra **89** (1984), no. 2, 375–396, DOI 10.1016/0021-8693(84)90224-2. MR751151

[BCDM] S. Breaz, G. Călugăreanu, P. Danchev, and T. Micu, *Nil-clean matrix rings*, Linear Algebra Appl. **439** (2013), no. 10, 3115–3119, DOI 10.1016/j.laa.2013.08.027. MR3116417

[BDZ] S. Breaz, P. Danchev, and Y. Zhou, *Rings in which every element is either a sum or a difference of a nilpotent and an idempotent*, J. Algebra Appl. **15** (2016), no. 8, 1650148, 11, DOI 10.1142/S0219498816501486. MR3528770

[C] H. Chen, *On uniquely clean rings*, Comm. Algebra **39** (2011), no. 1, 189–198, DOI 10.1080/00927870903451959. MR2770888

[CDN] V. Camillo, T. J. Dorsey, and P. P. Nielsen, *Dedekind-finite strongly clean rings*, Comm. Algebra **42** (2014), no. 4, 1619–1629, DOI 10.1080/00927872.2012.746976. MR3169656

[D] P. V. Danchev, *Invo-clean unital rings*, Commun. Korean Math. Soc. **32** (2017), no. 1, 19–27, DOI 10.4134/CKMS.c160054. MR3608475

[DL] P. V. Danchev and T.-Y. Lam, *Rings with unipotent units*, Publ. Math. Debrecen **88** (2016), no. 3-4, 449–466, DOI 10.5486/PMD.2016.7405. MR3491753

[DM] P. V. Danchev and W. Wm. McGovern, *Commutative weakly nil clean unital rings*, J. Algebra **425** (2015), 410–422, DOI 10.1016/j.jalgebra.2014.12.003. MR3295991

[Di] A. J. Diesl, *Nil clean rings*, J. Algebra **383** (2013), 197–211, DOI 10.1016/j.jalgebra.2013.02.020. MR3037975

[F] C. Faith, *Algebraic division ring extensions*, Proc. Amer. Math. Soc. **11** (1960), 43–53, DOI 10.2307/2032714. MR0111775

[KLM] P. Kanwar, A. Leroy, and J. Matczuk, *Idempotents in ring extensions*, J. Algebra **389** (2013), 128–136, DOI 10.1016/j.jalgebra.2013.05.010. MR3065996

[L] T. Y. Lam, *A first course in noncommutative rings*, 2nd ed., Graduate Texts in Mathematics, vol. 131, Springer-Verlag, New York, 2001. MR1838439

[M] J. Matczuk, *Conjugate (nil) clean rings and Köthe's problem*, J. Algebra Appl. **16** (2017), no. 4, 1750073, 14, DOI 10.1142/S0219498817500736. MR3635122

[N1] W. K. Nicholson, *Lifting idempotents and exchange rings*, Trans. Amer. Math. Soc. **229** (1977), 269–278, DOI 10.2307/1998510. MR0439876

[N2] W. K. Nicholson, *Strongly clean rings and Fitting's lemma*, Comm. Algebra **27** (1999), no. 8, 3583–3592, DOI 10.1080/00927879908826649. MR1699586

[NZ] W. K. Nicholson and Y. Zhou, *Rings in which elements are uniquely the sum of an idempotent and a unit*, Glasg. Math. J. **46** (2004), no. 2, 227–236, DOI 10.1017/S0017089504001727. MR2062606

[R] L. H. Rowen, *Polynomial identities in ring theory*, Pure and Applied Mathematics, vol. 84, Academic Press, Inc. [Harcourt Brace Jovanovich, Publishers], New York-London, 1980. MR576061

Institute of Mathematics and Informatics, Bulgarian Academy of Sciences, 1113 Sofia, Bulgaria

Email address: pvdanchev@yahoo.com, danchev@math.bas.bg

Institute of Mathematics, Warsaw University, Banacha 2, 02-097 Warsaw, Poland

Email address: jmatczuk@mimuw.edu.pl

Contemporary Mathematics
Volume **727**, 2019
https://doi.org/10.1090/conm/727/14626

Generating characters of non-commutative Frobenius rings

Steven T. Dougherty, Arda Kör, and André Leroy

ABSTRACT. We give constructions for the generating character of non-commutative Frobenius rings, which is used in constructing the MacWilliams relations for a given ring. We show how to construct this character directly for any finite Frobenius ring and give examples for various families of Frobenius rings.

1. Introduction

The MacWilliams relations are one of the foundational results of algebraic coding theory. They relate the weight distribution of a code and the weight distribution of its orthogonal. These relations were first proven by J. MacWilliams in [**8**] and [**9**] for binary codes and this proof naturally extended first to codes over finite fields and then to codes over Frobenius rings in [**11**]. This result is one of the major reasons why coding theorists generally restrict their alphabets to Frobenius rings. The key element of the MacWilliams relations for Frobenius rings is to find a generating character. In [**6**], the technique for finding this generating character when R is commutative was described. In the commutative case, the main tool is the generalized Chinese Remainder Theorem which decomposes arbitrary Frobenius rings into local Frobenius rings. Then a generating character is constructed for local rings and the generating character for the ambient ring is constructed via the Chinese Remainder Theorem. See [**4**] for a complete description of this technique and for a description of the use of Frobenius rings in algebraic coding theory. For non-commutative rings there is no such decomposition into easily handled subrings, which complicates the situation greatly. In other words, one cannot assume that an arbitrary finite ring has a decomposition into local rings or any other type of ring with an easily constructed generating character. In this work, we shall study the construction of this generating character for non-commutative Frobenius rings. We begin with the necessary definitions both for rings and for codes.

1.1. Rings. For any ring R, we define the Jacobson radical as the intersection of all maximal left ideals in R. It turns out that this is equal to the intersection of all maximal right ideals as well. We denote the Jacobson radical by $J(R)$ and note that it is a two sided ideal. The right socle of a ring R, denoted by $soc(R)$ is the

2010 *Mathematics Subject Classification.* Primary 16Z05, 11T71.
Key words and phrases. Frobenius rings, generating character.
The first and second authors are grateful to the Université d'Artois where they stayed while this work was completed.

sum of all minimal right ideals. For a complete description of these ideals and for a reference for the results from ring theory see [**7**].

Let M be a module. A character of M is a homomorphism from the additive group $(M,+)$ to the multiplicative group of nonzero complex numbers. Note that sometimes characters are understood to have images in $\mathbb{Q}/\mathbb{Z}$ but we prefer the image to be the nonzero complex numbers. These maps will also be referred to as "additive maps". For an R module M, let $\widehat{M}$ be the set of characters of M. This set is an abelian group: if $\sigma, \tau \in \widehat{M}$, then we define $(\sigma+\tau)(m)$ to be $\sigma(m)\tau(m)$. Notice that if M_R (respectively ${}_RM$) is a right (respectively left) R-module, then $\widehat{M}$ is left (respectively right) R module via $(r.\varphi)(m) = \varphi(mr)$ (respectively $(\varphi.r)(m) = \varphi(rm)$) where $\varphi \in \widehat{M}$ and $r \in R$.

Throughout the paper, we assume that a ring has a multiplicative unity and that all rings are finite. Rather than providing a definition for Frobenius ring we shall give its characterization as given in Theorem 3.10 in [**11**]. This states that the following are equivalent for a finite ring R.

- The ring R is a Frobenius ring.
- As a left module, $\widehat{R} \cong {}_RR$.
- As a right module $\widehat{R} \cong R_R$.

Hence, we say that a ring is Frobenius when the second or third condition is met. For a complete description of the classical definition see [**3**]. Moreover for a Frobenius ring R we have $\mathcal{L}(J) = soc({}_RR) = soc(R_R) = \mathcal{R}(J)$, where $\mathcal{L}(J)$ is the left annihilator of J and $\mathcal{R}(J)$ is the right annihilator of J. Note that for non-Frobenius rings this is not necessarily true.

A character χ of a ring R is a right generating character if the mapping $\phi : R \to \widehat{R}$ defined by $\phi(r) = \chi^r$ is an isomorphism of right R modules (where $(\chi^r)(x) = \chi(rx)$ for $r, x \in R$). A character χ of a ring R is a left generating character if the mapping $\phi : R \to \widehat{R}$ defined by $\phi(r) = {}^r\chi$ is an isomorphism of left R modules (where $({}^r\chi)(x) = \chi(xr)$ for $x, r \in R$). Theorem 4.3 in [**11**] gives that a character χ on R is a left generating character if and only if it is a right generating character. Therefore, throughout this paper we shall restrict ourselves to finding right generating characters and eliminate the adjectives right and left.

The following is Corollary 3.6 in [**2**].

LEMMA 1.1. [**2**] *Let χ be a character of a finite ring R. Then χ is a right generating character if and only if $ker(\chi)$ contains no non-zero right ideals.*

This lemma gives our most powerful tool in the construction of right generating characters and we shall use it throughout the paper when proving that a given character is in fact a generating character.

1.2. Codes. A code over a Frobenius ring R of length n is a subset of R^n. It is said to be a left linear code if it is a left submodule of ${}_RR^n$ and it is said to be a right linear code if it is a right submodule of R^n_R. The ambient space R^n has an inner-product defined by

$$[\mathbf{v}, \mathbf{w}] = \sum v_i w_i.$$

For a code C we define

$$\mathcal{L}(C) = \{\mathbf{v} \mid [\mathbf{v}, \mathbf{w}] = 0, \forall \mathbf{w} \in C\} \tag{1}$$

and

$$\mathcal{R}(C) = \{\mathbf{v} \mid [\mathbf{w}, \mathbf{v}] = 0, \forall \mathbf{w} \in C\}. \tag{2}$$

In [5], it is shown that $\mathcal{L}(C)$ is a left linear code and $\mathcal{R}(C)$ is a right linear code. When the ring is commutative these orthogonals coincide and are denoted by $C^{\perp}$.

For a code over $R = \{a_0, a_1, \ldots, a_{s-1}\}$, the complete weight enumerator is defined to be the following polynomial in commuting variables:

$$cwe_C(x_{a_0}, x_{a_1}, \ldots, x_{a_{s-1}}) = \sum_{\mathbf{c} \in C} \prod_{i=0}^{s-1} x_{a_i}^{n_i(\mathbf{c})} \tag{3}$$

where there are $n_i(\mathbf{c})$ occurrences of a_i in the vector $\mathbf{c}$.

If χ is a right generating character then $\chi^a(b) = \chi(ab)$, where χ^a is the character corresponding to the element a of R. We use this to define the matrix T. Let T be the $|R|$ by $|R|$ matrix given by:

$$(T)_{a,b} = (\chi(ab)) \tag{4}$$

where a and b are elements in R. It is precisely this matrix T that we use to obtain the MacWilliams relations for a given ring R. We note here that the ordering of the elements is arbitrary and must only match the ordering given in the complete weight enumerator.

If $(x_0, \ldots, x_{s-1})$ is a vector we let $T \cdot (x_0, \ldots, x_{s-1}) = (T(x_0, \ldots, x_{s-1})^t)^t$ where M^t is the transpose of the matrix M. The following are the MacWilliams relations as given in Corollary 8.2, [11].

THEOREM 1.2. [11] *(Generalized MacWilliams Relations) Let R be a Frobenius ring. If C is a left submodule of R^n, then*

$$cwe_C(x_0, x_1, \ldots, x_{s-1}) = \frac{1}{|\mathcal{R}(C)|} cwe_{\mathcal{R}(C)}(T^t \cdot (x_0, x_1, \ldots, x_{s-1})).$$

If C is a right submodule of R^n, then

$$cwe_C(x_0, x_1, \ldots, x_{s-1}) = \frac{1}{|\mathcal{L}(C)|} cwe_{\mathcal{L}(C)}(T \cdot (x_0, x_1, \ldots, x_{s-1})).$$

It follows from these relations for Frobenius rings, that if C is a left linear code then $|C||\mathcal{R}(C)| = |R|^n$ and if C is a right linear code then $|C||\mathcal{L}(C)| = |R|^n$. See [3] for a proof of these facts. This implies that if you can find a linear code and its proper orthogonal whose cardinalities do not satisfy this then the ring is not Frobenius. Since ideals are codes of length 1, it is enough to show that a left ideal and its right annihilator do not have the product of their cardinalities equal to the size of the ring to show that the ring is not Frobenius.

EXAMPLE 1. Consider the ring $R = \{\begin{pmatrix} a & b \\ 0 & c \end{pmatrix} \mid a, b, c \in \mathbb{F}_q\}$. Here $|R| = q^3$. Consider the right ideal $I = \{\begin{pmatrix} x & y \\ 0 & 0 \end{pmatrix} \mid x, y \in \mathbb{F}_q\}$. This ideal has cardinality q^2. Its left orthogonal $\mathcal{L}(I) = \{\begin{pmatrix} 0 & d \\ 0 & e \end{pmatrix} \mid d, e \in \mathbb{F}_q\}$ which has cardinality q^2. But $q^2 q^2 \neq q^3$. Hence this ring is not Frobenius. In general, the ring of upper triangular matrices is not Frobenius.

EXAMPLE 2. Consider the non-commutative ring $R = \{\begin{pmatrix} a & b \\ 0 & c \end{pmatrix} \mid a,b,c \in \mathbb{F}_2\}$. One might naively assume that the additive map $\chi(\begin{pmatrix} a & b \\ 0 & c \end{pmatrix}) = (-1)^{a+b+c}$ would be a generating character. However, $\{\begin{pmatrix} 0 & 0 \\ 0 & 0 \end{pmatrix}, \begin{pmatrix} 0 & 1 \\ 0 & 1 \end{pmatrix}\}$ is a right ideal of R. The map χ contains this ideal in its kernel, hence χ is not a generating character. In fact, this ring is not Frobenius and does not have a generating character.

The previous example shows that a generating character involves much more than the additive structure of the ring. More to the point, the fact that a ring is Frobenius and has a generating character, in some sense, says that the multiplication of the ring works in a compatible way to the addition in terms of orthogonality. The MacWilliams relations can be defined simply on the additive group structure, where the orthogonality is determined via a character table. The ring is not necessary at all. However, we want the orthogonal as defined by the group's character table to coincide with the orthogonal as defined canonically with the ring. When the ring is Frobenius the coincidence of the two orthogonals occurs, when it is not Frobenius it does not. Namely, the group orthogonal comes from characters alone whereas the ring orthogonality uses the multiplication in the ring.

Other weight enumerators can be defined over specific rings and MacWilliams relations can be found for some of these specialized weight enumerators. In general, by collapsing the elements with an equivalence relation denoted by $\equiv$, we can build a matrix S for the MacWilliams relations which apply to the weight enumerator based on this equivalence relation by

$$S_{\overline{a},\overline{b}} = \sum_{c \equiv b} T_{a,c}. \tag{5}$$

For this construction to be meaningful, we need

$$\sum_{c \equiv b} T_{a,c} = \sum_{c \equiv b} T_{a',c}$$

if $a \equiv a'$.

Notice that, except for the first row, the rows of the matrix T must sum to 0. Indeed if $a \neq 0$, since χ is a generating character, the ideal aR is not contained in $\ker(\chi)$. Therefore, there exists $b_0 \in R$ such that $\chi(ab_0) \neq 1$. We then have $\chi(ab_0)\sum_{b\in R}\chi(ab) = \sum_{b\in R}\chi(ab_0+ab) = \sum_{b\in R}\chi(a(b_0+b)) = \sum_{b\in R}\chi(ab)$. From this we conclude that $\sum_{b\in R}\chi(ab) = 0$, as claimed.

The rows of the matrix S must also sum to 0. For example, the Hamming weight enumerator equates all non-zero elements in the ring. Therefore, there are two equivalence classes, $\overline{0}$ and $\overline{1}$. Then we have that $\sum_{a\in R}\chi(0a) = |R|$ and $\sum_{a\in R}\chi(1a) = 0$. Hence, the matrix S for the MacWilliams relations for the Hamming weight enumerator is given by:

$$\begin{pmatrix} 1 & |R|-1 \\ 1 & -1 \end{pmatrix}. \tag{6}$$

As a small commutative example, consider the ring $\mathbb{F}_2[x]/\langle x^2\rangle$. Then the matrix T indexed by $0, 1, x, 1+x$ is given by

$$T = \begin{pmatrix} 1 & 1 & 1 & 1 \\ 1 & -1 & -1 & 1 \\ 1 & -1 & 1 & -1 \\ 1 & 1 & -1 & -1 \end{pmatrix}.$$

Equating 1 and $1+x$ you obtain the matrix

$$S = \begin{pmatrix} 1 & 2 & 1 \\ 1 & 0 & -1 \\ 1 & -2 & 1 \end{pmatrix}.$$

However, equating 1 and x would give two different results on the row corresponding to 1 and the row corresponding to x. Hence this is not an acceptable specialization. If the equivalence relation is that two elements are the same if they are equal by left multiplication of a unit then it is a simple matter to see that one can always construct the matrix S. This weight enumerator is known as the symmetrized weight enumerator. We shall denote the matrix S for this weight enumerator by S_1. It is also possible to construct the matrix S, where the equivalence relation is given by two elements a, b are equivalent if $a = \pm b$. This is known as the symmetric weight enumerator. We shall denote the matrix S for this weight enumerator by S_2.

2. Generating Characters

We begin our study of generating characters with the following lemma which appears as Example 4.9 (iii) in Wood's paper [**11**]. We include the proof since it is constructive and shows how to construct the generating character.

LEMMA 2.1. *Let R be a finite Frobenius ring where*

$$R \cong R_1 \times R_2 \times \cdots \times R_s$$

where R_i is Frobenius for all $1 \le i \le s$. Let χ_{R_i} be the right generating character for R_i and let $\phi : R \to R_1 \times R_2 \times \cdots \times R_s$ be the isomorphism, with $\phi(a) = (a_1, a_2, \ldots, a_s)$. Define $\chi : R \to \mathbb{C}^*$ by

$$\chi(a) = \prod_{i=1}^{s} \chi_{R_i}(a_i).$$

Then χ is a right generating character of R.

PROOF. Assume I is a nonzero right ideal of R. Then I is isomorphic to $I_1 \times I_2 \times \cdots \times I_s$ where I_i is a nonzero right ideal in R_i. Let $a_i \in I_i$ with $\chi_{R_i}(a_i) \neq 1$. We know such an a_i exists since χ_{R_i} is not trivial on any ideal of R_i. Let $a = \phi^{-1}(0, 0, \ldots, a_i, \ldots, 0)$. Then $\chi(a) = 1\chi_{R_i}(a_i) \neq 1$. Hence χ is not trivial on any right ideal I and hence no right ideal is contained in $ker(\chi)$ which gives that χ is a generating character for R. □

THEOREM 2.2. *Let R be a finite ring. Let $\theta : soc(R) \to \mathbb{C}^*$ be an additive map such that* $\ker(\theta)$ *contains no one sided ideals and χ be any additive extension of θ to R. Then χ is a generating character of R and R is a Frobenius ring.*

PROOF. Since R is finite any one sided ideal of R contains a minimal one sided ideal. So any extension χ of θ is such that $\ker(\chi)$ does not contain any nonzero one sided ideal. Lemma 1.1 concludes the proof. □

In the next examples, we consider local rings. Here $J(R)$ is the unique maximal ideal and so $\mathcal{R}(J(R))$ is the socle of the ring and is the minimal ideal of the ring.

EXAMPLE 3. Let $\mathbb{F}_4 = \{0, 1, \omega, 1+\omega\}$ be the field with 4 elements and $R = \mathbb{F}_4[X, \sigma]/\langle X^2\rangle$, where σ is the Frobenius map. An element of this ring is of the form $a_0 + a_1 x$ where $x = X + \langle X^2\rangle$ and $a_0, a_1 \in \mathbb{F}_4$. This ring is a local ring where the maximal ideal is the Jacobson radical with $J(R) = Rx = xR = \{0, 1x, \omega x, (1+\omega)x\}$. Then $R/J(R) \cong \mathbb{F}_4$. This ideal is the only non-trivial ideal of the ring and in fact $\mathcal{R}(J(R)) = J(R)$ and hence it is a self-dual code of length 1. See [**5**] for a detailed study of self-dual codes over non-commutative Frobenius rings. The additive structure of $J(R)$ is isomorphic to the additive structure of $\mathbb{F}_4$ which is $\mathbb{F}_2 \times \mathbb{F}_2$. Let τ be the character on $\mathbb{F}_4$ defined by $\tau(\alpha) = (-1)^{b_0+b_1}$ where $\alpha = b_0 + b_1\omega$. Then define θ on the minimal ideal by defining $\theta(\alpha x) = \tau(\alpha)$. We can extend θ to χ where $\chi(a_0 + a_1 x) = \tau(a_0)\tau(a_1)$. Then χ is a generating character of R by Theorem 2.2. In this ring there are 3 nonzero non units, namely $\overline{x} = \{x, \omega x, (1+\omega)x\}$. This leaves 12 units. Then the matrix S_1 for the symmetrized weight enumerator is indexed by $\overline{0}, \overline{1}, \overline{x}$ and is given by:

$$\begin{pmatrix} 1 & 12 & 3 \\ 1 & 0 & -1 \\ 1 & -4 & 3 \end{pmatrix}.$$

Since the ring has characteristic 2, the symmetric weight enumerator is equal to the complete weight enumerator so $S_2 = T$.

EXAMPLE 4. Let $R = \mathbb{F}_4[X, \sigma]/\langle X^4\rangle$. As in the above example we let $x = X + \langle X^4\rangle$. Then $J(R) = Rx = xR$ and $R/J(R) \cong \mathbb{F}_4$. The ring is a local ring as all right ideals are contained in the Jacobson radical. We note that $\mathcal{R}(J(R)) = Rx^3$ which is isomorphic to the additive group of $\mathbb{F}_4$. Let τ be the character on $\mathbb{F}_4$ defined by $\tau(\alpha) = (-1)^{b_0+b_1}$ where $\alpha = b_0 + b_1\omega$. Then $\theta(\alpha x) = \tau(\alpha)$. Then we extend this to χ to have $\chi(a_0 + a_1 x + a_2 x^2 + a_3 x^3) = \prod_{i=1}^{4} \tau(a_i)$ which is an extension of θ. Then χ is a generating character of R by Theorem 2.2. In this ring there are 192 units, and three additional equivalence classes where the equivalence relation is given by left multiplication by a unit. They are $\overline{x}, \overline{x^2}, \overline{x^3}$ with $|\overline{x}| = 48, |\overline{x^2}| = 12, |\overline{x^3}| = 3$. Hence the matrix S_1 is a 5 by 5 matrix which can be computed from the 256 by 256 matrix T. This ring has characteristic 2 so $S_2 = T$.

EXAMPLE 5. Let $A = \mathbb{F}_q[x;\sigma]$ where $q > 2$ and σ is an automorphism of $\mathbb{F}_q$ of order 2. Consider $R = \mathbb{F}_q[x;\sigma][y;\sigma']/(Rx^2 + Ry^2)$ where σ' extends σ via $\sigma'(x) = x$. The ring is a local ring with maximal ideal $xR + yR$ which is then the Jacobson radical. The ring has cardinality q^4. The maximal ideal has cardinality q^3. The unique minimal ideal is $(xy)R$ which has cardinality q. Here we can define a character

$$\chi(a_0 + a_1 x + a_2 y + a_3 xy) = \prod \chi_{\mathbb{F}_q}(a_i).$$

This character is non-trivial on the minimal ideal and hence is a generating character for R. In this case S_2 is not T as in the previous cases. It is a $(\lceil \frac{q}{2} \rceil)^4$ by $(\lceil \frac{q}{2} \rceil)^4$ matrix.

Of course, not all local rings are Frobenius. Consider the finite commutative ring $\mathbb{F}_2[x,y]/\langle x^2, y^2, xy\rangle$. This ring has 8 elements, namely $R = \{0, 1, x, y, 1+x, 1+y, 1+x+y, x+y\}$. The maximal ideal of this ring contains all non-units and is $M = \{0, x, y, x+y\}$. Here $M^\perp = M$, but $4(4) \neq 8$ and so the ring is not Frobenius. It is instructive to point out that if one were to naively take the character of the underlying additive structure of this ring you would have a character defined as follows:

(7)

a	0	1	x	y	$x+y$	$1+x$	$1+y$	$1+x+y$
$\chi(a)$	1	-1	-1	-1	1	1	1	-1

This does not qualify as a generating character because χ is trivial on the ideal $\{0, x+y\}$. Moreover there are three minimal ideals, namely $\{0,x\}$, $\{0,y\}$, $\{0,x+y\}$. For a character to be a generating character, one would need $\chi(x) = \chi(y) = \chi(x+y) = -1$. However, if $\chi(x) = \chi(y) = -1$ then $\chi(x+y)$ would have to be 1, contradicting the earlier statement. This gives an example of why a non-Frobenius ring does not have a generating character and the MacWilliams relations do not apply. In particular, the orthogonal of all three of the minimal ideals would be M if the ring were Frobenius.

2.1. Frobenius Algebras. We shall now determine the generating character for Frobenius algebras. For a complete description of Frobenius algebras, see [**7**]. Let A be a Frobenius $\mathbb{F}_q$ algebra with $q = p^l$. There is a nonsingular bilinear pairing $B : A \times A \to \mathbb{F}_q$ with the associative property. It is known that the functional $\lambda : A \to \mathbb{F}_q$ defined by $\lambda(a) = B(a,1)$ has a kernel which contains no nonzero right ideals, see page 438 of [**7**]. This gives that there exists a $\lambda \in Hom_{\mathbb{F}_q}(A, \mathbb{F}_q)$, such that no right ideal I has $\lambda(I) = \{0\}$. Let tr be the trace map from $\mathbb{F}_q$ to $\mathbb{F}_p$ given by $tr(x) = x + x^p + \cdots + x^{p^{l-1}}$ where $q = p^l$. Define the map $\mu : \mathbb{F}_q \to \mathbb{C}^*$, $\mu(x) = e^{\frac{2\pi i tr(x)}{p}}$, for $x \in \mathbb{F}_q$.

The following theorem also appears in [**12**].

THEOREM 2.3. *Let A be a Frobenius $\mathbb{F}_q$-algebra with $q = p^l$. Define a character for A as*

$$\chi(a) = \mu(\lambda(a)) = \mu(B(a,1)), \qquad \text{for } a \in A. \tag{8}$$

Then χ is a generating character for A.

This theorem implies that when the bilinear form is known the generating character follows immediately.

An important particular case of a Frobenius algebra is the case of finite dimensional Hopf algebras. Two prominent examples in this family are the group algebras of a finite group and the enveloping algebra of a finite dimensional Lie algebra. In order to get a generating character for a $\mathbb{F}_p$ finite dimensional Hopf algebra it is enough, according to the previous paragraph, to determine a bilinear pairing from $B : H \times H \longrightarrow \mathbb{F}_p$ which is non-degenerate and associative. This is classical but let us recall that an element $t \in H$ is called a left integral if for every $h \in H$ we have that $ht = \epsilon(h)t$ where $\epsilon : H \longrightarrow \mathbb{F}_p$ is the counity. Now $H^* = Hom(H, \mathbb{F}_p)$ is also a Hopf algebra and if $f \in H^*$ is a left integral it is well known that the map $B : H \times H \longrightarrow \mathbb{F}_p$ defined by $B(x,y) = f(xy)$ is such a pairing.

Let us now analyze the case of a group ring $R = kG$ where k is a finite field and G is a finite group. Let us recall that kG is a Hopf algera with comultiplication, counit and antipode given, for all $g \in G$, by

$$\Delta(g) = g \otimes g, \qquad \epsilon(g) = 1, \qquad S(g) = g^{-1}.$$

In kG, the one dimensional k-space of left (right) integrals is given by $k(\sum_{g\in G} g)$ and in kG^* it is kp_{1_G}, where p_{1_G} is the projection of kG on $k1_G$. The generating character χ of kG is thus given by $\chi(a) = \mu(p_{1_G}(a))$.

2.2. Taft Hopf Algebras. Let us now consider the well-known Taft Hopf algebra T which is neither commutative nor cocommutative. Let $\mathbb{F}_q$ be a finite field with $q = p^l$. For $n \in \mathbb{N}$, such that p does not divide n, we let $\zeta \in \mathbb{F}_q$ be a primitive n^{th}-root of unity. We define $T = \mathbb{F}_q\langle x, g\rangle/I$ where I is the ideal generated by $x^n, g^n - 1$ and $xg - \zeta gx$. This algebra can be given the structure of a Hopf algebra via:

$$\Delta(x) = x \otimes g + 1 \otimes x, \quad \Delta(g) = g \otimes g \quad \epsilon(g) = 1, \quad \text{and } \epsilon(x) = 0.$$

The antipode S of T is given by $S(g) = g^{-1}$ and $S(x) = -xg^{-1}$. It is known that the left (respectively right) integrals of T is the one dimensional space generated by $\sum_{i=0}^{n-1} g^i x^{n-1}$ (respectively $\sum_{i=0}^{n-1} \zeta^i g^i x^{n-1}$). The Taft algebra T is isomorphic to its dual T^*. The isomorphism of Hopf algebras ψ is given by $\psi(x) = X$ and $\psi(g) = G$ with X and G (notice that G is in fact a ring homomorphism) are defined by

$$X(g^i x^j) = \delta_{1,j} \quad \text{for } 0 \leq i, j \leq n-1 \quad \text{and } G(g) = \zeta^{-1},\ G(x) = 0,$$

where δ is the Kronecker symbol.

The left (respectively right) integrals of T^* is the one dimensional space generated by $\sum_{i=0}^{n-1} G^i X^{n-1}$ (respectively $\sum_{i=0}^{n-1} \zeta^i G^i X^{n-1}$). The generating character χ of T is thus given by $\chi(x) = \mu(\lambda(x)) = \mu(B(x, 1)) = \mu(\sum_{i=0}^{n-1} G^i X^{n-1}(x))$.

It is well known that a restricted enveloping k-algebra of a restricted finite dimensional Lie algebra over a field (also called u-algebra) is a Frobenius algebra. For more information about this enveloping algebra $u(L)$ we refer the reader to [**10**]. If the Lie algebra is $L = \sum_{i=1}^n ka_i$, where k is a field of characteristic $p \neq 0$, the restricted enveloping algebra $u(L)$ of L has a k-basis $a_I = a_1^{i_1} \ldots a_n^{i_n}$ where $0 \leq i_s < p$. It is shown in [**1**] that the element $\phi \in u(L)^*$ defined by $\phi(a_I) = 0$ for every subset $I \subset \mathbb{N}^n$, except for $I = p - 1$, has $\phi(a_1^{p-1} \cdots a_n^{p-1}) = 1$. We then have that $u(L)^* = \{a\phi \mid a \in u(L)\}$. This shows that ϕ is a generating character over k.

3. The Socle and Generating Characters

Let R be a finite Frobenius ring and let $J(R)$ be the Jacobson radical of R. It is well known that,

$$R/J(R) \cong M_{n_1}(\mathbb{F}_{q_1}) \times M_{n_2}(\mathbb{F}_{q_2}) \times \cdots \times M_{n_s}(\mathbb{F}_{q_s}), \tag{9}$$

where $M_{n_i}(\mathbb{F}_{q_i})$ is the ring of n_i by n_i matrices with entries from the field $\mathbb{F}_{q_i}$ (for more details see [**7**]). Let us recall that, for $q = p^e$ where p is a prime number, the generating character $\chi_{\mathbb{F}_q}$ of $\mathbb{F}_q$ is defined by $\chi_{\mathbb{F}_q}(a) = e^{\frac{2\pi i t}{p}}$ where, for $a \in \mathbb{F}_q$, $t = tr(a) = a + a^p + \cdots + a^{p^{e-1}} \in \mathbb{F}_p$.

We now define a generating character σ for R/J by

$$\sigma(M_1, M_2, \ldots, M_s) = \prod_{i=1}^{s} \chi_{\mathbb{F}_{q_i}}(Tr(M_i)) \tag{10}$$

where, for $M_i \in M_{n_i}(\mathbb{F}_{q_i})$, $Tr(M_i)$ is the trace of the matrix M_i. It is easily seen from Lemma 2.1 that this is a generating character for $M_{n_1}(\mathbb{F}_{q_1}) \times M_{n_2}(\mathbb{F}_{q_2}) \times \cdots \times M_{n_s}(\mathbb{F}_{q_s})$.

THEOREM 3.1. *Let R be a finite Frobenius ring. Let χ be any additive extension on*

$$\sigma : soc(R) \to \mathbb{C}^*$$

defined in Equation 10, *then χ is a generating character for R.*

PROOF. Since the ring is Frobenius we have that $R/J(R) \cong soc(R)$. We also have that

$$R/J(R) \cong M_{n_1}(\mathbb{F}_{q_1}) \times M_{n_2}(\mathbb{F}_{q_2}) \times \cdots \times M_{n_s}(\mathbb{F}_{q_s})$$

which gives that

$$soc(R) \cong M_{n_1}(\mathbb{F}_{q_1}) \times M_{n_2}(\mathbb{F}_{q_2}) \times \cdots \times M_{n_s}(\mathbb{F}_{q_s}).$$

Therefore, we can define σ as in Equation 10 on $soc(R)$. Then Theorem 2.2 gives that χ is a generating character. □

Note that this theorem gives the generating character for any finite Frobenius ring. Namely, we explicitly construct the generating character for the socle. Then the additive structure of the ring is a finite abelian additive group. Therefore, it is a trivial matter to extend the map σ to the entire ring additively.

COROLLARY 3.2. *Let R be a finite local Frobenius ring. Let $\chi_{\mathbb{F}_q}$ be the standard character on the field of order q. Let χ be any additive extension of $\chi_{\mathbb{F}_q} : soc(R) \to \mathbb{C}^*$ to R. Then χ is a generating character for R.*

PROOF. The proof is the same as Theorem 3.1, noting that $R/J(R)$ is a finite field of order q if R is a local ring. □

Note that if the ring is not Frobenius then it is not possible to apply the theorem since $R/J(R)$ is not necessarily the socle. Therefore, we cannot say that the socle is isomorphic to the direct product of matrix rings and so we cannot define σ on the socle. For example, consider the non-Frobenius ring $R = \{\begin{pmatrix} a & b \\ 0 & c \end{pmatrix}, a, b, c \in \mathbb{F}_2\}$. This ring is not Frobenius and therefore does not admit a generating character. However, $R/J(R) \cong \mathbb{F}_2 \times \mathbb{F}_2$ has a (unique) generating character defined via,

$$\sigma((0,0)) = 1, \sigma((0,1)) = \sigma((1,0)) = -1, \sigma((1,1)) = 1.$$

Concretely, $J(R) = \{\begin{pmatrix} 0 & 0 \\ 0 & 0 \end{pmatrix}, \begin{pmatrix} 0 & 1 \\ 0 & 0 \end{pmatrix}\}$ and

$$R/J(R) = \{\begin{pmatrix} 0 & 0 \\ 0 & 0 \end{pmatrix}, \begin{pmatrix} 1 & 0 \\ 0 & 0 \end{pmatrix}, \begin{pmatrix} 0 & 0 \\ 0 & 1 \end{pmatrix}, \begin{pmatrix} 1 & 0 \\ 0 & 1 \end{pmatrix}\}.$$

However, this is not the socle of the ring and the so the map cannot be extended. For example, if one were to define χ to be as given on $R/J(R)$ and extend it to the whole ring defined by

$$\chi(\begin{pmatrix} 0 & 1 \\ 0 & 0 \end{pmatrix}) = -1, \chi(\begin{pmatrix} 1 & 1 \\ 0 & 0 \end{pmatrix}) = 1, \chi(\begin{pmatrix} 0 & 1 \\ 0 & 1 \end{pmatrix}) = 1, \chi(\begin{pmatrix} 1 & 1 \\ 0 & 1 \end{pmatrix}) = -1.$$

Then χ would be trivial on the ideal

$$\{\begin{pmatrix} 0 & 1 \\ 0 & 0 \end{pmatrix}\begin{pmatrix} 1 & 1 \\ 0 & 0 \end{pmatrix}\begin{pmatrix} 0 & 1 \\ 0 & 1 \end{pmatrix}\begin{pmatrix} 1 & 1 \\ 0 & 1 \end{pmatrix}\}$$

and so would not be a generating character.

4. Conclusion

Generating characters for finite Frobenius rings are the key to finding the MacWilliams relations for codes over a specific ring. The MacWilliams relations are one of the foundational results of algebraic coding theory. In fact, we only study codes in spaces where the MacWilliams relations exist. In this work, we have shown how to construct a generating character for an arbitrary Frobenius ring and we have given numerous examples of finding generating characters for specific rings. We have also identified another ideal which allows us to find generating characters based on maps from this ideal.

References

[1] A. J. Berkson, *The u-algebra of a restricted Lie algebra is Frobenius*, Proc. Amer. Math. Soc. **15** (1964), 14–15, DOI 10.2307/2034339. MR0158910

[2] H. L. Claasen and R. W. Goldbach, *A field-like property of finite rings*, Indag. Math. (N.S.) **3** (1992), no. 1, 11–26, DOI 10.1016/0019-3577(92)90024-F. MR1157515

[3] S. T. Dougherty, *Foundations of algebraic coding theory*, Noncommutative rings and their applications, Contemp. Math., vol. 634, Amer. Math. Soc., Providence, RI, 2015, pp. 101–136, DOI 10.1090/conm/634/12693. MR3307392

[4] S. T. Dougherty, *Algebraic coding theory over finite commutative rings*, SpringerBriefs in Mathematics, Springer, Cham, 2017. MR3675844

[5] S. T. Dougherty and A. Leroy, *Euclidean self-dual codes over non-commutative Frobenius rings*, Appl. Algebra Engrg. Comm. Comput. **27** (2016), no. 3, 185–203, DOI 10.1007/s00200-015-0277-0. MR3500238

[6] S. T. Dougherty, E. Saltürk, and S. Szabo, *Codes over local rings of order 16 and binary codes*, Adv. Math. Commun. **10** (2016), no. 2, 379–391, DOI 10.3934/amc.2016012. MR3503678

[7] T. Y. Lam, *Lectures on modules and rings*, Graduate Texts in Mathematics, vol. 189, Springer-Verlag, New York, 1999. MR1653294

[8] F. J. MacWilliams, Combinatorial problems of elementary group theory, Ph.D. thesis, Harvard University, 1961.

[9] J. MacWilliams, *A theorem on the distribution of weights in a systematic code*, Bell System Tech. J. **42** (1963), 79–94, DOI 10.1002/j.1538-7305.1963.tb04003.x. MR0149978

[10] S. Montgomery, *Hopf algebras and their actions on rings*, CBMS Regional Conference Series in Mathematics, vol. 82, Published for the Conference Board of the Mathematical Sciences, Washington, DC; by the American Mathematical Society, Providence, RI, 1993. MR1243637

[11] J. A. Wood, *Duality for modules over finite rings and applications to coding theory*, Amer. J. Math. **121** (1999), no. 3, 555–575. MR1738408

[12] J. A. Wood, *Applications of finite Frobenius rings to the foundations of algebraic coding theory*, Proceedings of the 44th Symposium on Ring Theory and Representation Theory, Symp. Ring Theory Represent. Theory Organ. Comm., Nagoya, 2012, pp. 223–245. MR2975625

Department of Mathematics, University of Scranton, Scranton, Pennsylvania 18518
Email address: prof.steven.dougherty@gmail.com

Gebze Technical University, 41400 Gebze, Kocaeli, Turkey
Email address: ardakor07@gmail.com

Faculté Jean Perrin, Université d'Artois, 62307 Lens, France
Email address: andreleroy55@gmail.com

Contemporary Mathematics
Volume **727**, 2019
https://doi.org/10.1090/conm/727/14627

Constacyclic codes over local rings of order 16

Steven T. Dougherty and Esengül Saltürk

ABSTRACT. In this work, we study constacyclic codes of odd length over finite commutative local Frobenius non-chain rings of order 16. We give the structure of λ_i-constacyclic codes over each local Frobenius non-chain ring via cyclic codes over these rings where λ_i acts as a weight preserving unit in the corresponding ring. We also obtain binary optimal codes which are images of these constacyclic codes under a gray map.

1. Introduction

Codes over rings have become an increasingly important area of study beginning with the landmark paper [**8**] in 1994. The key part of this paper was establishing a gray map from $\mathbb{Z}_4$ to the binary Hamming space. Following this work, numerous extensions of gray maps have been made. It was shown in [**4**] that a family of gray maps could be defined for the local Frobenius rings of order 16. These rings were classified in [**10**]. This family of gray maps was used in [**5**] to consider images of cyclic codes over the local rings of order 16.

There are 12 local Frobenius rings of order 16. Of these, 7 are non-chain rings and 5 are chain rings. All of these rings have maximal ideal $\mathfrak{m}$, with $|\mathfrak{m}| = 8$ and socle with $|soc(r)| = 2$. In [**3**], a generator matrix has been defined for the local Frobenius non-chain rings of order 16 and generating characters for each ring have been described. Moreover, MacWilliams relations for the symmetrized and the complete weight enumerators have been given for codes over these rings. In [**4**], a weight preserving Gray map has been defined from these 12 rings to the binary space and self-dual and formally self-dual codes over these rings have been studied. The importance of the local Frobenius non-chain rings of order 16 comes from the fact that they are the smallest class of Frobenius rings which are both non-chain and local. In [**6**], codes over a larger class of local Frobenius rings, called X-rings, have been studied where the order is 2^{2^k}, k is a positive integer.

Cyclic codes and their generalizations have been one of the most studied classes of codes, due largely to the fact that they have a canonical algebraic representation in polynomial rings. Constacyclic codes are a generalization of both cyclic and negacyclic codes. Constacyclic codes can be viewed as ideals of the quotient ring $\mathbb{F}_q[x]/\langle x^n - \lambda\rangle$ where λ is a unit of the finite field $\mathbb{F}_q$. When $\lambda = 1$ we get cyclic codes and when $\lambda = -1$ we get negacyclic codes.

2010 *Mathematics Subject Classification.* Primary 11T71, 94B15.
Key words and phrases. Constacyclic code, local ring.

In this work, we study the structure of constacyclic codes of odd length over finite commutative local Frobenius rings of order 16. We first study the units in each ring that act as a swap map. We then define ring isomorphisms to characterize the constacyclic codes over each local Frobenius non-chain ring of order 16. We obtain some good optimal binary codes which are Gray images of those constacyclic codes.

2. Definitions and Notations

2.1. Rings and Codes. We begin with some definitions from ring theory. Characters of a module M are homomorphisms from M to $\mathbb{C}^*$. For a module M, let $\widehat{M}$ denote the character module of M. We define Frobenius rings via their characterization with respect to their character module. Specifically, we have the following. If R is a finite ring then the following statements are equivalent:

(1) R is a Frobenius ring;
(2) as a left module, $\widehat{R} \cong_R R$;
(3) as a right module $\widehat{R} \cong R_R$.

See [**2**] for a complete discussion of Frobenius rings in the context of coding theory.

In this paper, all the rings we study are finite, commutative and Frobenius. This simplifies many of the definitions. The Jacobson radical of a commutative ring is defined to be the intersection of all maximal ideals in the ring, which we denote as $J(R)$. The socle of the ring is the sum of all the minimal ideals of the ring (note that since the rings are commutative there is no reason to consider one sided ideals), which we denote as $soc(r)$. A ring is said to be a local ring if it has a unique maximal ideal. For a local ring, this unique maximal ideal is necessarily the Jacobson radical. If the ideals of a local ring are linearly ordered then it is said to be a chain ring.

We take the standard definitions from codes over rings. A code over a ring R of length n is a subset of R^n. If, additionally, it is a submodule, then we say that the code is linear. We attach to the ambient space R^n the standard inner-product, namely $[\mathbf{v}, \mathbf{w}] = \sum v_i w_i$. The orthogonal is defined in the usual way as $C^\perp = \{\mathbf{v} \mid \mathbf{v} \in R^n, [\mathbf{v}, \mathbf{w}] = 0, \forall \mathbf{w} \in C\}$. Since the ring is commutative we do not need to consider left and right orthogonals. Notice that an ideal in a ring R is a code of length 1 and hence we can speak of $\mathfrak{a}^\perp$ if $\mathfrak{a}$ is an ideal of R. Since the ring is a Frobenius ring we have that $|C||C^\perp| = |R|^n$ for all linear codes C. This fundamental result does not hold for codes over non-Frobenius rings.

Two codes C and C' are said to be equivalent if one can be obtained from the other by a combination of a permutation of the coordinates and multiplication of coordinates by a unit. For binary codes this means that two codes are equivalent if one can be obtained by a permutation of the coordinates alone since the only non-zero element is 1.

The Hamming weight of a vector over any ring is the number of non-zero coordinates in the vector. The Hamming weight enumerator of a code of length n is defined as

$$W_C(x, y) = \sum_{\mathbf{c} \in C} x^{n - w_H(\mathbf{c})} y^{w_H(\mathbf{c})}.$$

The minimum weight d_C of a code C is defined as $\min\{w_H(\mathbf{c}) \mid \mathbf{c} \in C\}$.

For any undefined terms or foundational results from the theory of codes over rings see [**2**].

2.2. Local Frobenius rings of order 16. The local Frobenius rings of order 16 were classified in [**10**]. We shall now describe their structure. Let R be a local Frobenius ring with maximal ideal $\mathfrak{m}$, with $|\mathfrak{m}| = 8$, such that $|R| = 16$. All local Frobenius rings of order 16 have two elements u and v such that $\mathfrak{m} = \langle u, v\rangle$. There exists an element w in these rings with $soc(r) = \langle w\rangle = \{0, w\}$. These rings are in fact local rings since they have $\mathfrak{m}$ as the unique maximal ideal which is not a principal ideal. The following diagram shows an ideal structure for any local Frobenius non-chain ring R of order 16.

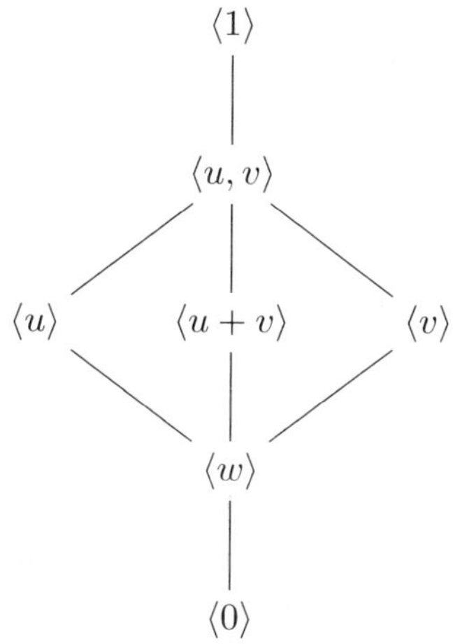

Since the ring R is Frobenius we have that $soc(r) = \mathfrak{m}^{\perp}$. Any local Frobenius non-chain ring of order 16 has 5 non-trivial ideals. The maximal ideal $\mathfrak{m} = \langle u, v\rangle$ is of size 8, the socle is of size 2 and the remaining 3 ideals are of size 4.

Given the ideal structure of the rings it is possible to write every element of the ring uniquely in the form $a + bu + cv + dw$ where $a, b, c, d \in \mathbb{F}_2$. Note that this does not imply that the additive structure of the ring is $\mathbb{Z}_2^4$. In fact, the possibilities for the additive structure are $\mathbb{Z}_2^4, \mathbb{Z}_4 \times \mathbb{Z}_2^2, \mathbb{Z}_8 \times \mathbb{Z}_2$ and $\mathbb{Z}_{16}$. We are simply saying that every element can be written in this way.

The group of units of the ring R is $U(R) = \{1, 1+u, 1+v, 1+u+v, 1+w, 1+u+w, 1+v+w, 1+u+v+w\}$. Specifically, any element $a + bu + cv + dw$ in R is a unit if and only if $a = 1$.

There are 5 local Frobenius chain rings which are given as follows: $\mathbb{F}_2[x]/\langle x^4\rangle, \mathbb{Z}_4[x]/\langle x^2 - 2\rangle, \mathbb{Z}_4[x]/\langle x^2 - 2x - 2\rangle, \mathbb{Z}_4[x]/\langle x^3 - 2, 2x\rangle, \mathbb{Z}_{16}$ and there are 7 local Frobenius non-chain rings which are given as follows: $\mathbb{F}_2[u, v]/\langle u^2, v^2\rangle, \mathbb{F}_2[u, v]/\langle u^2 + v^2, uv\rangle, \mathbb{Z}_4[x]/\langle x^2\rangle, \mathbb{Z}_4[x]/\langle x^2 - 2x\rangle, \mathbb{Z}_4[x, y]/\langle x^2, xy - 2, y^2, 2x, 2y\rangle, \mathbb{Z}_4[x, y]/\langle x^2 - 2, xy - 2, y^2, 2x, 2y\rangle, \mathbb{Z}_8[x]/\langle x^2 - 4, 2x\rangle$.

Note that, in this paper, we are interested in non-chain rings. This is because constacyclic codes over chain rings have already been studied.

We shall now define the canonical Gray map for these rings. The following map was defined in [**4**]:

$$\phi : R \to \mathbb{F}_2^4$$

with

$$\phi(a + bu + cv + dw) = (d, c + d, b + d, a + b + c + d). \tag{2.1}$$

This map is formed by taking the classical Gray map for local Frobenius rings of order 4 and applying it recursively. That is, for $\mathbb{Z}_4$ and $\mathbb{F}_2 + u\mathbb{F}_2$, we have the standard Gray map where

$$\phi_1(a + bx) = (b, a + b)$$

where $x = 2$ for $\mathbb{Z}_4$ and $x = u$ for $\mathbb{F}_2 + u\mathbb{F}_2$. Then, applying this map recursively, viewing a local Frobenius ring of order 16 as having $x = u$ for the first application of ϕ_1 and $x = v$ for the second application of ϕ_1, we have the map as defined above.

The following theorem appears in [**4**].

THEOREM 2.1. *If C is a linear code over R, where R is a local Frobenius ring of order 16, of length n, size 2^k and minimum Lee weight d, then $\phi(C)$ is a binary code with length $4n$, minimum Hamming distance d and size 2^k. The binary code is linear only if the characteristic of R is 2.*

For any vector $\mathbf{c} \in R^n$ we define $w_L(\mathbf{c}) = w_H(\phi(\mathbf{c}))$ where w_H denotes the Hamming weight. That is, the Lee weight of a vector is the Hamming weight of its image under the gray map. Note then that ϕ is a weight preserving map since the Lee weight is defined as the Hamming weight of the image under the map ϕ.

3. Constacyclic Codes Over Local Frobenius Rings of Order 16

We shall now consider constacyclic codes over rings. Let $\lambda \in R$ be a unit. The λ-constacyclic (λ-twisted) shift τ_λ on R^n acts as

$$\tau_\lambda(c_0, c_1, \ldots, c_{n-1}) = (\lambda c_{n-1}, c_0, \ldots, c_{n-2}). \tag{3.1}$$

A linear code C over a ring R is said to be λ-constacyclic if C is closed under the λ-constacyclic shift τ_λ, that is $\tau_\lambda(C) = C$.

As usual we associate a vector $(c_0, c_1, \ldots, c_{n-1})$ with the polynomial $c_0 + c_1x + c_2x^2 + \cdots + c_{n-1}x^{n-1}$. If we consider our polynomials as elements of the ring $R[x]/\langle x^n - \lambda\rangle$, then $xc(x)$ modulo $x^n - \lambda$ represents the constacyclic shift of c. The next proposition follows from this discussion.

PROPOSITION 3.1. *Let R be a local Frobenius ring of order 16. A subset C of R^n is a linear λ-constacyclic code of length n if and only if its polynomial representation is an ideal of $R[x]/\langle x^n - \lambda\rangle$ where λ is a unit in R.*

Hence, a λ-constacyclic code C of length n over R is an ideal of the ring $R[x]/\langle x^n - \lambda\rangle$ where R is a local Frobenius ring of order 16. The λ-constacyclic code C is called a cyclic code when $\lambda = 1$ and it is called a negacyclic code when $\lambda = -1$. Throughout the paper assume that the length n is an odd integer so that $gcd(n, char(R)) = 1$.

Note that the ring $\Re_n = R[x]/\langle x^n - \lambda\rangle$ is not a principal ideal ring in general, where R is a local Frobenius non-chain ring of order 16.

Theorem 6.1 in [**2**] gives the following. Let R be a finite commutative Frobenius ring. If C is a λ-constacyclic code over R of length n, then $C^\perp$ is a λ^{-1}-constacyclic code of length n over R. This is one reason why we require λ to be a unit.

DEFINITION 1. Let C be a code in $\mathbb{F}_2^{sn}$. If $(c_0, c_1, c_2, \ldots, c_{sn-1}) \in C$ implies that

$$(c_s, c_{s+1}, c_{s+2}, \ldots, c_{sn-1}, c_0, c_1, c_2, \ldots, c_{s-1}) \in C$$

then C is said to be a binary quasi-cyclic code of index s. If there is a permutation σ on s coordinates such that $(c_0, c_1, c_2, \ldots, c_{sn-1}) \in C$ implies that

$$(\sigma(c_s, c_{s+1}, c_{s+2}, \ldots, c_{sn-1}), c_0, c_1, c_2, \ldots, c_{s-1}) \in C$$

then C is said to be a binary skew quasi-cyclic code of index s.

We say that a unit λ in R^* acts as a swap map, if $\phi(\lambda a)$ and $\phi(a)$ are equivalent vectors for all elements $a \in R$ using the same permutation for each $a \in R$. It follows immediately that $w_L(a) = w_L(\lambda a)$ for all $a \in R$.

THEOREM 3.2. *Let C be a λ-constacyclic code of length n over a local ring of order 16 where λ has the property that the map $f : R \to R$, $f(a) = \lambda a$ is a swap map. Then $\phi(C)$ is a binary skew quasi-cyclic code of length $4n$ with index 4.*

PROOF. Let σ be the permutation of coordinates induced by multiplication by λ. Let σ act on the first four coordinates of $\mathbb{F}_2^{4n}$. Then $\mathbf{c} = (c_0, c_1, \ldots, c_{n-1}) \in C$ implies $(\lambda c_{n-1}, c_0, \ldots, c_{n-2}) \in C$. Then

$$\phi(\mathbf{c}) = (\phi(c_0), \phi(c_1), \ldots, \phi(c_{n-1})) \in \phi(C)$$

implies $(\sigma(\phi(c_{n-1})), \phi(c_0), \phi(c_1), \ldots, \phi(c_{n-2}))) \in \phi(C)$. □

In the next subsection, we shall determine which units in each of the non-chain local Frobenius rings of order 16 have the property that their multiplication induces a swap map.

3.1. Units in R When R Has Characteristic 2 or 4.

Let R be a local Frobenius non-chain ring of order 16 of characteristic 2 or 4. In the following we study the units for each ring.

(1) The ring $\mathbb{F}_2[u, v]/\langle u^2, v^2\rangle$.

Let $R_1 = \mathbb{F}_2[u, v]/\langle u^2, v^2\rangle$ where $w = uv$. The group of units is $U(R_1) = \{1, 1+u, 1+v, 1+u+v, 1+uv, 1+u+uv, 1+v+uv, 1+u+v+uv\}$. Only the units $1+u$ and $1+v$ act as swap maps so the weight is preserved:

$$\begin{aligned} w_L(\phi((1+u)(a+bu+cv+duv))) &= w_L((c+d, d, a+b+c+d, b+d)) \\ &= w_L((d, c+d, b+d, a+b+c+d)) \\ &= w_L(a+bu+cv+dw) \end{aligned}$$

and

$$\begin{aligned} w_L(\phi((1+v)(a+bu+cv+duv))) &= w_L((b+d, a+b+c+d, d, c+d)) \\ &= w_L((d, c+d, b+d, a+b+c+d)) \\ &= w_L(a+bu+cv+dw). \end{aligned}$$

A $(1+u)$-constacyclic shift τ_{1+u} and a $(1+v)$-constacyclic shift τ_{1+v} on R_1^n act as

$$\tau_{1+u}(c_0, c_1, \ldots, c_{n-1}) = ((1+u)c_{n-1}, c_0, \ldots, c_{n-2})$$

and

$$\tau_{1+v}(c_0, c_1, \ldots, c_{n-1}) = ((1+v)c_{n-1}, c_0, \ldots, c_{n-2}),$$

respectively. Note that $(1+v)$-constacyclic codes over R_1 were studied in **[9]**.

(2) The ring $\mathbb{F}_2[u,v]/\langle u^2+v^2, uv\rangle$.
Let $R_2 = \mathbb{F}_2[u,v]/\langle u^2+v^2, uv\rangle$ where $w = u^2$. The group of units is $U(R_2) = \{1, 1+u, 1+v, 1+u+v, 1+v^2, 1+u+v^2, 1+v+v^2, 1+u+v+v^2\}$. In R_2, only the unit $1+u+v+u^2$ acts as a swap map so the weight is preserved:

$$\begin{aligned} w_L(\phi((1+u+v+v^2)(a+bu+cv+dv^2))) &= w_L((a+b+c+d, b+d, c+d, d)) \\ &= w_L(a+bu+cv+dw). \end{aligned}$$

A $(1+u+v+v^2)$-constacyclic shift $\tau_{1+u+v+v^2}$ on R_2^n acts as

$$\tau_{1+u+v+v^2}(c_0, c_1, \ldots, c_{n-1}) = ((1+u+v+v^2)c_{n-1}, c_0, \ldots, c_{n-2}).$$

(3) The ring $R = \mathbb{Z}_4[x]/\langle x^2\rangle$.
Let $R_3 = \mathbb{Z}_4[x]/\langle x^2\rangle$ where $w = 2x$. The group of units is $U(R_3) = \{1, 3, 1+x, 3+x, 1+2x, 3+2x, 1+3x, 3+3x\}$. In R_3, only the unit 3 acts as a swap map so the weight is preserved:

$$\begin{aligned} w_L(\phi((3)(a+bx+c2+d2x))) &= w_L((b+d, a+b+c+d, d, c+d)) \\ &= w_L(a+bu+cv+dw). \end{aligned}$$

A 3-constacyclic shift τ_3 on R_3^n acts as

$$\tau_3(c_0, c_1, \ldots, c_{n-1}) = (3c_{n-1}, c_0, \ldots, c_{n-2}).$$

(4) The ring $R = \mathbb{Z}_4[x]/\langle x^2-2x\rangle$.
Let $R_4 = \mathbb{Z}_4[x]/\langle x^2-2x\rangle$ where $w = 2x$. The group of units is $U(R_4) = \{1, 3, 1+x, 3+x, 1+2x, 3+2x, 1+3x, 3+3x\}$. In R_4, only the unit 3 acts as a swap map so the weight is preserved:

$$\begin{aligned} w_L(\phi((3)(a+bx+c2+d2x))) &= w_L((b+d, a+b+c+d, d, c+d)) \\ &= w_L(a+bu+cv+dw). \end{aligned}$$

A 3-constacyclic shift τ_3 on R_4^n acts as

$$\tau_3(c_0, c_1, \ldots, c_{n-1}) = (3c_{n-1}, c_0, \ldots, c_{n-2}).$$

(5) The ring $\mathbb{Z}_4[x,y]/\langle x^2, xy-2, y^2, 2x, 2y\rangle$.
Let $R_5 = \mathbb{Z}_4[x,y]/\langle x^2, xy-2, y^2, 2x, 2y\rangle$ for which $w = xy = 2$. The group of units is $U(R_5) = \{1, 1+x, 1+y, 1+x+y, 1+xy, 1+x+xy, 1+y+xy, 1+x+y+xy\}$. In R_5, only the units $1+x$ and $1+y$ act as swap maps so the weight is preserved:

$$\begin{aligned} w_L(\phi((1+x)(a+bx+cy+d2))) &= w_L((c+d, d, a+b+c+d, b+d)) \\ &= w_L(a+bu+cv+dw) \end{aligned}$$

and

$$\begin{aligned} w_L(\phi((1+y)(a+bx+cy+d2))) &= w_L((b+d, a+b+c+d, d, c+d)) \\ &= w_L(a+bu+cv+dw). \end{aligned}$$

A $(1+x)$-constacyclic shift τ_{1+x} and a $(1+y)$-constacyclic shift τ_{1+y} on R_5^n act as

$$\tau_{1+x}(c_0, c_1, \ldots, c_{n-1}) = ((1+x)c_{n-1}, c_0, \ldots, c_{n-2})$$

and

$$\tau_{1+y}(c_0, c_1, \ldots, c_{n-1}) = ((1+y)c_{n-1}, c_0, \ldots, c_{n-2})$$

respectively.

(6) The ring $\mathbb{Z}_4[x, y]/\langle x^2 - 2, xy - 2, y^2, 2x, 2y\rangle$.

Let $R_6 = \mathbb{Z}_4[x, y]/\langle x^2, xy-2, y^2, 2x, 2y\rangle$ where $w = xy = 2$. The group of units is $U(R_6) = \{1, 1+x, 1+y, 1+x+y, 3, 3+x, 3+y, 3+x+y\}$. In R_6, only the unit $1+y$ acts as swap maps so the weight is preserved:

$$\begin{aligned} w_L(\phi((1+y)(a+bx+cy+d2))) &= w_L((b+d, a+b+c+d, d, c+d)) \\ &= w_L(a+bu+cv+dw). \end{aligned}$$

A $(1+y)$-constacyclic shift τ_{1+y} and on R_6^n acts as

$$\tau_{1+y}(c_0, c_1, \ldots, c_{n-1}) = ((1+y)c_{n-1}, c_0, \ldots, c_{n-2})$$

respectively.

3.2. λ_i-Constacyclic Codes. We study λ_i-constacyclic codes over R_i where $i \in \{1, 2, 3, 4, 5, 6\}$ and $\lambda_1 \in \{1+u, 1+v\}$, $\lambda_2 = 1+u+v+u^2$, $\lambda_3 = \lambda_4 = 3$, $\lambda_5 \in \{1+x, 1+y\}$, $\lambda_6 = 1+y$ as stated in the previous subsection.

To understand λ_i constacyclic codes over R_i we shall study the ring $R_i[x]/\langle x^n - \lambda_i\rangle$, $i \in \{1, 2, 3, 4, 5, 6\}$.

Note that in R_i, $\lambda_i^n = \lambda_i$ if n is odd and $\lambda_i^n = 1$ if n is even, $i \in \{1, 2, 3, 4, 5, 6\}$. The fact that $\lambda_i^2 = 1$ for each i is key in these cases. This means that the orthogonal $C^\perp$ is a λ_i-constacyclic code as well since $\lambda_i^{-1} = \lambda_i$. Moreover, the permutation of binary coordinates induced by multiplication by λ_i has order 2 since $\lambda_i(\lambda_i \mathbf{v}) = \mathbf{v}$.

We study the case when n is odd. In [**5**], cyclic codes over local Frobenius rings of order 16 of odd lengths were studied. We use this structure to study constacyclic codes by defining a ring isomorphism between the rings $\mathfrak{R}_n^i = R_i[x]/\langle x^n - 1\rangle$ and $\mathfrak{M}_n^i = R_i[x]/\langle x^n - \lambda_i\rangle$.

PROPOSITION 3.3. *Let n be odd. The map $\mu_i : \mathfrak{R}_n^i \to \mathfrak{M}_n^i$ is a ring isomorphism such that $\mu_i(a_i(x)) = a_i(\lambda_i x)$, for all $a_i(x) \in \mathfrak{R}_n^i$ where $i \in \{1, 2, 3, 4, 5, 6\}$ and λ_i is as stated above.*

PROOF. If n is odd, for polynomials $a_i(x), b_i(x) \in R_i[x]$,

$$a_i(x) \equiv b_i(x) \pmod{x^n - 1}$$

if and only if

$$a_i(x) - b_i(x) = (x^n - 1)q_i(x)$$

for some $q_i(x) \in R_i[x]$. Recall that $\lambda_i^2 = 1$ and when n is odd $\lambda_i^n = \lambda_i$ in R_i, $i \in \{1, 2, 3, 4, 5, 6\}$. On the other hand,

$$\begin{gathered} a_i(\lambda_i x) - b_i(\lambda_i x) = ((\lambda_i x)^n - 1)q_i(\lambda_i x) \Leftrightarrow \\ (\lambda_i x^n - 1))q_i(\lambda_i x) = \lambda_i(x^n - \lambda_i))q_i(\lambda_i x) \Leftrightarrow \\ a_i(\lambda_i x) \equiv b_i(\lambda_i x) \pmod{x^n - \lambda_i}. \end{gathered}$$

Hence μ_i is well-defined and injective. It is obvious that μ_i is surjective. The remaining part of the proof is straightforward. □

Hence we get the following:

COROLLARY 3.4. *For $i \in \{1, 2, 3, 4, 5, 6\}$, I_i is an ideal of $\mathfrak{R}_n^i$ if and only if $\mu_i(I_i)$ is an ideal of $\mathfrak{M}_n^i$.*

Recall the following theorem from [**5**]:

THEOREM 3.5. [**5**] *Let n be odd. Let $x^n - 1 = f_1(x)f_2(x)\cdots f_r(x)$ be the representation of $x^n - 1$ as a product of basic irreducible pairwise coprime polynomials in $R[x]$. Let $\widehat{f_i}(x)$ denote the product of all $f_j(x)$ except $f_i(x)$. Then any ideal in $\mathfrak{R}_n$ is a sum of the following ideals, $\langle \widehat{f_i}(x) + \langle x^n - 1\rangle\rangle$, $\langle u\widehat{f_i}(x) + \langle x^n - 1\rangle, v\widehat{f_i}(x) + \langle x^n - 1\rangle\rangle$ and $\langle \alpha\widehat{f_i}(x) + \langle x^n - 1\rangle\rangle$, $\alpha \in \{u, v, u+v, w\}$.*

By using Theorem 3.5, we get the structure of λ_i-constacyclic codes over R_i.

COROLLARY 3.6. *Let n be odd. Let R_i be a local Frobenius non-chain ring of order 16 of characteristic 2 or 4. Let $x^n - \lambda_i = f_1(x)f_2(x)\cdots f_r(x)$ be a representation of $x^n - \lambda_i$ as a product of basic irreducible pairwise coprime polynomials in $R_i[x]$ where $i \in \{1,2,3,4,5,6\}$. Let $\widehat{f_i}(x)$ denote the product of all $f_j(x)$ except $f_i(x)$. Then any ideal in $\mathfrak{M}_n^i$ is a sum of the following ideals, $\langle \widehat{f_i}(x) + \langle x^n - \lambda_i\rangle\rangle$, $\langle u\widehat{f_i}(x) + \langle x^n - \lambda_i\rangle, v\widehat{f_i}(x) + \langle x^n - \lambda_i\rangle\rangle$ and $\langle \alpha\widehat{f_i}(x) + \langle x^n - \lambda_i\rangle\rangle$, $\alpha \in \{u, v, u+v, w\}$.*

We also get the following corollary.

COROLLARY 3.7. *Let R_i be a local Frobenius non-chain ring of order 16 of characteristic 2 or 4. If $x^n - \lambda_i = f_1(x)f_2(x)\cdots f_r(x)$, $i \in \{1,2,3,4,5,6\}$, then the number of λ_i-constacyclic codes over R_i of length n is 7^r.*

We note that the number is 7^r for each λ_i, meaning if there is more than one unit in R_i inducing a swap map then there are 7^r constacyclic codes for each unit.

COROLLARY 3.8. *Let n be odd and λ_i, $i \in \{1,2,3,4,5,6\}$, be a unit as stated above. Let $D \subseteq R_i^n$ and let $\bar{\mu}$ be the permutation of R_i^n given by $\bar{\mu}_i(c_0,\ldots,c_{n-1}) = (c_0, \lambda_i c_1, \lambda_i^2 c_2, \ldots, \lambda_i^{n-1}c_{n-1})$. Then D is a cyclic code if and only if $\bar{\mu}_i(D)$ is a λ_i-constacyclic code.*

PROOF. Note that the permutation $\bar{\mu}_i$ corresponds to the ring homomorphism μ_i in the polynomial form. A cyclic code D over R_i is an ideal of $R_i[x]/\langle x^n - 1\rangle$. Its image $\mu_i(D)$ is an ideal of $R_i[x]/\langle x^n - \lambda_i\rangle$ which means it is a λ_i-constacyclic code of length n over R. □

Notice that since λ^i is a unit for all n, the Hamming weight of $(c_0, c_1, \ldots, c_{n-1})$ is the same as the Hamming weight of $\bar{\mu}_i(c_0,\ldots,c_{n-1}) = (c_0, \lambda_i c_1, \lambda_i^2 c_2, \ldots, \lambda_i^{n-1}c_{n-1})$. Using this result and Corollary 3.8, we have the following result.

COROLLARY 3.9. *Let n be odd and λ_i, $i \in \{1,2,3,4,5,6\}$, be a unit as stated above. Let C be a cyclic code over R_i with parameters $(n, 2^k, d)$ if and only if $\mu_i(C)$ is a λ_i-constacyclic code over R_i with parameters $(n, 2^k, d)$. Moreover, $W_C(x_0, x_1, \ldots, x_{s-1}) = W_{\mu_i(C)}(x_0, x_1, \ldots, x_{s-1})$.*

This corollary implies that a λ_i-constacyclic code has the same weight enumerator as a cyclic code in these cases.

3.3. Nechaev Permutation. We introduce the Nechaev permutation on $\mathbb{F}_2^{2n}$ given in [**11**] as in the following definition.

DEFINITION 2. Let n be a positive odd integer and let σ be the permutation on $\{0, 1, 2\ldots, 2n-1\}$ given by

$$\sigma = (1, n+1)(3, n+3)\cdots(2i+1, n+2i+1)\cdots(n-2, 2n-2).$$

The Nechaev permutation π on $\mathbb{F}_2^{2n}$ is defined as

$$\pi(a_0, a_1, \dots, a_{2n-1}) = (a_{\sigma(0)}, a_{\sigma(1)}, \dots, a_{\sigma(2n-1)}).$$

Next we define an extension of the Nechaev permutation on $(\mathbb{F}_2^{2n})^s$ where s is an integer.

DEFINITION 3. Let n be a positive odd integer and s any positive integer. The extension of Nechaev permutation π_s on $(\mathbb{F}_2^{2n})^s$ is defined as

$$\pi_s(\mathbf{r}_0|\mathbf{r}_1|\cdots|\mathbf{r}_s) = (\pi(\mathbf{r}_0)|\pi(\mathbf{r}_1)|\cdots|\pi(\mathbf{r}_s))$$

where $\mathbf{r}_i$ in $\mathbb{F}_2^{2n}$.

As a consequence we have the following proposition.

PROPOSITION 3.10. *Let n be odd. Let R_i be a local Frobenius ring of order 16 where $i = 1$ or 5 and the corresponding units be $1 + u$ for R_1 and $1 + x$ for R_5. Then for $i \in \{1, 5\}$, $\phi\bar{\mu}_i = \pi_2\phi$.*

PROOF. Consider the ring $R_1 = \mathbb{F}_2[u, v]/\langle u^2, v^2\rangle$ with the unit $\lambda_1 = 1 + u$. Recall that $(1 + u)^n = 1 + u$ if n is odd and $(1 + u)^n = 1$ if n is even. Let $\mathbf{r} = (r_0, r_1, \dots, r_{n-1}) \in R_1^n$ where $r_i = a_i + b_i u + c_i v + d_i uv$. Then $\bar{\mu}_1(\mathbf{r}) = (r_0, (1 + u)r_1, (1 + u)^2 r_2, \dots, (1 + u)^{n-1} r_{n-1})$. It follows that if

$$\phi\bar{\mu}_1(\mathbf{r}) = (s_1, s_2, \dots, s_{n-1}, s_n, \dots, s_{2n-1}, \dots, s_{4n-1}),$$

then for $0 \geq j \geq n - 1$.

If j is even: $s_j = d_j$, $s_{n+j} = c_j + d_j$, $s_{2n+j} = b_j + d_j$, $s_{3n+j} = a_j + b_j + c_j + d_j$. If j is odd: $s_j = c_j + d_j$, $s_{n+j} = d_j$, $s_{2n+j} = a_j + b_j + c_j + d_j$, $s_{3n+j} = b_j + d_j$.

On the other hand we have: $\phi(\mathbf{r}) = (d_0, d_1, \dots, d_{n-1}; c_0 + d_0, c_1 + d_1, \dots, c_{n-1} + d_{n-1}; b_0 + d_0, b_1 + d_1, \dots, b_{n-1} + d_{n-1}; a_0 + b_0 + c_0 + d_0, a_1 + b_1 + c_1 + d_1, \dots, a_{n-1} + b_{n-1} + c_{n-1} + d_{n-1})$. Applying π_s with $s = 2$, we get $\pi_2(\phi(\mathbf{r})) = (d_0, c_1 + d_1, d_2, \dots, d_{n-1}; c_0 + d_0, d_1, c_2 + d_2, \dots, d_{n-2}, c_{n-1} + d_{n-1}; b_0 + d_0, a_1 + b_1 + c_1 + d_1, \dots, b_{n-1} + d_{n-1}; a_0 + b_0 + c_0 + d_0, b_1 + d_1, \dots, a_{n-1} + b_{n-1} + c_{n-1} + d_{n-1})$.
Comparing the components s_i of $(\phi\bar{\mu}_1)(\mathbf{r})$ with the last equality $\pi_s(\phi(\mathbf{r}))$ we have that $(\phi\bar{\mu}_1)(\mathbf{r}) = (\pi_2\phi)(\mathbf{r})$ and therefore $\phi\bar{\mu}_i = \pi_2\phi$ for the unit $1 + u$. For the ring R_5, where the unit $1 + x$ acts on the elements, the proof is similar. □

COROLLARY 3.11. *Let π_s be the permutation defined above. If n is odd and Γ is the Gray image of a linear cyclic code over R_i^n with $i \in \{1, 5\}$, then $\pi_s(\Gamma)$ is a quasi-cyclic code.*

PROOF. Let $i = 1$ or 5. Let Γ be such that $\phi(D) = \Gamma$ where D is a linear cyclic code over R_i. From Proposition 3.10, $(\phi\bar{\mu}_i)(D) = (\pi_s\phi)(D) = \pi_s(\Gamma)$ where $s = 2$. From Corollary 3.8, we have $\bar{\mu}_i(D)$ is a λ_i-constacyclic code C where λ_i is $1 + u$ for R_1 and $1 + x$ for R_5. Thus $\phi\bar{\mu}_i(D) = \phi(C)$ and it follows from Theorem 3.2 that $\phi(C)$ is a binary quasi-cyclic code of length $4n$ with index 4. □

3.4. Constacyclic Codes over the local ring with characteristic 8. Next we study the units in $\mathbb{Z}_8[u]/\langle u^2 + 4, 2u\rangle$, which is the only local Frobenius ring of order 16 of characteristic 8.

The ring $R = \mathbb{Z}_8[u]/\langle u^2 + 4, 2u\rangle$.

Let $R_7 = \mathbb{Z}_8[u]/\langle u^2+4, 2u\rangle$ where $w = 4$. The units are $U(R_7) = \{1, 3, 5, 7, 1+u, 3+u, 5+u, 7+u\}$. In R_7, only the unit $7+u$ acts as swap maps so the weight is preserved:

$$\begin{aligned} w_L(\phi((7+u)(a+b2+cu+d4))) &= w_L((a+b+c+d, b+d, c+d, d)) \\ &= w_L(a+bu+cv+dw). \end{aligned}$$

A $(7+u)$-constacyclic shift τ_{7+u} and on R_7^n acts as

$$\tau_{7+u}(c_0, c_1, \ldots, c_{n-1}) = ((7+u)c_{n-1}, c_0, \ldots, c_{n-2})$$

respectively.

Note that in R_7, for some positive integer k,

$$(7+u)^n = \begin{cases} 7+u & \text{if } n = 4k+1, \\ 5 & \text{if } n = 4k+2, \\ 3+u & \text{if } n = 4k+3, \\ 1 & \text{if } n = 4k. \end{cases}$$

Let $\mathfrak{R}_n^7 = R_7[x]/\langle x^n - 1\rangle$ and $\mathfrak{M}_n^7 = R_7[x]/\langle x^n - (7+u)\rangle$.

The next proposition gives a ring isomorphism between the rings $\mathfrak{R}_n^7 = R_7[x]/\langle x^n - 1\rangle$ and $\mathfrak{M}_n^7 = R_7[x]/\langle x^n - (7+u)\rangle$.

PROPOSITION 3.12. *Let $n = 4k+1$. The map $\mu_7 : \mathfrak{R}_n^7 \to \mathfrak{M}_n^7$ is a ring isomorphism such that $\mu_7(a(x)) = a((7+u)x)$, for all $a(x) \in \mathfrak{R}_n^7$.*

PROOF. If $n = 4k+1$ is odd, for polynomials $a(x), b(x) \in R_7[x]$,

$$a(x) \equiv b(x) \pmod{x^n - 1}$$

if and only if

$$a(x) - b(x) = (x^n - 1)q(x)$$

for some $q(x) \in R_7[x]$.

On the other hand,

$$a((7+u)x) - b((7+u)x) = (((7+u)x)^n - 1)q((7+u)x) \Leftrightarrow$$

$$a((7+u)x) - b((7+u)x) = (7+u)^{-1}((7+u)^{n-1}x^n - (7+u))q((7+u)x) \Leftrightarrow$$

$$a((7+u)x) - b((7+u)x) = (3+u)(x^n - (7+u))q((7+u)x) \Leftrightarrow$$

$$a((7+u)x) \equiv b((7+u)x) \pmod{x^n - (7+u)}$$

noting that $7+u$ is a unit in R_7 so it has an inverse $3+u$ and $(7+u)^{n-1} = 1$ when $n = 4k+1$. Hence μ_7 is well-defined and injective. It is obvious that μ_7 is surjective. The remaining part of the proof is straightforward. □

Unlike the case for $i = 1, 2, \ldots, 6$ the orthogonal of a $7+u$-constacyclic code is a $3+u$-constacyclic code, but this unit does not induce a swap map. Hence the image of $C^\perp$ under the gray map is not a binary skew quasi-cyclic code.

Hence we get the following:

COROLLARY 3.13. *When $n = 4k+1$, I is an ideal of $\mathfrak{R}_n^7$ if and only if $\mu_7(I)$ is an ideal of $\mathfrak{M}_n^7$.*

Applying the ring $\mathbb{Z}_8[u]/\langle u^2+4, 2u\rangle$ and the isomorphism μ_7 in Corollaries 3.6, 3.7, 3.8, 3.9, we get similar results for the ring $\mathbb{Z}_8[u]/\langle u^2+4, 2u\rangle$.

Next we give a concrete example:

EXAMPLE 1. Let $R = \mathbb{Z}_4[\alpha]/\langle \alpha^2 \rangle$. We have

$$x^7 - 3 = (x+1)(x^3 + 2x^2 + x - 3)(x^3 - 3x^2 + 2x - 3) = f_1 f_2 f_3$$

where $f_1 = x+1$, $f_2 = x^3+2x^2+x-3$ and $f_3 = x^3-3x^2+2x-3$. Let $C = uf_2+vf_1f_3$. Then C is a 3-constacyclic code over R and so it is a negacyclic code over R.

In [**5**], the authors obtained tables of examples for cyclic codes over local Frobenius rings of order 16 whose binary images under the Gray map are optimal. Modifying the generators, given in [**5**], by μ_2, we get Table 1 and Table 2 for $(1 + u + v + v^2)$-constacyclic codes over $\mathbb{F}_2[u, v] / \langle u^2 + v^2, uv \rangle$.

In the Tables 1 and 2, a polynomial $f(x) = a_{n-1}x^{n-1} + a_{n-2}x^{n-2} + \cdots + a_2x^2 + a_1x + a_0$ is abbreviated as $(a_0, a_1, a_2, \ldots, a_{n-2}, a_{n-1})$. The ones marked with $*$ and b denote the corresponding binary code is optimal or the best known code with respect to the online database of linear codes [**7**].

In order to fit Table 1 and Table 2 regarding the results, we label the elements of $\mathbb{F}_2[u, v] / \langle u^2 + v^2, uv \rangle$ as follows:

y_1	1	y_5	$1+u+v$	z_1	0	z_5	$u+v$
y_2	$1+u$	y_6	$1+u+v^2$	z_2	u	z_6	$u+v^2$
y_3	$1+v$	y_7	$1+v+v^2$	z_3	v	z_7	$v+v^2$
y_4	$1+v^2$	y_8	$1+u+v+v^2$	z_4	v^2	z_8	$u+v+v^2$

REMARK 1. The codes with parameters $[92, 66, 8]_2^*$ and $[84, 58, 8]_2^b$ in Table 1 are the first quasi-cyclic codes of these parameters with respect to the online database [**1**].

REMARK 2. The codes with parameters $[84, 63, 6]_2^{*-2}$ and $[92, 78, 4]_2^{b-1}$ in Table 2 are the first quasi-cyclic codes of these parameters which are the best known quasi-cyclic code with respect to the online database [**1**].

TABLE 1. $(1+u+v+v^2)$-constacyclic codes over $\mathbb{F}_2[u,v]/\langle u^2+v^2, uv\rangle$

n	$f(x)$	$\phi(C)$
3	$(0, z_4, z_4)$	$[12,2,8]_2^*$
3	(z_7, z_6, z_8)	$[12,4,6]_2^*$
3	(z_1, z_8, z_6)	$[12,6,4]_2^*$
3	(y_7, y_4, y_2)	$[12,8,3]_2^*$
5	$(z_7, z_6, z_2, z_3, z_1)$	$[20,8,8]_2^*$
5	$(y_4, y_1, y_3, y_3, y_5)$	$[20,12,4]_2^*$
7	$(z_4, z_1, z_4, z_1, z_1, z_4, z_4)$	$[28,3,16]_2^*$
7	$(z_7, z_5, z_7, z_6, z_2, z_5, z_1)$	$[28,6,12]_2^*$
7	$(z_6, z_5, z_6, z_7, z_3, z_5, z_4)$	$[28,7,12]_2^*$
7	$(z_7, z_6, z_2, z_4, z_5, z_3, z_4)$	$[28,8,10]_2^{*-1}$
7	$(z_5, z_6, z_2, z_4, z_5, z_3, z_4)$	$[28,12,8]_2^*$
7	$(y_1, y_8, y_4, z_6, y_6, z_8, z_5)$	$[28,19,4]_2^*$
7	$(y_2, y_5, y_4, z_8, y_1, z_5, z_1)$	$[28,20,4]_2^*$
9	$(z_8, z_8, z_3, z_4, z_3, z_2, z_2, z_5, z_5)$	$[36,16,8]_2^{*-2}$
9	$(y_7, z_2, y_2, y_4, z_1, y_8, y_2, z_1, y_8)$	$[36,20,6]_2^{*-2}$
15	$(z_5, z_7, z_4, z_1, z_4, z_2, z_7, z_5, z_7, z_1, z_3, z_2, z_3, z_4, z_7)$	$[60,20,16]_2^{b-1}$
15	$(z_4, y_6, z_7, z_1, y_5, z_5, z_3, y_8, z_8, y_5, z_4, z_6, z_3, z_2, z_7)$	$[60,50,4]_2^*$
17	$(z_7, z_1, z_2, z_8, y_2, y_6, y_7, z_1, z_3, y_3, y_3, z_2, z_3, y_5, y_4, y_2, z_6)$	$[68,48,6]_2^{*-2}$
21	$(z_4, y_2, y_1, y_8, z_3, y_3, z_3, z_6, z_2, y_3, y_7, y_3, y_5, y_8, z_1, z_6, y_5, y_7, y_1, z_2, z_1)$	$[84,58,8]_2^b$
21	$(y_2, y_7, y_4, y_6, y_5, z_8, z_3, y_1, z_5, y_8, y_4, z_4, y_3, z_6, z_3, z_3, z_4, z_2, z_8, y_2, y_8)$	$[84,62,6]_2^{b-2}$
21	$(y_2, z_6, z_6, y_8, z_3, y_2, y_3, z_7, z_1, z_2, z_8, z_6, y_1, z_8, z_7, z_6, y_1, y_5, z_8, y_4, z_7)$	$[84,72,4]_2^b$
23	$(y_4, y_5, y_4, y_1, y_6, z_8, y_1, z_2, y_2, y_3, z_1, z_4, y_4, y_1, z_4, z_2, y_1, z_3, y_6, z_1, z_7, z_7, z_6)$	$[92,66,8]_2^*$

TABLE 2. $(1+u+v+v^2)$-constacyclic codes over $\mathbb{F}_2[u,v]/\langle u^2+v^2, uv\rangle$ with two generators

n	$f(x), g(x)$	$\phi(C)$
3	$(z_6, z_5, z_7), (z_8, z_8, z_8)$	$[12,6,4]_2^*$
3	$(z_8, z_2, z_7), (z_5, z_7, z_6)$	$[12,7,4]_2^*$
5	$(z_7, z_6, z_2, z_3, z_1), (z_3, z_7, z_3, z_3, z_1)$	$[20,13,4]_2^*$
7	$(z_7, z_4, z_4, z_2, z_2, z_4, z_7), (z_8, z_5, z_1, z_1, z_4, z_6, z_2)$	$[28,15,6]_2^*$
7	$(z_5, z_5, z_7, z_6, z_4, z_3, z_6), (z_5, y_3, z_8, y_4, y_1, y_1, z_7)$	$[28,21,4]_2^*$
7	$(z_5, z_5, z_7, z_6, z_4, z_3, z_6), (z_7, y_5, z_1, y_6, y_2, y_5, z_3)$	$[28,22,4]_2^*$
15	$(z_5, z_7, z_4, z_1, z_4, z_2, z_7, z_5, z_7, z_1, z_3, z_2, z_3, z_4, z_7)$, $(z_2, z_3, z_3, z_3, z_3, z_1, z_8, z_2, z_3, z_6, z_1, z_4, z_4, z_7, z_6)$	$[60,28,10]_2^{b-2}$
15	$(z_5, z_7, z_4, z_1, z_4, z_2, z_7, z_5, z_7, z_1, z_3, z_2, z_3, z_4, z_7)$, $(z_4, y_1, z_3, z_3, z_8, y_3, z_4, z_5, y_2, y_4, y_5, y_6, y_8, y_7, z_4)$	$[60,44,6]_2^b$
17	$(z_7, z_1, z_2, z_8, y_2, y_6, y_7, z_1, z_3, y_3, y_3, z_2, z_3, y_5, y_4, y_2, z_6)$, $(y_5, y_8, z_2, y_6, y_5, z_6, z_7, y_4, z_1, y_6, y_2, y_3, y_8, z_4, y_6, z_2, z_2)$	$[68,56,4]_2^b$
17	$(z_7, z_1, z_2, z_8, y_2, y_6, y_7, z_1, z_3, y_3, y_3, z_2, z_3, y_5, y_4, y_2, z_6)$, $(z_1, z_5, y_3, y_8, z_6, z_4, z_2, z_2, y_7, y_4, z_3, z_6, y_2, z_7, z_7, z_2, y_2)$	$[68,57,4]_2^b$
21	$(z_4, y_2, y_1, y_8, z_3, y_3, z_3, z_6, z_2, y_3, y_7, y_3, y_5, y_8, z_1, z_6, y_5, y_7, y_1, z_2, z_1)$, $(z_3, z_2, z_3, z_1, z_1, z_8, z_3, z_1, z_6, z_2, z_4, z_6, z_3, z_5, z_8, z_4, z_5, z_4, z_7, z_1, z_7)$	$[84,63,6]_2^{*-2}$
21	$(z_4, y_2, y_1, y_8, z_3, y_3, z_3, z_6, z_2, y_3, y_7, y_3, y_5, y_8, z_1, z_6, y_5, y_7, y_1, z_2, z_1)$, $(z_6, y_5, y_8, z_7, z_7, z_7, y_5, y_6, y_5, z_3, y_7, z_8, y_4, z_8, y_3, y_3, y_4, z_5, y_3, z_1, y_2)$	$[84,75,4]_2^*$
21	$(z_4, y_2, y_1, y_8, z_3, y_3, z_3, z_6, z_2, y_3, y_7, y_3, y_5, y_8, z_1, z_6, y_5, y_7, y_1, z_2, z_1)$, $(y_6, y_4, z_7, y_4, y_5, z_1, y_2, y_5, y_6, z_2, z_4, y_2, z_8, z_1, z_5, y_5, y_4, z_3, z_8, z_8, z_6)$	$[84,76,4]_2^*$
23	$(y_4, y_5, y_4, y_1, y_6, z_8, y_1, z_2, y_2, y_3, z_1, z_4, y_4, y_1, z_4, z_2, y_1, z_3, y_6, z_1, z_7, z_7, z_6)$, $(z_5, z_7, z_1, z_3, z_6, z_7, z_5, z_1, z_3, z_6, z_5, z_6, z_2, z_4, z_7, z_1, z_1, z_7, z_2, z_8, z_3, z_8, z_4)$	$[92,78,4]_2^{b-1}$

References

[1] E. Z. Chen, A database of quasi-twisted codes, available at `http://moodle.tec.hkr.se/~chen/research/codes`.

[2] S. T. Dougherty, *Algebraic coding theory over finite commutative rings*, SpringerBriefs in Mathematics, Springer, Cham, 2017. MR3675844

[3] S.T. Dougherty, E. Salturk and S. Szabo, On Codes over local rings: generator matrices, generating characters and MacWilliams identities, accepted for publication in Applicable Algebra in Engineering, Communication and Computing (AAECC).

[4] S. T. Dougherty, E. Saltürk, and S. Szabo, *Codes over local rings of order 16 and binary codes*, Adv. Math. Commun. **10** (2016), no. 2, 379–391, DOI 10.3934/amc.2016012. MR3503678

[5] S. T. Dougherty, A. Kaya, and E. Saltürk, *Cyclic codes over local Frobenius rings of order 16*, Adv. Math. Commun. **11** (2017), no. 1, 99–114, DOI 10.3934/amc.2017005. MR3613286

[6] S. T. Dougherty and E. Saltürk, *Codes over a family of local Frobenius rings, Gray maps and self-dual codes*. part 3, Discrete Appl. Math. **217** (2017), no. part 3, 512–524, DOI 10.1016/j.dam.2016.09.025. MR3579930

[7] M. Grassl, Bounds on the minimum distance of linear codes and quantum codes, available at http://codetables.de.

[8] A. R. Hammons Jr., P. V. Kumar, A. R. Calderbank, N. J. A. Sloane, and P. Solé, *The* $\mathbf{Z}_4$*-linearity of Kerdock, Preparata, Goethals, and related codes*, IEEE Trans. Inform. Theory **40** (1994), no. 2, 301–319, DOI 10.1109/18.312154. MR1294046

[9] S. Karadeniz and B. Yildiz, $(1+v)$*-constacyclic codes over* $\mathbb{F}_2+u\mathbb{F}_2+v\mathbb{F}_2+uv\mathbb{F}_2$, J. Franklin Inst. **348** (2011), no. 9, 2625–2632, DOI 10.1016/j.jfranklin.2011.08.005. MR2845342

[10] E. Martínez-Moro and S. Szabo, *On codes over local Frobenius non-chain rings of order 16*, Noncommutative rings and their applications, Contemp. Math., vol. 634, Amer. Math. Soc., Providence, RI, 2015, pp. 227–241, DOI 10.1090/conm/634/12702. MR3307401

[11] J. Wolfmann, *Negacyclic and cyclic codes over* Z_4, IEEE Trans. Inform. Theory **45** (1999), no. 7, 2527–2532, DOI 10.1109/18.796397. MR1725144

Department of Mathematics, University of Scranton, Scranton, Pennsylvania 18510
Email address: `prof.steven.dougherty@gmail.com`

Academic Department, TUBITAK TUSSIDE, 41401 Kocaeli, Turkey
Email address: `esengulsalturk@gmail.com`

Contemporary Mathematics
Volume **727**, 2019
https://doi.org/10.1090/conm/727/14628

Skew Reed-Muller codes

Willi Geiselmann and Felix Ulmer

ABSTRACT. We extend the classical Reed-Muller codes by using non-commutative iterated skew polynomial rings instead of classical commutative polynomial rings. This involves the construction of iterated skew polynomial rings and the definition of the notion of points and evaluation at those points for iterated skew polynomials. Our approach is based on the notion of a left module Gröbner basis in iterated skew polynomial rings.

1. Introduction

Let A be a ring and θ an automorphism of A. A ***θ-derivation*** is a map $\delta : A \to A$ such that $\delta(a+b) = \delta(a) + \delta(b)$ and $\delta(ab) = \delta(a)b + \theta(a)\delta(b)$ for all a and b in A. In the following we denote by $A^\theta \subset A$ the fixed field of θ and we will also use the notation a^θ for $\theta(a)$ and a^δ for $\delta(a)$.

Consider a ring A, an automorphism θ of A and a θ-derivation on A. On the set $\{a_n X^n + \ldots + a_1 X + a_0 \mid a_i \in A \text{ and } n \in \mathbb{N}\}$ we consider the usual addition of polynomials and define a multiplication by the basic rule $X\, a = \theta(a)\, X + \delta(a)$ for $a \in \mathbb{F}_q$ and extend this rule to all elements of R by associativity and distributivity. This defines the ***skew polynomial ring*** $A[X;\theta,\delta]$ (see [**9**]). The classical commutative polynomial ring corresponds to A commutative, $\theta = \text{id}$ and $\delta : a \mapsto 0$. By repeating this construction we obtain the ***iterated skew polynomial ring*** $R_\ell = (\cdots(A[X_1;\theta_1,\delta_1])\ldots)[X_\ell;\theta_\ell,\delta_\ell]$ in ℓ variables over A, which we simply note $R_\ell = A[X_1;\theta_1,\delta_1][X_2;\theta_2,\delta_2]\ldots[X_\ell;\theta_\ell,\delta_\ell]$. For a finite field $\mathbb{F}_q$ and an automorphism $\theta \in \text{Aut}(\mathbb{F}_q)$ the univariate skew polynomial ring $\mathbb{F}_q[X;\theta]$ is a left and right euclidean ring (see [**9**]).

DEFINITION 1.1. A ***code*** $\mathcal{C}$ of length $n \in \mathbb{N}$ over a finite commutative ring A is a nonempty subset of A^n. The elements of $\mathcal{C}$ are called codewords. The code $\mathcal{C}$ is a ***linear code*** if it is an A-submodule of A^n. If A is a finite field $\mathbb{F}_q$, then a linear code of length n and dimension k is a k-dimensional subspace of $\mathbb{F}_q^n$. The ***Hamming distance*** between two vectors of $\mathbb{F}_q^n$ is defined as the number of coordinates at which the two vectors differ. The ***minimal distance*** d of a k-dimensional linear code $\mathcal{C} \subset \mathbb{F}_q^n$ is defined to be the minimum Hamming distance between two distinct codewords of $\mathcal{C}$. In this case we say that $\mathcal{C}$ is a code with ***parameters*** $[n,k,d]_q$.

2010 *Mathematics Subject Classification.* Primary 94Bxx, 16Zxx; Secondary 13Pxx.
Key words and phrases. Linear codes, skew polynomial rings, Gröbner basis.

The ***Reed-Solomon codes*** over $\mathbb{F}_q$ that we will now define are examples of ***algebraic codes***, whose construction and properties result from the algebraic structure of the code. Reed-Solomon codes are constructed using evaluation of polynomials in $\mathbb{F}_q[X]$. In order to construct a Reed-Solomon code $\mathcal{C} \subset \mathbb{F}_q^n$ with parameters $[n, k, n-k+1]_q$ (where $n \leq q$) we start with the k-dimensional space of polynomials $\sum_{i=0}^{k-1} b_i X^i \in \mathbb{F}_q[X]$ of degree $< k$ and n distinct elements $\alpha_1, \ldots, \alpha_n$ of $\mathbb{F}_q$. The encoding of the message $(b_0, b_1 \ldots, b_{k-1})$ of length k corresponding to the polynomial $f = \sum_{i=0}^{k-1} b_i$ is the vector $(f(\alpha_1), f(\alpha_2), \ldots, f(\alpha_n)) \in \mathcal{C} \subset \mathbb{F}_q^n$. The minimal distance of this code is known to be best possible and the algebraic structure of the code can be used to efficiently correct up to $< \frac{n-k+1}{2}$ transmission errors. Note that the length of a Reed-Solomon code is bounded by the size q of the alphabet $\mathbb{F}_q$.

There exist two generalizations of Reed-Solomon codes to skew polynomial rings:

(1) In [**3**] the evaluation of a skew polynomial $f \in \mathbb{F}_q[X; \theta]$ at a point $b \in \mathbb{F}_q$ is defined as the remainder $f(b)$ of a right division $f = q(X-b) + f(b)$ of f by $X - b$ in $\mathbb{F}_q[X; \theta]$ (cf. [**8**]). This allows for a direct generalization of Reed-Solomon codes using univariate skew polynomial rings.

(2) Consider $q = p^m$ and $\theta : \mathbb{F}_q \to \mathbb{F}_q; y \mapsto y^p$ the Frobenius morphism. The map $\varphi : \mathbb{F}_q[X; \theta] \to \mathrm{End}(\mathbb{F}_q)$, $\sum_{i=0}^{m} a_i X^i \mapsto \sum_{i=0}^{m} a_i \theta^i$ is a ring morphism. One can define the evaluation of a skew polynomial $f \in \mathbb{F}_q[X; \theta]$ at a point $b \in \mathbb{F}_q$ as $\varphi(f)(b)$, which corresponds to the evaluation of the linearized polynomial $\sum_{i=0}^{m} a_i X^{q^i}$ at b. This evaluation leads to "Gabidulin codes" [**7**].

In both generalizations of Reed-Solomon codes the length of the resulting code is smaller than the length for the corresponding Reed-Solomon codes in the commutative case. For Gabidulin codes this follows from the fact that the solution space of an operator $\sum_{i=0}^{m} a_i \theta^i$ is a vector space over $\mathbb{F}_p$, therefore the evaluation points need to be linearly independent over $\mathbb{F}_p$. This reduces the number of possible evaluation points, i.e. the length of the code. The vector space structure of the solution of an operator $\sum_{i=0}^{m} a_i \theta^i$ is also the reason why the notion of a rank distance is more appropriated than the notion of a Hamming distance, when dealing with Gabidulin codes. A future project will be to generalize the notion of rank distance for the codes presented in this paper.

Reed-Muller codes are based on the evaluation of multivariate polynomials. A polynomial $f \in \mathbb{F}_q[X_1, \ldots, X_\ell]$ of total degree $s < q - 1$ contains $k = \binom{\ell+s}{s}$ coefficients $(b_0, \ldots, b_{k-1})$. In order to construct a Reed-Muller code $\mathcal{C} \subset \mathbb{F}_q^n$ we choose $n \leq q^\ell$ points $a_i = (\alpha_{i,1}, \ldots, \alpha_{i,\ell}) \in \mathbb{F}_q^\ell$. The encoding of the message $(b_0, b_1 \ldots, b_{k-1})$ of length k corresponding to the polynomial $f \in \mathbb{F}_q[X_1, \ldots, X_\ell]$ is the vector $(f(a_1), f(a_2), \ldots, f(a_n)) \in \mathcal{C} \subset \mathbb{F}_q^n$. Reed-Muller codes, unlike Reed-Solomon codes, are not optimal with respect to the minimal distance. However, the maximal length of the code is $n = q^\ell$, so that the alphabet size q can be exponentially smaller than the length of the code.

The paper is organized in the following way: In the first section we give some constructions of iterated skew polynomial rings. We then give a definition for an evaluation of a multivariate skew polynomial using left module Gröbner basis computations. In the third section we define skew Reed-Muller codes and provide some examples. In Section 4, we extend the notion of skew Reed-Muller code to skew polynomial rings over chain rings.

2. Iterated skew polynomial rings

The construction of iterated skew polynomial rings is a difficult problem because little is known about the automorphism ring and the derivations of

$$R_\ell = (\cdots(A[X_1;\theta_1,\delta_1])\ldots)[X_\ell;\theta_\ell,\delta_\ell].$$

Classical examples are quantum Weyl algebras where the ground ring A is central ([**6**], Section 2.3.3), iterated skew polynomial rings of derivation type ([**12**]) and iterated skew polynomial rings whose variables commute ([**5**]). None of those examples turned out to be sufficiently general. In this paper, our examples will be built using inner automorphisms and inner derivations.

EXAMPLE 2.1. Consider a ring A and an invertible element ν in A. Then $\theta_\nu^A : A \to A; a \mapsto \nu^{-1}a\nu$ is an ***inner automorphism*** of A. The automorphism θ_ν^A is the identity on A if and only if ν is a central invertible element in A.

EXAMPLE 2.2. Consider a ring A, an automorphism $\theta \in \mathrm{Aut}(A)$ and an element $\beta \in A$. The map $\delta_\beta^{A,\theta} : A \to A; a \mapsto \beta a - a^\theta\beta$ is an ***inner θ-derivation*** on A.

It is well known that skew polynomial rings that differ by inner derivations or inner automorphisms are isomorphic (see [**1**]), which explains why rings that differ by inner derivations or inner automorphisms will often lead to equivalent codes. A more general family of skew polynomial rings would probably lead to better codes.

EXAMPLE 2.3. Consider $\mathbb{F}_4 = \{0, 1, \alpha, \alpha^2\}$, $\theta_1 : \mathbb{F}_4 \to \mathbb{F}_4; y \to y^2$ the Frobenius automorphisms. We give an example of in iterated skew polynomial ring over $\mathbb{F}_4$ constructed using inner automorphisms and inner derivations. The parameters used are random but meet Definition 3.2 allowing to later compute a Gröbner basis over this ring.

(1) In the ring $R_1 = \mathbb{F}_4[X_1;\theta,\delta_1^{\mathbb{F}_4,\theta}]$ we have the commutation relation

$$X_1\alpha = \theta(\alpha)X_1 + \delta_1^{\mathbb{F}_4,\theta}(\alpha) = \alpha^2X_1 + (1\cdot\alpha - \theta(\alpha)\cdot 1) = \alpha^2X_1 + 1.$$

(2) In the ring $R_2 = R_1[X_2;\theta_\alpha^{R_1},\delta_{X_1+\alpha}^{R_1,\theta_\alpha^{R_1}}]$ we have the above commutation relation $X_1\alpha = \alpha^2X_1 + 1$ together with

$$\begin{aligned}
X_2\alpha &= \theta_\alpha^{R_1}(\alpha)X_2 + \delta_{X_1+\alpha}^{R_1,\theta_\alpha^{R_1}}(\alpha) = \alpha X_2 + ((X_1+\alpha)\alpha - \alpha^2(X_1+\alpha))\\
&= \alpha X_2 + X_1 + 1,\\
X_2X_1 &= \theta_\alpha^{R_1}(X_1)X_2 + \delta_{X_1+\alpha}^{R_1,\theta_\alpha^{R_1}}(X_1)\\
&= \alpha^2X_1\alpha X_2 + (X_1+\alpha)\alpha - \alpha^2X_1\alpha(X_1+\alpha))\\
&= \alpha^2X_1X_2 + \alpha X_2 + \alpha X_1^2 + \alpha X_1.
\end{aligned}$$

(3) In the ring $R_3 = R_2[X_3; \theta_\alpha^{R_2}, \delta_{\alpha X_1}^{R_2,\theta_\alpha^{R_2}}]$ we have the above commutation relations together with

$$\begin{aligned} X_3\alpha &= \theta_\alpha^{R_2}(\alpha)X_3 + \delta_{\alpha X_1}^{R_2,\theta_\alpha^{R_2}}(\alpha) = \alpha X_3 + \alpha X_1 + \alpha, \\ X_3X_1 &= \theta_\alpha^{R_2}(X_1)X_3 + \delta_{\alpha X_1}^{R_2,\theta_\alpha^{R_2}}(X_1) = \alpha^2 X_1X_3 + \alpha X_3, \\ X_3X_2 &= \theta_\alpha^{R_2}(X_2)X_3 + \delta_{\alpha X_1}^{R_2,\theta_\alpha^{R_2}}(X_2) \\ &= X_2X_3 + (\alpha X_1 + \alpha)X_3 + (\alpha^2X_1 + \alpha^2)X_2 + \alpha^2X_1^2 + \alpha^2X_1. \end{aligned}$$

3. Left ideal Gröbner bases and skew Reed-Muller codes

In order to generalize Reed-Muller codes we need to define the evaluation of an element of an iterated skew polynomial ring. The evaluation of a classical polynomial $f \in \mathbb{F}_q[X_1, \ldots, X_\ell]$ at the point $(\alpha_1, \ldots, \alpha_\ell)$ can be seen as the remainder of successive divisions of f by $X_1 - \alpha_1, \ldots, X_\ell - \alpha_\ell$, i.e.:

$$f = q_1(X_1 - \alpha_1) + \ldots + q_\ell(X_\ell - \alpha_\ell) + f(\alpha_1, \ldots, \alpha_\ell).$$

The result is independent of the order of the division, which corresponds to the fact that $\{X_1 - \alpha_1, \ldots, X_\ell - \alpha_\ell\}$ is a ***Gröbner basis*** for the ideal generated by this set. We refer to [**2**] for the definition of a Gröbner basis and a ***reduced Gröbner basis***.

There exist several generalizations of the notion of Gröbner basis to various types of iterated skew polynomial rings in the literature [**10, 11**].

We refer to [**10**] for the classical definition of a monomial ordering $<$ on $\mathbb{N}^m$. Classically a monomial ordering induces an ordering $\prec$ on the set of monomials $\mathcal{M} = \{X^\alpha = X_1^{\alpha_1}X_2^{\alpha_2}\cdots X_m^{\alpha_m} \mid \alpha_i \in \mathbb{N}\}$ (note that the variables need to be in a precise order when dealing with a non-commutative ring) via $X^\alpha \prec X^\beta$ if and only if $\alpha < \beta$. For any expression $f = \sum_{\alpha\in\mathbb{N}^m} c_\alpha X^\alpha$ where only finitely many constants c_α are nonzero, the monomial $X^\gamma = \max\{X^\alpha | c_\alpha \neq 0\}$ is the ***leading monomial*** of $f$and c_α is the ***leading coefficient*** of f, denoted respectively by $\mathrm{lm}(f)$ and $\mathrm{lc}(f)$. Then the least common multiple of X^α and X^β is defined as $\mathrm{lcm}(X^\alpha, X^\beta) = X^\gamma$ where $\gamma_i = \max(\alpha_i, \beta_i)$. We will be interested in left ideals I of skew polynomial rings $R_\ell = A[X_1; \theta_1, \delta_1][X_2; \theta_2, \delta_2] \ldots [X_\ell; \theta_\ell, \delta_\ell]$.

A Gröbner basis can be computed in a Poincaré-Birkhoff-Witt extension (PBW) (see also [**10**], Definition 1.2):

DEFINITION 3.1. (see [**11**], Definition 3.2.1) Let A and B be two associative rings with $A \subset B$. The ring B is called a (finite) Poincaré-Birkhoff-Witt PBW extension **(PBW extension)** of A if there exist $X_1, X_2, \ldots, X_\ell$ in B such that

(1) the monomials $X_1^{i_1}X_2^{i_1}\cdots X_\ell^{i_\ell}$ form a basis for B as a free left A-module, where $i_1, \ldots, i_\ell$ are in $\mathbb{N}$;
(2) $X_ia - aX_i = [X_i, a] \in A$ for each $i \in \{1, \ldots, n\}$ and any $a \in A$;
(3) $X_iX_j - X_jX_i = [X_i, X_j] \in A + AX_1 + \cdots + AX_\ell$ for all i, j in $\{1, \ldots, n\}$.

We write $B = A\langle X_1, \ldots, X_\ell\rangle$.

In [**10, 11**] algorithms are given for computing Gröbner bases of a left ideal I in solvable polynomial algebras and skew solvable polynomial rings. We will work with the following slight generalization of the last definition

DEFINITION 3.2. Let $R_\ell = (\cdots A[X_1;\theta_1,\delta_1])\ldots)[X_\ell;\theta_\ell,\delta_\ell]$ be an iterative skew polynomial ring in $n \in \mathbb{N}$. We call the ring ***left-lex-solvable***, for the lexicographical order $1 \prec X_1 \prec \ldots \prec X_\ell$, if

(1) for any $a \in A$ and any $i \in \{1,\ldots,n\}$, $X_i a = bX_i + p_{i,a}$ where $b \in A$ and $p_{i,a} \in R_{i-1}$;
(2) for all $j < i$ in $\{1,\ldots,n\}$, $X_iX_j = bX_jX_i + p_{i,j}$ where $b \in A$ and all monomials in $p_{i,j}$ are $\prec X_iX_j$.

Suppose now that R_ℓ is a left-lex-solvable iterated skew polynomial ring in the (non commuting) variables $X_1\ldots,X_m$. We say that $X^\alpha \in \mathcal{M}$ is divisible by X^β if $X^\alpha = \operatorname{lm}(X^\omega X^\beta)$ for some $X^\omega \in \mathcal{M}$ (note that $X^\omega X^\beta$ may no longer be a monomial, but that the non leading monomials of $X^\omega X^\beta$ are $\prec X^\alpha$).

We follow the definition of an ***S-polynomial*** given in ([**10**], Definition 2.5). If $X^\gamma = \operatorname{lm}(\operatorname{lcm}(X^\alpha X^\beta))$, $t_f = X^{\gamma-\alpha}$ and $t_g = X^{\gamma-\beta}$, then

$$\operatorname{SPoly}(f,g) = t_f\, f \;-\; c\, t_g\, g, \text{ where } c = \frac{\operatorname{lc}(t_f\, f)}{\operatorname{lc}(t_g\, g)}.$$

If the iterative skew polynomial ring R_ℓ is left-lex-solvable, then, according to ([**10**], Section 2.2) the classical Buchberger algorithm, applied to the above S-polynomials using a lexicographic order $X_1 \prec \ldots \prec X_\ell$, produces a left Gröbner basis of any left ideal $I \subset_\ell$. For a given Gröbner basis $\mathcal{B}_I = \{g_1,\ldots,g_s\}$ of a left ideal $I \subset R_\ell$, the ***right reduction*** of f by $\mathcal{B}_I$ is the unique polynomial $\overline{f}^{\mathcal{B}_I} \in R_\ell$ in the decomposition $f = \left(\sum_{j=1}^s q_j \cdot g_i\right) + \overline{f}^{\mathcal{B}_I}$ with the property that no leading monomial of any $g_i \in \mathcal{B}_I$ divides any monomial of $\overline{f}^{\mathcal{B}_I}$.

DEFINITION 3.3. Let F be a field, B a left-finitely generated algebra over F and $\prec$ an admissible monomial ordering on B. We call a left Gröbner basis $\mathcal{B}_I = \{g_1,\ldots,g_s\}$ of a left ideal $I \subset R$ an ***evaluation base*** if

(1) $I = (g_1,\ldots,g_s) \neq \{1\}$ (we exclude the "always zero" evaluation which is of no interest for Reed-Muller type codes).
(2) the right reduction $\overline{f}^{\mathcal{B}_I}$ of any $f \in B$ by $\mathcal{B}_I$ belongs to F.

PROPOSITION 3.4. *Let F be a field, R_ℓ a left-lex-solvable skew polynomial ring over F generated by $X_1,\ldots,X_\ell$ and $\prec$ an admissible monomial ordering. If the ordering $\prec$ is a well ordering, then any reduced evaluation basis $\mathcal{B}_I$ is of the form*

$$\{X_1-\alpha_1, X_2-\alpha_2,\ldots,X_\ell-\alpha_\ell\}$$

where $\alpha_i \in F$.

PROOF. For an evaluation basis $\mathcal{B}_I$ each generator X_i must reduce to $\alpha_i \in F$: $X_i = (\sum_{g_i\in\mathcal{B}_I} h_i g_i) + \alpha_i \in F$. Therefore $X_i - \alpha_i = (\sum_{g_i\in\mathcal{B}_I} h_i g_i) \in I$. We suppose that $X_1 \prec \cdots \prec X_\ell$ and proceed by induction on i:

(1) Since $\prec$ is a well ordering and $1 \notin \mathcal{B}_I$, the monomial X_1 is minimal among the leading monomials in $\mathcal{B}_I$ and therefore $X_1 - \alpha_1$ must belong to the Gröbner basis $\mathcal{B}_I$.
(2) Suppose that $X_1-\alpha_1,\ldots X_i-\alpha_i$ belongs to $\mathcal{B}_I$. A reduced evaluation basis $\mathcal{B}_I$ cannot contain any other monomial divisible by $X_1,\ldots,X_i$, and X_{i+1} is a minimal leading monomial among the other polynomials of $\mathcal{B}_I$. Suppose that $X_j \in \{X_1,\ldots,X_i\}$ divides the monomial X_{i+1}, then $X_{i+1} = MX_j$

for some monomial M which must contain a variable in $\{X_{i+1}, \ldots, X_\ell\}$. Reordering the variables in MX_j using (2) in Definition 3.2, we obtain a leading term containing X_j and therefore a contradiction. As a result, $X_{i+1} - \alpha_{i+1}$ belongs to $\mathcal{B}_I$.

Therefore $\mathcal{B}_I$ contains $X_1 - \alpha_1, \ldots, X_\ell - \alpha_\ell$. Since the basis is reduced it can only contain those polynomials, showing that for a lexicographic order a reduced Gröbner basis of an evaluation ideal is always of the form $(X_1-\alpha_1, X_2-\alpha_2, \ldots, X_\ell-\alpha_\ell)$. □

EXAMPLE 3.5. In Example 2.3 we constructed the iterate skew polynomial ring

$$R_3 = \mathbb{F}_4[X_1; \theta, \delta_1^{\mathbb{F}_4,\theta}][X_2; \theta_\alpha^{R_1}, \delta_{X_1+\alpha}^{R_1,\theta_\alpha^{R_1}}][X_3; \theta_\alpha^{R_2}, \delta_{\alpha X_1}^{R_2,\theta_\alpha^{R_2}}].$$

For this ring and the lexicographic order $X_3 > X_2 > X_1$ only 28 ideals of the form $(X_1 - \alpha_1, X_2 - \alpha_2, X_3 - \alpha_3)$ are distinct from $(1) = R_3$. For example (X_1, X_2, X_3) is a Gröbner basis, but for $(X_1, X_2, X_3 - 1)$ the Gröbner basis turns out to be (1) since $(\alpha^2 X_3 + \alpha)X_1 + (\alpha X_1 + 1)(X_3 - 1) = 1$.

DEFINITION 3.6. $\mathbb{F}_q$ Consider an iterated skew polynomial ring

$$R_\ell = \mathbb{F}_q[X_1; \theta_1, \delta_1][X_2; \theta_2, \delta_2] \ldots [X_\ell; \theta_\ell, \delta_\ell]$$

over a finite field $\mathbb{F}_q$, and $\prec$ an admissible monomial ordering on R and a list $\mathcal{B}_{I_1}, \ldots, \mathcal{B}_{I_n}$ of Gröbner bases which are evaluation bases for R_ℓ over $\mathbb{F}_q$. If an $\mathbb{F}_q$-subspace W of polynomials of R_ℓ is of dimension k, then a ***skew Reed-Muller encoding*** of length n of $f \in W$ is given by

$$\left(\overline{f}^{\mathcal{B}_{I_1}}, \ldots, \overline{f}^{\mathcal{B}_{I_n}}\right) \in \mathbb{F}_q^n.$$

The resulting code is a linear code with parameters $[n, k]$.

In order to verify that this gives an $\mathbb{F}_q$-linear code we need to show that

$$\mathcal{C} = \left\{\left(\overline{f}^{\mathcal{B}_{I_1}}, \ldots, \overline{f}^{\mathcal{B}_{I_n}}\right) \mid f \in W\right\}$$

is a subspace of $\mathbb{F}_q^n$. In order to see this we note that for all s in $\{1, \ldots, \ell\}$, for all f in R_ℓ with $f = (\sum_{g_i \in \mathcal{B}_{I_s}} h_i g_i) + \overline{f}^{\mathcal{B}_{I_s}}$, all $\tilde{f}$ in R_ℓ with $\tilde{f} = (\sum_{g_i \in \mathcal{B}_{I_s}} \tilde{h}_i g_i) + \overline{\tilde{f}}^{\mathcal{B}_{I_s}}$ and all λ in $\mathbb{F}_q$ we have:

$$\lambda f = (\sum_{g_i \in \mathcal{B}_{I_s}} \lambda h_i g_i) + \lambda \cdot \overline{f}^{\mathcal{B}_{I_s}} \tag{3.1}$$

$$\tilde{f} + f = (\sum_{g_i \in \mathcal{B}_{I_s}} (\tilde{h}_i + h_i) g_i) + \overline{\tilde{f}}^{\mathcal{B}_{I_s}} + \overline{f}^{\mathcal{B}_{I_s}}. \tag{3.2}$$

The result now follows from the uniqueness of the reduction by a Gröbner basis. We note that, even if the evaluation map is a ring homomorphism

$$R_\ell \to \mathbb{F}_q; f \mapsto \overline{f}^{\mathcal{B}_{I_s}},$$

the fact that we obtain a linear code over $\mathbb{F}_q$ relies only on the fact that the map $R_\ell \to \mathbb{F}_q; f \mapsto \overline{f}^{\mathcal{B}_{I_s}}$ is an $\mathbb{F}_q$-linear map.

EXAMPLE 3.7. In Example 2.3 we constructed the iterate skew polynomial ring

$$R_3 = \mathbb{F}_4[X_1; \theta, \delta_1^{\mathbb{F}_4,\theta}][X_2; \theta_\alpha^{R_1}, \delta_{X_1+\alpha}^{R_1,\theta_\alpha^{R_1}}][X_3; \theta_\alpha^{R_2}, \delta_{\alpha X_1}^{R_2,\theta_\alpha^{R_2}}].$$

While classical Reed-Muller codes are of length 64, for this ring and the lexicographic order $X_3 > X_2 > X_1$ we obtain 28 evaluation points with Gröbner bases of the form $(X_1-\alpha_1, X_2-\alpha_2, X_3-\alpha_3)$ distinct from $(1) = R_3$ (the indicated distance is the hamming distance):

(1) Considering the vector space of all polynomials of degree 1 we obtain a $[28, 4, 12]_4$.
(2) Considering the vector space of all polynomials of degree 2 we obtain a $[28, 10, 6]_4$.
(3) Considering the vector space of all polynomials of degree 3 we obtain a $[28, 17, 3]_4$.

4. Skew Reed-Muller codes over finite chain rings

A finite commutative ring with identity $1 \neq 0$ is called a ***finite chain ring*** if its ideals are linearly ordered by inclusion. In order to define skew Reed-Muller codes over finite chain rings we need to define the notion of a Gröbner basis to finite chain rings. We do that by following [**4**]. In a finite chain ring A the unique maximal ideal is generated by an element ω of nilpotency index m (i.e. m is the smallest integer such that $\omega^m = 0$) any element a can be written in the form $\mu_a \omega^{m_a}$ where $0 \leq m_a \leq m-1$ and μ_a is a invertible element of A, unique modulo ω^{m-m_a}.

EXAMPLE 4.1. The ring $A = \mathbb{F}_4[z]/(z^2)$ is a chain ring of order 16. The unique maximal ideal of A is (z) whose nilpotency index is $m = 2$. The invertible elements of A are

$$\{\alpha, z+\alpha, \alpha z+\alpha, \alpha^2 z+\alpha, 1, z+1, \alpha z+1, \alpha^2 z+1, \alpha^2, z+\alpha^2, \alpha z+\alpha^2, \alpha^2 z+\alpha^2\}.$$

We can construct an iterated skew polynomial ring using inner automorphisms and inner derivations in the same way than in Example 2.3. Two simple examples that we will use later are

(1) The ring homomorphism $\theta_3 : A \to A$ defined by $a \mapsto a$ and $z \mapsto \alpha^2 z$ is an automorphism of order 3 of A. In the ring $R_{2,3} = A[X_1; \theta_3][X_2]$, we have the commutation relation
$$X_1\alpha = \theta_3(\alpha)X_1 = \alpha X_1;\ X_1 z = \theta_3(z)X_1 = \alpha^2 z X_1$$
and X_2 is a central element.
(2) The ring homomorphism $\theta_2 : A \to A$ defined by $a \mapsto a^2$ and $z \mapsto \alpha z$ is an automorphism of order 2 of A. In the ring $R_{2,2} = A[X_1; \theta][X_2]$, we have the commutation relation
$$X_1\alpha = \theta_2(\alpha)X_1 = \alpha^2 X_1;\ X_1 z = \theta_2(z)X_1 = \alpha z X_1$$
and X_2 is a central element.

We follow the definition of an S***-polynomial*** in ([**4**], Definition 3.4). Suppose now that R is a left-lex-solvable iterated polynomial ring in the (non commuting) variables $X_1 \ldots, X_m$ over a finite commutative chain ring A. Recall that $\operatorname{lcm}(X^\alpha, X^\beta) = X^\gamma$ where $\gamma_i = \max(\alpha_i, \beta_i)$ and that $X^\alpha \in \mathcal{M}$ is divisible by X^β if $X^\alpha = \operatorname{lm}(X^\omega X^\beta)$ for some $X^\omega \in \mathcal{M}$ (note that the product of monomial $X^\omega X^\beta$ may no longer be a monomial, but that the non leading monomials of $X^\omega X^\beta$ are $\prec X^\alpha$ in a left-lex-solvable iterated polynomial ring).

Adapting ([**4**], Definition 3.2) to the non-commutative situation we define that the polynomial g ***reduces*** the monomial $\mu_f \omega^{m_f} X^\alpha$ if

(1) $\mathrm{lm}(g) = X^\beta$ with $\beta < \alpha$;
(2) if $\mathrm{lc}(X^{\alpha-\beta} g) = \mu_g \omega^{m_g}$ with $m_g < m_f$.

The corresponding reduction step is the result of

$$\mu_f \omega^{m_f} X^\alpha - \mu_f \mu_g^{-1} \omega^{m_f - m_g} X^{\alpha-\beta} g.$$

We say that a polynomial g reduces a polynomial f if g reduces $\mathrm{lt}(f)$.

Let f and g be two non zero polynomials whose leading monomials are X^α and X^β. If $X^\gamma = \mathrm{lm}(\mathrm{lcm}(X^\alpha, X^\beta))$, $a_f = \mathrm{lc}(X^{\gamma-\alpha} f) = \mu_f \omega^{m_f}$, $b_g = \mathrm{lc}(X^{\gamma-\beta} g) = \mu_g \omega^{m_g}$ and $m_{f,g} = \max\{m_f, m_g\}$. We follow the definition of an S-polynomial of ([**4**], Definition 3.4):

$$\mathrm{SPoly}(f,g) = \mu_f^{-1} \omega^{m_{f,g} - m_f} X^{\gamma-\alpha} f - \mu_g{}^{-1} \omega^{m_{f,g} - m_g} X^{\gamma-\beta} g.$$

According to [**4**] we also need the notion of an ***A-polynomial***. Consider f with $\mathrm{lc}(f) = \mu_a \omega^{m_a}$ with $m_a > 0$ (i.e. a non invertible zero divisor):

$$\mathrm{APoly}(f) = \omega^{m-m_a} f$$

A basis of an ideal is a Gröbner basis if all S-polynomials between the elements of the basis and all A-polynomials of the elements of the basis reduce to zero.

Similarly to $\mathbb{Z}[X]$ where $(X, 2)$ is a maximal ideal, the maximal ideals in an iterated skew polynomial ring are no longer all of the form $(X_1 - \alpha_1, \ldots, X_\ell - \alpha_\ell)$.

EXAMPLE 4.2. Consider $A = \mathbb{F}_4[z]/(z^2)$ and $R_{2,2} = A[X_1; \theta_2][X_2]$ as defined in Example 4.1. With lex order $X_2 > X_1$ there are 88 evaluation points with Gröbner bases of the form $(X_1 - \alpha_1, X_2 - \alpha_2)$ and 12 additional evaluation points with Gröbner bases $(X_1 - \alpha_1, X_2 - \alpha_2, z)$. For example the Gröbner basis of $(X_1 - 1, X_2 - z)$ is $(X_1 - 1, X_2, z)$.
Using only the 88 evaluation points whose Gröbner bases are of the special form $(X_1 - \alpha_1, X_2 - \alpha_2)$, we obtain the following skew Reed-Muller codes (the indicated distance is the hamming distance):

(1) Considering the evaluation of all $|A|^3$ polynomials of total degree 1, we obtain a linear code of length 88 over A which maps, using the mapping $A \to \mathbb{F}_4^2$ given by $(\alpha_1 + \alpha_2 z) \mapsto (\alpha_2, \alpha_1 + \alpha_2)$, to a $[176, 6, 88]_4$ code.
(2) Considering the evaluation of all $|A|^6$ polynomials of total degree 2, we obtain a linear code of length 88 over A which maps to a $[176, 12, 16]_4$ code.
(3) Considering the evaluation of all $|A|^9$ polynomials of total degree 3, we obtain a linear code of length 88 over A which maps to a $[176, 18, 8]_4$ code.

Using all 96 evaluation points, including those with Gröbner bases of the form $(X_1 - \alpha_1, X_2 - \alpha_2, z)$, we obtain the following skew Reed-Muller codes (the indicated distance is the hamming distance):

(1) Considering the evaluation of all $|A|^3$ polynomials of total degree 1, we obtain a linear code of length 96 over A which maps, under the above mapping, to a $[192, 6, 96]_4$ code.
(2) Considering the evaluation of all $|A|^6$ polynomials of total degree 2, we obtain a linear code of length 96 over A which maps to a $[192, 12, 16]_4$ code.

(3) Considering the evaluation of all $|A|^9$ polynomials of total degree 3, we obtain a linear code of length 96 over A which maps to a $[192, 18, 8]_4$ code.

EXAMPLE 4.3. Consider $A = \mathbb{F}_4[z]/(z^2)$ and $R_{2,3} = A[X_1; \theta_3][X_2]$ as defined in Example 4.1. With lex order $X_2 > X_1$ there are 112 evaluation points with Gröbner basis of the form $(X_1 - \alpha_1, X_2 - \alpha_2)$ and 12 additional evaluation points with Gröbner bases of the form $(X_1 - \alpha_1, X_2 - \alpha_2, z)$. For example the points $(X_1 - a, X_2 - a, z)$ or $(X_1 - 1, X_2, z)$. Using only the 112 evaluation points with Gröbner bases of the form $(X_1 - \alpha_1, X_2 - \alpha_2)$, we obtain the following skew Reed-Muller codes (the indicated distance is the hamming distance):

(1) Considering the evaluation of all $|A|^3$ polynomials of total degree 1, we obtain a linear code of length 112 over A which, under the mapping $A \to \mathbb{F}_4^2$ with $(\alpha_1 + \alpha_2 z) \mapsto (\alpha_2, \alpha_1 + \alpha_2)$, map to a $[224, 6, 96]_4$ code.
(2) Considering the evaluation of all $|A|^6$ polynomials of total degree 2, we obtain a linear code of length 112 over A which under the mapping $A \to \mathbb{F}_4^2$ with $(\alpha_1 + \alpha_2 z) \mapsto (\alpha_2, \alpha_1 + \alpha_2)$, map to a $[224, 12, 64]_4$ code.
(3) Considering the evaluation of all $|A|^9$ polynomials of total degree 3, we obtain a linear code of length 112 over A which under the mapping $A \to \mathbb{F}_4^2$ with $(\alpha_1 + \alpha_2 z) \mapsto (\alpha_2, \alpha_1 + \alpha_2)$, map to a $[224, 20, 32]_4$ code.

References

[1] S. A. Abramov, Kh. K. Le, and Z. Li, *Ore polynomial rings in one variable in computer algebra* (Russian, with Russian summary), Sovrem. Mat. Prilozh. **13, Algebra** (2004), 24–39, DOI 10.1007/s10958-005-0449-8; English transl., J. Math. Sci. (N.Y.) **131** (2005), no. 5, 5885–5903. MR2153691

[2] W. W. Adams and P. Loustaunau, *An introduction to Gröbner bases*, Graduate Studies in Mathematics, vol. 3, American Mathematical Society, Providence, RI, 1994. MR1287608

[3] D. Boucher and F. Ulmer, *Linear codes using skew polynomials with automorphisms and derivations*, Des. Codes Cryptogr. **70** (2014), no. 3, 405–431, DOI 10.1007/s10623-012-9704-4. MR3160738

[4] A. Hashemi and P. Alvandi, *Applying Buchberger's criteria for computing Gröbner bases over finite-chain rings*, J. Algebra Appl. **12** (2013), no. 7, 1350034, 15, DOI 10.1142/S0219498813500345. MR3063473

[5] L. Chaussade, *Codes correcteurs avec les polynômes tordus*, Thèse Université de Rennes 1, novembre 2010.

[6] F. Dumas, *An introduction to noncommutative polynomial invariants*, CIMPA course "Homological methods and representations of non-commutative algebras", Argentina, 2006

[7] È. M. Gabidulin, *Theory of codes with maximum rank distance* (Russian), Problemy Peredachi Informatsii **21** (1985), no. 1, 3–16. MR791529

[8] T. Y. Lam and A. Leroy, *Vandermonde and Wronskian matrices over division rings*, J. Algebra **119** (1988), no. 2, 308–336, DOI 10.1016/0021-8693(88)90063-4. MR971137

[9] O. Ore, *Theory of non-commutative polynomials*, Ann. of Math. (2) **34** (1933), no. 3, 480–508, DOI 10.2307/1968173. MR1503119

[10] X. Zhao and Y. Zhang, *A signature-based algorithm for computing Gröbner-Shirshov bases in skew solvable polynomial rings*, Open Math. **13** (2015), 298–307, DOI 10.1515/math-2015-0028. MR3346488

[11] Y. Zhang, *Algorithms for noncommutative differential operators*, ProQuest LLC, Ann Arbor, MI, 2004. Thesis (Ph.D.)–The University of Western Ontario (Canada). MR2706667

[12] M. Gr. Voskoglou, *Derivations and Iterated Skew Polynomial Rings*, International journal of applied mathematics and informatics, Issue 2, Volume 5, 2011

KIT, Institut für Theoretische Informatik (ITI), Am Fasanengarten 5, D-76131 Karlsruhe, Germany

Email address: willi.geiselmann@kit.edu

Univ Rennes, CNRS, IRMAR - UMR 6625, F-35000 Rennes, France

Email address: felix.ulmer@univ-rennes1.fr

Contemporary Mathematics
Volume **727**, 2019
https://doi.org/10.1090/conm/727/14629

The extension theorem for bi-invariant weights over Frobenius rings and Frobenius bimodules

Oliver W. Gnilke, Marcus Greferath, Thomas Honold, Jay A. Wood, and Jens Zumbrägel

In memory of our colleague and friend Aleksandr A. Nechaev, 1945–2014

Abstract. We give a sufficient condition for a bi-invariant weight on a Frobenius bimodule to satisfy the extension property. This condition applies to bi-invariant weights on a finite Frobenius ring as a special case. The complex-valued functions on a Frobenius bimodule are viewed as a module over the complex monoid algebra of the multiplicative monoid of the coefficient ring.

1. Introduction

When coding theory was first developed in the 1940s and 50s, linear codes were defined over finite fields using the Hamming weight. Coding theory has evolved since those early years, and linear codes are now often defined over alphabets that are finite modules over a finite ring using a more general weight on the module.

A fundamental question about linear codes, regardless of the level of abstraction, is: *When should two linear codes be considered equivalent?* There are two competing notions of equivalence. One notion is that two linear codes are equivalent if there exists a weight-preserving monomial transformation of the ambient space that takes one code to the other. The other notion is that two linear codes are equivalent if there exists a weight-preserving linear isomorphism between the two codes. If two linear codes are equivalent in the first sense, then they are equivalent in the second sense. Indeed, the restriction of the weight-preserving monomial transformation to the codes defines a weight-preserving linear isomorphism between them.

The converse, i.e., if two linear codes are equivalent in the second sense, then they are equivalent in the first sense, is called the *extension problem*, because the question is whether every weight-preserving linear isomorphism between two linear codes extends to a weight-preserving monomial transformation of the ambient space. When the extension problem for a given alphabet and a given weight has a positive solution, i.e., when every weight-preserving linear isomorphism extends to a weight-preserving monomial transformation, we often say that the *extension theorem holds* for that alphabet and weight.

The extension theorem is known to hold in a number of situations. The earliest version, over finite fields with the Hamming weight, is due to MacWilliams [**10**, **11**]. The extension theorem holds when the alphabet is a finite Frobenius ring using the

Hamming weight [**13**] or the homogeneous weight [**7**], and the extension theorem holds over Frobenius bimodules over any finite ring using either the Hamming or the homogeneous weight [**6**].

There has been considerable work on determining tractable conditions on a general weight in order that the extension theorem hold for that weight. Some of the earliest work involved a nonsingularity condition on a matrix determined by both the alphabet and the weight that allowed one to reduce the problem to the known extension properties of a so-called symmetrized weight composition [**14**].

This paper, drawing on ideas introduced in [**2**], [**3**] and [**5**], describes conditions on a weight with maximal symmetry (called a *bi-invariant* weight) over a Frobenius bimodule so that the extension problem reduces to the known extension properties of the homogeneous weight. The conditions found generalize those for weights defined over a finite principal ideal ring in [**4**, Theorem 4.4]. The reduction process is accomplished by viewing any weight on a Frobenius bimodule as an element of a complex vector space which is itself a module over the complex monoid algebra of the multiplicative monoid of the coefficient ring of the Frobenius bimodule. The additional algebraic structure of modules over monoid algebras allows us to compare the extension properties of weights that are scalar multiples over the monoid algebra. The key observation, dating from [**2**], is that if w' is a scalar multiple of w, both having maximal symmetry, and if the extension theorem holds for w', then the extension theorem holds for w. We apply this idea with w' equaling the homogeneous weight, and the conditions we determine are those that allow us to find a scalar γ such that $w\gamma = w'$.

Here is a short guide to the rest of the paper. Monoid algebras and certain modules are introduced in Section 2. Detailed computations, especially solving $w\gamma = w'$ for γ, are carried out in Section 3. The proofs of the main results are in Section 4. The main theorem is illustrated by an example in Section 5. The paper concludes in Section 6 with an extension of the main theorem to alphabets that can be viewed as pseudo-injective submodules of a Frobenius bimodule.

We thank the referee for helpful comments, especially for suggesting what became Lemma 3.7.

2. Preliminaries

Let R be a finite ring with 1. Denote the group of units of R by $\mathcal{U}$. Let $\widehat{R}$ be the group of complex characters of the additive group of R; i.e., $\widehat{R} = \{\pi : R \to \mathbb{C}^\times : \pi(r_1 + r_2) = \pi(r_1)\pi(r_2), r_1, r_2 \in R\}$. Then $\widehat{R}$ is an abelian group under point-wise multiplication of functions, and $\widehat{R}$ is a bimodule over R under the following scalar multiplications: $({}^s\pi)(r) = \pi(rs)$ and $(\pi^s)(r) = \pi(sr)$, for all $r, s \in R$.

Let A be a Frobenius bimodule over R; i.e., A is a bimodule over R such that ${}_RA \cong {}_R\widehat{R}$ and $A_R \cong \widehat{R}_R$. In particular, $|A| = |R|$. The Frobenius bimodule A admits a character χ that generates $\widehat{A}$ both as a left R-module and as a right R-module. This character χ also has the equivalent properties that $\ker\chi$ contains no nonzero left nor right R-submodules. We will refer to χ as a *generating character* for A. Details concerning these results may be found in [**15**, Section 5.2].

LEMMA 2.1. *If S is a nonzero left or right submodule of A, then $\sum_{s\in S} \chi(s) = 0$.*

PROOF. Because $\ker\chi$ contains no nonzero left or right submodules, there is some $s_0 \in S$ with $\chi(s_0) \neq 1$. Let $t = s + s_0$ be a change of variables in the given sum:

$$\sum_{t\in S}\chi(t)=\sum_{s\in S}\chi(s+s_0)=\sum_{s\in S}\chi(s)\chi(s_0)=\chi(s_0)\sum_{s\in S}\chi(s).$$

Thus, $(1-\chi(s_0))\sum_{s\in S}\chi(s)=0$, so that $\sum_{s\in S}\chi(s)=0$. □

Let $\mathcal{R}$ denote the complex monoid algebra of the multiplicative monoid of R. Explicitly, $\mathcal{R}$ is the vector space of all functions $\mathcal{R}=\{\alpha : R\to\mathbb{C}\}$ from R to the complex numbers $\mathbb{C}$, equipped with the multiplicative convolution product $*$:

$$(\alpha*\beta)(r)=\sum_{st=r}\alpha(s)\beta(t),$$

where the sum is over all pairs $(s,t)\in R\times R$ satisfying $st=r$.

Let $\mathcal{A}=\{w : A\to\mathbb{C}\}$ be the vector space of all functions from A to $\mathbb{C}$. For $w\in\mathcal{A}$ and $\alpha\in\mathcal{R}$, define the (right) multiplicative correlation $w\circledast\alpha\in\mathcal{A}$ by $(w\circledast\alpha)(a)=\sum_{r\in R}w(ra)\alpha(r)$, for $a\in A$. In a similar fashion, one can define a left multiplicative correlation; we will not need the left version in this paper.

Define a *Fourier transform* $\widehat{\ }:\mathcal{R}\to\mathcal{A}$ by $\widehat{\alpha}=\chi\circledast\alpha$; that is, for $a\in A$,

$$\widehat{\alpha}(a)=\sum_{r\in R}\chi(ra)\alpha(r).$$

LEMMA 2.2. *The spaces $\mathcal{R}$ and $\mathcal{A}$ satisfy the following, with $w\in\mathcal{A}$, $\alpha,\beta\in\mathcal{R}$:*

(1) *$\mathcal{R}$ is a finite-dimensional algebra over $\mathbb{C}$; $\dim\mathcal{R}=|R|$.*
(2) *$\mathcal{A}$ is a right $\mathcal{R}$-module under $\circledast$: $w\circledast(\alpha*\beta)=(w\circledast\alpha)\circledast\beta$.*
(3) *$\widehat{\ }:\mathcal{R}\to\mathcal{A}$ is an isomorphism of right $\mathcal{R}$-modules; in particular, $\widehat{\alpha*\beta}=\widehat{\alpha}\circledast\beta$. The inverse sends $w\in\mathcal{A}$ to $\widetilde{w}(r)=(1/|A|)\sum_{a\in A}w(a)\chi(-ra)$.*

PROOF. These are routine verifications. We will show one of them.

$$\begin{aligned}((w\circledast\alpha)\circledast\beta)(a)&=\sum_{t\in R}(w\circledast\alpha)(ta)\beta(t)\\&=\sum_{t\in R}(\sum_{s\in R}w(sta)\alpha(s))\beta(t)\\&=\sum_{r\in R}\sum_{st=r}w(ra)\alpha(s)\beta(t)=(w\circledast(\alpha*\beta))(a)\end{aligned}$$

For the inverse transform claim, see Lemma 3.1. □

We define subspaces of $\mathcal{R}$ and $\mathcal{A}$: let $\mathcal{R}_0=\{\alpha\in\mathcal{R}:\sum_{r\in R}\alpha(r)=0\}$ and $\mathcal{A}_0=\{w\in\mathcal{A}:w(0)=0\}$.

LEMMA 2.3. (1) *For $\alpha,\beta\in\mathcal{R}$,*

$$\sum_{r\in R}(\alpha*\beta)(r)=\left(\sum_{s\in R}\alpha(s)\right)\left(\sum_{t\in R}\beta(t)\right).$$

(2) *For $w\in\mathcal{A}$ and $\alpha\in\mathcal{R}$, $(w\circledast\alpha)(0)=w(0)\sum_{r\in R}\alpha(r)$.*
(3) *For $\alpha\in\mathcal{R}$, $\widehat{\alpha}(0)=\sum_{r\in R}\alpha(r)$.*
(4) *$\mathcal{R}_0$ is a two-sided ideal in $\mathcal{R}$.*
(5) *$\mathcal{A}_0$ is a right $\mathcal{R}$-submodule of $\mathcal{A}$.*
(6) *$\widehat{\ }:\mathcal{R}_0\to\mathcal{A}_0$ is an isomorphism of right $\mathcal{R}$-modules.*

PROOF. Again, these are routine verifications. □

An element $\alpha \in \mathcal{R}$ is *bi-invariant* if $\alpha(urv) = \alpha(r)$ for all $r \in R$ and $u, v \in \mathcal{U}$. Similarly, an element $w \in \mathcal{A}$ is *bi-invariant* if $w(uav) = w(a)$ for all $a \in A$ and $u, v \in \mathcal{U}$.

LEMMA 2.4. *Let $\alpha, \beta \in \mathcal{R}$ and $w \in \mathcal{A}$ all be bi-invariant. Then $\alpha * \beta$, $\widehat{\alpha}$, and $w \circledast \alpha$ are bi-invariant.*

PROOF. We will verify the claim for $w \circledast \alpha$; the others are similar.

$$\begin{aligned}(w \circledast \alpha)(uav) &= \sum_{r\in R} w(ruav)\alpha(r) = \sum_{r\in R} w(rua)\alpha(r) \\ &= \sum_{s\in R} w(sa)\alpha(su^{-1}) = \sum_{s\in R} w(sa)\alpha(s) = (w \circledast \alpha)(a),\end{aligned}$$

where we made use of the change of variables $s = ru$, $u \in \mathcal{U}$. □

3. Some Computations

Suppose $w \in \mathcal{A}_0$ is bi-invariant. By making the change of variable $a \leftrightarrow -a$ and using bi-invariance, the inverse Fourier transform of w is

$$\widetilde{w}(r) = \frac{1}{|A|}\sum_{a\in A} w(a)\chi(ra).$$

LEMMA 3.1. *Suppose $w \in \mathcal{A}_0$ is bi-invariant. Then $\widetilde{w} \in \mathcal{R}_0$ is bi-invariant and $\chi \circledast \widetilde{w} = w$.*

PROOF. Let us begin by verifying that $\widetilde{w} \in \mathcal{R}_0$:

$$\begin{aligned}|A|\sum_{r\in R}\widetilde{w}(r) &= \sum_{r\in R}\sum_{a\in A} w(a)\chi(ra) \\ &= \sum_{a\in A} w(a)\sum_{r\in R}\chi(ra) = |R|w(0) = 0,\end{aligned}$$

where we simplified the summation using Lemma 2.1 for $S = Ra$:

$$\sum_{r\in R}\chi(ra) = \begin{cases}|R|, & a = 0,\\ 0, & a \neq 0.\end{cases}$$

By the change of variable $b = va$, $v \in \mathcal{U}$, one sees that $\widetilde{w}(rv) = \widetilde{w}(r)$. For $u \in \mathcal{U}$, $|A|\widetilde{w}(ur) = \sum_{a\in A} w(a)\chi^u(ra)$. Because u is a unit, χ^u is another generating character for A. As such, it must equal ${}^{u'}\chi$ for some unit $u' \in \mathcal{U}$. Then $|A|\widetilde{w}(ur) = \sum_{a\in A} w(a)\chi(rau')$. Using the change of variable $b = au'$, we conclude that $\widetilde{w}(ur) = \widetilde{w}(r)$.

Finally, using bi-invariance of w so that $w(-b) = w(b)$,

$$\begin{aligned}|A|(\chi \circledast \widetilde{w})(a) &= \sum_{r\in R}\chi(ra)\sum_{b\in A} w(b)\chi(-rb) \\ &= \sum_{b\in A} w(b)\sum_{r\in R}\chi(r(a-b)) = |A|w(a),\end{aligned}$$

where we used $|A| = |R|$ and Lemma 2.1 with $S = R(a-b)$. □

A particular example of a bi-invariant weight on a Frobenius bimodule is the homogeneous weight w_{Hom}. The existence of the homogeneous weight on any Frobenius bimodule was proved in [**6**] and [**9**]. We will need the features of the homogeneous weight listed in the next theorem.

THEOREM 3.2 ([**8**]). *Suppose w_{Hom} is the homogeneous weight on a Frobenius bimodule A. Then*

(1) *w_{Hom} is expressible in terms of the generating character χ of A:*

$$w_{\mathrm{Hom}}(a) = 1 - \frac{1}{|\mathcal{U}|} \sum_{u \in \mathcal{U}} \chi(ua);$$

(2) *w_{Hom} is bi-invariant.*

Define $\varepsilon \in \mathcal{R}_0$ by

$$\varepsilon(r) = \begin{cases} -1/|\mathcal{U}|, & r \in \mathcal{U}, \\ 1, & r = 0, \\ 0, & \text{otherwise.} \end{cases}$$

LEMMA 3.3. *The Fourier transform of ε is w_{Hom}, the homogeneous weight of A. That is, $\chi \circledast \varepsilon = w_{\mathrm{Hom}}$.*

PROOF. Calculate and make use of Theorem 3.2:

$$\begin{aligned}(\chi \circledast \varepsilon)(a) &= \sum_{r \in R} \chi(ra)\varepsilon(r) \\ &= 1 - (1/|\mathcal{U}|) \sum_{u \in \mathcal{U}} \chi(ua) = w_{\mathrm{Hom}}(a).\end{aligned}$$ □

We quote the following lemma:

LEMMA 3.4 ([**13**, Prop. 5.1]). *For cyclic right submodules of A, $aR = bR$ if and only if $a\mathcal{U} = b\mathcal{U}$.*

The set $P = \{aR : a \in A\}$ of all cyclic right submodules of A is a partially ordered set (poset) under inclusion $\subseteq$. Denote the Möbius function of P by μ. Then μ is characterized by the following properties:

- $\mu(aR, cR) = 0$ when $aR \not\subseteq cR$;
- $\mu(aR, aR) = 1$;
- $\sum_{aR \subseteq bR \subseteq cR} \mu(aR, bR) = 0$, where bR varies for fixed $aR \subsetneq cR$.

For further details about posets and their Möbius functions, see [**12**].

The next result appears in [**8**].

LEMMA 3.5. *The Möbius function of P has $\mu(0, aR) = \sum_{b \in a\mathcal{U}} \chi(b)$.*

PROOF. Define $g : P \to \mathbb{C}$ by $g(aR) = \sum_{b \in a\mathcal{U}} \chi(b)$. We verify that g satisfies the properties of $\mu(0, \cdot)$.

Certainly $g(0) = \chi(0) = 1$. For fixed $cR \neq 0$, use Lemma 2.1:

$$\sum_{bR \subseteq cR} g(bR) = \sum_{b\mathcal{U} \subseteq cR} \sum_{a \in b\mathcal{U}} \chi(a) = \sum_{a \in cR} \chi(a) = 0.$$ □

Suppose $w \in \mathcal{A}_0$ is bi-invariant. Consider w that satisfy:

$$\sum_{aR \subseteq S} w(a)\mu(0, aR) \neq 0, \tag{3.1}$$

for all nonzero right submodules $S \subseteq A$.

LEMMA 3.6. *Let $w \in \mathcal{A}_0$ be bi-invariant. Then* (3.1) *is equivalent to*

$$\sum_{a \in S} w(a)\chi(a) \neq 0, \quad \textit{for all nonzero right submodules } S \subseteq A.$$

PROOF. The group of units $\mathcal{U}$ acts on A on the right by scalar multiplication. Any right submodule S of A is invariant under this action. Thus S is the disjoint union of right $\mathcal{U}$-orbits. Break up the sum over S into the sum over these right $\mathcal{U}$-orbits. Then, by Lemma 3.5,

$$\begin{aligned}\sum_{a \in S} w(a)\chi(a) &= \sum_{a\mathcal{U} \subseteq S} \sum_{b \in a\mathcal{U}} w(b)\chi(b) = \sum_{a\mathcal{U} \subseteq S} w(a) \sum_{b \in a\mathcal{U}} \chi(b) \\ &= \sum_{aR \subseteq S} w(a)\mu(0, aR).\end{aligned}$$

□

For later use in the proof of Lemma 3.8, we examine factorizations $st = r$ with $Rr = Rt$. Fix $r \in R$. Suppose $st = r$ and $Rr = Rt$. Lemma 3.4 implies that $t = ur$ for some $u \in \mathcal{U}$. Then $r = st = sur$, so that $(su - 1)r = 0$. That is, $su - 1 \in \mathrm{ann}_{\mathrm{lt}}(r)$, where $\mathrm{ann}_{\mathrm{lt}}(r) = \{q \in R : qr = 0\}$ is the left annihilator of r in R; $\mathrm{ann}_{\mathrm{lt}}(r)$ is a left ideal in R. In summary, if $st = r$ and $Rr = Rt$, then there exist $u \in \mathcal{U}$ and $q \in \mathrm{ann}_{\mathrm{lt}}(r)$ with $s = (1+q)u^{-1}$ and $t = ur$. Conversely, if $u \in \mathcal{U}$ and $q \in \mathrm{ann}_{\mathrm{lt}}(r)$, then setting $s = (1+q)u^{-1}$ and $t = ur$ yields a factorization $st = r$ with $Rr = Rt$.

It is possible that different choices of u and q may yield the same factors s and t. The next lemma gives a precise description of the possibilities. Fix $r \in R$. Define $\mathcal{F}(r) = \{(s,t) \in R \times R : st = r \text{ and } Rr = Rt\}$, the set of all factorizations of r with $Rr = Rt$. Define $\mathrm{Stab}(r) = \{u \in \mathcal{U} : ur = r\}$, the stabilizer subgroup of r under the left action of $\mathcal{U}$ on R. Define

$$F : \mathrm{ann}_{\mathrm{lt}}(r) \times \mathcal{U} \to \mathcal{F}(r), \quad F(q, u) = ((1+q)u^{-1}, ur).$$

LEMMA 3.7. *The function $F : \mathrm{ann}_{\mathrm{lt}}(r) \times \mathcal{U} \to \mathcal{F}(r)$ is onto and, for every $(s,t) \in \mathcal{F}(r)$, $|F^{-1}(s,t)| = |\mathrm{Stab}(r)|$.*

PROOF. The discussion in the preceding paragraphs shows that F is onto. For any $(s,t) \in \mathcal{F}(r)$, pick one point $(q_0, u_0) \in F^{-1}(s,t)$. If (q,u) is any other point in $F^{-1}(s,t)$, then $s = (1+q_0)u_0^{-1} = (1+q)u^{-1}$ and $t = u_0 r = ur$. From the latter equation, we see that $u_0^{-1}u \in \mathrm{Stab}(r)$.

Define $\phi : F^{-1}(s,t) \to \mathrm{Stab}(r)$ by $\phi(q,u) = u_0^{-1}u$, and define $\psi : \mathrm{Stab}(r) \to F^{-1}(s,t)$ by $\psi(z) = ((1+q_0)z - 1, u_0 z)$. One readily verifies that ϕ and ψ are inverse functions. □

The next computation is the key technical result in the paper.

LEMMA 3.8. *Suppose $w \in \mathcal{A}_0$ is bi-invariant and satisfies* (3.1) *for all nonzero right submodules $S \subseteq A$. Then there exists $\gamma \in \mathcal{R}_0$ that is bi-invariant and satisfies $\widetilde{w} * \gamma = \varepsilon$.*

PROOF. We define γ in stages, starting on the group of units $\mathcal{U}$ in R, by solving the equation

$$\varepsilon(r) = \sum_{st=r} \widetilde{w}(s)\gamma(t). \tag{3.2}$$

If $r = u \in \mathcal{U}$, then $st = u$ implies that s and t are in $\mathcal{U}$. Equation (3.2) then becomes $-1/|\mathcal{U}| = \sum_{t\in\mathcal{U}} \widetilde{w}(rt^{-1})\gamma(t) = |\mathcal{U}|\widetilde{w}(1)\gamma(1)$. We have used the bi-invariance of $\widetilde{w}$ and the desired bi-invariance of γ. Note that $\widetilde{w}(1) \neq 0$, this being the $S = A$ case of (3.1). Thus, by defining $\gamma(t) = \gamma(1) = -1/(|\mathcal{U}|^2\widetilde{w}(1))$, we solve (3.2) when $r = u$, and that portion of γ that has been defined is bi-invariant.

As a recursive step, suppose a solution γ has been defined for various $r \in R$ such that γ is bi-invariant; i.e., if $\gamma(r)$ is defined, then $\gamma(ur)$ and $\gamma(ru)$ are also defined for all $u \in \mathcal{U}$, and $\gamma(ur) = \gamma(ru) = \gamma(r)$ for all $u \in \mathcal{U}$. Now let $r \in R$ be any element, neither a unit nor 0, such that Rr is maximal among all principal left ideals of R where γ is not defined on $\mathcal{U}r$. Consider (3.2) for this choice of r.

Any solution of the equation $st = r$ has the property that $Rr \subseteq Rt$. If $Rr \subsetneq Rt$, then the maximality property of Rr implies that $\gamma(t)$ is already defined. Then (3.2) becomes

$$0 = \sum_{\substack{st=r\\ Rr\subsetneq Rt}} \widetilde{w}(s)\gamma(t) + \sum_{\substack{st=r\\ Rr=Rt}} \widetilde{w}(s)\gamma(t).$$

We analyze the sum where $Rr = Rt$. Using Lemma 3.7 and the bi-invariance of $\widetilde{w}$ and γ, we see that

$$\begin{aligned}|\mathrm{Stab}(r)| \sum_{\substack{st=r\\ Rr=Rt}} \widetilde{w}(s)\gamma(t) &= \sum_{q\in \mathrm{ann}_{\mathrm{lt}}(r), u\in\mathcal{U}} \widetilde{w}((1+q)u^{-1})\gamma(ur)\\ &= |\mathcal{U}|\gamma(r) \sum_{q\in \mathrm{ann}_{\mathrm{lt}}(r)} \widetilde{w}(1+q).\end{aligned}$$

The last summation simplifies in the following way:

$$\begin{aligned}|A| \sum_{q\in \mathrm{ann}_{\mathrm{lt}}(r)} \widetilde{w}(1+q) &= \sum_{q\in \mathrm{ann}_{\mathrm{lt}}(r)} \sum_{a\in A} w(a)\chi((1+q)a)\\ &= \sum_{a\in A} w(a)\chi(a) \sum_{q\in \mathrm{ann}_{\mathrm{lt}}(r)} \chi(qa).\end{aligned}$$

The summation $\sum_{q\in \mathrm{ann}_{\mathrm{lt}}(r)} \chi(qa)$ is a sum over the left submodule $\mathrm{ann}_{\mathrm{lt}}(r)a \subseteq A$. Because χ is a generating character for A, this sum vanishes (Lemma 2.1) unless the submodule $\mathrm{ann}_{\mathrm{lt}}(r)a = 0$, in which case the sum equals $|\mathrm{ann}_{\mathrm{lt}}(r)|$.

For $r \in R$, define $S_r = \{a \in A : \mathrm{ann}_{\mathrm{lt}}(r)a = 0\}$; S_r is a right submodule of A. Continuing the simplification from above, we see that

$$|A| \sum_{q\in \mathrm{ann}_{\mathrm{lt}}(r)} \widetilde{w}(1+q) = |\mathrm{ann}_{\mathrm{lt}}(r)| \sum_{a\in S_r} w(a)\chi(a).$$

By the $S = S_r$ case of (3.1) and Lemma 3.6, we see that $\sum_{a\in S_r} w(a)\chi(a) \neq 0$ and hence $\sum_{q\in \mathrm{ann}_{\mathrm{lt}}(r)} \widetilde{w}(1+q) \neq 0$. This allows us to solve for $\gamma(r)$ in terms of previously defined values of γ:

$$\gamma(r) = -\left(|\mathrm{Stab}(r)| \sum_{\substack{st=r\\ Rr\subsetneq Rt}} \widetilde{w}(s)\gamma(t)\right) \Bigg/ \left(|\mathcal{U}| \sum_{q\in \mathrm{ann}_{\mathrm{lt}}(r)} \widetilde{w}(1+q)\right).$$

This formula gives the same value for $\gamma(ru)$, $u \in \mathcal{U}$, and the entire derivation is the same if r is replaced by ur. Thus γ is bi-invariant.

Proceeding recursively, we eventually arrive at the case $r = 0$. Then (3.2) becomes

$$1 = \sum_{\substack{st=0 \\ t \neq 0}} \widetilde{w}(s)\gamma(t) + \sum_{s \in R} \widetilde{w}(s)\gamma(0).$$

Because $\widetilde{w} \in \mathcal{R}_0$, the coefficient $\sum_{s\in R} \widetilde{w}(s)$ of $\gamma(0)$ vanishes. Thus, the equation puts no restriction on $\gamma(0)$. By setting $\gamma(0) = -\sum_{r\neq 0} \gamma(r)$, we have $\gamma \in \mathcal{R}_0$.

Why does $(\widetilde{w} * \gamma)(0) = 1$? We know $\widetilde{w} \in \mathcal{R}_0$, and we have already shown that $\gamma \in \mathcal{R}_0$ and $(\widetilde{w} * \gamma)(r) = \varepsilon(r)$ for all $r \neq 0$. Since $\widetilde{w} * \gamma$ and ε belong to $\mathcal{R}_0$, we conclude that their values at $r = 0$ also agree. □

COROLLARY 3.9. *Suppose $w \in \mathcal{A}_0$ is bi-invariant and satisfies* (3.1) *for all nonzero right submodules $S \subseteq A$. Then $w \circledast \gamma = w_{\mathrm{Hom}}$.*

PROOF. Use Lemmas 2.2, 3.1, 3.3, and 3.8. Then, $w \circledast \gamma = (\chi \circledast \widetilde{w}) \circledast \gamma = \chi \circledast (\widetilde{w} * \gamma) = \chi \circledast \varepsilon = w_{\mathrm{Hom}}$. □

4. The Main Theorem

As in earlier sections, R is a finite ring with 1 and A is a Frobenius bimodule over R. An *R-linear code* over A of length n is a left R-submodule $C \subseteq A^n$. A function $w \in \mathcal{A}_0$ defines a *weight* on A^n (and, by restriction, on C) by $w(x) = \sum_{i=1}^n w(x_i)$, where $x = (x_1, x_2, \ldots, x_n) \in A^n$. An injective R-linear homomorphism $f : C \to A^n$ is a *w-isometry* if $w(f(x)) = w(x)$ for all $x \in C$.

PROPOSITION 4.1. *Suppose $f : C \to A^n$ is a w-isometry. Then f is a $w \circledast \alpha$-isometry for any $\alpha \in \mathcal{R}$.*

PROOF. For $x \in C$,

$$\begin{aligned}(w \circledast \alpha)(f(x)) &= \sum_{r\in R} w(rf(x))\alpha(r) = \sum_{r\in R} w(f(rx))\alpha(r) \\ &= \sum_{r\in R} w(rx)\alpha(r) = (w \circledast \alpha)(x).\end{aligned}$$ □

For units $u_1, u_2, \ldots, u_n \in \mathcal{U}$ and a permutation σ of $\{1, 2, \ldots, n\}$, the left R-module homomorphism $T : A^n \to A^n$ defined by

$$T(x_1, x_2, \ldots, x_n) = (x_{\sigma(1)}u_1, x_{\sigma(2)}u_2, \ldots, x_{\sigma(n)}u_n),$$

$(x_1, x_2, \ldots, x_n) \in A^n$, is called a *monomial transformation* of A^n. If w is bi-invariant, then any monomial transformation is a w-isometry.

DEFINITION 4.2. A bi-invariant weight w on a Frobenius bimodule A has the *extension property* if the following holds: for any linear code $C \subseteq A^n$ and any w-isometry $f : C \to A^n$, the isometry f extends to a monomial transformation of A^n.

An example of a bi-invariant weight that has the extension property is the homogeneous weight w_{Hom}:

THEOREM 4.3 ([**6**, Theorem 4.10]). *Any Frobenius bimodule A has the extension property with respect to w_{Hom}.*

THEOREM 4.4. *Suppose A is a Frobenius bimodule over R and w is a bi-invariant weight on A. If w satisfies (3.1) for all nonzero right submodules $S \subseteq A$, then w has the extension property.*

PROOF. Suppose $C \subseteq A^n$ is a linear code and $f : C \to A^n$ is a w-isometry. Then f is an isometry with respect to $w \circledast \gamma = w_{\mathrm{Hom}}$, Corollary 3.9 and Proposition 4.1. By Theorem 4.3, w_{Hom} has the extension property, so f extends to a monomial transformation, as desired. □

COROLLARY 4.5. *Suppose R is a finite Frobenius ring and w is a bi-invariant weight on R. If w satisfies (3.1) for all nonzero right ideals $S \subseteq R$, then w has the extension property.*

REMARK 4.6. Corollary 4.5 generalizes to all finite Frobenius rings the condition proved for finite principal ideal rings in [**4**, Theorem 4.4].

5. Example

We illustrate Theorem 4.4 with an example.

EXAMPLE 5.1. Let $R = U_2(\mathbb{F}_2)$ be the ring of 2×2 upper-triangular matrices over $\mathbb{F}_2$. The ring R is an algebra over $\mathbb{F}_2$ of dimension 3, so that $|R| = 8$. When $r \in R$ is the matrix

$$r = \begin{bmatrix} a & b \\ 0 & c \end{bmatrix},$$

we will denote the element r with the notation $r = [a, b, c]$. It is known that the ring R is not Frobenius.

The Frobenius bimodule $A = \widehat{R}$ is a vector space over $\mathbb{F}_2$ of dimension 3. For $\pi \in A$, we will use the notation $\pi = (x, y, z)$, where the evaluation $\pi(r)$, $r = [a, b, c] \in R$, is

$$\pi(r) = (-1)^{ax+by+cz}.$$

The following formulas describe the scalar multiplications on the bimodule A over R. As above, $r = [a, b, c]$ and $\pi = (x, y, z)$:

$$\begin{aligned} {}^r\pi &= (ax + by, cy, cz), \\ \pi^r &= (ax, ay, by + cz). \end{aligned}$$

In order to understand (3.1), we need to know the right submodules of A. For $\pi = (x, y, z) \in A$, we will denote the right submodule generated by π as $(x, y, z)R$. By straight-forward computations using the formulas above, we have the following list of right submodules of A. The module A is not a cyclic module, while all its proper submodules are cyclic.

dim 0: $0 = (0, 0, 0)R$;
dim 1: $(0, 0, 1)R$, $(1, 0, 0)R$;
dim 2: $(0, 1, 0)R = (0, 1, 1)R$, $(1, 0, 1)R$, $(1, 1, 0)R = (1, 1, 1)R$;
dim 3: A itself.

Also, $(1, 0, 0)R \subset (1, 0, 1)R$, and $(0, 0, 1)R$ is contained in every submodule of dimension 2. See Figure 5.1, where some containments are dotted because A is not itself a cyclic right module.

FIGURE 5.1. The cyclic right submodules of A.

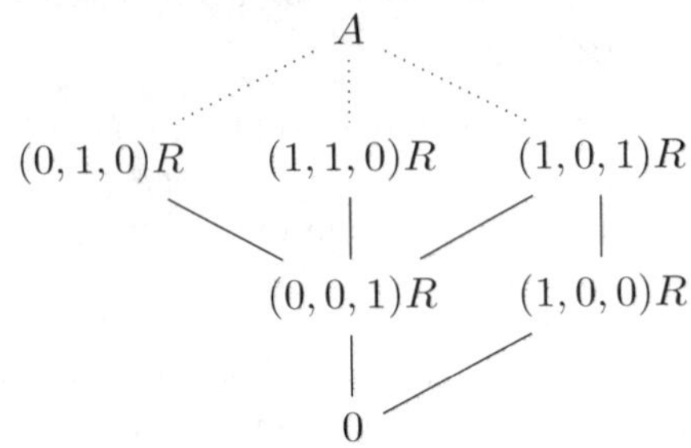

By using the containments just described, we calculate that the values of the Möbius function $\mu(0, aR)$ are as follows:

a	0	$(0,0,1)$	$(1,0,0)$	$(0,1,0)$	$(1,0,1)$	$(1,1,0)$
$\mu(0,aR)$	1	-1	-1	0	1	0

Theorem 4.4 applies to bi-invariant weights. The description above of the right cyclic submodules implies that $(0,1,0)\mathcal{U} = (0,1,1)\mathcal{U}$ and $(1,1,0)\mathcal{U} = (1,1,1)\mathcal{U}$, with all other right $\mathcal{U}$-orbits being distinct. Similar computations of left $\mathcal{U}$-orbits show that all the left $\mathcal{U}$-orbits are distinct except for $\mathcal{U}(0,1,0) = \mathcal{U}(1,1,0)$ and $\mathcal{U}(0,1,1) = \mathcal{U}(1,1,1)$. Thus, a weight w on A is bi-invariant if and only if

$$w(0,1,0) = w(0,1,1) = w(1,1,0) = w(1,1,1).$$

Theorem 4.4 says that a bi-invariant weight w has the extension property when the following expressions are nonzero:

right module S	expression in (3.1)
$(0,0,1)R$	$-w(0,0,1)$
$(1,0,0)R$	$-w(1,0,0)$
$(0,1,0)R$	$-w(0,0,1)$
$(1,0,1)R$	$-w(0,0,1) - w(1,0,0) + w(1,0,1)$
$(1,1,0)R$	$-w(0,0,1)$
A	$-w(0,0,1) - w(1,0,0) + w(1,0,1)$

In particular, the value of $w(0,1,0)$ is irrelevant.

6. Other Alphabets

In this section we consider the extension problem for certain other alphabets.

Suppose R is a finite ring with 1 and A is a finite unital left R-module. Let w be a weight on A, i.e., a function $w : A \to \mathbb{C}$ with $w(0) = 0$. Then w extends to a weight on A^n by $w(x) = \sum_{i=1}^{n} w(x_i)$, where $x = (x_1, x_2, \ldots, x_n) \in A^n$. The group of all invertible homomorphisms $\phi : A \to A$ is denoted $\mathrm{GL}_R(A)$, and ϕ will be written with inputs on the left, so that $(ra)\phi = r(a\phi)$ for all $r \in R$ and $a \in A$. The *right symmetry group* of w on A is

$$G_{\mathrm{rt}} = \{\phi \in \mathrm{GL}_R(A) : w(a\phi) = w(a), a \in A\}.$$

A G_{rt}*-monomial transformation* of A^n is any invertible homomorphism $T : A^n \to A^n$ of the form

$$(x_1, x_2, \ldots, x_n)T = (x_{\sigma(1)}\phi_1, x_{\sigma(2)}\phi_2, \ldots, x_{\sigma(n)}\phi_n),$$

for $(x_1, x_2, \ldots, x_n) \in A^n$, where σ is a permutation of $\{1, 2, \ldots, n\}$ and $\phi_i \in G_{\mathrm{rt}}$. Observe that a G_{rt}-monomial transformation preserves w: $w(xT) = w(x)$ for all $x \in A^n$.

A weight w on A has the *extension property* if, for any left R-linear code (submodule) $C \subseteq A^n$ and any injective homomorphism $f : C \to A^n$ of left R-modules that preserves w, $w(c) = w(f(c))$, $c \in C$, it follows that f extends to a G_{rt}-monomial transformation of A^n.

We will prove that the extension property holds for certain alphabets. Let A be a finite unital left R-module that is pseudo-injective and such that its socle $\mathrm{soc}(A)$ is a cyclic left R-module. Recall that a left R-module is *pseudo-injective* if for any left R-submodule $B \subseteq A$ and any injective homomorphism $\phi : B \to A$ of left R-modules, ϕ extends to a homomorphism $\tilde{\phi} : A \to A$ of left R-modules. It follows from [**1**, Proposition 3.2] that there exists an extension $\tilde{\phi}$ that is injective, hence invertible. The *socle* $\mathrm{soc}(A)$ of A is the left submodule generated by all simple left R-submodules of A. The hypothesis that $\mathrm{soc}(A)$ is cyclic implies that A injects into $\widehat{R}$ as a left R-module [**15**, Proposition 5.3].

THEOREM 6.1. *Suppose R is a finite ring with 1 and A is a finite unital left R-module that is pseudo-injective and has $\mathrm{soc}(A)$ cyclic. Fix an injection of A into $\widehat{R}$, and view A as a left submodule of $\widehat{R}$. Let w be a weight on A, and assume that w is the restriction to A of a bi-invariant weight on $\widehat{R}$ that satisfies* (3.1). *Then the weight w on the alphabet A satisfies the extension property.*

Before proving Theorem 6.1 we state the following lemma.

LEMMA 6.2. *Suppose R is a finite ring with 1, I is a finite injective left module over R, and $A \subseteq I$ is a left R-submodule. Then every element of $\mathrm{GL}_R(A)$ extends to an element of $\mathrm{GL}_A(I)$.*

PROOF. This is essentially [**1**, Proposition 3.2], discussed above. Any injective module is, a fortiori, pseudo-injective. Injections $A \to A \subseteq I$ then extend to injections $I \to I$. □

PROOF OF THEOREM 6.1. Suppose $C \subseteq A^n$ is a left R-linear code and $f : C \to A^n$ is an injective homomorphism that preserves w. Because $A \subseteq \widehat{R}$, we can view $C \subseteq \widehat{R}^n$ as an R-linear code over the alphabet $\widehat{R}$ and $f : C \to \widehat{R}^n$; f still preserves w, now viewed as a bi-invariant weight on $\widehat{R}$ that satisfies (3.1). By Theorem 4.4, f extends to a monomial transformation $T : \widehat{R}^n \to \widehat{R}^n$.

Write the components of $f : C \to A^n$ as $f = (f_1, f_2 \ldots, f_n)$, $f_i : C \to A$. The fact that f extends to the monomial transformation T means there exist a permutation σ of $\{1, 2, \ldots, n\}$ and units $u_i \in \mathcal{U}(R)$ such that $f_i(x) = x_{\sigma(i)} u_i$, for $x = (x_1, x_2, \ldots, x_n) \in C$. Because $f : C \to A^n$, this means that $x_{\sigma(i)} u_i \in A$ for $x \in C$.

Let $C_i = \{x_i : x \in C\}$ be the projection of $C \subseteq A^n$ on the ith coordinate; $C_i \subseteq A$ is an R-submodule. The extendability of f implies that right multiplication by u_i injects $C_{\sigma(i)} \to A$. Because A is pseudo-injective, this injection extends to an injection $A \to A$, i.e., to an element $\phi_i \in \mathrm{GL}_R(A)$. Thus, $f_i(x) = x_{\sigma(i)} \phi_i$, $x \in C$.

The last thing we need to prove is that $\phi_i \in G_{\mathrm{rt}}$, i.e., that ϕ_i preserves w on A. This follows from Lemma 6.2. Indeed, $\widehat{R}$ is always an injective module, so $\phi_i \in \mathrm{GL}_R(A)$ extends to an element of $\mathrm{GL}_R(\widehat{R})$. But $\mathrm{GL}_R(\widehat{R}) = \mathcal{U}(R)$, and

w is bi-invariant as a weight on $\widehat{R}$, so every element of $\mathrm{GL}_R(\widehat{R})$ preserves w. By restriction, every element of $\mathrm{GL}_R(A)$ preserves w on A. $\square$

References

[1] H. Q. Dinh and S. R. López-Permouth, *On the equivalence of codes over finite rings*, Appl. Algebra Engrg. Comm. Comput. **15** (2004), no. 1, 37–50, DOI 10.1007/s00200-004-0149-5. MR2142429

[2] M. Greferath and T. Honold, *On weights allowing for MacWilliams equivalence theorem*, Proceedings of the Fourth International Workshop in Optimal Codes and Related Topics (Pamporovo, Bulgaria), 2005, pp. 182–192.

[3] M. Greferath and T. Honold, *Monomial extensions of isometries of linear codes II: Invariant weight functions on Z_m*, Proceedings of the Tenth International Workshop in Algebraic and Combinatorial Coding Theory (ACCT-10) (Zvenigorod, Russia), 2006, pp. 106–111.

[4] M. Greferath, T. Honold, C. Mc Fadden, J. A. Wood, and J. Zumbrägel, *MacWilliams' extension theorem for bi-invariant weights over finite principal ideal rings*, J. Combin. Theory Ser. A **125** (2014), 177–193, DOI 10.1016/j.jcta.2014.03.005. MR3207471

[5] M. Greferath, C. Mc Fadden, and J. Zumbrägel, *Characteristics of invariant weights related to code equivalence over rings*, Des. Codes Cryptogr. **66** (2013), no. 1-3, 145–156, DOI 10.1007/s10623-012-9671-9. MR3016561

[6] M. Greferath, A. Nechaev, and R. Wisbauer, *Finite quasi-Frobenius modules and linear codes*, J. Algebra Appl. **3** (2004), no. 3, 247–272, DOI 10.1142/S0219498804000873. MR2096449

[7] M. Greferath and S. E. Schmidt, *Finite-ring combinatorics and MacWilliams' equivalence theorem*, J. Combin. Theory Ser. A **92** (2000), no. 1, 17–28, DOI 10.1006/jcta.1999.3033. MR1783936

[8] T. Honold, *Characterization of finite Frobenius rings*, Arch. Math. (Basel) **76** (2001), no. 6, 406–415, DOI 10.1007/PL00000451. MR1831096

[9] A. A. Nechaev and T. Khonol′d, *Fully weighted modules and representations of codes* (Russian, with Russian summary), Problemy Peredachi Informatsii **35** (1999), no. 3, 18–39; English transl., Problems Inform. Transmission **35** (1999), no. 3, 205–223. MR1730800

[10] J. MacWilliams, *Error-correcting codes for multiple-level transmission*, Bell System Tech. J. **40** (1961), 281–308, DOI 10.1002/j.1538-7305.1961.tb03986.x. MR0141541

[11] F. J. MacWilliams, *COMBINATORIAL PROBLEMS OF ELEMENTARY ABELIAN GROUPS*, ProQuest LLC, Ann Arbor, MI, 1962. Thesis (Ph.D.)–Radcliffe College. MR2939359

[12] R. P. Stanley, *Enumerative combinatorics. Vol. I*, The Wadsworth & Brooks/Cole Mathematics Series, Wadsworth & Brooks/Cole Advanced Books & Software, Monterey, CA, 1986. With a foreword by Gian-Carlo Rota. MR847717

[13] J. A. Wood, *Duality for modules over finite rings and applications to coding theory*, Amer. J. Math. **121** (1999), no. 3, 555–575. MR1738408

[14] J. A. Wood, *Weight functions and the extension theorem for linear codes over finite rings*, Finite fields: theory, applications, and algorithms (Waterloo, ON, 1997), Contemp. Math., vol. 225, Amer. Math. Soc., Providence, RI, 1999, pp. 231–243, DOI 10.1090/conm/225/03225. MR1650644

[15] J. A. Wood, *Foundations of linear codes defined over finite modules: the extension theorem and the MacWilliams identities*, Codes over rings, Ser. Coding Theory Cryptol., vol. 6, World Sci. Publ., Hackensack, NJ, 2009, pp. 124–190, DOI 10.1142/9789812837691_0004. MR2850303

Department of Mathematics and Systems Analysis, Aalto University, P.O. Box 11100, FI-00076 Aalto, Finland
Email address: oliver.gnilke@aalto.fi

Department of Mathematics and Systems Analysis, Aalto University, P.O. Box 11100, FI-00076 Aalto, Finland
Email address: marcus.greferath@aalto.fi

ZJU-UIUC Institute, Zhejiang University, 718 East Haizhou Road, Haining, 314400, China
Email address: honold@zju.edu.cn

Department of Mathematics, Western Michigan University, 1903 W. Michigan Avenue, Kalamazoo, Michigan 49008-5248
Email address: jay.wood@wmich.edu

Faculty of Computer Science and Mathematics, University of Passau, Innstrasse 33, 94032 Passau, Germany
Email address: jens.zumbraegel@uni-passau.de

Contemporary Mathematics
Volume **727**, 2019
https://doi.org/10.1090/conm/727/14630

Dual skew codes from annihilators: Transpose Hamming ring extensions

José Gómez-Torrecillas, F. J. Lobillo, and Gabriel Navarro

Introduction

Linear codes may be endowed with cyclic structures by means of skew polynomial rings. This is the case of Piret cyclic convolutional codes [**26**] and the subsequent generalizations and alternatives (see [**27**], [**14**], [**11**], [**25**], [**16**], [**20**]). Non commutative cyclic structures of this kind have been also considered for block linear codes ([**7**], [**5**], [**4**], [**12**], [**1**]), and for linear codes over commutative rings ([**6**], [**22**], [**10**]).

A desirable property of any class of linear codes is to be stable under duals. This property has been already studied in several of the aforementioned references. A common feature of many of these approaches to duality is the presence of a suitable anti-isomorphism of rings that encodes, more or less explicitly, the transfer of the cyclic structure from the code to its dual. Our aim is to present a systematization of this method, besides some relevant examples where it successfully applies.

The strategy is to establish a formal framework, called transpose Hamming ring extension, designed to derive that the dual of every cyclic code is cyclic. Cyclic codes will be, from an algebraic point of view, identified as left ideals of suitable (non-commutative) ring extensions of a given commutative ring C, well understood that such an "identification" has to be made explicit by an isomorphism of C–modules from the ring to C^m, where m is the length of the C–linear code. The transposition will be an anti-isomorphism of rings which allows to transform annihilators into duals. Details are to be found in Section 1

In Section 2 we apply our general approach to left ideal convolutional codes in the sense of [**25**], extending to a more general setting, and improving, results from [**25**] and [**16**] on the description of dual codes in this setting (Theorem 11). The case of a simple word-ambient algebra is analyzed in detail (Theorem 23).

Section 3 is devoted to dual codes of skew constacyclic codes over a commutative ring. Several results from [**7**], [**6**], [**22**], [**10**] on these codes are covered by our general result (Theorem 29).

Finally, in Section 4, we compute (Theorem 37) the dual of a skew Reed Solomon code over a general field in the sense of [**18**], and, as a consequence, these codes are shown to be evaluation codes (Theorem 39).

Research supported by grants MTM2013-41992-P from Ministerio de Economía y Competitividad and FEDER, and MTM2016-78364-P from Agencia Estatal de Investigación and FEDER.

1. Transpose ring extensions and dual codes

Let C be a commutative ring. A C–linear code of length m is, by definition, a C–submodule $\mathcal{C}$ of C^m. The *dual* of $\mathcal{C}$ is defined as

$$\mathcal{C}^\perp = \{w \in C^m : wv^{\mathsf{T}} = 0, \forall v \in \mathcal{C}\},$$

where M^{T} denotes the transpose of a matrix M with coefficients in C.

DEFINITION 1. A *Hamming ring extension of rank* m is a three-tuple $(C, R, \mathfrak{v})$, where R is a ring, C is a commutative subring of R and $\mathfrak{v} : R \to C^m$ is an isomorphism of C–modules. Here, the C–module structure of R is given by left multiplication.

Hamming ring extensions encode cyclic structures on some C–linear codes, according to the following definition. Precise examples will be examined later.

DEFINITION 2. Every $f \in R$ leads to a C–linear code $\mathcal{C} = \mathfrak{v}(Rf)$. We say then that $\mathcal{C}$ is the $(C, R, \mathfrak{v})$*–cyclic code generated by* f. We will also say that $\mathcal{C}$ is R*–cyclic.*

Given an R–cyclic code $\mathcal{C} = \mathfrak{v}(Rf)$, we have the multiplication map $\cdot f : R \to R$. There is a unique square matrix $M_R(f) \in \mathcal{M}_m(C)$ making commute the following diagram of C–module morphisms

$$\begin{array}{ccc} R & \xrightarrow{\cdot f} & R \\ {\scriptstyle \mathfrak{v}}\downarrow & & \downarrow{\scriptstyle \mathfrak{v}} \\ C^m & \xrightarrow[\cdot M_R(f)]{} & C^m, \end{array}$$

so

$$\mathcal{C} = \operatorname{im}(\cdot M_R(f)).$$

A straightforward computation shows that

$$\mathcal{C}^\perp = \ker(\cdot M_R(f)^{\mathsf{T}}). \tag{1}$$

The map $M_R : R \to \mathcal{M}_m(C)$ sending f onto $M_R(f)$ is an injective homomorphism of rings.

DEFINITION 3. Two Haming ring extensions $(C, R, \mathfrak{v})$ and $(C, \widehat{R}, \widehat{\mathfrak{v}})$ are said to be *transposed* if there exists an anti-isomorphism of rings $\Theta : R \to \widehat{R}$ such that the diagram

$$\begin{array}{ccc} R & \xrightarrow{M_R} & \mathcal{M}_m(C)\ , \\ {\scriptstyle \Theta}\downarrow & & \downarrow{\scriptstyle (-)^{\mathsf{T}}} \\ \widehat{R} & \xrightarrow{M_{\widehat{R}}} & \mathcal{M}_m(C) \end{array}$$

is commutative. Equivalently if,

$$M_{\widehat{R}}(\Theta(f)) = M_R(f)^{\mathsf{T}},\ \forall f \in R.$$

We say also that Θ is a *transposition* from $(C, R, \mathfrak{v})$ to $(C, \widehat{R}, \widehat{\mathfrak{v}})$.

Given a subset X of a ring S, the *left annihilator* and the *right annihilator* of X are defined, respectively, by

$$\mathsf{l}_S(X) = \{s \in S : sx = 0, \forall x \in X\}, \quad \mathsf{r}_S(X) = \{s \in S : xs = 0, \forall x \in X\}.$$

The first of these sets is a left ideal of S, while the second one is a right ideal.

THEOREM 4. *Let $(C, R, \mathfrak{v})$ and $(C, \widehat{R}, \widehat{\mathfrak{v}})$ be transpose Hamming ring extensions with transposition $\Theta : R \to \widehat{R}$. Let $f, h \in R$ and $\mathcal{C} = \mathfrak{v}(Rf)$. Then $hR = \mathsf{r}_R(Rf)$ if and only if $\mathcal{C}^\perp = \widehat{\mathfrak{v}}(\widehat{R}\Theta(h))$. In this case, $\mathcal{C}^\perp$ is generated by the rows of $M_R(h)^\intercal$.*

PROOF. Assume $\mathsf{r}_R(Rf) = hR$. Then we get $\mathsf{l}_{\widehat{R}}(\Theta(f)\widehat{R}) = \widehat{R}\Theta(h)$. This implies that the top row of the commutative diagram

$$\begin{array}{ccccc}
\widehat{R} & \xrightarrow{\cdot\Theta(h)} & \widehat{R} & \xrightarrow{\cdot\Theta(f)} & \widehat{R} \\
\downarrow{\scriptstyle\widehat{\mathfrak{v}}} & & \downarrow{\scriptstyle\widehat{\mathfrak{v}}} & & \downarrow{\scriptstyle\widehat{\mathfrak{v}}} \\
C^m & \xrightarrow[\cdot M_{\widehat{R}}(\Theta(h))]{} & C^m & \xrightarrow[\cdot M_{\widehat{R}}(\Theta(f))]{} & C^m
\end{array}$$

is exact, so is the bottom row. Therefore,

$$\widehat{\mathfrak{v}}(\widehat{R}\Theta(h)) = \mathrm{im}(\cdot M_{\widehat{R}}(\Theta(h)) = \ker(\cdot M_{\widehat{R}}(\Theta(f)) = \ker(\cdot M_R(f)^\intercal) = \mathcal{C}^\perp,$$

where the last equality is (1).

Conversely, assume $\mathcal{C}^\perp = \widehat{\mathfrak{v}}(\widehat{R}\Theta(h))$. Then

$$\mathcal{C}^\perp = \mathrm{im}(\cdot M_{\widehat{R}}(\Theta(h))) = \mathrm{im}(\cdot M_R(h)^\intercal),$$

and therefore $M_R(f)M_R(h) = 0$. So $fh = 0$ and $hR \subseteq \mathsf{r}_R(Rf)$. Let $h' \in \mathsf{r}_R(Rf)$, i.e. $fh' = 0$. Then $M_R(f)M_R(h') = 0$ and so

$$M_R(h')^\intercal M_R(f)^\intercal = 0.$$

Since $M_R(h')^\intercal = M_{\widehat{R}}(\Theta(h'))$ and $\mathcal{C}^\perp = \ker(\cdot M_R(f)^\intercal)$, it follows that $\widehat{\mathfrak{v}}(\widehat{R}\Theta(h')) = \mathrm{im}(\cdot M_{\widehat{R}}(\Theta(h'))) \subseteq \mathcal{C}^\perp = \widehat{\mathfrak{v}}(\widehat{R}\Theta(h))$. Hence $\Theta(h') \in \widehat{R}\Theta(h)$, which implies $h' \in hR$. Therefore $\mathsf{r}_R(Rf) \subseteq hR$ and the equality holds.

Finally, the equality $\widehat{\mathfrak{v}}(\widehat{R}\Theta(h)) = \mathrm{im}(\cdot M_R(h)^\intercal)$, implies that $\mathcal{C}^\perp$ is generated by the rows of $M_R(h)^\intercal$. □

COROLLARY 5. *If $\mathcal{C}^{\perp\perp} = \mathcal{C}$ and $hR = \mathsf{r}_R(Rf)$, then $Rf = \mathsf{l}_R(hR)$.*

PROOF. We have $\mathcal{C}^\perp = \widehat{\mathfrak{v}}(\widehat{R}\Theta(h))$. Let $\widehat{\Theta} : \widehat{R} \to R$ be the inverse of Θ. Since

$$\mathcal{C}^{\perp\perp} = \mathcal{C} = \mathfrak{v}(Rf) = \mathfrak{v}(R\widehat{\Theta}\Theta(f)),$$

we get from Theorem 4, applied to the transposition $\widehat{\Theta} : \widehat{R} \to R$, that $\mathsf{r}_{\widehat{R}}(\widehat{R}\Theta(h)) = \Theta(f)\widehat{R}$. But this implies that $\mathsf{l}_R(hR) = Rf$. □

REMARK 6. It is well known that, if C is a field, then $\mathcal{C}^{\perp\perp} = \mathcal{C}$ for every C–linear code $\mathcal{C}$. This equality also holds true in more general situations of interest. This is the case, for instance, if C is a principal ideal domain and $\mathcal{C}$ is a direct summand of C^m, or $\mathcal{C}$ is any C–linear code over a Frobenius ring C.

2. Dual left ideal convolutional codes

The method described in Section 1 is abstracted from the study of the dual of a group convolutional code developed in [**25**, Section 4]. The ideas from [**25**, Section 4] were adapted in [**16**] to $\mathbb{F}$–linear cyclic convolutional codes when the word ambient algebra is a matrix algebra $\mathcal{M}_n(\mathbb{F})$. The aim of this section is to extend the results on duality from [**16**] to the matrix $\mathbb{F}$–algebra $\mathcal{M}_n(\mathbb{K})$, where $\mathbb{K}$ is a finite field extension of $\mathbb{F}$. To this end, we first work, in the spirit of Section 1, in a more general setting, and then apply the general results to the more concrete situation.

2.1. Left ideal convolutional codes of automorphism type. Let A be a ring and $\sigma : A \to A$ a ring automorphism. The skew *right* polynomial ring $[z;\sigma]A$ is defined as the free right A–module with basis $\{z^i : i \in \mathbb{N}\}$ with the multiplication determined by the rules $z^i z^j = z^{i+j}$, and $az = z\sigma(a)$, for all $i,j \in \mathbb{N}$ and $a \in A$. If A is a finite algebra over a finite field $\mathbb{F}$, and σ is an $\mathbb{F}$–automorphism, then $R = [z;\sigma]A$ becomes the sentence ambient algebra for some cyclic convolutional codes as follows. First, observe that $\mathbb{F}[z]$, the commutative polynomial ring in the variable z, is a subring of R. Moreover, each $\mathbb{F}$–basis $B = \{v_0, v_1, \dots, v_{m-1}\}$ of A leads to the associated coordinate map $\mathfrak{v} : A \to \mathbb{F}^m$. Since B becomes a basis of R as a (left) $\mathbb{F}[z]$–module, we get that $\mathfrak{v}$ extends to an $\mathbb{F}[z]$–module isomorphism $\mathfrak{v} : R \to \mathbb{F}[z]^m$. That is, $(\mathbb{F}[z], R, \mathfrak{v})$ becomes a Hamming ring extension of rank m.

Convolutional codes may be understood as $\mathbb{F}[z]$–submodules of $\mathbb{F}[z]^m$, so, they may be considered as $\mathbb{F}[z]$–linear codes. The variable z is interpreted as the delay operator [**13**]. Convolutional codes are often required to be, in addition, direct summands of $\mathbb{F}[z]^m$.

Definition 7. [**25**] A direct summand $\mathbb{F}[z]$–submodule $\mathcal{C}$ of $\mathbb{F}[z]^m$ is said to be a *left ideal convolutional code* if there exits a left ideal I of $[z;\sigma]A$ such that $\mathcal{C} = \mathfrak{v}(I)$.

Our next aim is to prove that, under suitable conditions, it is possible to construct a transpose Hamming ring extension to R. For $a \in A$, we use the notation $M_a = M_R(a)$. Observe that $M_a \in \mathcal{M}_m(\mathbb{F})$. On the other hand, for every $\mathbb{F}$–linear map $\lambda : A \to A$, let $M_\lambda \in \mathcal{M}_m(\mathbb{F})$ be the unique matrix such that the diagram

$$\begin{array}{ccc} A & \xrightarrow{\lambda} & A \\ \downarrow{\scriptstyle \mathfrak{v}} & & \downarrow{\scriptstyle \mathfrak{v}} \\ \mathbb{F}^m & \xrightarrow{\cdot M_\lambda} & \mathbb{F}^m \end{array}$$

is commutative. A straightforward computation shows that, if λ is an algebra map, then, for all $a \in A$,

$$M_a M_\lambda = M_\lambda M_{\lambda(a)} \tag{2}$$

Proposition 8. *For every* $f = \sum_k z^k f_k \in R$ *we have:*

$$M_R(f) = \sum_k z^k M_\sigma^k M_{f_k}.$$

Proof. This proof is adapted and simplified from [**25**, Proposition 4.7] and [**16**, Proposition 2.4]. Since $\mathfrak{v} : R \to \mathbb{F}[z]^m$ is an $\mathbb{F}[z]$–linear map, we have, for every

$g \in R$,

$$\begin{aligned}\mathfrak{v}(g)\textstyle\sum_k z^k M_\sigma^k M_{f_k} &= \mathfrak{v}\left(\textstyle\sum_l z^l g_l\right)\textstyle\sum_k z^k M_\sigma^k M_{f_k}\\ &= \textstyle\sum_{l,k} z^l z^k \mathfrak{v}(g_l) M_{\sigma^k} M_{f_k}\\ &= \textstyle\sum_{l,k} z^l z^k \mathfrak{v}(\sigma^k(g_l) f_k)\\ &= \mathfrak{v}\left(\textstyle\sum_l z^l \sum_k z^k \sigma^k(g_l) f_k\right)\\ &= \mathfrak{v}\left(\textstyle\sum_l z^l \sum_k g_l z^k f_k\right)\\ &= \mathfrak{v}(gf).\end{aligned}$$

□

Consider now an involution $\theta : A \to A$ (that is, an anti-algebra homomorphism such that $\theta^2 = \mathrm{id}_A$). Let us denote $\widehat{\sigma} = \theta\sigma^{-1}\theta$. Define the map

$$\begin{aligned}\Theta : [z;\sigma]A &\to [z;\widehat{\sigma}]A\\ \textstyle\sum_k z^k a_k &\mapsto \Theta(\textstyle\sum_k z^k a_k) = \textstyle\sum_k z^k \theta\sigma^{-k}(a_k)\end{aligned} \tag{3}$$

Note that, as left $\mathbb{F}[z]$–modules, both rings are equal, and, once the basis B of A is fixed, we have that $(\mathbb{F}[z], [z;\widehat{\sigma}]A, \mathfrak{v})$ is a Hamming ring extension.

PROPOSITION 9. *The map $\Theta : [z;\sigma]A \to [z;\widehat{\sigma}]A$ is an anti-isomorphism of $\mathbb{F}$–algebras. Moreover, if $M_{\widehat{\sigma}} = M_\sigma^{\mathsf{T}}$ and $M_{\theta(a)} = M_a^{\mathsf{T}}$ for all $a \in A$, then $(\mathbb{F}[z], [z;\sigma]A, \mathfrak{v})$ and $(\mathbb{F}[z], [z;\widehat{\sigma}]A, \mathfrak{v})$ are transpose Hamming ring extensions via Θ.*

PROOF. Since θ and σ are $\mathbb{F}$–linear and bijective, Θ is also $\mathbb{F}$–linear and bijective. Let us check that Θ is anti-multiplicative. Obviously, the restriction of Θ to A is just θ. Since $[z;\sigma]A$ is generated as an $\mathbb{F}$–algebra by A and z, it is enough to consider $a \in A$ and compute

$$\Theta(a)\Theta(z) = \theta(a)z = z\theta\sigma^{-1}\theta\theta(a) = z\theta\sigma^{-1}(a) = \Theta(za)$$

and

$$\Theta(az) = \Theta(z\sigma(a)) = z\theta\sigma^{-1}\sigma(a) = z\theta(a) = \Theta(z)\Theta(a).$$

In order to check that Θ is a transposition, let us compute

$$\begin{aligned}M_R(f)^{\mathsf{T}} &= \left[\sum_k z^k M_\sigma^k M_{f_k}\right]^{\mathsf{T}} && \text{by Proposition 8}\\ &= \left[\sum_k z^k M_{\sigma^{-k}(f_k)} M_\sigma^k\right]^{\mathsf{T}} && \text{by (2)}\\ &= \sum_k z^k (M_\sigma^k)^{\mathsf{T}} (M_{\sigma^{-k}(f_k)})^{\mathsf{T}}\\ &= \sum_k z^k (M_\sigma^{\mathsf{T}})^k M_{\theta\sigma^{-k}(f_k)} && \text{by assumption}\\ &= \sum_k z^k (M_{\widehat{\sigma}})^k M_{\theta\sigma^{-k}(f_k)} && \text{by assumption}\\ &= M_{\widehat{R}}(\Theta(f)) && \text{by Proposition 8,}\end{aligned}$$

as desired. □

REMARK 10. If $A = \mathbb{F}G$ is the group algebra of a finite group G, and θ is the involution defined on G as $\theta(g) = g^{-1}$, then Proposition 9 gives [**25**, Proposition 4.18] in the case where the skew derivation used in [**25**] is zero.

THEOREM 11. *Assume that $M_{\widehat{\sigma}} = M_{\sigma}^{\mathrm{T}}$ and $M_{\theta(a)} = M_a^{\mathrm{T}}$ for all $a \in A$. Let $f \in R = [z;\sigma]A$ such that $\mathsf{r}_R(Rf) = hR$ for some $h \in R$. If $\mathcal{C} = \mathfrak{v}(Rf)$, then $\mathcal{C}^{\perp} = \mathfrak{v}(\widehat{R}\Theta(h))$, and both $\mathcal{C}$ and $\mathcal{C}^{\perp}$ are left ideal convolutional codes. Moreover, $Rf = \mathsf{l}_R(hR)$.*

PROOF. The equality $\mathcal{C}^{\perp} = \mathfrak{v}(\widehat{R}\Theta(h))$ follows from Proposition 4 and Proposition 9. As observed in the proof of Theorem 4, $\widehat{R}\Theta(h) = \mathsf{l}_{\widehat{R}}(\Theta(f))$. Then $\mathcal{C}^{\perp}$ is a left ideal convolutional code by [**16**, Lemma 2.1]. Corollary 5, by virtue of Remark 6, gives the equality $Rf = \mathsf{l}_R(hR)$. Therefore, $\mathcal{C}$ is a left ideal convolutional code, by [**16**, Lemma 2.1]. □

REMARK 12. If $\mathbb{F}[z] \subseteq R$ is a separable ring extension, then every left ideal convolutional code is of the form $\mathfrak{v}(Re)$ for some idempotent $e \in R$ (see [**17**]). Therefore, $\mathsf{r}_R(Re) = (1-e)R$.

REMARK 13. If $A = \mathbb{F}G$, the group algebra of a finite group G, then Theorem 11 gives, in this particular case, a stronger statement than [**25**, Theorem 4.21], for the case where the σ–derivation considered there is zero.

2.2. Convolutional codes with a simple word-ambient algebra. In [**16**], dual codes of left ideal convolutional codes, when A is the matrix algebra $\mathcal{M}_n(\mathbb{F})$, are studied. Concretely, [**16**, propositions 2.9 and 2.10] provide a transposition of Hamming ring extensions in this special case, and, henceforth, Theorem 11 generalizes and improves [**16**, Theorem 2.12].

Our next aim is to give sufficient conditions to apply Proposition 9 and Theorem 11 to $A = \mathcal{M}_n(\mathbb{K})$, where $\mathbb{K} = \mathbb{F}_{q^t}$ is a finite field extension of $\mathbb{F} = \mathbb{F}_q$ of degree t. Let $D = \{\alpha_0, \dots, \alpha_{t-1}\}$ be a basis of $\mathbb{K}$ as an $\mathbb{F}$–vector space. We have a monomorphism of $\mathbb{F}$–algebras

$$\begin{aligned} \mathfrak{m} : \mathbb{K} &\to \mathcal{M}_t(\mathbb{F}) \\ \gamma &\mapsto \mathfrak{m}(\gamma), \end{aligned}$$

where $\mathfrak{m}(\gamma)$ is the matrix that represents the multiplication map $\cdot\gamma : \mathbb{K} \to \mathbb{K}$ for $\gamma \in \mathbb{K}$ with respect to D.

Let V, V' be finite-dimensional $\mathbb{K}$–vector spaces with bases $B = \{v_0, \dots, v_{r-1}\}$ and $B' = \{v'_0, \dots, v'_{s-1}\}$, respectively. Define the $\mathbb{F}$–basis

$$B_{\mathbb{F}} = \{\alpha_i v_j : 0 \leq i \leq t, 0 \leq j \leq r\}$$

of V ordered by the condition that $\alpha_i v_j$ is before than $\alpha_k v_{j+1}$ for all i, j, k. The $\mathbb{F}$–basis $B'_{\mathbb{F}}$ of V' is defined analogously.

LEMMA 14. *Let $M = (m_{ij}) \in \mathcal{M}_{r\times s}(\mathbb{K})$ be the matrix associated with respect to the bases B and B' to a $\mathbb{K}$–linear map $\lambda : V \to V'$. The matrix associated to λ considered as an $\mathbb{F}$–linear map with respect to the bases $B_{\mathbb{F}}$ and $B'_{\mathbb{F}}$ is*

$$\mathfrak{m}(M) = \Big(\mathfrak{m}(m_{ij})\Big)_{\substack{0 \leq i < r \\ 0 \leq j < s}} \in \mathcal{M}_{rt\times st}(\mathbb{F}).$$

PROOF. Straightforward. □

Consider the $\mathbb{K}$–basis $\mathcal{B} = \{E_{ij} \mid 0 \leq i,j < n, 0 \leq k < t\}$ of $A = \mathcal{M}_n(\mathbb{K})$, where E_{ij} is the matrix with 1 in the position corresponding to ith row and jth column, and 0 elsewhere. Order $\mathcal{B}$ in such a way that the corresponding coordinate map $\mathcal{M}_n(\mathbb{K}) \to \mathbb{K}^{n^2}$ writes each matrix as the concatenation of its rows. We thus get the $\mathbb{F}$–basis $\mathcal{B}_{\mathbb{F}} = \{\alpha_k E_{ij} : 0 \leq k \leq t-1, 0 \leq i,j \leq n-1\}$ ordered as described above. This last basis leads to a coordinate map $\mathfrak{v} : A \to \mathbb{F}^{tn^2}$ which, for any $\mathbb{F}$–automorphism σ of A, extends to an $\mathbb{F}[z]$–linear isomorphism $\mathfrak{v} : [z;\sigma]A \to \mathbb{F}[z]^{tn^2}$. Our aim is to prove that, if the basis D of $\mathbb{K}$ over $\mathbb{F}$ is a normal self-dual basis, then a transposition Θ fulfilling the hypotheses of Proposition 9 can be constructed.

Let $D^* = \{\beta_0, \ldots, \beta_{t-1}\}$ the dual basis of $D = \{\alpha_0, \ldots, \alpha_{t-1}\}$. Recall that D is normal if there exists $\alpha \in \mathbb{K}$ such that $\alpha_k = \alpha^{q^k}$ for all $0 \leq k < t$. Also recall that D is self-dual if $\alpha_k = \beta_k$ for all $0 \leq k < t$.

The coordinate map with respect to D is given by

$$\mathbb{K} \to \mathbb{F}^t, \quad \gamma \mapsto (\mathrm{Tr}(\beta_0\gamma), \ldots, \mathrm{Tr}(\beta_{t-1}\gamma))$$

where Tr denotes the trace map of the field extension $\mathbb{F} \subseteq \mathbb{K}$. By the properties of the trace function,

$$\mathfrak{m}(\gamma) = \Big(\mathrm{Tr}(\beta_i\gamma\alpha_j) \Big)_{0 \leq i,j < t}.$$

As involution on A, we use the natural one, that is, $\theta : A \to A$ be the involution given by $\theta(a) = a^{\mathsf{T}}$. In order to show how θ fulfills the Proposition 9, we need to use some properties of the Kronecker product of matrices over a field L.

Recall that if $M \in \mathcal{M}_{r\times s}(L)$ and $N \in \mathcal{M}_{r'\times s'}(L)$, the Kronecker product of M and N is defined as

$$M \boxtimes N = \begin{pmatrix} m_{0,0}N & m_{0,1}N & \cdots & m_{0,s-1}N \\ m_{1,0}N & m_{1,1}N & \cdots & m_{1,s-1}N \\ \vdots & \vdots & \ddots & \vdots \\ m_{r-1,0}N & m_{r-1,1}N & \cdots & m_{r-1,s-1}N \end{pmatrix},$$

where $M = \big(m_{i,j}\big)_{0\leq i<r, 0\leq j<s}$. Properties of the Kronecker product can be seen in [**21**, Chapter 4].

LEMMA 15. *The following properties hold for all $a \in A$:*

(i) $M_a = \mathfrak{m}(I \boxtimes a)$.
(ii) *If B is self-dual,* $M_a^{\mathsf{T}} = M_{a^{\mathsf{T}}}$.

PROOF. (i) follows from Lemma 14 and [**16**, Lemma 2.2]. If D is self-dual then $\mathfrak{m}(\gamma)$ is symmetric for all $\gamma \in \mathbb{K}$, hence (ii) follows from (i) and [**16**, Lemma 2.3(iii)]. □

Lemma 15 ensures the first hypothesis of Proposition 9. As for the second one concerns, namely the equality $M_{\widehat{\sigma}} = M_{\sigma}^{\mathsf{T}}$, it will be obtained with the help of a suitable decomposition of σ. Let $\tau : \mathbb{K} \to \mathbb{K}$ denote the Frobenius $\mathbb{F}$–automorphism of $\mathbb{K}$, i.e. $\tau(\gamma) = \gamma^q$. By [**8**, Theorem 2.4], for every $\mathbb{F}$–automorphism σ of the matrix algebra A, there exist a regular matrix $U \in A$ and $0 \leq h \leq t-1$ such that $\sigma = \sigma_U \circ \sigma_{\tau^h}$, where σ_U is the inner automorphism associated to U, i.e. $\sigma_U(a) = UaU^{-1}$, and σ_{τ^h} is the componentwise extension of τ^h to A, i.e. $\sigma_{\tau^h}\big(m_{ij}\big) = \big(\tau^h(m_{ij})\big)$.

We are going to analyze inner automorphisms and extensions of field automorphisms independently, and later we will join both results.

For each regular matrix $U \in A$, a straightforward computation shows that $\widehat{\sigma_U} = \theta\sigma_U^{-1}\theta = \sigma_{U^{\mathrm{T}}}$.

LEMMA 16. *The following properties hold:*

(i) $M_{\sigma_U} = \mathfrak{m}(U^{\mathrm{T}} \boxtimes U^{-1})$.

(ii) *If D is self-dual, then* $M_{\widehat{\sigma_U}} = M_{\sigma_U}^{\mathrm{T}}$.

PROOF. (i) follows from Lemma 14 and [**16**, Lemma 2.3(ii)]. If D is self-dual then $\mathfrak{m}(\gamma)$ is symmetric for all $\gamma \in \mathbb{K}$, hence (ii) follows from (i). □

PROPOSITION 17. *Let $R = [z;\sigma_U]A$ and $\widehat{R} = [z;\widehat{\sigma_U}]A$. If D is a self-dual basis, then $(\mathbb{F}[z], R, \mathfrak{v})$ and $(\mathbb{F}[z], \widehat{R}, \mathfrak{v})$ are transposed Hamming extensions.*

PROOF. By Lemma 15, $M_{a^{\mathrm{T}}} = M_a^{\mathrm{T}}$ for all $a \in A$. By Lemma 16, $M_{\widehat{\sigma_U}} = M_{\sigma_U}^{\mathrm{T}}$. Hence the result follows from Proposition 9. □

Now we focus on extensions of field automorphisms.

LEMMA 18. *Let τ be the Frobenius automorphism of $\mathbb{K}$ over $\mathbb{F}$, and let $0 \leq h < t$. If D is a normal basis, then the automorphism τ^h is represented as $\mathbb{F}$–linear map by the matrix*

$$P_h = \begin{pmatrix} 0 & I_{t-h} \\ I_h & 0 \end{pmatrix} \in \mathcal{M}_t(\mathbb{F}).$$

The automorphism $\sigma_{\tau^h} : A \to A$ is represented as $\mathbb{F}$–linear map by the matrix

$$M_{\sigma_{\tau^h}} = I_{n^2} \boxtimes P_h = \begin{pmatrix} P_h & & \\ & \ddots & \\ & & P_h \end{pmatrix} \in \mathcal{M}_{n^2t}(\mathbb{F}).$$

PROOF. Straightforward since τ performs a cyclic permutation of one position to the right of the elements of D. □

LEMMA 19. *Assume D to be a normal basis. The following properties hold:*

(i) $M_{\sigma_{\tau^h}}^{-1} = M_{\sigma_{\tau^{t-h}}} = M_{\sigma_{\tau^h}}^{\mathrm{T}}$.

(ii) $M_{\widehat{\sigma_{\tau^h}}} = M_{\sigma_{\tau^h}}^{\mathrm{T}}$,

PROOF. (i) is a direct consequence of Lemma 18. Since σ_{τ^h} acts componentwise, it commutes with θ, hence $\widehat{\sigma_{\tau^h}} = \theta\sigma_{\tau^h}^{-1}\theta = \theta^2\sigma_{\tau^h}^{-1} = \sigma_{\tau^h}^{-1} = \sigma_{\tau^{t-h}}$, hence (ii) follows from (i). □

PROPOSITION 20. *Let $R = [z;\sigma_{\tau^h}]A$ and $\widehat{R} = [z;\widehat{\sigma_{\tau^h}}]A$. If D is a normal basis then $(\mathbb{F}[z], R, \mathfrak{v})$ and $(\mathbb{F}[z], \widehat{R}, \mathfrak{v})$ are transposed Hamming extensions.*

PROOF. Again, Lemma 15 implies $M_{a^{\mathrm{T}}} = M_a^{\mathrm{T}}$ for all $a \in A$. Lemma 19 implies $M_{\widehat{\sigma_{\tau^h}}} = M_{\sigma_{\tau^h}}^{\mathrm{T}}$. So Proposition 9 gives the result. □

Finally we get to general automorphisms $\sigma = \sigma_U\sigma_{\tau^h} \in \mathrm{Aut}_{\mathbb{F}}(A)$ where $U \in A$ is a regular matrix and $0 \leq h < t$.

LEMMA 21. *Let $\sigma = \sigma_U\sigma_{\tau^h} \in \mathrm{Aut}_{\mathbb{F}}(A)$. If D is a self-dual normal basis, we have*

$$M_{\widehat{\sigma}} = M_{\sigma}^{\mathrm{T}}$$

PROOF. The decomposition $\sigma = \sigma_U \sigma_{\tau^h}$ implies $M_\sigma = M_{\sigma_{\tau^h}} M_{\sigma_U}$. Since

$$\widehat{\sigma} = \widehat{\sigma_U \sigma_{\tau^h}} = \theta \sigma_{\tau^h}^{-1} \sigma_U^{-1} \theta = \theta \sigma_{\tau^h}^{-1} \theta \theta \sigma_U^{-1} \theta = \widehat{\sigma_{\tau^h}} \widehat{\sigma_U},$$

we get $M_{\widehat{\sigma}} = M_{\widehat{\sigma_U}} M_{\widehat{\sigma_{\tau^h}}}$. Hence the result follows from Lemma 16 and Lemma 19. □

PROPOSITION 22. *Let $R = [z;\sigma]A$ and $\widehat{R} = [z;\widehat{\sigma}]A$. If D is a self-dual normal basis, then $(\mathbb{F}[z], R, \mathfrak{v})$ and $(\mathbb{F}[z], \widehat{R}, \mathfrak{v})$ are transposed Hamming extensions.*

PROOF. Decompose $\sigma = \sigma_U \sigma_{\tau^h}$ according to [**8**, Theorem 2.4]. The proof is now completely analogous to Proposition 17 and Proposition 20 by using Lemma 15, Lemma 21 and Proposition 9. □

We are now in position to state the main result of this subsection, which is a consequence of Theorem 11 and Proposition 22. Field extensions which have a self-dual normal basis are characterized by results of Lempel, Weinberger, Seroussi, Imamura and Morii, see [**23**] and its references.

THEOREM 23. *Let σ be any $\mathbb{F}$–automorphism of $A = \mathcal{M}_n(\mathbb{K})$. Assume there exists a self-dual normal basis D of $\mathbb{K}$ over $\mathbb{F}$. Let $f \in R = [z;\sigma]A$ such that $\mathsf{r}_R(Rf) = hR$ for some $h \in R$. If $\mathcal{C} = \mathfrak{v}(Rf)$, then $\mathcal{C}^\perp = \mathfrak{v}(\widehat{R}\Theta(h))$, and $\mathcal{C}$ and $\mathcal{C}^\perp$ are both left ideal convolutional codes. Moreover, $Rf = \mathsf{l}_R(hR)$.*

PROOF. The result follows from Proposition 22 and Theorem 11. □

REMARK 24. Taking $\mathbb{K} = \mathbb{F}$ in Theorem 23 leads to a strong form of [**16**, Theorem 2.12].

REMARK 25. By [**23**, Theorems 1 and 2], if t is odd, or q is even and $t \equiv 2 \mod 4$, then there exists a self-dual normal basis D of $\mathbb{K}$ over $\mathbb{F}$, and Theorem 23 applies.

3. Dual of skew constacyclic codes

Let C be a commutative ring and σ an automorphism of C. Assume that σ has finite order, and let $n \geq 1$ such that $\sigma^n = \mathrm{id}_C$. If $u \in C$ is a unit such that $\sigma(u) = u$, then the both $x^n - u$ and $x^n - u^{-1}$ are central elements in the skew left polynomial ring $C[x;\sigma]$, where the multiplication rule is now $xa = \sigma(a)x$. Consider the factor rings

$$\mathcal{R} = \frac{C[x;\sigma]}{\langle x^n - u\rangle}, \qquad \widehat{\mathcal{R}} = \frac{C[x;\sigma]}{\langle x^n - u^{-1}\rangle},$$

which contain C as a subring. Since the ideal $\langle x^n - u\rangle$ coincides with the left ideal generated by $x^n - u$, and this polynomial is monic, we easily get that $\{1, x, \dots, x^{n-1}\}$ is a basis of $\mathcal{R}$ as a left C–module (we are identifying, as usual, each equivalent class modulo $x^n - u$ with its unique representative of degree less than n). We have a Hamming ring extension $(C, \mathcal{R}, \mathfrak{v})$, where $\mathfrak{v} : \mathcal{R} \to C^n$ is the coordinate map with respect to the aforementioned basis.

Since x is a unit in both rings, we may define

$$\Theta : \mathcal{R} \to \widehat{\mathcal{R}}, \qquad \sum_{i=0}^{n-1} a_i x^i \mapsto \sum_{i=0}^{n-1} \sigma^{-i}(a_i) x^{-i}.$$

LEMMA 26. *The map Θ is an anti-isomorphim of rings.*

PROOF. Since Θ is clearly additive, in order to prove that it is an anti-homomorphism of rings, it suffices to check that it is anti-multiplicative on monomials. To this end, let us first observe that, since $x^n = u^{-1}$ in $\widehat{\mathcal{R}}$, we have for all integers i, j, k with $k = in + j$

$$x^{-k} = u^i x^{-j} \tag{4}$$

Now, given any integer k, write $k = in + j$ for some integers i, j with $0 \leq j < n$. Then, for any $a \in C$, we have

$$\Theta(ax^k) = \Theta(au^i x^j) = u^i \sigma^{-j}(a) x^{-j} = \sigma^{-k}(a) x^{-k},$$

where, in the last equality, we used (4) and that $\sigma^n = \mathrm{id}_C$. Therefore, for any $a, b \in C$ and integers i, j we have

$$\Theta(ax^i bx^j) = \Theta(a\sigma^i(b)x^{i+j}) = \sigma^{-i-j}(a\sigma^i(b))x^{-i-j} = \sigma^{-i-j}(a)\sigma^{-j}(b)x^{-i-j},$$

while

$$\Theta(bx^j)\Theta(ax^i) = \sigma^{-j}(b)x^{-j}\sigma^{-i}(a)x^{-i} = \sigma^{-j}(b)\sigma^{-j-i}(a)x^{-i-j}.$$

Therefore, Θ is anti-multiplicative. The map Θ is proved to be bijective by defining a map $\widehat{\Theta} : \widehat{\mathcal{R}} \to \mathcal{R}$ analogous to Θ which turns out to be its inverse. $\square$

The set $\{1, x, \ldots, x^{n-1}\}$ provides bases as left C–modules of $\mathcal{R} = C[x;\sigma]/\langle x^n - u\rangle$ and of $\widehat{\mathcal{R}} = C[x;\sigma]/\langle x^n - u^{-1}\rangle$.

PROPOSITION 27. *Consider the basis $\{1, x, \ldots, x^{n-1}\}$ of $\mathcal{R}$ and $\widehat{\mathcal{R}}$, with corresponding coordinate isomorphisms $\mathfrak{v} : \mathcal{R} \to C^n$ and $\widehat{\mathfrak{v}} : \widehat{\mathcal{R}} \to C^n$. The anti-isomorphism $\Theta : \mathcal{R} \to \widehat{\mathcal{R}}$ is a transposition from $(C, \mathcal{R}, \mathfrak{v})$ to $(C, \widehat{\mathcal{R}}, \widehat{\mathfrak{v}})$.*

PROOF. First, observe that if $f = \sum_{i=0}^{n-1} a_i x^i \in \mathcal{R}$ then

$$M_{\mathcal{R}}(f) = \begin{pmatrix} a_0 & a_1 & \cdots & a_{n-1} \\ u\sigma(a_{n-1}) & \sigma(a_0) & \cdots & \sigma(a_{n-2}) \\ \vdots & \vdots & \ddots & \vdots \\ u\sigma^{n-1}(a_1) & u\sigma^{n-1}(a_2) & \cdots & \sigma^{n-1}(a_0) \end{pmatrix}.$$

Now, since $x^{-1} = ux^{n-1} \in \widehat{\mathcal{R}}$, we get that $x^{-i} = ux^{n-i}$ for $i = 0, 1, \ldots, n-1$. Hence,

$$\Theta(f) = \sum_{i=0}^{n-1} \sigma^{n-i}(a_i)ux^{n-i} = a_0 + \sum_{j=1}^{n-1} u\sigma^j(a_{n-j})x^j.$$

Therefore,

$$M_{\widehat{\mathcal{R}}}(\Theta(f)) = \begin{pmatrix} a_0 & u\sigma(a_{n-1}) & \cdots & u\sigma^{n-1}(a_1) \\ a_1 & \sigma(a_0) & \cdots & u\sigma^{n-1}(a_2) \\ \vdots & \vdots & \ddots & \vdots \\ a_{n-1} & \sigma(a_{n-2}) & \cdots & \sigma^{n-1}(a_0) \end{pmatrix} = M_{\mathcal{R}}(f)^{\mathrm{T}}$$

because $\sigma^n = \mathrm{id}_C$. $\square$

DEFINITION 28. Let $(C, \mathcal{R}, \mathfrak{v})$ be the Hamming ring extension where $\mathcal{R} = \frac{C[x;\sigma]}{\langle x^n - u\rangle}$ with $\sigma^n = \mathrm{id}_C$ and $u \in C$ is a unit such that $\sigma(u) = u$. A $(C, \mathcal{R}, \mathfrak{v})$-cyclic code is called (u, σ)-constacyclic code.

THEOREM 29. *Let $f \in C[x;\sigma]$ be monic such that $x^n - u = fh$ for some $h \in C[x;\sigma]$. Let $\mathcal{C} = \mathfrak{v}(\mathcal{R}f)$ be the (u,σ)–constacyclic C–linear code generated by f. Then $\mathcal{C}^\perp = \widehat{\mathfrak{v}}(\widehat{\mathcal{R}}\Theta(h))$, and it is thus a (u^{-1},σ)–constacyclic code.*

PROOF. Let us first check that $\mathsf{r}_\mathcal{R}(\mathcal{R}f) = h\mathcal{R}$. By hypotesis, $h \in \mathsf{r}_\mathcal{R}(\mathcal{R}f)$ so that $\mathsf{r}_\mathcal{R}(\mathcal{R}f) \supseteq h\mathcal{R}$. To see the converse inclusion, let $h' \in \mathsf{r}_\mathcal{R}(\mathcal{R}f)$. Then $fh' = (x^n - u)g = fhg$ for some $g \in C[x;\sigma]$. Since monic polynomials are nonzero divisors in $C[x;\sigma]$, we get that $h' = hg$ as desired.

By virtue of Proposition 27 we may apply Theorem 4 and thus get $\mathcal{C}^\perp = \widehat{\mathfrak{v}}(\widehat{\mathcal{R}}\Theta(h))$. □

REMARK 30. If $u^2 = 1$, we get $\mathcal{R} = \widehat{\mathcal{R}}$ and Θ becomes an involution in $\mathcal{R}$, i.e. an anti-algebra automorphism such that $\Theta^2 = \mathrm{id}_\mathcal{R}$.

COROLLARY 31. *If $u^2 = 1$, then $\mathcal{C}^\perp = \mathfrak{v}(\mathcal{R}\Theta(h))$.*

EXAMPLE 32. Cyclic and negacyclic codes over finite chain rings defined in [**9**] fit in our construction. Let C be a commutative chain ring and $\mathcal{C}$ a C-linear code of length n. Then $\mathcal{C}$ is cyclic if and only if $\mathfrak{v}^{-1}(\mathcal{C})$ is an ideal of $\mathcal{R} = C[x]/\langle x^n - 1\rangle = \widehat{\mathcal{R}}$. The ring $\mathcal{R}$ is proven to be a principal ideal ring, [**9**, Corollary 3.7], and generators F and G for a cyclic code $\mathcal{C}$ and its dual $\mathcal{C}^\perp$ are computed, [**9**, theorems 3.6 and 3.10]. Theorem 4 implies that $\Theta(G)\mathcal{R} = \mathsf{r}_\mathcal{R}(\mathcal{R}F)$. The same can be said for negacyclic C-codes, i.e. ideals of $C[x]/\langle x^n + 1\rangle$, if F and G are the generators computed in [**9**, Theorem 5.7 and 5.12] then $\Theta(G)$ generates the annihilator of F.

EXAMPLE 33. Since, in Theorem 29, $x^n - u$ is a central element of $C[x;\sigma]$, the equality $fh = x^n - u$ implies, by a standard argument, that $hf = x^n - u$. Now, x is a unit in $\widehat{\mathcal{R}}$, which implies that $x^k\Theta(h)$ generates $\widehat{\mathcal{R}}\Theta(h)$ for every integer k. Setting k equal to the degree of h, we get that $x^k\Theta(h)$ is a generator polynomial of the dual code $\mathcal{C}^\perp$ which takes the form of the given in [**7**, Corollary 18] for σ–codes over a finite field. We also get the generator of the dual of any skew constacyclic code computed in [**6**, Theorem 4.4], [**22**, Lemma 3.1] and [**10**, Proposition 3] in the realm of codes over finite commutative rings.

4. Dual of skew Reed-Solomon codes

Let σ be an automorphism of finite order n of a field L. Since L is a field, all left and right ideals in the skew left polynomial ring $L[x;\sigma]$ are principal. So greatest common left and right divisors and least common left and right multiples exist and they can be computed with the corresponding versions of the extended Euclidean algorithm, see e.g. [**15**]. Concretely, given $f, g \in L[x;\sigma]$, the least common left multiple of f and g, denoted by $[f,g]_\ell$, is the monic generator of $L[x;\sigma]f \cap L[x;\sigma]g$, and the greatest common right divisor, denoted by $(f,g)_r$, is the monic generator of $L[x;\sigma]f + L[x;\sigma]g$. Analogously, the least common right multiple is denoted by $[f,g]_r$ and the greatest common left divisor by $(f,g)_\ell$.

The theory developed in Section 3 can be applied to $\mathcal{R} = L[x;\sigma]/\langle x^n - 1\rangle$ with $u = 1$. In this case, $\mathcal{R} = \widehat{\mathcal{R}}$ and $(1,\sigma)$-constacyclic codes are called σ-cyclic codes. Theorem 31 says that the dual of a σ-cyclic code is σ-cyclic. In this σ-cyclic setting the morphism Θ becomes an involution, i.e.

$$\Theta : \mathcal{R} \to \mathcal{R}, \qquad \sum_{i=0}^{n-1} a_i x^i \mapsto \sum_{i=0}^{n-1} \sigma^{-i}(a_i) x^{-i} = a_0 + \sum_{j=1}^{n-1} \sigma^j(a_{n-j}) x^j,$$

satisfies $\Theta^2 = \mathrm{id}_{\mathcal{R}}$.

Skew Reed-Solomon codes are σ-cyclic codes generated by some special polynomials which we describe below. Let $\{\alpha, \sigma(\alpha), \dots, \sigma^{n-1}(\alpha)\}$ be a normal basis of L/K where $K = L^\sigma$ denotes de invariant subfield. Let $\beta = \sigma(\alpha)\alpha^{-1}$. By [**18**, Lemma 3.1],

$$x^n - 1 = \left[x - \beta, x - \sigma(\beta), \dots, x - \sigma^{n-1}(\beta)\right]_\ell . \tag{5}$$

Let t such that $2t < n$, $\delta = 2t + 1$. Let

$$g = \left[x - \beta, x - \sigma(\beta), \dots, x - \sigma^{\delta-2}(\beta)\right]_\ell .$$

The code $\mathcal{C} = \mathfrak{v}(\mathcal{R}g)$ is called a skew RS code of designed distance δ. If $k = n - \delta + 1$, $\mathcal{C}$ is an $[n, k, \delta]$-code over L, hence it is MDS with respect to the Hamming distance (see [**18**, Definition 2, Theorem 3.4]). Skew Reed-Solomon codes can be efficiently decoded, see [**18, 19**]

We will prove that the dual of a skew Reed-Solomon code is again a skew Reed-Solomon code, and describe explicitly its generator polynomial as a least common left multiple of linear polynomials. Some preliminary results are needed. Observe that σ extends to an automorphism of $\mathcal{R}$, which acts on x as the identity.

LEMMA 34. *The morphisms Θ and σ commute, i.e. $\Theta\sigma = \sigma\Theta$.*

PROOF. It is a straightforward computation. □

Let $\gamma \in L$ such that $x - \gamma$ left divides $x^n - 1$. Then $(x - \gamma)\mathcal{R}$ is a maximal right ideal, and therefore $\mathcal{R}\Theta(x - \gamma)$ is a maximal left ideal. Hence there exists $\gamma' \in L$ such that

$$\mathcal{R}(x - \gamma') = \mathcal{R}\Theta(x - \gamma).$$

In fact, since $\Theta(x - \gamma) = x^{-1} - \gamma$, we get that

$$\gamma' = \sigma(\gamma)^{-1}.$$

LEMMA 35. *If $h = \left[\{x - \sigma^i(\gamma) \mid 0 \leq i \leq k-1\}\right]_r$, then $\mathcal{R}\Theta(h) = \mathcal{R}h'$, where*

$$h' = \left[\{x - \sigma^i(\gamma') \mid 0 \leq i \leq k-1\}\right]_\ell .$$

PROOF. Since $h = \left[\{x - \sigma^i(\gamma) \mid 0 \leq i \leq k-1\}\right]_r$,

$$h\mathcal{R} = \bigcap_{i=0}^{k-1} (x - \sigma^i(\gamma))\mathcal{R} = \bigcap_{i=0}^{k-1} \sigma^i\left((x - \gamma)\mathcal{R}\right).$$

So, by Lemma 34,

$$\begin{aligned}
\mathcal{R}\Theta(h) &= \Theta\left(h\mathcal{R}\right) \\
&= \Theta\left(\bigcap_{i=0}^{k-1} \sigma^i\left((x-\gamma)\mathcal{R}\right)\right) \\
&= \bigcap_{i=0}^{k-1} \Theta\left(\sigma^i\left((x-\gamma)\mathcal{R}\right)\right) \\
&= \bigcap_{i=0}^{k-1} \sigma^i\left(\Theta\left((x-\gamma)\mathcal{R}\right)\right) \\
&= \bigcap_{i=0}^{k-1} \sigma^i\left(\mathcal{R}\Theta(x-\gamma)\right) \\
&= \bigcap_{i=0}^{k-1} \sigma^i\left(\mathcal{R}(x-\gamma')\right) \\
&= \bigcap_{i=0}^{k-1} \mathcal{R}(x-\sigma^i(\gamma')) \\
&= \mathcal{R}\left[\{x-\sigma^i(\gamma') \mid 0 \le i \le k-1\}\right]_\ell.
\end{aligned}$$

□

LEMMA 36. *Let $\gamma \in L$ such that $(x-\gamma)\left[x-\sigma(\beta), \dots, x-\sigma^{n-1}(\beta)\right]_\ell = x^n - 1$. Then*

$$\left[x-\gamma, \dots, x-\sigma^k(\gamma)\right]_r \left[x-\sigma^{k+1}(\beta), \dots, x-\sigma^{n-1}(\beta)\right]_\ell = x^n - 1$$

for all $0 \le k \le n-1$.

PROOF. Denote $N_i = \left[\{x-\sigma^j(\beta) \mid j \neq i\}\right]_\ell$ for each $i = 0, \dots, n-1$. Then N_i has degree $n-1$ by (5). So, there exist $\gamma_0, \gamma_1, \dots, \gamma_{n-1} \in L$ such that

$$(x-\gamma_i)N_i = x^n - 1$$

for $i = 0, \dots, n-1$. Let $\gamma = \gamma_0$. Then $(x-\gamma)N_0 = x^n - 1$, and therefore $(x-\sigma^k(\gamma))\sigma^k(N_0) = x^n - 1$ for $k = 1, \dots, n-1$. On the other hand, observe that $\sigma^k(N_0) = N_k$ for any $k = 1, \dots, n-1$. Thus, for $k = 1, \dots, n-1$, $(x-\sigma^k(\gamma))N_k = (x-\gamma_k)N_k$, and then $\gamma_k = \sigma^k(\gamma)$ for all $k = 1, \dots, n-1$.

Let us now prove our thesis, i.e.

$$\left[x-\gamma, \dots, x-\sigma^k(\gamma)\right]_r \left[x-\sigma^{k+1}(\beta), \dots, x-\sigma^{n-1}(\beta)\right]_\ell = x^n - 1$$

for $k = 0, \dots, n-1$, by induction on k. The case $k = 0$ is trivial. Assume that it holds for $j \le k$ and we prove it for $k+1$. Firstly, the polynomial $\left[x-\gamma, \dots, x-\sigma^{k+1}(\gamma)\right]_r$ has degree $k+2$. Indeed, otherwise, by hypothesis, $\left[x-\gamma, \dots, x-\sigma^k(\gamma)\right]_r = \left[x-\gamma, \dots, x-\sigma^{k+1}(\gamma)\right]_r$ and then $(x-\sigma^{k+1}(\gamma))p = \left[x-\gamma, \dots, x-\sigma^k(\gamma)\right]_r$. So,

$$(x-\sigma^{k+1}(\gamma))p\left[x-\sigma^{k+1}(\beta), \dots, x-\sigma^{n-1}(\beta)\right]_\ell = x^n - 1.$$

In particular, this implies that

$$p\left[x-\sigma^{k+1}(\beta), \dots, x-\sigma^{n-1}(\beta)\right]_\ell = N_{k+1}$$

and, consequently, $x-\sigma^{k+1}(\beta)$ right divides N_{k+1}, which contradicts (5).

Now, by hypothesis,

$$\left[x-\gamma,\dots,x-\sigma^{k}(\gamma)\right]_r\left[x-\sigma^{k+1}(\beta),\dots,x-\sigma^{n-1}(\beta)\right]_\ell = x^n-1$$

so, applying σ,

$$\left[x-\sigma(\gamma),\dots,x-\sigma^{k+1}(\gamma)\right]_r\left[x-\sigma^{k+2}(\beta),\dots,x-\sigma^{n-1}(\beta),x-\beta\right]_\ell = x^n-1$$

so

$$\left[x-\sigma(\gamma),\dots,x-\sigma^{k+1}(\gamma)\right]_r q\left[x-\sigma^{k+2}(\beta),\dots,x-\sigma^{n-1}(\beta)\right]_\ell = x^n-1$$

for some polynomial q. That is, if h is the monic polynomial verifying that

$$h\left[x-\sigma^{k+2}(\beta),\dots,x-\sigma^{n-1}(\beta)\right]_\ell = x^n-1,$$

then $\left[x-\sigma(\gamma),\dots,x-\sigma^{k+1}(\gamma)\right]_r$ left divides h. On the other hand, the case $k=0$ provides that

$$(x-\gamma)\left[x-\sigma(\beta),\dots,x-\sigma^{n-1}(\beta)\right]_\ell = x^n-1$$

and therefore $x-\gamma$ left divides h as well. So $\left[x-\gamma,x-\sigma(\gamma),\dots,x-\sigma^{k+1}(\gamma)\right]_r$ left divides h. Since both have degree $k+2$, the result follows. □

THEOREM 37. *Consider a skew RS code $\mathcal{C}=\mathfrak{v}(\mathcal{R}g)$, where*

$$g=\left[x-\beta,x-\sigma(\beta),\dots,x-\sigma^{\delta-2}(\beta)\right]_\ell.$$

If $\gamma\in L$ is such that

$$(x-\gamma)\left[x-\sigma(\beta),\dots,x-\sigma^{n-1}(\beta)\right]_\ell = x^n-1,$$

then $\mathcal{C}^\perp$ is the skew RS code generated by

$$\left[x-\sigma^{\delta}(\gamma)^{-1},\dots,x-\sigma^{n}(\gamma)^{-1}\right]_\ell.$$

PROOF. By Lemma 36

$$\left[x-\gamma,\dots,x-\sigma^{n-\delta}(\gamma)\right]_r\left[x-\sigma^{n-\delta+1}(\beta),\dots,x-\sigma^{n-1}(\beta)\right]_\ell = x^n-1,$$

and, by applying $\sigma^{\delta-1}$, we obtain

$$\left[x-\sigma^{\delta-1}(\gamma),\dots,x-\sigma^{n-1}(\gamma)\right]_r\left[x-\beta,\dots,x-\sigma^{\delta-2}(\beta)\right]_\ell = x^n-1.$$

Therefore, $hg=x^n-1$, where $h=\left[x-\sigma^{\delta-1}(\gamma),\dots,x-\sigma^{n-1}(\gamma)\right]_r$. A standard argument, which uses that x^n-1 is central, proves that $gh=x^n-1$.

By Corollary 31, we get that $\mathcal{C}^\perp=\mathcal{R}\Theta(h)$. Lemma 35 gives then that $\mathcal{R}\Theta(h)$ is generated by

$$\left[x-\sigma(\sigma^{\delta-1}(\gamma))^{-1},\dots,\sigma^{n-\delta+1}(\sigma^{\delta-1}(\gamma))^{-1}\right]_\ell,$$

which finishes the proof. □

We finish by proving that, as a consequence of Theorem 37, skew RS codes can be seen as evaluation codes. The right evaluation of a skew polynomial $f=\sum_i f_i x^i\in L[x;\sigma]$ by $a\in L$ is the remainder of the right division of f by $x-a$, i.e. the unique element $f(a)\in L$ such that $f(x)=q(x)(x-a)+f(a)$. Then

$$f(a)=\sum_i f_iN_i(a)$$

where

$$N_i(a)=a\sigma(a)\dots\sigma^{i-1}(a).$$

DEFINITION 38. Let $\overline{\alpha} = (\alpha_0, \alpha_1, \ldots, \alpha_{m-1}) \in L^m$ and $\overline{v} = (v_0, v_1, \ldots, v_{m-1}) \in (L \setminus \{0\})^m$. The skew Generalized Evaluation code associated to $(\overline{\alpha}, \overline{v})$ is

$$\mathrm{sGE}^k_{(\overline{\alpha},\overline{v})} = \{(v_0 f(\alpha_0), \ldots, v_{m-1} f(\alpha_{m-1})) \mid f \in L[x;\sigma], \deg f < k\}.$$

It is straightforward to check that $\mathrm{sGE}^k_{(\overline{\alpha},\overline{v})}$ is an L–linear code. In fact a generator matrix for it is

$$\begin{pmatrix} v_0 & v_1 & \cdots & v_{m-1} \\ v_0\alpha_0 & v_1\alpha_1 & \cdots & v_{m-1}\alpha_{m-1} \\ \vdots & \vdots & \ddots & \vdots \\ v_0 N_{k-1}(\alpha_0) & v_1 N_{k-1}(\alpha_1) & \cdots & v_{n-1} N_{k-1}(\alpha_{m-1}) \end{pmatrix}$$
$$= \begin{pmatrix} 1 & 1 & \cdots & 1 \\ \alpha_0 & \alpha_1 & \cdots & \alpha_{m-1} \\ \vdots & \vdots & \ddots & \vdots \\ N_{k-1}(\alpha_0) & N_{k-1}(\alpha_1) & \cdots & N_{k-1}(\alpha_{m-1}) \end{pmatrix} \begin{pmatrix} v_0 & 0 & \cdots & 0 \\ 0 & v_1 & \cdots & 0 \\ \vdots & \vdots & \ddots & \vdots \\ 0 & 0 & \cdots & v_{m-1} \end{pmatrix}$$

This definition is an extension of [**24**, Definition 9] to arbitrary fields, including therefore the convolutional case when $L = \mathbb{F}(t)$. A different approach to evaluation codes, where the norms are replaced by powers of the automorphism, can be found in [**2**, **3**].

Let us now prove that a skew RS code is also a sGE code.

THEOREM 39. *Let $\mathcal{C} = \mathfrak{v}(\mathcal{R}g)$ be a skew RS code where*

$$g = \left[x - \beta, \ldots, x - \sigma^{\delta-2}(\beta)\right]_\ell.$$

Then there exist $\mu, \nu \in L \setminus \{0\}$ such that $\mathcal{C} = \mathrm{sGE}^{n-\delta+1}_{(\overline{\mu},\overline{\nu})}$, where $\overline{\mu} = (\mu, \sigma(\mu), \ldots, \sigma^{n-1}(\mu))$ and $\overline{\nu} = (\nu, \sigma(\nu), \ldots, \sigma^{n-1}(\nu))$.

PROOF. By Theorem 37 there exists $\mu \in L$ such that $\mathcal{C}^\perp = \mathfrak{v}(\mathcal{R}g')$ where

$$g' = \left[x - \mu, \ldots, x - \sigma^{n-\delta}(\mu)\right]_\ell.$$

Since $\sigma^i(\mu)$ is a right root of g' for $0 \leq i \leq n - \delta$, it follows that a parity check matrix of $\mathcal{C}^\perp$ is

$$\begin{pmatrix} 1 & 1 & \cdots & 1 \\ \mu & \sigma(\mu) & \cdots & \sigma^{n-\delta}(\mu) \\ \vdots & \vdots & \ddots & \vdots \\ N_{n-1}(\mu) & N_{n-1}(\sigma(\mu)) & \cdots & N_{n-1}(\sigma^{n-\delta}(\mu)) \end{pmatrix}.$$

Since μ is also a left root of $x^n - 1$, it follows that $N_n(\mu) = 1$ and, by Hilbert's 90 Theorem, there exists $\nu \in L$ such that $\mu = \sigma(\nu)\nu^{-1}$. So, up to multiply each column by the corresponding scalar, a new parity check matrix for $\mathcal{C}^\perp$ is

$$H = \begin{pmatrix} \nu & \sigma(\nu) & \cdots & \sigma^{n-\delta}(\nu) \\ \sigma(\nu) & \sigma^2(\nu) & \cdots & \sigma^{n-\delta+1}(\nu) \\ \vdots & \vdots & \ddots & \vdots \\ \sigma^{n-1}(\nu) & \nu & \cdots & \sigma^{n-\delta-1}(\nu) \end{pmatrix}.$$

Therefore $\mathcal{C}$ is generated by the rows of the matrix

$$H^{\mathrm{T}} = \begin{pmatrix} \nu & \sigma(\nu) & \cdots & \sigma^{n-1}(\nu) \\ \sigma(\nu) & \sigma^2(\nu) & \cdots & \nu \\ \vdots & \vdots & \ddots & \vdots \\ \sigma^{n-\delta}(\nu) & \sigma^{n-\delta+1}(\nu) & \cdots & \sigma^{n-\delta-1}(\nu) \end{pmatrix}.$$

We have

$$\begin{aligned} H^{\mathrm{T}} &= \begin{pmatrix} \nu & \sigma(\nu) & \cdots & \sigma^{n-1}(\nu) \\ \sigma(\nu) & \sigma^2(\nu) & \cdots & \nu \\ \vdots & \vdots & \ddots & \vdots \\ \sigma^{n-\delta}(\nu) & \sigma^{n-\delta+1}(\nu) & \cdots & \sigma^{n-\delta-1}(\nu) \end{pmatrix} \\ &= \begin{pmatrix} 1 & 1 & \cdots & 1 \\ \sigma(\nu)\nu^{-1} & \sigma^2(\nu)\sigma(\nu)^{-1} & \cdots & \nu\sigma^{n-1}(\nu)^{-1} \\ \vdots & \vdots & \ddots & \vdots \\ \sigma^{n-\delta}(\nu)\nu^{-1} & \sigma^{n-\delta+1}(\nu)\sigma(\nu)^{-1} & \cdots & \sigma^{n-\delta-1}(\nu)\sigma^{n-1}(\nu)^{-1} \end{pmatrix} \\ &\quad \cdot \begin{pmatrix} \nu & 0 & \cdots & 0 \\ 0 & \sigma(\nu) & \cdots & 0 \\ \vdots & \vdots & \ddots & \vdots \\ 0 & 0 & \cdots & \sigma^{n-1}(\nu) \end{pmatrix} \\ &= \begin{pmatrix} 1 & 1 & \cdots & 1 \\ \mu & \sigma(\mu) & \cdots & \sigma^{n-1}(\mu) \\ \vdots & \vdots & \ddots & \vdots \\ N_{n-\delta}(\mu) & N_{n-\delta}(\sigma(\mu)) & \cdots & N_{n-\delta}(\sigma^{n-1}(\mu)) \end{pmatrix} \\ &\quad \cdot \begin{pmatrix} \nu & 0 & \cdots & 0 \\ 0 & \sigma(\nu) & \cdots & 0 \\ \vdots & \vdots & \ddots & \vdots \\ 0 & 0 & \cdots & \sigma^{n-1}(\nu) \end{pmatrix}, \end{aligned}$$

hence $\mathcal{C} = \mathrm{sGE}^{n-\delta+1}_{(\overline{\gamma},\overline{\nu})}$. □

References

[1] A. Alahmadi, A. Leroy, and P. Solé, *Long module skew codes are good*, Discrete Math. **339** (2016), no. 5, 1624–1627, DOI 10.1016/j.disc.2016.01.008. MR3475578

[2] D. Augot, P. Loidreau, and G. Robert. Rank metric and Gabidulin codes in characteristic zero. In *ISIT 2013 IEEE International Symposium on Information Theory, Jul 2013, Istanbul, Turkey.*, 2013.

[3] D. Augot, P. Loidreau, and G. Robert, *Generalized Gabidulin codes over fields of any characteristic*, Des. Codes Cryptogr. **86** (2018), no. 8, 1807–1848, DOI 10.1007/s10623-017-0425-6. MR3810530

[4] D. Boucher and F. Ulmer, *A note on the dual codes of module skew codes*, Cryptography and coding, Lecture Notes in Comput. Sci., vol. 7089, Springer, Heidelberg, 2011, pp. 230–243, DOI 10.1007/978-3-642-25516-8_14. MR2945261

[5] D. Boucher, W. Geiselmann, and F. Ulmer, *Skew-cyclic codes*, Appl. Algebra Engrg. Comm. Comput. **18** (2007), no. 4, 379–389, DOI 10.1007/s00200-007-0043-z. MR2322946

[6] D. Boucher, P. Solé, and F. Ulmer, *Skew constacyclic codes over Galois rings*, Adv. Math. Commun. **2** (2008), no. 3, 273–292, DOI 10.3934/amc.2008.2.273. MR2429458

[7] D. Boucher and F. Ulmer, *Coding with skew polynomial rings*, J. Symbolic Comput. **44** (2009), no. 12, 1644–1656, DOI 10.1016/j.jsc.2007.11.008. MR2553570

[8] G. Cauchon and J. C. Robson, *Endomorphisms, derivations, and polynomial rings*, J. Algebra **53** (1978), no. 1, 227–238, DOI 10.1016/0021-8693(78)90213-2. MR0486004

[9] H. Q. Dinh and S. R. López-Permouth, *Cyclic and negacyclic codes over finite chain rings*, IEEE Trans. Inform. Theory **50** (2004), no. 8, 1728–1744, DOI 10.1109/TIT.2004.831789. MR2096841

[10] J. Ducoat and F. Oggier, *On skew polynomial codes and lattices from quotients of cyclic division algebras*, Adv. Math. Commun. **10** (2016), no. 1, 79–94, DOI 10.3934/amc.2016.10.79. MR3471062

[11] S. Estrada, J. R. García-Rozas, J. Peralta, and E. Sánchez-García, *Group convolutional codes*, Adv. Math. Commun. **2** (2008), no. 1, 83–94, DOI 10.3934/amc.2008.2.83. MR2377237

[12] N. Fogarty and H. Gluesing-Luerssen, *A circulant approach to skew-constacyclic codes*, Finite Fields Appl. **35** (2015), 92–114, DOI 10.1016/j.ffa.2015.03.008. MR3368804

[13] G. D. Forney Jr., *Convolutional codes. I. Algebraic structure*, IEEE Trans. Information Theory **IT-16** (1970), 720–738. MR0275988

[14] H. Gluesing-Luerssen and W. Schmale, *On cyclic convolutional codes*, Acta Appl. Math. **82** (2004), no. 2, 183–237, DOI 10.1023/B:ACAP.0000027534.61242.09. MR2066259

[15] J. Gómez-Torrecillas, *Basic module theory over non-commutative rings with computational aspects of operator algebras*, Algebraic and algorithmic aspects of differential and integral operators, Lecture Notes in Comput. Sci., vol. 8372, Springer, Heidelberg, 2014, pp. 23–82, DOI 10.1007/978-3-642-54479-8_2. With an appendix by V. Levandovskyy. MR3188538

[16] J. Gómez-Torrecillas, F. J. Lobillo, and G. Navarro, *Convolutional codes with a matrix-algebra word-ambient*, Adv. Math. Commun. **10** (2016), no. 1, 29–43, DOI 10.3934/amc.2016.10.29. MR3471058

[17] J. Gómez-Torrecillas, F. J. Lobillo, and G. Navarro, *Ideal codes over separable ring extensions*, IEEE Trans. Inform. Theory **63** (2017), no. 5, 2796–2813. MR3649761

[18] J. Gómez-Torrecillas, F. J. Lobillo, and G. Navarro, *Peterson-Gorenstein-Zierler algorithm for skew RS codes*, Linear Multilinear Algebra **66** (2018), no. 3, 469–487, DOI 10.1080/03081087.2017.1301364. MR3760381

[19] J. Gómez-Torrecillas, F. J. Lobillo, and G. Navarro, *A Sugiyama-like decoding algorithm for convolutional codes*, IEEE Trans. Inform. Theory **63** (2017), no. 10, 6216–6226, DOI 10.1109/TIT.2017.2731774. MR3707532

[20] J. Gómez-Torrecillas, F. J. Lobillo, and G. Navarro, *A new perspective of cyclicity in convolutional codes*, IEEE Trans. Inform. Theory **62** (2016), no. 5, 2702–2706, DOI 10.1109/TIT.2016.2538264. MR3493871

[21] R. A. Horn and C. R. Johnson, *Topics in matrix analysis*, Cambridge University Press, Cambridge, 1994. Corrected reprint of the 1991 original. MR1288752

[22] S. Jitman, S. Ling, and P. Udomkavanich, *Skew constacyclic codes over finite chain rings*, Adv. Math. Commun. **6** (2012), no. 1, 39–63, DOI 10.3934/amc.2012.6.39. MR2885847

[23] A. Lempel and M. J. Weinberger, *Self-complementary normal bases in finite fields*, SIAM J. Discrete Math. **1** (1988), no. 2, 193–198, DOI 10.1137/0401021. MR941350

[24] S. Liu, F. Manganiello, and F. R. Kschischang. Construction and decoding of generalized skew-evaluation codes. In *2015 IEEE 14th CanadianWorkshop on Information Theory (CWIT), St. John's, NL*, pages 9–13, 2015.

[25] S. R. López-Permouth and S. Szabo, *Convolutional codes with additional algebraic structure*, J. Pure Appl. Algebra **217** (2013), no. 5, 958–972, DOI 10.1016/j.jpaa.2012.09.017. MR3003319

[26] P. Piret, *Structure and constructions of cyclic convolutional codes*, IEEE Trans. Information Theory **IT-22** (1976), no. 2, 147–155. MR0416746

[27] C. Roos, *On the structure of convolutional and cyclic convolutional codes*, IEEE Trans. Inform. Theory **25** (1979), no. 6, 676–683, DOI 10.1109/TIT.1979.1056108. MR551266

CITIC and Department of Algebra, University of Granada, 18071 Granada, Spain

Email address: gomezj@ugr.es

CITIC and Department of Algebra, University of Granada, 18071 Granada, Spain

Email address: jlobillo@ugr.es

CITIC and Department of Computer Sciences and AI, University of Granada, 18071 Granada, Spain

Email address: gnavarro@ugr.es

Contemporary Mathematics
Volume **727**, 2019
https://doi.org/10.1090/conm/727/14631

Injective hulls of simple modules over nilpotent Lie color algebras

Can Hatipoğlu

ABSTRACT. Using cocycle twists for associative graded algebras, we characterize finite dimensional nilpotent Lie color algebras L graded by arbitrary abelian groups whose enveloping algebras $U(L)$ have the property that the injective hulls of simple right $U(L)$-modules are locally Artinian.

1. Introduction

One of the famous open problems in ring theory is Jacobson's Conjecture, which asks for a Noetherian ring R with a Jacobson radical J whether it is true that $\cap_{i=1}^{\infty} J^i = 0$. When Jategaonkar proved in [**10**] that fully bounded Noetherian rings satisfy Jacobson's Conjecture, a key step in his proof was to observe that over such rings finitely generated essential extensions of simple modules were Artinian. This created an interest in this finiteness property, and a number of Noetherian rings have been studied in relation with it. The obvious question whether this property was shared by every Noetherian ring was answered negatively by Musson in [**12, 13**].

Recall that a module M is said to be *locally Artinian* if every finitely generated submodule of it is Artinian. For a Noetherian ring A, finitely generated essential extensions of simple right modules are Artinian if and only if the property

($\diamond$) Injective hulls of simple right A-modules are locally Artinian

is satisfied. We should stress that property ($\diamond$) is sufficient for a Noetherian ring to satisfy Jacobson's conjecture.

Interest in property ($\diamond$) has increased in recent years, with Noetherian down-up algebras having this property have been characterized in a series of papers by Carvalho *et al.* [**4**], Carvalho and Musson [**6**] and Musson [**14**]. Later, finite dimensional solvable Lie superalgebras over algebraically closed fields whose enveloping algebras have property ($\diamond$) have been characterized in [**9**] and then a characterization of Ore extensions $k[x][y; \alpha, d]$ of $k[x]$ having property ($\diamond$) has been obtained in [**3**]. A characterization involving torsion theories and some sufficient conditions have been given by the author in [**8**]. Most recently, ($\diamond$) property of semiprimary commutative quasi-local rings and skew polynomial rings have been investigated [**2, 5**].

2010 *Mathematics Subject Classification.* Primary 17B75.

Lie color algebras are natural generalizations of Lie superalgebras. In this note we show using a technique, which is sometimes called "Scheunert's trick", that the result on nilpotent Lie superalgebras could easily be generalized to nilpotent Lie color algebras. The characterization of nilpotent Lie superalgebras L with property ($\diamond$) has shown that it only depends on the even part of the Lie superalgebra. We will see that the same holds in this more general setting of Lie color algebras.

In the rest of the paper, we assume that k is an algebraically closed field of characteristic zero, G is an additive abelian group and all the graded rings and modules are graded by G. The group of nonzero elements of k will be denoted by k^*.

2. Cocycle Twists of Associative Graded Algebras and Property ($\diamond$)

Let A be a G-graded algebra over k. That is, A is a G-graded k-vector space and there is a family $\{A_g \mid g \in G\}$ of subspaces of A such that $A = \oplus_{g\in G} A_g$ and $A_g A_h \subseteq A_{g+h}$ for every $g, h \in G$. The set of all $g \in G$ such that $A_g \neq 0$ is called the *support* of the grading. A nonzero element a of A_g is said to be a *homogeneous element of degree* g, and in this case we will write $|a| = g$ to indicate the degree of a.

A graded right A-module is a right A-module M such that there exists a family $\{M_g \mid g \in G\}$ of k-subspaces of M such that $M = \oplus_{g\in G} M_g$ and for every $g, h \in G$ we have $M_g A_h \subseteq M_{g+h}$. For a graded ring A, we will denote the category of graded right A-modules by *Gr-A* and Mod-A will denote the category of right A-modules.

By a *2-cocycle* on G we mean a map $\sigma : G \times G \longrightarrow k^*$ which satisfies

$$\sigma(a,b)\sigma(a+b,c) = \sigma(b,c)\sigma(a,b+c) \tag{2.1}$$

for all a, b, and $c \in G$. For our purposes, we will also assume that a 2-cocycle σ satisfies the normalization condition $\sigma(0,0) = 1$. It follows from this assumption that $\sigma(a,0) = \sigma(0,a) = \sigma(0,0) = 1$. Note that if σ is a 2-cocycle, then the map $\sigma^{-1} : G \times G \longrightarrow k^*$ defined by

$$\sigma^{-1}(a,b) = \frac{1}{\sigma(a,b)}$$

is also a 2-cocycle.

Let σ be a 2-cocycle on G. The *cocycle twist* of the algebra A is the algebra A^σ whose underlying G-graded vector space is the same as A, and whose multiplication is defined as

$$a \cdot_\sigma b = \sigma(|a|,|b|)ab \tag{2.2}$$

for every homogeneous elements $a, b \in A$. Observe that the defining relation (2.1) of a 2-cocycle is exactly what is needed for the twisted algebra A^σ to be an associative algebra, because, for homogeneous elements $a, b, c \in A$ we have

$$\begin{aligned} a \cdot_\sigma (b \cdot_\sigma c) &= a \cdot_\sigma (\sigma(|b|,|c|)bc) \\ &= \sigma(|a|,|b|+|c|)\sigma(|b|,|c|)abc \\ &= \sigma(|a|,|b|)\sigma(|a|+|b|,|c|)abc \\ &= (a \cdot_\sigma b) \cdot_\sigma c \end{aligned}$$

There is a corresponding relation between the categories of graded modules over A and its cocycle twist A^σ. Let M be a graded right A-module. We define a right A^σ-module structure on M as follows. For homogeneous elements $a \in A^\sigma$ and $m \in M$, we define the action of a on m by

$$m *_\sigma a = \sigma(|a|, |m|)ma.$$

where the action on the right hand side is the right A-action on M. We will shortly see that this defines an equivalence of categories between the categories of graded right modules *Gr-A* and *Gr-*A^σ but first we need to introduce more terminology.

For a graded right module M over A and $g \in G$, the *g-suspension* of M is defined to be the graded right A-module $M(g) = \oplus_{h\in G} M(g)_h$ where $M(g)_h = M_{g+h}$. This defines a functor T_g : *Gr-A* $\longrightarrow$ *Gr-A* by letting $T_g(M) = M(g)$. For G-graded rings A and B, Gordon and Green [**7**] call a functor F : *Gr-A* $\longrightarrow$ *Gr-B* a *graded functor* if it commutes (in the appropriate categories) with the g-suspension functor for every $g \in G$. A graded functor U : *Gr-A* $\longrightarrow$ *Gr-B* is called a *graded equivalence* if there is a graded functor V : *Gr-B* $\longrightarrow$ *Gr-A* such that $VU \simeq 1_{Gr\text{-}A}$ and $UV \simeq 1_{Gr\text{-}B}$. If this is the case, the categories *Gr-A* and *Gr-B* are said to be *graded equivalent.* We are ready to state the following lemma.

LEMMA 2.1. *Let A be a graded ring and σ a 2-cocycle on G. Then Gr-A is graded equivalent to Gr-A^σ.*

PROOF. Let $(-)^\sigma$: *Gr-A* $\longrightarrow$ *Gr-*A^σ denote the functor which assigns to each graded right A-module M the twisted module M^σ, leaving homomorphisms unchanged. Since the functor does not change the underlying G-grading on M, it is clear that it commutes with the suspension functor T_g for every $g \in G$. Moreover, if we let $\sigma^{-1} : G \times G \longrightarrow k^*$ be the 2-cocycle defined by $\sigma^{-1}(a,b) = \frac{1}{\sigma(a,b)}$, then we have $(M^\sigma)^{\sigma^{-1}} = M$ and hence $(-)^{\sigma^{-1}} \circ (-)^\sigma \simeq 1_{Gr\text{-}A}$ and similarly $(-)^\sigma \circ (-)^{\sigma^{-1}} \simeq 1_{Gr\text{-}A^\sigma}$. So it follows that the categories *Gr-A* and *Gr-*A^σ are graded equivalent. □

We would like to connect ($\diamond$) property of A to that of A^σ. We already know that there is an equivalence of categories between *Gr-A* and *Gr-*A^σ, so that the ($\diamond$) property for graded injective hulls passes from *Gr-A* to *Gr-*A^σ. The question is whether the same is true for the categories Mod-A and Mod-A^σ. It turns out that this is indeed the case, as the following result shows that graded equivalences between graded module categories give rise to Morita equivalences between module categories.

THEOREM 2.2 ([**11**], Theorem 1.3. See also [**7**] Theorem 5.4.). *Let A and B be G-graded rings. Let Φ_A (resp. Φ_B) denote the forgetful functor Gr-A $\longrightarrow$ Mod-A (resp. Gr-B $\longrightarrow$ Mod-B). Then the following statements are equivalent.*

(i) *The categories Gr-A and Gr-B are graded equivalent;*
(ii) *There is a Morita equivalence L : Mod-A $\longrightarrow$ Mod-B and a graded functor F : Gr-A $\longrightarrow$ Gr-B such that $\Phi_B \circ F = L \circ \Phi_A$;*
(iii) *There exists an object $P \in$ Gr-A such that $\Phi_A(P)$ is a finitely generated projective generator in Mod-A and the graded ring $\mathrm{End}_A(P)$ is isomorphic to B as graded rings.*

Since ($\diamond$) is a Morita invariant property, the next result follows from the above theorem.

COROLLARY 2.3. *Let A be a G-graded associative k-algebra and let σ be a 2-cocycle on G. Then A has property ($\diamond$) if and only if A^σ does.*

3. Cocycle twists of Lie color algebras

A *commutation factor* ε on G with values in k^* is a mapping $\varepsilon : G \times G \longrightarrow k^*$ such that

$$\varepsilon(a,b)\varepsilon(b,a) = 1, \quad \varepsilon(a,b+c) = \varepsilon(a,b)\varepsilon(a,c), \quad \varepsilon(a+b,c) = \varepsilon(a,c)\varepsilon(b,c)$$

for all $a,b,c \in G$. It is easy to see from the definition that $\varepsilon(a,a) = \pm 1$ for every $a \in G$ and accordingly we define two sets $G_+ = \{a \in G \mid \varepsilon(a,a) = 1\}$ and $G_- = \{a \in G \mid \varepsilon(a,a) = -1\}$. Here, G_+ is a subgroup of G of index at most two and G_- is the other coset when the index of G_+ is two.

A G-graded *Lie color algebra* L with a commutation factor ε is a G-graded vector space $L = \oplus_{g\in G} L_g$ with a bracket operation $[,] : L \times L \longrightarrow L$ which satisfies $[L_g, L_h] \subseteq L_{g+h}$ for all $g, h \in G$, along with color skew symmetry and color Jacobi identities:

$$[x,y] = -\varepsilon(|x|,|y|)[y,x], \tag{3.1}$$

$$[[x,y],z] = [x,[y,z]] - \varepsilon(|x|,|y|)[y,[x,z]] \tag{3.2}$$

for all homogeneous elements x, y, z of L. For example, if ε is chosen to be the trivial commutation factor defined by $\varepsilon(a,b) = 1$ for all $a, b \in G$, then L is nothing but a G-graded ordinary Lie algebra. If $G = \mathbb{Z}_2$ and ε is defined as $\varepsilon(a,b) = (-1)^{ab}$ for all $a, b \in \mathbb{Z}_2$, then L is a Lie superalgebra.

Like the superalgebra case, we have the notion of even and odd elements for Lie color algebras. For a Lie color algebra L, we define two sets $L_+ = \oplus_{g\in G_+} L_g$ and $L_- = \oplus_{g\in G_-} L_g$. The elements of L_+ are called *even* while the elements of L_- are called the *odd* elements of L.

The *universal enveloping algebra* $U(L)$ of a Lie color algebra L is defined as

$$U(L) = T(L)/J(L)$$

where $T(L)$ is the tensor algebra of L and $J(L)$ is the ideal of $T(L)$ which is generated by the elements $x \otimes y - \varepsilon(|x|,|y|)y \otimes x - [x,y]$ for every homogeneous elements x, y of L. The enveloping algebra $U(L)$ of a Lie color algebra is a G-graded associative algebra and it is well-known that when L is finite dimensional $U(L)$ is Noetherian [**15**, Theorem 1].

Let L be a G-graded Lie color algebra with a commutation factor ε. For a 2-cocycle σ on G, if we define a new multiplication on L by

$$[x,y]^\sigma = \sigma(|x|,|y|)[x,y] \tag{3.3}$$

for every homogeneous elements x and y of L, then with this multiplication L^σ becomes a G-graded Lie color algebra with commutation factor $\varepsilon' = \varepsilon\delta$ where $\varepsilon'(a,b) = \varepsilon(a,b)\delta(a,b)$ and $\delta(a,b) = \sigma(a,b)/\sigma(b,a)$ for all $a, b \in G$.

The *descending central sequence* of a Lie color algebra L is defined as

$$\mathcal{C}^0(L) = L, \quad \mathcal{C}^{n+1}(L) = [\mathcal{C}^n(L), L], \ \forall n \geq 0.$$

L is called *nilpotent* if $\mathcal{C}^n(L) = 0$ for some n. Before we proceed any further, we note the following easy fact that being nilpotent is preserved under cocycle twists.

LEMMA 3.1. *Let L be a G-graded Lie color algebra and let σ be a 2-cocycle on G. Then, L is nilpotent if and only if L^σ is nilpotent.*

PROOF. Obviously, $\mathcal{C}^0(L^\sigma) = L = \mathcal{C}(L)$. It is also clear that as k-spaces $\mathcal{C}^{i+1}(L^\sigma) = \mathcal{C}^{i+1}(L)$ for any $i \geq 0$. Hence, $\mathcal{C}^n(L) = 0$ if and only if $\mathcal{C}^n(L^\sigma) = 0$, *i.e.* L is nilpotent if and only if L^σ is nilpotent. □

Cocycle twists can be applied to pass from a Lie color algebra to a Lie superalgebra and we briefly explain this process now. Let L be a G-graded Lie color algebra with a commutation factor ε. If σ is a 2-cocycle on G, we know that L^σ is a G-graded Lie color algebra with a commutation factor $\varepsilon' = \varepsilon\delta$, where δ is as in the lemma. Let $\varepsilon_0 : G \times G \longrightarrow k^*$ denote the superalgebra commutation factor, which is defined by the rule

$$\varepsilon_0(a, b) = 1$$

if $a \in G_+$ or $b \in G_+$ and

$$\varepsilon_0(a, b) = -1$$

when both $a, b \in G_-$. The following result is the crucial step in the process of passing from a Lie color algebra to a Lie superalgebra.

THEOREM 3.2. [**1**, Theorem 2.3] *Let G be an arbitrary abelian group, and k an arbitrary commutative ring with 1 and with group of units k^*. Then for any commutation factor $\varepsilon : G \times G \longrightarrow k^*$ there exists a 2-cocycle $\sigma : G \times G \longrightarrow k^*$ such that if we set $\delta(g, h) = \sigma(g, h)/\sigma(h, g)$, then $\varepsilon\delta = \varepsilon_0$.*

That is, for any G-graded Lie color algebra L with a commutation factor ε, we can find a 2-cocycle σ such that the twisted Lie color algebra L^σ is a Lie superalgebra $L^\sigma = L^\sigma_{\bar{0}} \oplus L^\sigma_{\bar{1}}$ where $L^\sigma_{\bar{0}} = L_+$ and $L^\sigma_{\bar{1}} = L_-$.

Hence, the discussions of the previous section apply to Lie color algebras as well and there is a corresponding equivalence of categories between the graded representations of a Lie color algebra L and the graded representations of its cocycle twist L^σ. While this correspondence only exists between the graded modules over Lie color algebras and Lie superalgebras, we are interested in modules over the enveloping algebras of such algebras. Fortunately, the cocycle twists of Lie color algebras are connected to their enveloping algebras in the following way.

LEMMA 3.3. [**15**, Theorem 2] *Let L be a G-graded Lie color algebra and let σ be a 2-cocycle on G. Then there is an algebra isomorphism*

$$U(L^\sigma) \cong U(L)^\sigma$$

where $U(L)$ is considered as a G-graded associative algebra and $U(L)^\sigma$ is its cocycle twist.

Together with Corollary 2.3, we have:

LEMMA 3.4. *Let L be a G-graded Lie color algebra and let σ be a 2-cocycle on G. Then $U(L)$ has property $(\diamond)$ if and only if $U(L^\sigma)$ does.*

The previous lemma and the Morita equivalence between $U(L)$ and its cocycle twists allow us to reduce our study to that of the enveloping algebra of a Lie superalgebra. Recall that a *central abelian direct factor* of a Lie algebra L is an abelian Lie subalgebra A of L such that $L = A \times B$ for some Lie subalgebra B of L. We can now state the main result of this note.

THEOREM 3.5. *Let L be a finite dimensional nilpotent G-graded Lie color algebra with a commutation factor ε. Let σ be the 2-cocycle such that the twisted* *algebra L^σ is a G-graded Lie superalgebra with $L^\sigma_{\bar 0} = L_+$ and $L^\sigma_{\bar 1} = L_-$. The following statements are equivalent.*

(i) *$U(L)$-Mod has property $(\diamond)$;*
(ii) *$U(L^\sigma)$-Mod has property $(\diamond)$;*
(iii) *$U(L^\sigma_{\bar 0})$-Mod has property $(\diamond)$;*
(iv) *Up to a central abelian direct factor, $L^\sigma_{\bar 0}$ is isomorphic to either*
  (a) *a nilpotent Lie algebra with an abelian ideal of codimension one,*
  (b) *the five dimensional Lie algebra $\mathfrak{g}_5$ with basis $\{e_1, \dots, e_5\}$ and nonzero brackets $[e_1, e_2] = e_3$, $[e_1, e_3] = e_4$, $[e_2, e_3] = e_5$,*
  (c) *the six dimensional Lie algebra $\mathfrak{g}_6$ with basis $\{e_1, \dots, e_6\}$ and nonzero brackets $[e_1, e_3] = e_4$, $[e_2, e_3] = e_5$, $[e_1, e_2] = e_6$.*

PROOF. The equivalence of (i) and (ii) follows from Lemma 3.4. (ii) $\Leftrightarrow$ (iii) $\Leftrightarrow$ (iv) from Lemma 3.1 and [**9**, Theorem 1.1]. □

We remark that the nilpotency of the Lie algebra in Theorem 3.5 is fundamental in the sense that a solvable non-nilpotent Lie algebra does not satisfy property $(\diamond)$ by a result of Musson [**13**, Theorem 2]. In fact, Musson shows that the enveloping algebra $U(L)$ of the two dimensional nonabelian Lie algebra L has simple modules with non-Artinian finitely generated essential extensions, hence $U(L)$ does not satisfy property $(\diamond)$. On the other hand, L is an epimorphic image of every finite dimensional solvable non-nilpotent Lie algebra, and since $(\diamond)$ property is preserved under homomorphic images, the enveloping algebras of such Lie algebras cannot satisfy $(\diamond)$.

4. Nilpotent Lie algebras of almost maximal index

If L is a Lie algebra and $f \in L^*$, we denote the alternating bilinear form $(x, y) \mapsto f([x, y])$ by B_f. The *index* of L is defined to be the number $\inf_{f \in L^*} \dim L^f$ where $L^f = \{x \in L \mid f([x, y]) = 0 \text{ for all } y \in L\}$ denotes the kernel of the bilinear form B_f.

A Lie algebra L is said to have *almost maximal index* if $\operatorname{ind}(L) = \dim(L) - 2$. According to [**9**, Proposition 5.3], a finite dimensional nilpotent Lie algebra has almost maximal index if and only if it is one of the Lie algebras appearing in Theorem 3.5(iv).

Finite dimensional Lie algebras with an abelian ideal of codimension one are in one-to-one correspondence with finite dimensional vector spaces V and nilpotent endomorphisms $f : V \longrightarrow V$. Given such a vector space and a nilpotent endomorphism, one defines a Lie algebra structure on the vector space $L = kx \oplus V$ by defining $[x, v] = f(v)$. A typical example of this construction is the n-dimensional standard filiform Lie algebra, which is the Lie algebra L_n with a basis $\{e_1, e_2, \dots, e_n\}$ and whose nonzero brackets are $[e_1, e_i] = e_{i+1}$ for $i = 2, 3, \dots, n-1$. For instance the three dimensional Heisenberg Lie algebra is the three dimensional standard filiform Lie algebra.

The n-dimensional standard filiform Lie algebra L_n is realized in the above sense as a vector space $L_n = kx_1 \oplus V$ where V is the span of the elements $x_2, x_3, \dots, x_n$ with a nilpotent endomorphism $f : V \longrightarrow V$ given by $f(x_i) = x_{i+1}$ for all $i = 2, 3, \dots, n-1$.

5. Acknowledgments

The author wishes to thank the referee for his/her comments and suggestions.

References

[1] Y. Bahturin and S. Montgomery, *PI-envelopes of Lie superalgebras*, Proc. Amer. Math. Soc. **127** (1999), no. 10, 2829–2839, DOI 10.1090/S0002-9939-99-04825-X. MR1600092

[2] K. Brown, P.A.A.B. Carvalho, and J. Matczuk, *Simple Modules and Their Essential Extensions for Skew Polynomial Rings*, arXiv:1705.06596.

[3] P. A. A. B. Carvalho, C. Hatipoğlu, and C. Lomp, *Injective hulls of simple modules over differential operator rings*, Comm. Algebra **43** (2015), no. 10, 4221–4230, DOI 10.1080/00927872.2014.941469. MR3366571

[4] P. A. A. B. Carvalho, C. Lomp, and D. Pusat-Yilmaz, *Injective modules over down-up algebras*, Glasg. Math. J. **52** (2010), no. A, 53–59, DOI 10.1017/S0017089510000261. MR2669095

[5] P.A.A.B. Carvalho, C. Lomp, and P.F. Smith, *A Note on Simple Modules Over Quasi-Local Rings*, arXiv:1711.10580.

[6] P. A. A. B. Carvalho and I. M. Musson, *Monolithic modules over Noetherian rings*, Glasg. Math. J. **53** (2011), no. 3, 683–692, DOI 10.1017/S0017089511000267. MR2822809

[7] R. Gordon and E. L. Green, *Graded Artin algebras*, J. Algebra **76** (1982), no. 1, 111–137, DOI 10.1016/0021-8693(82)90240-X. MR659212

[8] C. Hatipoğlu, *Stable torsion theories and the injective hulls of simple modules*, Int. Electron. J. Algebra **16** (2014), 89–98, DOI 10.24330/ieja.266229. MR3239073

[9] C. Hatipoğlu and C. Lomp, *Injective hulls of simple modules over finite dimensional nilpotent complex Lie superalgebras*, J. Algebra **361** (2012), 79–91, DOI 10.1016/j.jalgebra.2012.03.033. MR2921612

[10] A. V. Jategaonkar, *Jacobson's conjecture and modules over fully bounded Noetherian rings*, J. Algebra **30** (1974), 103–121, DOI 10.1016/0021-8693(74)90195-1. MR0352170

[11] C. Menini and C. Năstăsescu, *When is R-***gr** *equivalent to the category of modules?*, J. Pure Appl. Algebra **51** (1988), no. 3, 277–291, DOI 10.1016/0022-4049(88)90067-9. MR946579

[12] I. M. Musson, *Injective modules for group rings of polycyclic groups. I, II*, Quart. J. Math. Oxford Ser. (2) **31** (1980), no. 124, 429–448, 449–466, DOI 10.1093/qmath/31.4.429. MR596979

[13] I. M. Musson, *Some examples of modules over Noetherian rings*, Glasgow Math. J. **23** (1982), no. 1, 9–13, DOI 10.1017/S0017089500004730. MR641613

[14] I. M. Musson, *Finitely generated, non-Artinian monolithic modules*, New trends in noncommutative algebra, Contemp. Math., vol. 562, Amer. Math. Soc., Providence, RI, 2012, pp. 211–220, DOI 10.1090/conm/562/11138. MR2905561

[15] M. Scheunert, *Generalized Lie algebras*, J. Math. Phys. **20** (1979), no. 4, 712–720, DOI 10.1063/1.524113. MR529734

Department of Mathematics, College of Engineering and Technology, American University of the Middle East, Kuwait

Email address: osman.hatipoglu@aum.edu.kw

Contemporary Mathematics
Volume **727**, 2019
https://doi.org/10.1090/conm/727/14632

U-rings generated by its idempotents

Yasser Ibrahim and Mohamed Yousif

ABSTRACT. According to [Comm. Algebra 46 (2018), no. 2, 870–886], a ring R is called right Utumi-ring (U-ring) if, whenever A and B are right ideals of R with $A \cong B$ and $A \cap B = 0$, there exist idempotents e, f and g of R such that $A \subseteq^{ess} eR$, $B \subseteq^{ess} fR$ and $eR \oplus fR = gR$. The class of U-rings is a strict and simultaneous generalization of quasi-continuous, square-free and automorphism-invariant rings. In this paper, we prove that every right U-ring that is totally non-abelian (i.e. every non-zero right ideal of R contains a non-central idempotent) is generated as a ring by its idempotents. Moreover, under these assumptions R turns out to be regular and right self-injective. This result extends the work of Utumi who proved that every totally non-abelian regular right self-injective ring is generated as a ring by its idempotents.

1. Introduction

In [**4**], Erdös showed that the linear transformations of a finite-dimensional vector space over a division ring which are products of proper idempotents are precisely the singular ones. Erdös' results were subsequently extended by Reynolds and Sullivan [**17**] to linear transformations of an arbitrary dimension vector space over a division ring. Subsequently, O'Meara [**16**] provided a complete characterization of the elements of a prime, regular, right self-injective ring which can be written as a product of idempotents. In [**9**], Hannah and O'Meara extended the work in [**16**] to regular right self-injective rings. Subsequently, more work on the decomposition of singular matrices as a product of idempotents was carried out by Alahmedi, Jain, Lam and Leroy in [**1**], Alahmedi, Jain, and Leroy in [**2**], and by Facchini and Leroy in [**6**].

In the mean time it has been known that many of the above classes of regular self-injective rings are generated as rings by their idempotents, that is each element is a sum of products of idempotents. Indeed, Wolfson [**20**] proved that every right full linear ring that is not a division ring is generated as a ring by its idempotents. Right full linear rings R are known to be prime, regular, and right self-injective, and if the ring R is not a division ring then R is also totally non-abelian, where in this paper a ring R (not necessarily regular) is called totally non-abelian if every non-zero right ideal of R contains a non-central idempotent. Consequently, Wolfson's

2010 *Mathematics Subject Classification.* Primary 16D40, 16D50, 16D60; Secondary 16L30, 16L60, 16P20, 16P40, 16P60.

Key words and phrases. Injective, quasi-injective, quasi-continuous, square-free and automorphism-invariant modules, self-injective and regular rings.

result was extended by Utumi in [**19**], who proved that if R is a regular right self-injective ring that is totally non abelian, then R is generated as a ring by its idempotents (see Goodearl [**11**, Theorem 2.37] for an excellent proof of this result).

The aim of this paper is to extend the aforementioned result of Utumi to a larger class of rings. Indeed, we will show that if R is a totally non abelian right U-ring, then R is generated as a ring by its idempotents. Moreover, under these assumptions R turns out to be regular and right self-injective. The class of U-rings is quite large and was introduced in [**10**] as a non-trivial and simultaneous generalization of the quasi-continuous, the square-free and the automorphism-invariant rings (i.e. R_R is invariant under any automorphism of its injective hull [**14**]). As an immediate application we provide a different argument for one of the main results in [**5**], namely every prime right non-singular right auto-invariant ring is right self-injective. In particular, every simple right auto-invariant ring is right self-injective.

Throughout this paper, rings R are associative with unity and modules are unitary. If M is an R-module, we denote by $rad(M)$, $soc(M)$, and $Z(M)$ the Jacobson radical, the socle, and the singular submodule of M, respectively. If $M = R$, we write $J = J(R) = rad(R)$. We also write $N \subseteq M$ if N is a submodule of M, $N \subseteq^{ess} M$ if N is an essential submodule of M, $N \subseteq^{\oplus} M$ if N is a (direct) summand of M, and $A \cong B$ if A is isomorphic to B as R-modules.

2. Preliminaries

For R-modules M and N, M is called N-*injective* if every R-homomorphism from a submodule of N into M can be extended to an R-homomorphism from N into M. The module M is called quasi-injective if M is M-injective. Quasi-injective modules are described in the literature as those modules which are invariant under any homomorphism of its injective hull. Modules that are invariant under any automorphism of their injective hulls are called auto-invariant modules by Lee and Zhou in [**14**]. Every quasi-injective module satisfies what is known in the literature by the Ci-conditions (continuity conditions), $1 \leq i \leq 3$. A module M is called a $C1$-module (CS-module) if every submodule is essential in a (direct) summand, M is called a $C2$-module if every submodule isomorphic to a summand of M is itself a summand of M. Every $C2$-module is a $C3$-module, i.e. the sum of any two summands of M with zero intersection is a summand of M. A module M is called (quasi-)continuous if M is both a CS- and a $C2$-module ($C3$-module).

By modifying and combining the above continuity conditions in one single definition, the notion of a U-module was introduced and thoroughly investigated in [**10**].

DEFINITION 2.1. A right R-module M is called a Utumi-module (U-module) if for any two non-zero submodules A and B of M with $A \cong B$ and $A \cap B = 0$, there exist two summands K and L of M such that $A \subseteq^{ess} K$, $B \subseteq^{ess} L$ and $K \oplus L \subseteq^{\oplus} M$. A ring R is called a right U-ring if R is a U-module as a right R-module.

Clearly, every quasi-continuous module is a U-module, and it was shown in [**10**] that every auto-invariant module is a U-module. Another interesting class of U-modules is that of square-free modules, where a module is called square-free if it contains no non-zero submodules isomorphic to a square $A \oplus A$.

The next four lemmas are needed for the proof of our main result, the first three of which have already been established in [**10**]. But first, recall that a module M is called pseudo-injective relative to another module N (pseudo-N-*injective*) if every monomorphism $f : K \to M$, where $K \subseteq N$, can be extended to a homomorphism from N into M. A module M is called pseudo-injective if it is pseudo-M-*injective*. For a full account on the subject of relative pseudo-injectivity, see [**18**]. It was shown in [**5**] that the class of pseudo-injective modules coincide with that of auto-invariant modules.

LEMMA 2.2. [**10**, Propositions 2.11 and 2.12] *If $X =: A \oplus B$ is a U-module, then A and B are relatively pseudo-injective.*

LEMMA 2.3. [**10**, Lemma 2.8] *and* [**3**, Theorem 2.2] *If M is a U-module and A and B are two submodules of M with $A \subseteq^{\oplus} M$, $A \cap B = 0$ and $B \cong A$, then $B \subseteq^{\oplus} M$.*

LEMMA 2.4. [**10**, Lemma 3.20] *If e is a non-central idempotent, then either eR or $(1-e)R$ is not a two-sided ideal of R.*

LEMMA 2.5. *Every right non-singular right square-free ring R is abelian.*

PROOF. Let e be a non-central idempotent of the ring R. By Lemma 2.4, either eR or $(1-e)R$ is not a two-sided ideal of R. Without loss of generality assume that eR is not a two sided ideal of R. Therefore, $Re \nsubseteq eR$ and $(1-e)Re \neq 0$. Let $0 \neq (1-e)ae \in (1-e)Re$ and define $h : eR \longrightarrow (1-e)R$ by $h(er) = (1-e)aer$. Clearly h is a non-zero map and $\ker h$ is not essential in eR. If T is a non-zero submodule of eR with $\ker(h) \cap T = 0$, then $T \cong h(T)$ with $T \cap h(T) \subseteq eR \cap (1-e)R = 0$, a contradiction. □

3. The Results

We will say that a right ideal T of R is square-free if T_R is square-free as a right R-module.

THEOREM 3.1. *Let R be a right U-ring. If every non-zero square-free right ideal of R contains a non-central idempotent, then $R = A \oplus B \oplus K$ with $A \cong B$ and K is a square-free module isomorphic to a submodule of $A \oplus B$. Moreover, R is a right self-injective ring and is generated as a ring by its idempotents.*

PROOF. Let $\digamma = \left\{ (A, B, f) : A, B \subseteq R_R,\ A \cap B = 0, \text{ and } A \overset{f}{\cong} B \right\}$ with the following order:

$$(A, B, f) \leq (A_1, B_1, f_1) \text{ if } A \subseteq A_1, B \subseteq B_1 \text{ and } f_1 \text{ extends } f.$$

Clearly, $\digamma$ is a non-empty inductive set. As in the proof of [**10**, Theorem 3.11], if (A, B, f) is a maximal element of $\digamma$, then $R = A \oplus B \oplus K$ where $A_R \cong B_R$, $A \oplus B$ is quasi-injective, K_R is a square-free module and K is $(A \oplus B)$-injective. If $K = 0$, then R is right self-injective. Otherwise, assume that $K \neq 0$. Since every non-zero square-free right ideal of R contains a non-central idempotent, $Z(K_R) = 0$, i.e. K_R is non-singular. By Lemma 2.5, $A \oplus B \neq 0$. Now let

$$\Gamma = \{X \subseteq K : X \text{ is isomorphic to a submodule of } C = A \oplus B\}.$$

Claim: every non-zero submodule M of K contains a non-zero element in Γ.

By the hypothesis, M contains a non-central idempotent e of R and, by Lemma 2.4, either eR or $(1-e)R$ is not a two-sided ideal of R.

Case 1: eR is not a two sided ideal of R. Therefore, $Re \nsubseteq eR$ and so $(1-e)Re \neq 0$. Let $0 \neq (1-e)ae \in (1-e)Re$ and define $h : eR \longrightarrow (1-e)R$ by $h(er) = (1-e)aer$. Clearly, h is a non-zero well-defined map. If $\gamma : R \longrightarrow A \oplus B$ is the projection map with $\ker \gamma = K$, then either $\gamma(\mathrm{Im}h) = 0$ or $\gamma(\mathrm{Im}h) \neq 0$.

Subcase 1a: $\gamma(\mathrm{Im}h) = 0$. In this case, $\mathrm{Im}h \subseteq \ker \gamma = K$. Since K is non-singular, $\ker h$ is not essential in eR. Let T be a submodule of eR with $T \cap \ker h = 0$. Now $T \cong h(T)$ with $T \cap h(T) \subseteq eR \cap (1-e)R = 0$. Inasmuch as K is square-free, we infer that $h(T) \oplus T \nsubseteq K$ and $\gamma(h(T)) \neq 0$, a contradiction.

Subcase 1b: $\gamma(\mathrm{Im}h) \neq 0$. In this case, the restriction $g =: \gamma|_{\mathrm{Im}h} : \mathrm{Im}h \longrightarrow A \oplus B$ is a non-zero map with $\ker g = \mathrm{Im}h \cap \ker \gamma = \mathrm{Im}h \cap K$. We claim that g is a monomorphism. For, suppose that $\ker g \neq 0$. Then there exists $0 \neq h(x) \in K$ with $x \in eR$ and, consequently, the restriction $h' =: h|_{xR} : xR \longrightarrow (1-e)R$ is a non-zero map with $\mathrm{Im}h' = h(xR) \subseteq K$ is nonsingular. Thus $\ker h'$ is not essential in xR and there is a non-zero submodule $T' \subseteq xR$ with $\ker h' \cap T' = 0$. Consequently, $T' \cong h'(T')$ with $T' \cap h'(T') \subseteq xR \cap (1-e)R \subseteq eR \cap (1-e)R = 0$. This is a contradiction since both T' and $h'(T')$ are contained in K and K is square-free. Therefore $\ker g = 0$ and the map $\gamma \circ h : eR \longrightarrow A \oplus B$ is a monomorphism, proving the claim in this subcase.

Case 2: $(1-e)R$ is not a two sided ideal of R. Therefore, $R(1-e) \nsubseteq (1-e)R$, $eR(1-e) \neq 0$ and as in **case 1,** there exists a non-zero map $\theta : (1-e)R \longrightarrow eR$. Since eR is a non-singular module, $\ker \theta$ is not essential in $(1-e)R$. Let T be a non-zero submodule of $(1-e)R$ with $\ker(\theta) \cap T = 0$. Clearly, $T \cap \theta(T) = 0$ and we claim that $(\theta(T) \oplus T) \cap (A \oplus B) \neq 0$. Assume to the contrary that $(\theta(T) \oplus T) \cap (A \oplus B) = 0$ and define the map $f_1 : A \oplus \theta(T) \to B \oplus T$ by $f_1(a+x) = f(a) + \sigma(x)$, for all $a \in A$ and $x \in \theta(T)$, where $\theta(T) \overset{\sigma}{\cong} T$. Clearly, f_1 is an extension of f and $A \oplus \theta(T) \overset{f_1}{\cong} B \oplus T$. This means $(A, B, f) \lneqq (A \oplus \theta(T), B \oplus T, f_1)$, a contradiction to the maximality of (A, B, f). Thus $(\theta(T) \oplus T) \cap (A \oplus B) \neq 0$. If $\pi : R \longrightarrow (1-e)R$ denotes the obvious projection, then the restriction $h =: \pi|_{(A \oplus B)} : A \oplus B \longrightarrow (1-e)R$ is a non-zero map with $\ker h = eR \cap (A \oplus B) = 0$. Thus h is a monomorphism with $A \oplus B \cong \mathrm{Im}h \subseteq (1-e)R$.

We claim that if $\beta =: \theta|_{(\mathrm{Im}h)} : \mathrm{Im}h \longrightarrow eR$, then β is a non-zero map. For, suppose to the contrary that $\beta = 0$ and let $z = x + y \in (\theta(T) \oplus T) \cap (A \oplus B)$, where $x \in \theta(T)$ and $y \in T$. Then $0 = \beta(h(z)) = \theta(\pi(z)) = \theta(\pi(x+y)) = \theta(y)$, and so $y \in T \cap \ker \theta = 0$. Therefore $z = x \in (A \oplus B) \cap \theta(T) = 0$, and hence $(\theta(T) \oplus T) \cap (A \oplus B) = 0$, a contradiction.

Now the map $\beta \circ h : A \oplus B \longrightarrow eR$ is a non-zero R-homomorphism and since eR is non-singular, $\ker(\beta \circ h)$ is not essential in $A \oplus B$. If $0 \neq L \subseteq A \oplus B$ with $\ker(\beta \circ h) \cap L = 0$, then $L \cong \beta \circ h(L) \in \Gamma$ as required, proving the claim in this case.

Now if we order the set Γ by inclusion, then clearly Γ is a non-empty inductive set. Let H be a maximal element of Γ. We will show that $H \subseteq^{ess} K$. Suppose that $H \cap E = 0$ for a submodule $E \subseteq K$. If $E \neq 0$, then by the above claim E contains a non-zero element $Y \in \Gamma$. Now, since $\{H, Y\} \subseteq \Gamma$, there exist sumodules H', Y' such that $H \cong H' \subseteq A \oplus B$ and $Y \cong Y' \subseteq A \oplus B$. We claim that $H' \cap Y' = 0$. Otherwise if $Z' = H' \cap Y' \neq 0$, then $0 \neq Z' \cong X_1$ and $0 \neq Z' \cong X_2$ where $X_1 \subseteq H \subseteq K$, $X_2 \subseteq Y \subseteq K$ and $X_1 \cap X_2 = 0$. This is a contradiction since K is a square-free module. Therefore $H \oplus Y \cong H' \oplus Y' \subseteq A \oplus B$ with $H \oplus Y \subseteq K$, a clear contradiction to the choice of H. Therefore, $E = 0$ and $H \subseteq^{ess} K$.

Next, we will show that K is isomorphic to a submodule of $A \oplus B$. To see this, let $\alpha : H \longrightarrow A \oplus B$ be an R-monomorphism. Inasmuch as $R = A \oplus B \oplus K$ is a U-module, we infer from Lemma 2.2 that $A \oplus B$ is pseudo-K-injective. Let $\lambda : K \longrightarrow A \oplus B$ be an extension of α. Since $H \subseteq^{ess} K$, λ is a monomorphism and $K \cong L \subseteq A \oplus B$.

Now we are in a position to show that R is right self-injective. Since $L \cong K \subseteq^{\oplus} R$, $L \cap K = 0$ and R_R is a U-module, it follows from Lemma 2.3 that $L \subseteq^{\oplus} R_R$ and consequently $L \subseteq^{\oplus} A \oplus B$. Since $A \oplus B$ is quasi-injective, L is quasi-injective and consequently K is both quasi-injective and $A \oplus B$-injective. This means that K is $(A \oplus B \oplus K)$-injective, i.e. K is R-injective and hence injective. Similarly since $L \subseteq A \oplus B$ and $A \oplus B$ is quasi-injective, $A \oplus B$ is L-injective. Thus $A \oplus B$ is K-injective and hence $A \oplus B$ is $(A \oplus B \oplus K)$-injective, i.e. $A \oplus B$ is injective and $R = A \oplus B \oplus K$ is right self-injective.

Finally we will show that R is generated as a ring by its idempotents. Let E denote the subring of R generated by its idempotents. Write $R = e_1R \oplus e_2R \oplus e_3R$ with $e_1R = A$, $e_2R = B$, $e_3R = K$ and e_1, e_2, e_3 are orthogonal idempotents of R. If $r \in e_iRe_j$ with $i \neq j$, then clearly $e_i + r$ is an idempotent of R and so $r = e_i + r - e_i \in E$. Hence $e_iRe_j \subseteq E$ whenever $i \neq j$. We claim that the isomorphism $e_1R \cong e_2R$ is given by left multiplication by some $x \in e_2Re_1$. For, if $f : e_1R \to e_2R$ is an isomorphism and $r \in R$, then $f(e_1r) = f(e_1)r = f(e_1)e_1r = e_2te_1r = (e_2te_1)e_1r = xe_1r$ where $t \in R$ and $x = e_2te_1$. Similarly, the inverse function $f^{-1} : e_2R \to e_1R$ is given by left multiplication by some $y \in e_1Re_2$. So $e_1 = f^{-1}f(e_1) = f^{-1}(xe_1) = f^{-1}(x) = yx$. Now if $r \in e_1Re_1$, then $r = e_1r = yxr = (y)(xr) \in (e_1Re_2)(e_2Re_1) \subseteq E$. Consequently, $e_1Re_1 \subseteq E$. Similarly, $e_2Re_2 \subseteq E$. Now, since $A = e_1Re_1 + e_1Re_2 + e_1Re_3$ and $B = e_2Re_2 + e_2Re_1 + e_2Re_3$, we infer that $A \subseteq E$ and $B \subseteq E$. To complete the proof, we only need to show that $K \subseteq E$. To see this, let $\lambda : K \to A \oplus B$ be the above monomorphism with $\lambda(K) = L \subseteq^{\oplus} A \oplus B$. By Lemma 2.3, since $K \subseteq^{\oplus} R, L \subseteq^{\oplus} R$ and $L \cap K = 0$, we infer that $K \oplus L \subseteq^{\oplus} R$. If we write $R = K \oplus L \oplus N$ for a submodule $N \subseteq R_R$, then we may proceed as before and show that $K \subseteq E$. Hence $R = A \oplus B \oplus K \subseteq E$, completing the proof. □

We will call a ring R *totally non-abelian* if every non-zero right ideal of R contains a non-central idempotent. This definition goes back to Utumi [**19**] who used a different terminology. The terminology "*totally non-abelian*" was introduced and used, strictly for regular rings, by Goodearl in [**11**, Page 62]. Indeed, this notion is left-right symmetric if R is a regular ring, and it is not difficult to see that if R is a prime regular ring that is not a division ring, then R is totally non-abelian. In particular every right full linear ring that is not a division ring is an example of a regular right self-injective ring that is totally non-abelian. Therefore, by [**19**], every right full linear ring that is not a division ring is generated as a ring by its idempotents. The later statement extends the main result of Wolfson in [**20**]. In the next theorem we obtain a far-reaching result for the class of U-rings.

THEOREM 3.2. *If R is a totally non-abelian right U-ring, then R is a regular right self-injective ring and is generated as a ring by its idempotents.*

PROOF. By Theorem 3.1, R is right self-injective and is generated as a ring by its idempotents. Since R is right self-injective, $Z(R_R) = J(R)$ and $R/J(R)$ is a regular ring. Inasmuch as R is totally non-abelian, $Z(R_R) = 0$, and hence R is regular. □

COROLLARY 3.3. *If R is a totally non-abelian right quasi-continuous ring, then R is a regular right self-injective ring and is generated as a ring by its idempotents.*

COROLLARY 3.4. *If R is a totally non-abelian right automorphism-invariant ring, then R is a regular right self-injective ring and is generated as a ring by its idempotents.*

In the next corollary, we provide a different argument for one of the main results in [**5**, Theorem 8].

COROLLARY 3.5. [**5**, Theorem 8 and Corollary 10] *If R is a prime right non-singular right auto-invariant ring, then R is right self-injective. In particular, every simple right auto-invariant ring is right self-injective.*

PROOF. If R is a division ring, then clearly R is self-injective. Otherwise suppose that R is not a division ring. By [**7**, Proposition 1], since R is non-singular, R is a regular ring. Now, inasmuch as R is a prime regular ring that is not a division ring, we infer that R is totally non-abelian. By Corollary 3.4, R is right self-injective. □

Recall that, [**15**, Definition 2.34], a submodule T of a module M is called a square-root in M if $T^2 =: T \oplus T$ embeds in M. A module M is called square-full if every non-zero submodule of M contains a non-zero square-root in M.

COROLLARY 3.6. *Every square-full right U-ring is right self-injective and is generated as a ring by its idempotents.*

PROOF. By [**10**, Theorem 3.13], $R_R = Q \oplus T$ where Q is a quasi-injective right ideal, T is a square-free right ideal, Q_R and T_R are orthogonal as right R-modules, and $Q = A \oplus B \oplus K$ such that $A_R \cong B_R$ and K_R is isomorphic to a submodule of $A \oplus B$. We claim that $T = 0$. Assume to the contrary that $T \neq 0$, and let $X_R \subseteq T_R$ be a non-zero square-root in R_R. This means there exist right ideals W and Z of R such that $W \cap Z = 0$ and $W \cong Z \cong X$. Now, we will establish the claim by considering the following two cases:

Case1: $Q \cap (W \oplus Z) = 0$. In this case, if $\eta : Q \oplus T \longrightarrow T$ is the natural projection, then the restriction map $\eta|_{W \oplus Z} : W \oplus Z \longrightarrow T$ is a monomorphism. This means T contains the two right ideals $\eta(W)$ and $\eta(Z)$ that satisfy $\eta(W) \cap \eta(Z) = 0$ and $\eta(W) \cong \eta(Z)$, a contradiction since T is square-free.

Case2: $Q \cap (W \oplus Z) \neq 0$. In this case, it follows from [**15**, Lemma 1.31] that either Q and W have non-zero ismomorphic submodules or Q and Z have non-zero ismomorphic submodules. Assume that Q and W have non-zero ismomorphic submodules Q' and W'. Since $W \cong X$, there exists a submodule $X' \subseteq X$ such that $Q \supseteq Q' \cong W' \cong X' \subseteq X \subseteq T$, a contradiction since Q and T are orthogonal.

Thus $T = 0$, proving the claim, and $R = Q = A \oplus B \oplus K$ is right self-injective. Now, by proceeding with an argument similar to the one used in the last paragraph of the proof of Theorem 3.1, we can show that R is generated as a ring by its idempotents. □

4. Examples

The following is an example of a totally non-abelian regular ring that is neither left nor right U-ring.

EXAMPLE 4.1. Let A be an infinite set and consider the column finite matrix ring $S = \mathbb{CFM}_A(F)$ over a division ring F. Let T be the set of all matrices in S with only a finite number of non-zero entries and view F as the subring of all scalar matrices of S. If $R = T + F$, then R is a prime regular ring with $soc(R) = T$. Inasmuch as R is a prime regular ring, R is totally non-abelian. Since R is not self-injective it follows, from Theorem 3.2, that R is neither left nor right U-ring.

REMARK 4.2. We should point out that on the other extreme of Utumi's result on totally non-abelian regular self-injective rings, a remarkable result of G.M. Bergman [**8**, Corollary 4.3] asserts that if R is a commutative ring generated by its idempotents, then the additive group $(R, +)$ is a direct sum of cyclic groups.

Next we give an example of a commutative U-ring (indeed it is self-injective, square-free ring) which is not generated as a ring by its idempotents.

EXAMPLE 4.3. Let $\mathcal{P}$ be the set of all primes $p \in \mathbb{Z}$, and set $R = \prod_{p\in\mathcal{P}} \mathbb{Z}/p\mathbb{Z}$. Clearly, R is commutative, regular, self-injective, and square-free ring. We claim that R is not generated as a ring by its idempotents. Otherwise, by Bergman's result [**8**, Corollary 4.3] the additive group R^+ will be a direct sum of cyclic groups, a contradiction.

The following is an example of an artinian, square-free, self-injective ring (hence U-ring) that is neither abelian nor totally non-abelian.

EXAMPLE 4.4. Let F be a field and R be the F-algebra consisting of all 4×4 matrices of the form $\begin{pmatrix} a & x & 0 & 0 \\ 0 & b & 0 & 0 \\ 0 & 0 & b & y \\ 0 & 0 & 0 & a \end{pmatrix}$, where $a, b, x, y \in F$. It is shown in [**13**, 16.19 (4)] that the ring R is quasi-Frobenius, in particular self-injective, with $J^2(R) = 0$. As in [**12**], if K consists of the matrices of the form $\begin{pmatrix} 0 & b & 0 & 0 \\ 0 & b & 0 & 0 \\ 0 & 0 & b & y \\ 0 & 0 & 0 & 0 \end{pmatrix}$ and L consists of the matrices of the form $\begin{pmatrix} 0 & 0 & 0 & 0 \\ 0 & b & 0 & 0 \\ 0 & 0 & b & y \\ 0 & 0 & 0 & 0 \end{pmatrix}$, then both K and L are direct summands of R_R. However, $(K + L) \subsetneqq^{ess} R_R$, and hence $K + L$ is not a summand of R_R. Therefore, R is not abelian. By Theorem 3.2, R can not be totally non-abelian. Note that the right ideals of the form $S_1 = \begin{pmatrix} 0 & x & 0 & 0 \\ 0 & 0 & 0 & 0 \\ 0 & 0 & 0 & 0 \\ 0 & 0 & 0 & 0 \end{pmatrix}$ and $S_2 = \begin{pmatrix} 0 & 0 & 0 & 0 \\ 0 & 0 & 0 & 0 \\ 0 & 0 & 0 & y \\ 0 & 0 & 0 & 0 \end{pmatrix}$ are the only minimal right ideals of R with $S_1 \not\cong S_2$ and

$J(R) = Z(R) = soc(R_R) = (S_1 \oplus S_2) \subseteq^{ess} R_R$. We claim that R is square-free as a right R-module. If A and B are any two non-zero right ideals of R with $A \cap B = 0$, then without loss of generality we may assume that $S_1 \subseteq A$ and $S_2 \subseteq B$. Since $S_1 \not\cong S_2$, we infer that $A \not\cong B$, proving the claim.

Acknowledgments

Part of this work was carried out while the first author was visiting the Mathematics Department of The Ohio State University. He would like to express his gratitude to the members of the mathematics department at OSU for the financial support and the warm reception. The research of the second author was supported by the Mathematics Research Institute of the Ohio State University.

References

[1] A. Alahmadi, S. K. Jain, T. Y. Lam, and A. Leroy, *Euclidean pairs and quasi-Euclidean rings*, J. Algebra **406** (2014), 154–170, DOI 10.1016/j.jalgebra.2014.02.009. MR3188333

[2] A. Alahmadi, S. K. Jain, and A. Leroy, *Decomposition of singular matrices into idempotents*, Linear Multilinear Algebra **62** (2014), no. 1, 13–27, DOI 10.1080/03081087.2012.754439. MR3175397

[3] N. Ding, Y. Ibrahim, M. Yousif, and Y. Zhou, *C4-modules*, Comm. Algebra **45** (2017), no. 4, 1727–1740, DOI 10.1080/00927872.2016.1222412. MR3576690

[4] J. A. Erdos, *On products of idempotent matrices*, Glasgow Math. J. **8** (1967), 118–122, DOI 10.1017/S0017089500000173. MR0220751

[5] N. Er, S. Singh, and A. K. Srivastava, *Rings and modules which are stable under automorphisms of their injective hulls*, J. Algebra **379** (2013), 223–229, DOI 10.1016/j.jalgebra.2013.01.021. MR3019253

[6] A. Facchini and A. Leroy, *Elementary matrices and products of idempotents*, Linear Multilinear Algebra **64** (2016), no. 10, 1916–1935, DOI 10.1080/03081087.2015.1127885. MR3521148

[7] P. A. Guil Asensio and A. K. Srivastava, *Automorphism-invariant modules satisfy the exchange property*, J. Algebra **388** (2013), 101–106, DOI 10.1016/j.jalgebra.2013.05.003. MR3061680

[8] G. M. Bergman, *Boolean rings of projection maps*, J. London Math. Soc. (2) **4** (1972), 593–598, DOI 10.1112/jlms/s2-4.4.593. MR0311531

[9] J. Hannah and K. C. O'Meara, *Products of idempotents in regular rings. II*, J. Algebra **123** (1989), no. 1, 223–239, DOI 10.1016/0021-8693(89)90044-6. MR1000485

[10] Y. Ibrahim and M. Yousif, *Utumi modules*, Comm. Algebra **46** (2018), no. 2, 870–886, DOI 10.1080/00927872.2017.1339064. MR3764904

[11] K. R. Goodearl, *Ring theory*, Marcel Dekker, Inc., New York-Basel, 1976. Nonsingular rings and modules; Pure and Applied Mathematics, No. 33. MR0429962

[12] A. Hamdouni, A. Ç. Özcan, and A. Harmancı, *Characterization of modules and rings by the summand intersection property and the summand sum property*, JP J. Algebra Number Theory Appl. **5** (2005), no. 3, 469–490. MR2189962

[13] T. Y. Lam, *Lectures on modules and rings*, Graduate Texts in Mathematics, vol. 189, Springer-Verlag, New York, 1999. MR1653294

[14] T.-K. Lee and Y. Zhou, *Modules which are invariant under automorphisms of their injective hulls*, J. Algebra Appl. **12** (2013), no. 2, 1250159, 9, DOI 10.1142/S0219498812501599. MR3005608

[15] S. H. Mohamed and B. J. Müller, *Continuous and discrete modules*, London Mathematical Society Lecture Note Series, vol. 147, Cambridge University Press, Cambridge, 1990. MR1084376

[16] K. C. O'Meara, *Products of idempotents in regular rings*, Glasgow Math. J. **28** (1986), no. 2, 143–152, DOI 10.1017/S0017089500006467. MR848421

[17] M. A. Reynolds and R. P. Sullivan, *Products of idempotent linear transformations*, Proc. Roy. Soc. Edinburgh Sect. A **100** (1985), no. 1-2, 123–138, DOI 10.1017/S0308210500013688. MR801849

[18] S. Singh and S. K. Jain, *On pseudo injective modules and self pseudo injective rings*, J. Math. Sci. **2** (1967), 23–31. MR0219569

[19] Y. Utumi, *On continuous regular rings and semi-simple self injective rings*, Canad. J. Math. **12** (1960), 597–605, DOI 10.4153/CJM-1960-053-9. MR0117250

[20] K. G. Wolfson, *An ideal-theoretic characterization of the ring of all linear transformations*, Amer. J. Math. **75** (1953), 358–386, DOI 10.2307/2372458. MR0053080

Department of Mathematics, Faculty of Science, Taibah University, Madina 42353, Saudi Arabia; and Department of Mathematics, Faculty of Science, Cairo University, Giza, 12613, Egypt

Email address: yfibrahim@sci.cu.edu.eg

Department of Mathematics, The Ohio State University, Lima, Ohio 45804

Email address: yousif.1@osu.edu

Contemporary Mathematics
Volume **727**, 2019
https://doi.org/10.1090/conm/727/14633

On radicals of graded ring constructions

Emil Ilić-Georgijević

ABSTRACT. In this paper by a graded ring we mean a ring which is a direct sum of a family of its additive subgroups such that the product of homogeneous elements is again homogeneous. We extend already known construction of a graded polynomial ring over a graded ring to other graded ring constructions such as graded semigroup rings and incidence rings of group automata over graded rings. We also establish graded analogues of results on the prime radical of polynomial rings and incidence rings of group automata.

1. Introduction

Radicals and graded radicals of graded rings and graded ring constructions have been studied by many authors (see [**14**, **31**] and for more recent results, see [**27**, **28**, **33**], and references therein). Also, there have been investigations on radicals of rings graded in the sense of the following definition.

DEFINITION 1.1 ([**14**, **19**]). Let R be a ring, and Δ a partial groupoid, that is, a set with a partial binary operation. Also, let $\{R_\delta\}_{\delta\in\Delta}$ be a family of additive subgroups of R. We say that $R=\bigoplus_{\delta\in\Delta} R_\delta$ is Δ-*graded* and R *induces* Δ (or R is a Δ-*graded ring inducing* Δ) if the following two conditions hold:

i) $R_\xi R_\eta \subseteq R_{\xi\eta}$ whenever $\xi\eta$ is defined;
ii) $R_\xi R_\eta \neq 0$ implies that the product $\xi\eta$ is defined.

A Δ-graded ring inducing Δ from above is also called a *homogeneous sum* [**14**, **19**]. However, we will call it simply a *graded ring* in the rest of the paper. For examples of these rings such as a semidirect extension of a ring (in particular, the Dorroh extension), path algebras, and Morita contexts, see [**14**] and [**9**], respectively. A notion of a graded ring, equivalent to that in Definition 1.1, was studied in [**2**, **7**, **21**] from a different point of view, origins of which go back to [**20**]. We will say more details about it in the next section. Definition 1.1 applies to both associative and nonassociative rings. However, all rings in this paper are assumed to be associative.

The homogeneity of the Jacobson radical of a graded ring has been studied in [**19**], the famous Bergman's Lemma for graded rings, to be more precise. In [**13**], one may find results concerning connections between the properties of a graded ring and homogeneous components corresponding to an idempotent. The graded Jacobson, Brown–McCoy, Thierrin, and, generally, graded special radicals which

2010 *Mathematics Subject Classification.* Primary 16W50, 16S36, 16N80.

contain the graded Jacobson radical have been studied in [**6**, **7**, **9**–**11**] respectively. The prime and the graded prime radical of a graded ring, as well as the question of whether a homogeneous subring of a Jacobson radical graded ring is Jacobson radical are investigated in [**8**].

In the classical ring theory, it is natural to study radicals of various ring constructions such as radicals of matrix rings, polynomial rings or semigroup rings. Many significant results are obtained, including some characterizations of the famous Koethe's Conjecture (see [**23**, **25**]). For more information on this, one may consult [**5**, **14**]. It is also natural to consider graded cases of the aformentioned results, including the gradings in the sense of Definition 1.1. In [**2**] (see also [**21**]), the question of how to construct a graded polynomial ring over a graded ring is resolved. After preliminaries, we first present that construction by extending it to the semigroup ring case. Afterwards, we focus on the graded prime radical of a graded polynomial ring over a graded ring and establish a graded analogue of a known fact that the classical prime radical satisfies the Amitsur property (see [**1**, **29**]). Recently, one more ring construction is introduced in [**15**, **16**], known as the *incidence ring* of a group automaton over a ring with identity. It represents a generalization of the notion of an incidence algebra of a graph, and for more information, see [**18**]. Also, an incidence ring of a group automaton is a group graded ring, but here we show how to construct the so-called *graded incidence ring* (graded in the sense of Definition 1.1), and establish a graded analogue of the result from [**18**] on the classical prime radical of an incidence ring.

2. Preliminaries

In this section we gather all known notions and results necessary for the rest of the article. More details can be found in [**2**, **7**, **13**, **14**, **19**, **21**]. If $R = \bigoplus_{\delta\in\Delta} R_\delta$ is a graded ring, the set $H = \bigcup_{\delta\in\Delta} R_\delta$ is called the *homogeneous part of* R, and elements of H are called *homogeneous elements of* R. *The degree* $\delta(a)$ of a nonzero homogeneous element a of R is defined to be the unique $\xi \in \Delta$ for which $a \in R_\xi$. If we define $\xi\eta = 0$ for those pairs (ξ, η) for which we have $R_\xi R_\eta = 0$, then we put $\delta(0) = 0$. This makes the *grading set* Δ a groupoid with zero element 0. In the rest of the article, we assume that grading sets are groupoids in the previous sense. Also, sometimes, the grading set of a graded ring R will be denoted by Δ_R and its zero by 0_R.

We have already mentioned that there is a notion of a graded ring in [**2**, **7**, **21**] equivalent to that from Definition 1.1. Namely, if R is a graded ring and H its homogeneous part, then we may consider H with induced partial addition and induced multiplication from R. Such a structure with a partial addition and everywhere defined multiplication is called an *anneid* [**2**, **7**, **21**]. For a survey on anneids, one may also consult [**22**, **34**]. An anneid H determines a graded ring whose homogeneous part it is up to an H-isomorphism [**2**, **7**, **21**]. Therefore, all results obtained for anneids can be translated to the language of graded rings and vice versa. In this paper we use the language of graded rings.

An ideal (two-sided) I of a graded ring $R = \bigoplus_{\delta\in\Delta} R_\delta$ is said to be *homogeneous* if $I = \bigoplus_{\delta\in\Delta} R_\delta \cap I$. Also recall that if I is a homogeneous ideal of a graded ring $R = \bigoplus_{\delta\in\Delta} R_\delta$ and $I_\delta = R_\delta \cap I$, then $R/I = \bigoplus_{\delta\in\Delta} R_\delta/I_\delta$ is a graded ring [**2**, **7**, **14**, **21**] (see also [**12**]).

If R and R' are two graded rings, then a homomorphism of rings $f : R \to R'$

is called a *homogeneous homomorphism of graded rings* [**2**, **7**, **21**] if the image of a homogeneous element is again a homogeneous element, and if the fact that $f(x)$ is a nonzero homogeneous element of R' implies that x is a homogeneous element of R.

As it was mentioned in the previous section, a question of how to construct a graded polynomial ring over a graded ring is resolved in [**2**] (see also [**21**]). Actually, the truth is that it is resolved more generally. Namely, a *quasi-graded polynomial ring* in several indeterminates over a quasi-graded ring is constructed. Since we will extend this construction to the semigroup ring case, we recall the notion of a quasi-graded ring.

DEFINITION 2.1 ([**2**, **21**]). Let R be a ring and $\{R_\delta\}_{\delta\in\Delta}$ a family of its additive subgroups. Then R is called a *quasi-graded ring* if the following conditions are satisfied:

i) There exists an ideal I of R, called the *heart of* R such that $R_\xi \cap R_\eta = I$ for all distinct $\xi, \eta \in \Delta$;
ii) R/I is a graded ring $\bigoplus_{\delta\in\Delta} R_\delta/I$.

With the abuse of language, Δ is again called the *grading set* of R.

We also need to recall the notion of the *graded prime radical* $P^g(R)$ of a graded ring R from [**8**]. A graded ring is called *graded prime* if for every two homogeneous ideals I, J of R we have that $IJ = 0$ implies $I = 0$ or $J = 0$. A homogeneous ideal I of a graded ring R is called *prime* if R/I is a graded prime ring. The graded prime radical is then naturally defined to be the intersection of all homogeneous prime ideals of a graded ring. Also, a graded ring R is called *graded semiprime* [**8**] if $I^2 = 0$ implies $I = 0$ for any homogeneous ideal I of R. A homogeneous ideal I of a graded ring R is called *semiprime* if R/I is a graded semiprime ring.

Since we will follow the exposition from [**5**] in order to establish a graded analogue of the fact that the classical prime radical has the Amitsur property, we need to set the stage for the general radical theory of graded rings. As is explained in [**11**], the general Kurosh–Amitsur radical theory holds also for the category $\mathcal{R}$ whose objects are graded rings and morphisms are homogeneous homomorphisms. We refer to [**5**] and [**11**] for more details. However, let us just emphasize that $\mathcal{S}$ stands for a *graded semisimple operator*, that is, if γ is a graded radical class, then $\mathcal{S}\gamma = \{R \in \mathcal{R} \mid \gamma(R) = 0\}$.

3. Graded semigroup rings

At the begining of this section, let us fix some notation. If S is a set (in our case it is either a semigroup or a singleton $\{x\}$, where x is an indeterminate), Δ a groupoid with the zero element 0, and $\bar{\Delta}$ an overgroupoid of Δ such that $0\bar{\delta} = \bar{\delta}0 = 0$ for all $\bar{\delta} \in \bar{\Delta}$, then a mapping

$$S \to Z_\Delta(\bar{\Delta}) = \{\bar{\delta} \in \bar{\Delta} \mid \bar{\delta}\delta = \delta\bar{\delta}\ (\delta \in \Delta)\},$$

such that the images mutually commute, will always be denoted by θ or by θ with additional subscripts or superscripts in order to distinguish several such muppings at the time.

Let $R = \bigoplus_{\delta\in\Delta} R_\delta$ be a graded ring, H its homogeneous part, S a semigroup, $\bar{\Delta}$ an overgroupoid of Δ such that $0\bar{\delta} = \bar{\delta}0 = 0$ for all $\bar{\delta} \in \bar{\Delta}$, where 0 is the zero element of Δ, and let $\theta : S \to Z_\Delta(\bar{\Delta})$.

DEFINITION 3.1 ([**21**]). Let $r \in R$ be a homogeneous element. A homogeneous element $h \in R$ is called *central divisor of* r if there exist homogeneous elements $x, y \in R$ such that $r = xhy$. Homogeneous elements $h_1, h_2 \in R$ are called *adjacent central divisors of* r if there exist homogeneous elements $x, y \in R$ such that $r = xh_1h_2y$.

As usual, let $R[S]$ denote the semigroup ring of R with respect to S. Let $r \in R$ be a homogeneous element, $s \in S$, and assume that s can be written as a product of finitely many elements $s_i \in S$, not necessarily distinct. Now, following conditions imposed on monomials in the case of quasi-graded polynomial rings in several indeterminates from [**21**] (see also [**2**]), we have the following definition.

DEFINITION 3.2. Element $rs = r\prod_i s_i \in R[S]$ is said to be *degree-associable* if $\delta(r) \neq 0$ and if the following conditions are satisfied:

i) For every central divisor h of r and for all s_i, s_j, s_k we have

$$\begin{aligned}((\delta(h)\theta(s_i))\theta(s_j))\theta(s_k) &= \delta(h)(\theta(s_i)(\theta(s_j)\theta(s_k))),\\ (\theta(s_i)(\delta(h)\theta(s_j)))\theta(s_k) &= \theta(s_i)((\delta(h)\theta(s_j))\theta(s_k));\end{aligned}$$

ii) For every central divisor h of r and for all s_i, s_j we have

$$\begin{aligned}(\delta(h)\theta(s_i))\theta(s_j) &= \delta(h)(\theta(s_i)\theta(s_j)),\\ (\theta(s_i)\delta(h))\theta(s_j) &= \theta(s_i)(\delta(h)\theta(s_j));\end{aligned}$$

iii) For all adjacent central divisors h_1, h_2 of r and for all s_i we have

$$\begin{aligned}(\theta(s_i)\delta(h_1))\delta(h_2) &= \theta(s_i)(\delta(h_1)\delta(h_2)),\\ \delta(h_1)(\theta(s_i)\delta(h_2)) &= (\delta(h_1)\theta(s_i))\delta(h_2).\end{aligned}$$

It is easy to see that the following lemma holds.

LEMMA 3.3. *Let $r \in R$ be a homogeneous element, and $s \in S$ such that $r = h_1 \dots h_p$ and $s = s_1 \dots s_q$, where $h_i \in H$, $s_j \in S$, $i = 1, \dots, p$, $j = 1, \dots, q$, $p, q \in \mathbb{N}$. If rs is degree-associable, then $\bar{\delta}(rs) := \delta(h_1)\dots\delta(h_p)\theta(s_1)\dots\theta(s_q)$ is an element of $\bar{\Delta}$, which is invariant under the order of multiplication, and it is called the* pre-degree *of the element rs. Also, if r_is_i, $r_i \in H$, $s_i \in S$, $i = 1, \dots, p$, $p \in \mathbb{N}$, are degree-associable, then the pre-degree of $\prod_{i=1}^p r_is_i$ equals $\prod_{i=1}^p \bar{\delta}(r_is_i)$.*

LEMMA 3.4. *Let I be an additive subgroup of $R[S]$ generated by elements $rs \in R[S]$, $r \in H$, which are not degree-associable. Then I is an ideal of $R[S]$.*

PROOF. Let $x = \sum_{i=1}^p r_is_i$, $y = \sum_{j=1}^q r'_js'_j \in I$, $p, q \in \mathbb{N}$. Then, obviously, $x - y \in I$, since r_is_i, $r'_js'_j$ are not degree-associable, and I is generated by such elements. Also, if $z \in R[S]$, we have that $xz, zx \in I$, since if r_is_i is not degree-associable and $r''s'' \in R[S]$, $r'' \in H$, then $(r_is_i)(r''s'') = r_ir''s_is''$ is not a degree-associable element, which completes the proof. □

THEOREM 3.5. *Under the notation of Lemma 3.4, $R[S]$ is a quasi-graded ring with the heart I and the grading set $\bar{\Delta}$ under the usual semigroup ring addition and multiplication. It is denoted by $R_\theta[S]$.*

PROOF. Let $\bar{\xi} \in \bar{\Delta}$ and let $(R_\theta[S])_{\bar{\xi}}$ be the set of elements of $R[S]$ which consists of all the finite sums of elements which are degree-associable elements of $R[S]$ which have the same pre-degree $\bar{\xi}$ or they belong to I. Then $(R_\theta[S])_{\bar{\xi}}$ is an additive subgroup of $R[S]$, according to Lemma 3.4. Clearly, for distinct $\bar{\xi}, \bar{\eta} \in \bar{\Delta}$, we have $(R_\theta[S])_{\bar\xi}\cap(R_\theta[S])_{\bar\eta} = I$, and $R[S]/I = \bigoplus_{\bar\delta\in\bar\Delta}(R_\theta[S])_{\bar\delta}/I$. Also, for $\bar\xi,\bar\eta\in\bar\Delta$, let $x = \sum_{i=1}^{p} r_i s_i \in (R_\theta[S])_{\bar\xi}$, $y = \sum_{j=1}^{q} r'_j s'_j \in (R_\theta[S])_{\bar\eta}$, $p, q \in \mathbb{N}$. Then xy will be a sum of elements which are from I or are degree-associable elements of the same pre-degree, according to Lemma 3.3. Hence, there exists $\bar\zeta \in \bar\Delta$ such that $((R_\theta[S])_{\bar\xi}/I)((R_\theta[S])_{\bar\eta}/I) \subseteq (R_\theta[S])_{\bar\zeta}/I$. Therefore, $R[S]$ is a quasi-graded ring with I as its heart. □

REMARK 3.6. If S is a semigroup with zero z, then we may obtain a quasi-graded contracted semigroup ring as a quotient quasi-graded ring modulo a homogeneous ideal Rz. For more details on homogeneous ideals and quotients of quasi-graded rings, we refer to [**2**] (see also [**21**]).

The next theorem generalizes Proposition 6.5 in [**30**].

THEOREM 3.7. *Let G be a group and $R_\theta[G]$ a quasi-graded group ring with respect to some $\theta : G \to Z_\Delta(\bar\Delta)$. If $R_\theta[G]$ is a graded group ring, then we have:*

i) A is a homogeneous prime ideal of R if and only if there exists a homogeneous prime ideal B of $R_\theta[G]$ such that $A = B \cap R_\theta[G]$;

ii) $P^g(R) = P^g(R_\theta[G]) \cap R$.

PROOF. cf. [**30**]. □

3.1. On graded radicals of graded polynomial rings. Let us now investigate the case of polynomial rings which motivated this research in the first place. In this subsection we will show how the theory of radicals of polynomial rings can be extended to the case of graded polynomial rings over graded rings. Also, throughout this subsection, if $R = \bigoplus_{\delta\in\Delta} R_\delta$ is a graded ring, x an indeterminate, $\bar\Delta$ an overgroupoid of Δ, such that $0\bar\delta = \bar\delta 0 = 0$ for all $\bar\delta \in \bar\Delta$, where 0 is the zero element of Δ, and $\theta : \{x\} \to Z_\Delta(\bar\Delta)$, then we will assume that $\theta(x)$ is not a zero divisor in $\bar\Delta$. Under this assumption, according to [**2**], the quasi-graded polynomial ring $R_\theta[x]$ is a graded ring.

Let $\mathcal{R}$ be the category whose objects are graded rings and morphisms are homogeneous homomorphisms, and let γ be a graded radical class. Following [**4**] (see also [**5**]), let us define

$$\gamma_x = \{R \in \mathcal{R} \mid (\forall\theta : \{x\} \to Z_{\Delta_R}(\bar\Delta_R))\ R_\theta[x] \in \gamma\},$$

where Δ_R denotes the grading set of $R \in \mathcal{R}$ and $\bar\Delta_R$ an overgroupoid of Δ_R such that $0_R\bar\delta = \bar\delta 0_R = 0_R$ for every $\bar\delta \in \bar\Delta_R$. We have the following theorem.

THEOREM 3.8. *γ_x is a graded radical class.*

PROOF. Let $R \in \gamma_x$ and let S be a homomorphic image of R under a homogeneous homomorphism. Then S belongs to $\mathcal{R}$ and $\Delta_S = \Delta_R$ (see [**2, 7, 21**]). Let $\bar\Delta_R$ be an overgroupoid of Δ_R such that $0_R\bar\delta_R = \bar\delta_R 0_R = 0_R$ for all $\bar\delta_R \in \bar\Delta_R$, and let $\theta : \{x\} \to Z_{\Delta_R}(\bar\Delta_R)$. Then $S_\theta[x]$ is a homomorphic image of $R_\theta[x]$ under a homogeneous homomorphism. Hence, $S_\theta[x] \in \gamma$, as $R_\theta[x] \in \gamma$. Now, let $\{I_\lambda \mid \lambda \in \Lambda\}$ be an ascending chain of homogeneous ideals of a graded ring $R \in \mathcal{R}$ such that $R = \bigcup_{\lambda\in\Lambda} I_\lambda$ and $I_\lambda \in \gamma_x$ $(\lambda \in \Lambda)$. Similarly to the classical case (see [**4**] or [**5**]), we have $R_\theta[x] = \bigcup_{\lambda\in\Lambda}(I_\lambda)_\theta[x]$ for any θ. Hence, $R \in \gamma_x$. Finally, let I be a homogeneous ideal of R such that I and R/I belong to γ_x. Then $I_\theta[x]$ and $(R/I)_{\theta'}[x]$ belong to γ for all the mappings $\theta : \{x\} \to Z_{\Delta_R}(\bar\Delta_R)$, $\theta' : \{x\} \to Z_{\Delta_R}(\bar\Delta_R)$. In particular, this holds for $\theta = \theta'$. Since, as in the classical case we have $R_\theta[x]/I_\theta[x] \cong (R/I)_\theta[x]$, it follows that $R \in \gamma_x$, which completes the proof. □

Following [**32**] (see also [**5**]), given a graded ring $R = \bigoplus_{\delta\in\Delta} R_\delta$ and an overgroupoid $\bar{\Delta}$ of Δ such that $0\bar{\delta} = \bar{\delta}0 = 0$ for all $\bar{\delta} \in \bar{\Delta}$, we may consider the homogeneous ideal

$$\rho_\theta(R) = \gamma(R_\theta[x]) \cap R$$

of R, for a mapping $\theta : \{x\} \to Z_\Delta(\bar{\Delta})$.

DEFINITION 3.9. We say that a graded radical class γ has the *graded Amitsur property* if

$$\gamma(R_\theta[x]) = \rho_\theta(R)_\theta[x] = (\gamma(R_\theta[x]) \cap R)_\theta[x]$$

for all $\theta : \{x\} \to Z_\Delta(\bar{\Delta})$.

It is known that the classical prime radical has the classical Amitsur property (see for instance [**5**]). In order to establish a graded analogue, we gather the preparatory facts in the next theorem which can be proved analogously to the classical case with the appropriate modifications (see [**5**]). However, in order to go through these modifications, one needs to introduce the *graded Dorroh extension* R^1 of a graded ring $R = \bigoplus_{\delta\in\Delta} R_\delta$. Let, as usual, R^1 be the ring $R \times \mathbb{Z}$ with componentwise addition and multiplication $(r, m) \cdot (s, n) = (rs + nr + ms, mn)$. Then R^1 is a graded ring with the grading $R^1_0 = \{(0, n) \mid n \in \mathbb{Z}\}$, $R^1_\delta = \{(r, 0) \mid r \in R_\delta\}$ $(\delta \in \Delta)$. It is important to emphasize that this construction works if we do not require R^1 to have a cancellative partial groupoid as the grading set. Otherwise, we need to assume that Δ contains only a finite number of nonzero idempotents, since, as we know from [**7**], an identity element of a ring is then the sum of identities of components which correspond to nonzero idempotent elements of Δ. However, we will not encounter such a case here.

THEOREM 3.10. *Let $R \in \mathcal{R}$ and let γ be a graded radical class. Then:*

i) $\gamma_x \subseteq \gamma$ *and* $\rho_\theta(R)_\theta[x] \subseteq \gamma(R_\theta[x])$ *for all the mappings* θ*;*

ii) If γ is graded hereditary, then $\rho_\theta(R) = \gamma(R)$ if and only if $\gamma_x = \gamma$ if and only if $R \in \gamma \Rightarrow R_\theta[x] \in \gamma$ if and only if $R \in \gamma \Leftrightarrow R_\theta[x] \in \gamma$ for all the mappings θ.

Using the technique of the proof of Theorem 4.9.22 in [**5**], with the help of Theorem 3.10*i*), we establish the following theorem.

THEOREM 3.11 (cf. [**24**]). *A graded radical γ has the graded Amitsur property if and only if $\rho_\theta(R) = 0$ implies $\gamma(R_\theta[x]) = 0$ for all $R \in \mathcal{R}$ and all the mappings θ.*

THEOREM 3.12. *The graded prime radical of a graded ring has the graded Amitsur property.*

PROOF. As in the classical case (see [**5**]), if R is a graded prime ring, then $R_\theta[x]$ is a graded prime ring for all θ. Also, the graded prime radical is graded hereditary, which can be easily seen from the fact that the graded prime radical coincides with the lower graded radical class of all nilpotent graded rings (for the proof, see for instance [**3**] and [**8**]). Then, according to Theorem 3.10*ii*), we have that $\gamma(R) = \rho_\theta(R)$ for all θ. Let us now assume that R belongs to $\mathcal{SP}^g$. We claim that $R_\theta[x]$ then belongs to $\mathcal{SP}^g$ for all θ. If we assume the opposite, then, for some θ, there exists a nonzero homogeneous ideal A of $P^g(R_\theta[x])$ such that $A^2 = 0$. Now, as in the classical case (see [**5**]), it can be easily seen that the set I, which consists of 0 and the leading coefficients of polynomials from A, is a homogeneous ideal of

R such that $I^2 = 0$, contradicting the fact that R belongs to $\mathcal{S}P^g$. Hence, $R \in \mathcal{S}P^g$ implies $R_\theta[x] \in \mathcal{S}P^g$ for all θ. This and the fact that $\gamma(R) = \rho_\theta(R)$ for all θ imply that P^g has the graded Amitsur property, which can be proved analogously to the proof of Theorem 4.9.23 in [**5**] with the help of Theorem 3.11. □

REMARK 3.13. It would also be interesting to test the other graded radicals for the graded Amitsur property such as the graded and the large graded Jacobson [**6**], Brown–McCoy [**9**] and Thierrin [**10**] radical.

3.2. Graded incidence rings of group automata. Recently, the notion of an *incidence ring of a group automaton* is introduced and its structure is obtained (see [**15–18**]). Let us briefly recall these notions. We use the exposition from [**18**].

DEFINITION 3.14. A *group automaton* is a triple $\mathcal{A} = (X, G, f)$, where

i) X is the *set of states*,

ii) G is the *group of input symbols*,

iii) $f : X \times G \to X$ is the *transition function*, denoted by $f(x, g) = xg$, such that
$$(xg)h = x(gh)$$
for all $x \in X$ and for all $g, h \in G$.

Let F be a ring with identity and $\mathcal{A} = (X, G, f)$ a group automaton. The *incidence ring* of the automaton $\mathcal{A} = (X, G, f)$ over F is the ring $I_\mathcal{A} = I_\mathcal{A}(F)$ that is spanned as a free left F-module by the set $T_\mathcal{A} = \{(x, g, xg) \mid x \in X, g \in G\}$ with multiplication defined by the distributive law and the rules

$$\begin{aligned} (x_1, g_1, x_1g_1) \cdot (x_2, g_2, x_2g_2) &= \begin{cases} (x_1, g_1g_2, x_1g_1g_2), & \text{if } x_1g_1 = x_2, \\ 0, & \text{otherwise}, \end{cases} \\ (x, g, xg) \cdot r &= r \cdot (x, g, xg), \end{aligned}$$

for all $x, x_1, x_2 \in X$, $g, g_1, g_2 \in G$, $r \in F$.

Clearly, $T^0_\mathcal{A} = T_\mathcal{A} \cup 0$ is a semigroup, and $I_\mathcal{A}$ is isomorphic to the contracted semigroup ring $F_0[T^0_\mathcal{A}]$ (see [**14, 15**]). Following [**18**], let $\bar{\mathcal{A}}$ be the group automaton (XG, G, f), and $I_{\bar{\mathcal{A}}}$ its incidence ring. Also, let $X' = X \setminus XG$, and $N_\mathcal{A}$ an F-subalgebra of $I_\mathcal{A}$ generated by all the triples (x, g, xg), $x \in X'$, $g \in G$. According to [**18**], the incidence ring $I_\mathcal{A}$ can be written as the direct sum of $N_\mathcal{A}$ and $I_{\bar{\mathcal{A}}}$, and $I_{\bar{\mathcal{A}}}$ is on the other hand the direct sum of incidence rings $I_{\mathcal{A}(\mathcal{O})}$ which correspond to subautomata of $\bar{\mathcal{A}}$ which correspond to orbits $\mathcal{O}$ of XG with respect to G, $\mathcal{O} \in \Omega_G$.

Also (see [**18**]), if T is any subset of $T_\mathcal{A}$, then $FT \subseteq I_\mathcal{A}$ denotes the F-linear span of T. Hence, if $(X', G, X'G) = \{(x, g, xg) \mid x \in X', g \in G\}$ and $(\mathcal{O}, G, \mathcal{O}) = \{(x, g, xg) \mid x \in \mathcal{O}, g \in G\}$, then $N_\mathcal{A} = F(X', G, X'G)$ and $I_{\mathcal{A}(\mathcal{O})} = F(\mathcal{O}, G, \mathcal{O})$.

Let now $F = \bigoplus_{\delta \in \Delta} F_\delta$ be a graded ring with identity, and let $\bar{\Delta}$ be an overgroupoid of Δ such that $0_F \in \Delta$ is also the zero element of $\bar{\Delta}$. Let us observe quasi-graded contracted semigroup rings $N^{\theta'}_\mathcal{A} = F_{\theta'}(X', G, X'G)$, $I^{\theta_\mathcal{O}}_{\mathcal{A}(\mathcal{O})} = F_{\theta_\mathcal{O}}(\mathcal{O}, G, \mathcal{O})$, $\mathcal{O} \in \Omega_G$, for some $\theta' : (X', G, X'G) \cup 0 \to Z_\Delta(\bar{\Delta})$ and $\theta_\mathcal{O} : (\mathcal{O}, G, \mathcal{O}) \cup 0 \to Z_\Delta(\bar{\Delta})$, where $\mathcal{O} \in \Omega_G$.

DEFINITION 3.15. A *graded incidence ring* $I^g_\mathcal{A}(F)$ *of a group automaton* $\mathcal{A} = (X, G, f)$ over a graded ring F is a direct sum of factor rings of $N^{\theta'}_\mathcal{A}$, $I^{\theta_\mathcal{O}}_{\mathcal{A}(\mathcal{O})}$, modulo their hearts C_N, $C_\mathcal{O}$, respectively, for some $\theta' : (X', G, X'G) \cup 0 \to Z_\Delta(\bar{\Delta})$, $\theta_\mathcal{O} : (\mathcal{O}, G, \mathcal{O}) \cup 0 \to Z_\Delta(\bar{\Delta})$, where $\mathcal{O} \in \Omega_G$.

REMARK 3.16. The notion from the previous definition is not unique. First of all, it depends on the choices of the mappings θ, and, secondly, a direct sum of graded rings is not unique (see for instance [**21**]).

As usual, if n is a cardinal number, $\mathbb{M}_n(R)$ denotes the ring of finite $n \times n$ matrices over a ring R (see [**18**]).

THEOREM 3.17. *If $I^g_{\mathcal{A}}(F)$ is a graded incidence ring, then*

$$P^g(I^g_{\mathcal{A}}(F)) = N^{\theta'}_{\mathcal{A}}/C_N \oplus \sum_{\mathcal{O}\in\Omega_G} P^g(I^{\theta_{\mathcal{O}}}_{\mathcal{A}(\mathcal{O})}/C_{\mathcal{O}}),$$

where the sum on the right is direct. If $x \in \mathcal{O}$ and if $H = G_x$ is the stabilizer subgroup of x in G, then $P^g(I^{\theta_{\mathcal{O}}}_{\mathcal{A}(\mathcal{O})}/C_{\mathcal{O}}) \cong \mathbb{M}_{|\mathcal{O}|}(P^g(F[H]/C))$, where C is the heart of $F[H]$ regarded as an arbitrary quasi-graded ring over the graded ring F.

LEMMA 3.18. *Let R be a graded ring with identity and $n \geq 1$ a natural number. Then, if $\mathbb{M}_n(R)$ is a graded ring (not necessarily with the same grading set as of the ring R), then $P^g(\mathbb{M}_n(R)) = \mathbb{M}_n(P^g(R))$.*

PROOF. First of all, as in the classical case, one may prove that if K is a homogeneous ideal of $\mathbb{M}_n(R)$, then $K = \mathbb{M}_n(I)$ for some homogeneous ideal I of R and vice versa. Also, as in the classical case, we establish that R is graded semiprime if and only if $\mathbb{M}_n(R)$ is graded semiprime (see for instance Proposition 10.20 in [**26**]). Now, we know that $R/P^g(R)$ is a graded semiprime ring (see Theorem 8.7 in [**8**]), and so, $\mathbb{M}_n(R/P^g(R))$ is graded semiprime. It follows that $\mathbb{M}_n(P^g(R))$ is a homogeneous semiprime ideal of $\mathbb{M}_n(R)$. Hence $\mathbb{M}_n(P^g(R)) \supseteq P^g(\mathbb{M}_n(R))$. Now, let K be a homogeneous prime ideal of $\mathbb{M}_n(R)$. Then $K = \mathbb{M}_n(P)$ for some homogeneous ideal P of R. One easily verifies that P is a homogeneous prime ideal of R. Therefore $\mathbb{M}_n(P^g(R)) \subseteq K$. Hence $\mathbb{M}_n(P^g(R)) \subseteq P^g(\mathbb{M}_n(R))$. □

PROOF OF THEOREM 3.17. The first part of this theorem can be proved analogously to the proof of Lemma 3.2 in [**18**]. For the second part, for the sake of simplicity, let us assume that $I^{\theta_{\mathcal{O}}}_{\mathcal{A}(\mathcal{O})}$ and $F[H]$ are graded rings. Now, according to Theorem 2.1(ii) in [**18**], $I_{\mathcal{A}(\mathcal{O})}$ is isomorphic to $\mathbb{M}_{|\mathcal{O}|}(F[H])$. This means that $\mathbb{M}_{|\mathcal{O}|}(F[H])$ is a graded ring with the grading set $\bar{\Delta}$. In order to finish the proof, it is enough to prove that $P^g(\mathbb{M}_{|\mathcal{O}|}(F[H])) = \mathbb{M}_{|\mathcal{O}|}(P^g(F[H]))$. However, this follows from Lemma 3.18. □

References

[1] S. A. Amitsur, *Radicals of polynomial rings*, Canad. J. Math. **8** (1956), 355–361, DOI 10.4153/CJM-1956-040-9. MR0078345

[2] M. Chadeyras, *Essai d'une théorie noethérienne homogène pour les anneaux commutatifs dont la graduation est aussi générale que possible* (French), Bull. Soc. Math. France Suppl. Mém. **22** (1970), 143. MR0276210

[3] M. Cohen and S. Montgomery, *Group-graded rings, smash products, and group actions*, Trans. Amer. Math. Soc. **282** (1984), no. 1, 237–258, DOI 10.2307/1999586. MR728711

[4] B. J. Gardner, *A note on radicals and polynomial rings*, Math. Scand. **31** (1972), 83–88, DOI 10.7146/math.scand.a-11414. MR0318206

[5] B. J. Gardner and R. Wiegandt, *Radical theory of rings*, Monographs and Textbooks in Pure and Applied Mathematics, vol. 261, Marcel Dekker, Inc., New York, 2004. MR2015465

[6] E. Halberstadt, *Le radical d'un annéide régulier* (French), C. R. Acad. Sci. Paris Sér. A-B **270** (1970), A361–A363. MR0262305

[7] E. Halberstadt, Théorie artinienne homogène des anneaux gradués à grades non commutatifs réguliers, PhD Thesis, University of Pierre and Marie Currie, 1971.

[8] E. Ilić-Georgijević, *On graded Ω-groups*, Filomat **29** (2015), no. 10, 2167–2183, DOI 10.2298/FIL1510167I. MR3434158

[9] E. Ilić-Georgijević, *On graded Brown-McCoy radicals of graded rings*, J. Algebra Appl. **15** (2016), no. 8, 1650143, 13, DOI 10.1142/S0219498816501437. MR3528765

[10] E. Ilić-Georgijević, *On graded Thierrin radicals of graded rings*, Comm. Algebra **45** (2017), no. 9, 3886–3891, DOI 10.1080/00927872.2016.1248243. MR3627636

[11] E. Ilić-Georgijević, *On graded special radicals of graded rings*, J. Algebra Appl. **17** (2018), no. 6, 1850109, 10, DOI 10.1142/S0219498818501098. MR3805720

[12] A. V. Kelarev, *Combinatorial properties and homomorphisms of semigroups*, Internat. J. Algebra Comput. **4** (1994), no. 3, 443–450, DOI 10.1142/S0218196794000087. MR1297150

[13] A. V. Kelarev, *On groupoid graded rings*, J. Algebra **178** (1995), no. 2, 391–399, DOI 10.1006/jabr.1995.1355. MR1359892

[14] A. V. Kelarev, *Ring constructions and applications*, Series in Algebra, vol. 9, World Scientific Publishing Co., Inc., River Edge, NJ, 2002. MR1875643

[15] A. V. Kelarev, *On incidence rings of group automata*, Bull. Austral. Math. Soc. **67** (2003), no. 3, 407–411, DOI 10.1017/S0004972700037217. MR1983873

[16] A. V. Kelarev, *On the structure of incidence rings of group automata*, Internat. J. Algebra Comput. **14** (2004), no. 4, 505–511, DOI 10.1142/S0218196704001888. MR2084383

[17] A. V. Kelarev, *On group automata over finite fields*, Finite Fields Appl. **10** (2004), no. 1, 65–71, DOI 10.1016/S1071-5797(03)00043-1. MR2024603

[18] A. V. Kelarev and D. S. Passman, *A description of incidence rings of group automata*, Noncommutative rings, group rings, diagram algebras and their applications, Contemp. Math., vol. 456, Amer. Math. Soc., Providence, RI, 2008, pp. 27–33, DOI 10.1090/conm/456/08882. MR2416142

[19] A. V. Kelarev and A. Plant, *Bergman's lemma for graded rings*, Comm. Algebra **23** (1995), no. 12, 4613–4624, DOI 10.1080/00927879508825490. MR1352559

[20] M. Krasner, *Une généralisation de la notion de corps—corpoïde. Un corpoïde remarquable de la théorie des corps valués* (French), C. R. Acad. Sci. Paris **219** (1944), 345–347. MR0015066

[21] M. Krasner, *Anneaux gradués généraux* (French), Algebra Colloquium (Rennes, 1980), Univ. Rennes, Rennes, 1980, pp. 209–308. MR645222

[22] M. Krasner and M. Vuković, *Structures paragraduées (groupes, anneaux, modules)* (French), Queen's Papers in Pure and Applied Mathematics, vol. 77, Queen's University, Kingston, ON, 1987. MR911680

[23] J. Krempa, *Logical connections between some open problems concerning nil rings*, Fund. Math. **76** (1972), no. 2, 121–130, DOI 10.4064/fm-76-2-121-130. MR0306251

[24] J. Krempa, *On radical properties of polynomial rings* (English, with Russian summary), Bull. Acad. Polon. Sci. Sér. Sci. Math. Astronom. Phys. **20** (1972), 545–548. MR0357470

[25] J. Krempa, *Radicals of semi-group rings*, Fund. Math. **85** (1974), no. 1, 57–71, DOI 10.4064/fm-85-1-57-71. MR0360898

[26] T. Y. Lam, *A first course in noncommutative rings*, Graduate Texts in Mathematics, vol. 131, Springer-Verlag, New York, 1991. MR1125071

[27] P.-H. Lee and E. R. Puczyłowski, *On prime ideals and radicals of polynomial rings and graded rings*, J. Pure Appl. Algebra **218** (2014), no. 2, 323–332, DOI 10.1016/j.jpaa.2013.06.004. MR3120633

[28] R. Mazurek, P. P. Nielsen, and M. Ziembowski, *The upper nilradical and Jacobson radical of semigroup graded rings*, J. Pure Appl. Algebra **219** (2015), no. 4, 1082–1094, DOI 10.1016/j.jpaa.2014.05.036. MR3282127

[29] N. H. McCoy, *The prime radical of a polynomial ring*, Publ. Math. Debrecen **4** (1956), 161–162. MR0079578

[30] C. Năstăsescu, *Group rings of graded rings. Applications*, J. Pure Appl. Algebra **33** (1984), no. 3, 313–335, DOI 10.1016/0022-4049(84)90065-3. MR761636

[31] C. Năstăsescu and F. Van Oystaeyen, *Methods of graded rings*, Lecture Notes in Mathematics, vol. 1836, Springer-Verlag, Berlin, 2004. MR2046303

[32] A. H. Ortiz, *A construction in general radical theory*, Canad. J. Math. **22** (1970), 1097–1100, DOI 10.4153/CJM-1970-126-x. MR0277563

[33] A. Smoktunowicz, *A note on nil Jacobson radicals in graded rings*, J. Algebra Appl. **13** (2014), no. 4, 1350121, 8, DOI 10.1142/S0219498813501211. MR3153856

[34] M. Vuković, Structures graduées et paragraduées, Prepublication de l'Institut Fourier, Université de Grenoble I, No. **536** (2001), pp. 1-40.

FACULTY OF CIVIL ENGINEERING, UNIVERSITY OF SARAJEVO, PATRIOTSKE LIGE 30, 71000 SARAJEVO, BOSNIA AND HERZEGOVINA

Email address: emil.ilic.georgijevic@gmail.com

Contemporary Mathematics
Volume **727**, 2019
https://doi.org/10.1090/conm/727/14634

Simple ambiskew polynomial rings II: Non-bijective endomorphisms

David A. Jordan

ABSTRACT. We study simplicity of the iterated skew polynomial rings in two indeterminates known as ambiskew polynomial rings. Simplicity criteria for the case where the defining endomorphisms are automorphisms were given in a previous joint paper with Wells, together with several examples and results on simplicity of related rings. In order to accommodate the non-Noetherian downup algebras of Benkart and Roby, the construction was amended to allow for non-bijective endomorphisms. The amendment involves one skew polynomial extension with coefficients on the left, followed by another with coefficients on the right. The main purpose of this paper is to extend the study of simplicity to the general case and to give examples of simple ambiskew polynomial rings for which the defining endomorphisms are not bijective. We also study primality of ambiskew polynomial rings and simplicity of analogues of generalized Weyl algebras.

1. Introduction

Throughout this paper, A denotes a unital algebra over a field $\mathbb{K}$, all endomorphisms are $\mathbb{K}$-endomorphisms and all right or left derivations are right or left $\mathbb{K}$-derivations. The centre and group of units of a ring S will be denoted $Z(S)$ and $U(S)$ respectively. In the case of the field $\mathbb{K}$, $U(\mathbb{K})$ will be written $\mathbb{K}^*$. We denote the identity map on any set S by $\mathrm{id}\,|_S$.

Constructing an ambiskew polynomial ring R over A involves the choice of an element ρ of $\mathbb{K}$, an element v of A such that $va = \gamma(a)v$ for some automorphism γ of A and an endomorphism α of A that commutes with γ. Two indeterminates y and x are adjoined subject to the relations

$$xy - \rho yx = v \text{ and, } \forall a \in A,\ ay = y\alpha(a) \text{ and } xa = \beta(a)x,$$

where $\beta = \alpha\gamma = \gamma\alpha$. Numerous motivating examples are documented in [**12**] and the references therein. In most of these examples α is an automorphism, and in the study of simplicity in [**12**] it was assumed to be so. However, in [**11**, Section 7], the construction was extended to cover arbitrary endomorphisms in order to incorporate the non-noetherian down-up algebras of Benkart and Roby [**3**]. Further examples of ambiskew polynomial rings with non-bijective endomorphisms are the non-noetherian generalized down-up algebras of Cassidy and Shelton [**4**, p.407]. In

2010 *Mathematics Subject Classification.* Primary 16D30; Secondary 16S36, 16W20, 16W25.
Key words and phrases. Skew polynomial ring, simple ring.

the automorphism case, simplicity criteria, depending on the characteristic of $\mathbb{K}$, were given by [**12**, Theorems 3.10 and 3.15]. The main purpose of this paper is to extend the study of simplicity to the general case and to construct examples of simple ambiskew polynomial rings in which the defining endomorphisms are not bijective. For example, we shall see that the ring extension of the rational function field $\mathbb{C}(t)$ generated by x and y subject to the relations

$$xy - \rho yx = t^2, \quad ty = yt^2, \quad xt = t^2x.$$

is simple for all $\rho \in \mathbb{K}^*$. The results on simplicity and examples appear in Section 3. We first establish necessary conditions for simplicity, analogous to those in [**12**], and then prove that they are also sufficient when A is commutative noetherian of characteristic 0.

Bavula [**2**] has introduced a class of iterated skew polynomial ring, the *diskew* polynomial rings, that contains those ambiskew polynomial rings that are constructed using bijective endomorphisms. In contrast to the general ambiskew polynomial rings described above, the indeterminates x and y act from the same side. See Definition 2.10 for the details. We shall see that the simple ambiskew polynomial ring mentioned above cannot be expressed as a diskew polynomial ring over any base ring.

Section 4 is concerned with primality and generalizes results from [**11**] where it was shown that, with one exceptional choice of parameters, non-noetherian down-up algebras are prime. If α is not injective then, as $\ker(\alpha)Y = 0$, R cannot be a domain but it can be prime. The results in this section will be applied to determine which ambiskew polynomial rings over the polynomial ring $\mathbb{K}[t]$ are prime.

In Section 5, we consider certain quotients T of non-simple ambiskew polynomial rings analogous to the generalized Weyl algebras (GWAs) of [**1**]. The general definition is given in Section 5 but, when A is commutative, α is injective and $u \in A$, the algebra $T = T(A, \alpha, u)$ is the extension of A generated by X and Y subject to the relations

$$XY = \alpha(u), YX = u, \text{ and, } \forall a \in A,\ aY = Y\alpha(a) \text{ and } Xa = \alpha(a)X.$$

This is different to the generalization of GWA to general endomorphisms given by Bavula [**2**], where Y and X act from the same side using different endomorphisms, so we shall use the name *hyperbolic algebra* which was used, in the automorphism case, by Rosenberg [**14**]. When $u = 1$ and α is injective, T is the skew Laurent polynomial ring $A[X^{\pm 1}; \alpha]$ as constructed in [**7**]. We begin Section 5 by establishing necessary conditions for simplicity in a general setting and then prove their sufficiency when A is commutative noetherian. One issue that arises is that, even when α is injective, Y and X need not be regular in T. As regularity of X and Y is a necessary condition for simplicity this influences the simplicity criterion. Related to this issue of regularity is the interesting but surprising possibility that if B is a commutative algebra with an injective endomorphism α and a subalgebra A, such that $\alpha(A) \subseteq A$, and an element $u \in Z(A)$ then $T(A, \alpha, u)$ need not embed in $T(B, \alpha, u)$.

2. Preliminaries

In [**11**], where the focus was on the relationship between ambiskew polynomial rings over the commutative polynomial ring $\mathbb{K}[t]$ and down-up algebras, the base ring was assumed to be commutative whereas in [**12**], where the focus was on simplicity and examples include simple ambiskew polynomial rings with noncommutative base rings, the endomorphisms were assumed to be bijective. Definition 2.1 below is the minimal common generalization of the ambiskew polynomial rings considered in [**11**] and in [**12**]. We shall need to distinguish between skew polynomial rings with coefficients on the left or right. We shall use the usual notation $A[x;\alpha,\delta]$ for the left skew polynomial ring over the ring A in the indeterminate x, given by an endomorphism α of A and a left α-derivation δ of A. Thus, coefficients appear to the left of powers of x and $xa=\alpha(a)x+\delta(a)$ for all $a\in A$. On the other hand, if δ is a right α-derivation δ of A, we shall write $[\delta,\alpha;y]A$ for the right skew polynomial ring in y in which $ay=y\alpha(a)+\delta(a)$ for all $a\in A$. In either case, if $\delta=0$ then it is omitted from the notation, giving $A[x;\alpha]$ or $[\alpha;y]A$ as appropriate.

An element v of a ring A is *normal* if $vA=Av$ and is γ-*normal* for an automorphism γ of A if $\gamma(v)=v$ and $va=\gamma(v)a$ for all $a\in A$. If a normal element v of A is regular then there is a uniquely determined automorphism γ of A such that v is γ-normal. The condition $\gamma(v)=v$ holds because $v^2=\gamma(v)v$.

DEFINITION 2.1. Let A be a $\mathbb{K}$-algebra with an automorphism γ, a γ-normal element v and an endomorphism α such that $\alpha\gamma=\gamma\alpha$. Let $\rho\in\mathbb{K}$ and let $\beta=\alpha\gamma$. The *ambiskew polynomial ring* $R=R(A,\alpha,v,\rho)$ is the iterated skew polynomial ring $([\alpha;y]A)[x;\beta,\delta]$ where β is extended to $[\alpha;y]A$ by setting $\beta(y)=\rho y$ and δ is the β-derivation of $[y;\alpha]A$ such that $\delta(y)=v$ and $\delta(A)=0$. Thus,

$$xy-\rho yx=v \text{ and, } \forall a\in A,\ ay=y\alpha(a) \text{ and } xa=\beta(a)x.$$

To justify the existence of the β-derivation δ above we shall make use of the universal mapping property for right skew polynomial rings and later we shall need universal mapping properties on both sides. The universal mapping property for left skew polynomial rings is stated in [**6**, Proposition 2.4] and the right analogue is as follows.

PROPOSITION 2.2. *Let α be an endomorphism of a ring R and let δ be a right α-derivation of R. Let S be the skew polynomial ring $[\delta,\alpha;x]R$. Let T be a ring with a ring homomorphism $\phi:R\to T$ and an element t such that $\phi(r)t=t\phi\alpha(r)+\phi(\delta(r))$ for all $r\in R$. Then there is a unique ring homomorphism $\psi:S\to T$ such that $\psi|_R=\phi$ and $\phi(x)=t$.*

With a view to establishing the existence of δ in the context of Definition 2.1, let $S=[\alpha;y]A$ and let T be the ring of 2×2 matrices over S. Specify an element $t\in T$ and a ring homomorphism $\phi:A\to T$ as follows:

$$t=\begin{pmatrix}\rho y & v\\ 0 & y\end{pmatrix} \text{ and, for } r\in A,\ \phi(r)=\begin{pmatrix}\beta(r) & 0\\ 0 & r\end{pmatrix}.$$

Then

$$\phi(r)t=\begin{pmatrix}\beta(r) & 0\\ 0 & r\end{pmatrix}\begin{pmatrix}\rho y & v\\ 0 & y\end{pmatrix}=\begin{pmatrix}\rho\beta(r)y & \beta(r)v\\ 0 & ry\end{pmatrix}$$

whereas

$$t\phi(\alpha(r))=\begin{pmatrix}\rho y & v\\ 0 & y\end{pmatrix}\begin{pmatrix}\beta\alpha(r) & 0\\ 0 & \alpha(r)\end{pmatrix}=\begin{pmatrix}\rho y\beta\alpha(r) & v\alpha(r)\\ 0 & y\alpha(r)\end{pmatrix}.$$

Here $\rho\beta(r)y=\rho\alpha\beta(r)=\rho\beta\alpha(r)$, $v\alpha(r)=\gamma\alpha(r)v=\beta(r)v$ and $y\alpha(r)=ry$ so $t\phi(\alpha(r))=\phi(r)t$ for all $r\in A$. By the universal mapping property for right skew polynomial rings, there exists a unique ring homomorphism $\psi:S\to T$ such that

$\psi|_A = \phi$ and $\psi(y) = t$. It then follows from [**5**, 0.8, p. 41], or [**6**, Exercise 2ZC], that there is a left β-derivation δ of S such that $\delta(A) = 0$ and $\delta(y) = v$.

REMARK 2.3. An alternative approach to justifying the existence of the left β-derivation δ of S above is to check that, when the Leibniz rules are applied on the assumption that δ is a left β-derivation, it preserves the defining relations for S as a ring extension of A. To be complete, this approach should include writing S as a quotient F/I, where F is the free product of the $\mathbb{K}$-algebras A and the polynomial algebra $\mathbb{K}[y]$ and β is extended appropriately to F, and establishing the existence of a left β-derivation Δ of F such that $\Delta(A) = 0$, $\Delta(y) = v$ and $\Delta(I) \subseteq I$.

REMARK 2.4. Definition 2.1 involves a change in notation from [**12**] where α was an automorphism and $R(A, \alpha, v, \rho)$ satisfied the relations $ya = \alpha(a)y$ and $xa = \gamma\alpha^{-1}(a)x$ for all $a \in A$ and $xy - \rho yx = v$. In the current noatation this would be $A(R, \alpha^{-1}, v, \rho)$.

DEFINITION 2.5. In any ambiskew polynomial ring $R(A, \alpha, v, \rho)$, we shall refer to elements of the form $y^i a x^j$, where $i, j \in \mathbb{N}_0$ and $a \in A$, as *monomials*. As R is a left skew polynomial ring in x over a right skew polynomial ring in y over A, every element of R is a unique sum of monomials.

REMARK 2.6. Note that, in Definition 2.1, the endomorphism β of $[\alpha; y]A$ is injective if and only if the original endomorphism β of A is injective and $\rho \neq 0$.

REMARK 2.7. Definition 2.1 is symmetric in the roles of x and y in the sense that $R(A, \alpha, v, \rho)$ has the alternative form $[\delta', \alpha; y](A[x; \beta])$, where $\alpha = \beta\gamma^{-1}$ is extended to $A[x; \beta]$ by setting $\alpha(x) = \rho x$ and the right α-derivation δ' is such that $\delta'(y) = v$ and $\delta'(A) = 0$. More formally, the opposite ring $R(A, \alpha, v, \rho)^{\mathrm{op}}$ is isomorphic to $R(A^{\mathrm{op}}, \beta, v, \rho)$ under an isomorphism that interchanges x and y.

DEFINITION 2.8. Let $R = R(A, \alpha, v, \rho)$ be an ambiskew polynomial ring. If $v = \alpha(u) - \rho u$ for some γ-normal $u \in A$ we say that the 4-tuple (A, α, v, ρ) is *conformal* or that R is a *conformal ambiskew polynomial ring*. We then refer to u as a *splitting element* and to the element $z = xy - \alpha(u) = \rho(yx - u)$ as a *Casimir element*.

Here neither u nor z need be unique. For example whenever $v \in Z(A)$, in which case $\gamma = \mathrm{id}\,|_A$, if u is splitting then so too is $u - \lambda$ for all $\lambda \in \mathbb{K}$.

If (A, α, v, ρ) is not conformal we say that it is *singular* or that R is a *singular ambiskew polynomial ring*.

LEMMA 2.9. *Let $R = R(A, \alpha, v, \rho)$ be a conformal ambiskew polynomial ring with splitting element u and Casimir element z. Then $za = \gamma(a)z$ for all $a \in A$, $xz = \rho zx$ and $zy = \rho yz$. Consequently, if $\rho \neq 0$ then z is a normal element of R.*

PROOF. For $a \in A$,

$$(yx - u)a = y\beta(a)x - ua = y\alpha\gamma(a)x - \gamma(a)u = \gamma(a)(yx - u)$$

so $za = \gamma(a)z$.

Also,

$$xz = \rho xyx - \rho xu = \rho(z + \alpha(u))x - \rho\alpha(u)x = \rho zx$$

and

$$zy = \rho(yxy - uy) = \rho y(z + \alpha(u) - \alpha(u)) = \rho yz.$$

If $\rho \neq 0$ then, as R is generated by A, x and y and $\rho \in \mathbb{K}^*$, it follows that $zR = Rz$ so the Casimir element z is normal in R. □

DEFINITION 2.10. [**2**] Let D be a ring with two endomorphisms σ and τ and two elements b and ρ such that, for all $d \in D$, $\sigma\tau(d)\rho = \rho\tau\sigma(d)$ and $\sigma\tau(d)b = bd$. The *diskew polynomial ring* $D[x, y; \sigma, \tau, b, \rho]$ is generated by D, x and y subject to the defining relations $xy - \rho yx = b$ and, for all $d \in D$, $xd = \sigma(d)x$ and $yd = \tau(d)y$. [**2**, Theorem 5.1] says that $D[x, y; \sigma, \tau, b, \rho]$ is an iterated skew polynomial ring $D[y; \tau][x; \sigma, \partial]$ where σ is extended to $D[y; \tau]$ by setting $\sigma(y) = \rho y$ and ∂ is the σ-derivation of $D[y; \tau]$ such that $\partial(D) = 0$ and $\partial(y) = b$.

NOTATION 2.11. The abbreviations ann, lann and rann will be used for annihilators, left annihilators and right annihilators respectively. The positive integers and the non-negative integers will be written $\mathbb{N}$ and $\mathbb{N}_0$ respectively.

3. Simplicity

This section is a study of simplicity in ambiskew polynomial rings with a view to finding examples of simple ambiskew polynomial rings $R(A, \alpha, v, \rho)$ in which the defining endomorphism α is not injective. We begin with some preliminary results and then establish some necessary conditions for simplicity. There are technical challenges in determining whether these necessary conditions are sufficient for the simplicity of $R(A, \alpha, v, \rho)$. However, by restricting the base ring A to be commutative noetherian we can establish a simplicity criterion analogous to [**12**, Theorem 3.10] and thereby we obtain examples of simple ambiskew polynomial algebras, with α non-injective.

DEFINITION 3.1. Let α be an endomorphism of a ring A. An ideal I of A is said to be an *α-ideal* if $\alpha(I) \subseteq I$, equivalently $I \subseteq \alpha^{-1}(I)$, and to be a *strong α-ideal* if $\alpha^{-1}(I) = I$. Note that $\alpha^{-1}(I)$ is always an ideal of A but $\alpha(I)$ need not be so. Also, a strong α-ideal is clearly an α-ideal.

The ring A is said to be *α-simple* if no non-zero proper ideal is an α-ideal and to be *weakly α-simple* if no non-zero proper ideal is a strong α-ideal. Thus, α-simplicity implies weak α-simplicity.

Let β be a second endomorphism of A. An ideal of A is said to be an *$\{\alpha, \beta\}$-ideal* if it is both an α-ideal and a β-ideal and to be a *strong $\{\alpha, \beta\}$-ideal* if $\alpha^{-1}(I) = I = \beta^{-1}(I)$. The ring A is said to be *$\{\alpha, \beta\}$-simple* if no non-zero proper ideal of A is an $\{\alpha, \beta\}$-ideal and to be *weakly $\{\alpha, \beta\}$-simple* if no non-zero proper ideal of A is a strong $\{\alpha, \beta\}$-ideal.

NOTATION 3.2. Let $R = R(A, \alpha, v, \rho)$ be an ambiskew polynomial ring. Let $v_1 = v$, let $w_1 = w = xy = \rho yx + v$ and, for $j \geq 2$, define $w_j \in R$ and $v_j \in A$ recursively by setting $w_{i+1} = \rho w_i + \alpha^i(v)$ and $v_{i+1} = \rho^i v + \alpha(v_i)$. It is a routine matter to check that each $v_{i+1} = \rho v_i + \alpha^i(v) = \rho^i v + \rho^{i-1}\alpha(v) + \cdots + \rho\alpha^{i-1}(v) + \alpha^i(v)$ and each $w_i = \rho^i(w - v) + v_i$.

LEMMA 3.3. *Let $R = R(A, \alpha, v, \rho)$ be an ambiskew polynomial ring.*

(i) *For $a \in A$, $wa = \gamma(a)w$.*
(ii) $xw = (\rho w + \beta(v))x$, $\rho wx = x(w - v)$, $wy = y(\rho w + \alpha(v))$ *and* $\rho yw = (w - v)y$.
(iii) *For $n \geq 1$, $w_n y = y w_{n+1}$ and $x w_n = w_{n+1} x$.*
(iv) *For $n \geq 1$, $x^n y^n = w_n \dots w_2 w_1$.*
(v) *For $n \geq 1$, $x^n y - \rho^n y x^n = v_n x^{n-1}$ and $x y^n = \rho^n y^n x + y^{n-1} v_n$.*

PROOF. Note that, for $k \geq 1$, $\beta^k(v) = \alpha^k\gamma^k(v) = \alpha^k(v)$. Hence, $\beta(v_i) = \alpha(v_i)$ for all i.

(i) For $a \in A$,

$$wa = (\rho yx + v)a = \rho y\alpha\gamma(a)x + va = \rho\gamma(a)yx + \gamma(a)v = \gamma(a)w.$$

(ii) $xw = \rho xyx + xv = \rho wx + xv = \rho wx + \beta(v)x$, so $\rho wx = x(w - v)$ and $xw = (\rho w+\beta(v))x$. Also, $\rho yw = \rho yxy = (w-v)y$ so $wy = \rho yw+vy = y(\rho w+\alpha(v))$.

(iii) By (ii), $w_1y = yw_2$ and $xw_1 = w_2x$. Let $k \geq 1$ and suppose that $w_ky = yw_{k+1}$ and $xw_k = w_{k+1}x$. Then

$$\begin{aligned} yw_{k+2} &= y(\rho w_{k+1} + \alpha^{k+1}(v)) \\ &= \rho w_k y + \alpha^k(v)y = w_{k+1}y \end{aligned}$$

and

$$\begin{aligned} w_{k+2}x &= (\rho w_{k+1} + \alpha^{k+1}(v))x \\ &= (\rho w_{k+1} + \beta^{k+1}(v))x \\ &= \rho xw_k + x\beta^k(v) = xw_{k+1}. \end{aligned}$$

By induction, $w_ny = yw_{n+1}$ and $xw_n = w_{n+1}x$ for all $n \geq 1$.

(iv) This clearly holds when $n = 1$. Let $k \geq 1$ and suppose that $x^ky^k = w_k \dots w_2w_1$. Then, using (iii),

$$\begin{aligned} x^{k+1}y^{k+1} &= xw_k \dots w_2w_1y \\ &= w_{k+1}w_k \dots w_3w_2xy = w_{k+1}w_k \dots w_3w_2w_1 \end{aligned}$$

and so the result holds by induction.

(v) When $n = 1$ both identities reduce to the defining relation $xy - \rho yx = 1$. Assuming that both hold when $n = k$ for some $k \geq 1$,

$$\begin{aligned} x^{k+1}y &= x(\rho^k yx^k + v_kx^{k-1}) \\ &= \rho^k xyx^k + \beta(v_k)x^k \\ &= \rho^k xyx^k + \alpha(v_k)x^k \\ &= \rho^{k+1}yx^{k+1} + (\rho^k v + \alpha(v_k))x^k \\ &= \rho^{k+1}yx^{k+1} + v_{k+1}x^k \end{aligned}$$

and

$$\begin{aligned} xy^{k+1} &= \rho^k y^k xy + y^{k-1}v_ky \\ &= \rho^k y^k(\rho yx + v) + y^k\alpha(v_k) \\ &= \rho^{k+1}y^{k+1}x + y^k(\rho^k v + \alpha(v_k)) \\ &= \rho^{k+1}y^{k+1}x + y^kv_{k+1}. \end{aligned}$$

By induction, both identities hold for all $n \geq 1$. □

NOTE 3.4. An ambiskew polynomial ring $R = R(A, \alpha, v, \rho)$ has a $\mathbb{Z}$-grading with $\deg(a) = 0$ for all $a \in A$, $\deg(x) = 1$ and $\deg(y) = -1$. The ℓ-th component consists of all finite sums $\sum y^i a_{ij} x^j$ where $j - i = \ell$ and each $a_{ij} \in A$. In particular, R_0 consists of all finite sums $\sum y^i a_i x^i$ where each $a_i \in A$.

If $\rho = 0$ then $w = v \in A$ but if $\rho \neq 0$ then, by Lemma 3.3(v) and as each w^n has the form $\rho^n y^n x^n + r$, where r has degree $< n$ in both x and y, the subalgebra of R generated by A and w is a skew polynomial ring $A[w; \gamma]$ over A and $A[w; \gamma] \subseteq R_0$.

Also, α, γ and β extend to endomorphisms of $A[w;\gamma]$ such that $\gamma(w) = w$ and $\alpha(w) = \beta(w) = \rho w + \alpha(v)$. By Lemma 3.3(ii), for all $f \in A[w;\gamma]$, $fy = y\alpha(f)$ and $xf = \alpha(f)x$.

PROPOSITION 3.5. *If an ambiskew polynomial ring $R(A, \alpha, v, \rho)$ over a $\mathbb{K}$-algebra A is simple then*

(i) *A is $\{\alpha, \beta\}$-simple.*
(ii) *α and β are injective.*
(iii) *x and y are regular in R.*
(iv) *$\rho \neq 0$.*
(v) *$Av_mA = A$ for all $m \geq 1$ and v is a unit.*
(vi) *A is α-simple.*
(vii) *R is singular.*

PROOF. Suppose that R is simple.

(i) Let I be a non-zero $\{\alpha, \beta\}$-ideal of A and let J denote the set of finite sums of monomials of the form y^iax^j where $a \in I$ and $i, j \in \mathbb{N}_0$. Let $m = y^iax^j$ be such a monomial and let $b \in A$. Clearly $ym \in J$ and $mx \in J$. If $i = 0$ then $xm = \beta(a)x^{j+1} \in J$ and if $i \neq 0$ then, by Lemma 3.3(v), $xm = \rho^iy^ixax^j + y^{i-1}v_iax^j = \rho^iy^i\beta(a)x^{j+1} + y^{i-1}v_iax^j \in J$. Similarly, if $j = 0$ then $my = y^{i+1}\alpha(a) \in J$ and if $j > 0$ then, by Lemma 3.3(v), $y^iax^jy = \rho^jy^iayx^j + y^iav_ix^{j-1} = \rho^jy^{i+1}\alpha(a)x^j + y^iav_ix^{j-1} \in J$. For $b \in A$, $bm = y^i\alpha^i(b)ax^j \in J$ and $mb = y^ia\beta^j(b)x^j \in J$. As R is generated by A, x and y, it follows that J is a non-zero ideal of R so $1 \in J$. Hence, $1 \in J \cap A = I$ and I is not proper.

(ii) As $\ker(\alpha)$ is both an α-ideal of A and, as $\alpha\beta = \beta\alpha$, a β-ideal of A, it follows from (i) that $\ker(\alpha) = 0$ and that α is injective. Similarly, β is injective.

(iii) If $\rho \neq 0$ then x and y are regular in R as both α and the extension of β to $[\alpha; y]A$ are injective by (ii). So we may suppose that $\rho = 0$. Thus, $xy = v$. Clearly x is right regular in R and y is left regular in R. Let $D = \cup_{j\geq 1} \operatorname{rann}(x^d)$ which is a right ideal of R. Let $d \in D$ and let $j \geq 1$ be such that $x^jd = 0$. Then $x^{j+1}yd = x^jvd = vx^jd = 0$ so $yd \in J$. Clearly $xd \in D$ and, if $a \in A$ then $x^jad = \beta^j(a)x^jd = 0$ so $ad \in D$. Thus, D, which cannot be R, is an ideal of R and must be zero. Therefore x is regular. Similarly $\cup_{j\geq 1} \operatorname{lann}(y^d)$ is an ideal of R and y is regular.

(iv) Suppose that $\rho = 0$. Recall that v is normal in A so vA is an ideal of A. Let J be the set of all finite sums of monomials of the form y^iax^j where $i > 0$ or $j > 0$ or $a \in vA$. Then J is clearly closed under left multiplication by y and right multiplication by x. For $b \in A$, $by^iax^j = y^j\alpha^j(b)ax^i$ and $y^iax^jb = y^ia\beta^j(b)x^j$ so J is closed under right and left multiplication by elements of A. If $i \geq 2$ then $xy^iax^j = vy^{i-1}ax^j = y^{i-1}\alpha^{i-1}(v)ax^j \in J$, if $i = 1$ then $xy^iax^j = vax^j \in J$ and if $i = 0$ then $xy^iax^j = \alpha(a)x^{j+1} \in J$. Thus, $yJ \subseteq J$ and similarly $Jx \subseteq J$. It now follows that J is an ideal of R and, by simplicity, $1 \in J$. But $J \cap A = vA$ so $1 \in vA = Av$ and v is a unit in A, with inverse t say. Let $d = ytx - 1 \in R\backslash\{0\}$. Then $xd = xytx - x = vtx - x = 0$ but, by (iii), x is regular so $d = 0$ and we have a contradiction. Thus, $\rho \neq 0$.

(v) We amend the corresponding argument from [**12**, Lemma 3.2] as, although v is normal, we do not know that v_m is normal. Let I be the ideal Av_mA and suppose that $I \neq A$. Let J be the $\mathbb{K}$-linear span of the monomials of the forms y^iax^j where either $j > 0$, $i \geq m$ or $a \in I$ and note that $J \neq R$ as $A\backslash I \not\subseteq J$. We claim that J is a left ideal of R. It is clear that $yJ \subseteq J$ and, as $by^iax^j = y^i\alpha^i(b)ax^j$, that $AJ \subseteq J$.

Also, by Lemma 3.3(v), if $i > 0$ then $xy^iax^j = \rho^iy^i\beta(a)x^{j+1} + y^{i-1}v_iax^j$. Hence, $xy^iax^j \in J$ if $j > 0$ or if $i \geq m+1$ and, as $v_ma \in I$, if $i = m$. Finally, if $j = 0$ and $i < m$ then $xy^ia = \rho^iy^i\beta(a)x + y^{i-1}v_ia \in J$ and if $i = 0$ then, as $j + 1 > 0$, $xax^j = \beta(a)x^{j+1} \in J$. Hence, J is a proper left ideal of R and $\operatorname{lann}_R R/J = \{r \in R : rR \subseteq J\}$ is a proper ideal of R. But $y^m(y^iax^j) = y^{m+i}ax^j \in J$ for all $i, j \geq 0$ and all $a \in A$ so $y^m \in \operatorname{lann}_R R/J$, contradicting the simplicity of R. Hence, $Av_mA = A$. In particular, $vA = Av = Av_1A = A$ so v is a unit.

(vi) By (v), v is a unit so, for any α-ideal I of A, $\gamma(a) = vav^{-1} \in I$ and $\beta(a) = \alpha\gamma(a) \in I$ for all $a \in I$. It follows from (i) that A is α-simple.

(vii) Suppose that R is conformal with Casimir element $z = \rho(yx - u)$. As $\rho \neq 0$ and α is injective, the leading coefficient ρy of z, as a polynomial in x, is regular in $[\alpha; y]A$. Hence, z has degree ≥ 1 in x and cannot be a unit, contradicting the simplicity of R as $zR = Rz$ and $z \neq 0$. □

We now aim to establish a simplicity criterion for R, analogous to [**12**, Theorem 3.10], in the case where A is commutative noetherian and $\operatorname{char}\mathbb{K} = 0$ and to apply it to obtain examples of simple ambiskew polynomial algebras in which α is non-injective. Note that, when A is commutative, $\beta = \alpha$, $\gamma = \operatorname{id}|_A$ and, if $\rho \neq 0$, $A[w; \gamma] = A[w]$.

LEMMA 3.6. *Let $R = R(A, \alpha, v, \rho)$ be an ambiskew polynomial ring for which A is commutative, α is injective, but not an automorphism of finite order, and $\rho \neq 0$. Let I be a non-zero ideal of R. If A is α-simple or if A is a domain then $I \cap A[w] \neq 0$.*

PROOF. We first show that $I \cap R_0 \neq 0$. For an element r of the form $\sum y^ir_{ij}x^j$, where each $r_{ij} \in A$, let the *weight* of r be the number of coefficients r_{ij} that are non-zero and choose $r \in I\backslash\{0\}$ with minimal weight. Multiplying on the left by an appropriate power of y or on the right by an appropriate power of x, we can assume that $r_{ii} \neq 0$ for at least one i. Let $a \in A$. Then $ar - ra \in I$ and

$$ar - ra = \sum y^i(\alpha^i(a) - \alpha^j(a))r_{ij}x^j.$$

As $a_{ii} \neq 0$ for some i, $ar - ra$ has lower weight than that of r so, by minimality, $ar - ra = 0$ and $(\alpha^i(a) - \alpha^j(a))r_{ij} = 0$ whenever $i \neq j$.

Suppose that $i \neq j$ are such that $r_{ij} \neq 0$. If A is a domain, $\alpha^i(a) - \alpha^j(a) = 0$ for all $a \in A$ and, as α is injective, $\alpha^{|j-i|} = \operatorname{id}_A$ whence α is an automorphism of finite order, contradicting our hypotheses. Hence, $r_{ij} = 0$. Suppose now that A is α-simple, let S_{ij} be the set $\{\alpha^i(a) - \alpha^j(a) : a \in A\}$ and let I_{ij} be the ideal of A generated by S_{ij}. As $\alpha(S_{ij}) \subseteq S_{ij}$, I_{ij} is an α-ideal of A so either $I_{ij} = 0$, in which case, as above, α is an automorphism of finite order $|j - i|$, or $I_{ij} = A$, in which case we again deduce that $r_{ij} = 0$.

As $r_{ij} = 0$ whenever $i \neq j$, it follows that $r \in I \cap R_0$ so $I \cap R_0 \neq 0$. Let $J = I \cap R_0$.

Let $0 \neq \sum_0^n y^ia_ix^i = r \in J$, where $a_n \neq 0$. By Lemma 3.3(iv), $x^iy^ia_ix^iy^i \in A[w]$ for $0 \leq i \leq n$. For $f(w) \in A[w]$, $xf(w)y = \alpha(f(w))xy \in A[w]$ so $xA[w]y \subseteq A[w]$. Hence, $x^tA[w]y^t \subseteq A[w]$ for $t \geq 1$. For $0 \leq i \leq n$, $x^ny^ia_ix^iy^n = x^{n-i}x^iy^ia_ix^iy^iy^{n-i} \in A[w]$ and so $x^nry^n \in J \cap A[w]$. Let $S = [\alpha; y]A$. Suppose that $x^nry^n = 0$ and, for $0 \leq i \leq n$, let $b_i = y^ia_i \in S$. Thus, $b_n \neq 0$, as y is regular in S by the injectivity of α, $r = b_nx^n + b_{n-1} + \ldots b_0$ and x^nr has degree at most $2n$ in x and its coefficient of x^{2n} is $\alpha^n(b_n)$. As α is injective on A and $\rho \neq 0$, α is injective on S so $x^n r$ has degree $2n$ and has the form $c_{2n}x^{2n}+c_{2n-1}x^{2n-1}+\cdots+c_0$, for some $c_{2n}, c_{2n-1}, \ldots, c_0 \in S$ with $c_{2n} = \alpha^n(b_n) \neq 0$. The coefficient of x^{2n} in $x^n r y^n$ is $c_{2n}\alpha^{2n}(y^n) = \rho^{2n}c_{2n}y^n$ so, as $\rho \neq 0$, $c_{2n}y^n = 0$ in S. As y is regular in S, this is a contradiction. Thus, $0 \neq x^n r y^n \in J \cap A[w] = I \cap A[w]$. □

LEMMA 3.7. *Suppose that* $\operatorname{char} \mathbb{K} = 0$. *Let* $R = R(A, \alpha, v, \rho)$ *be an ambiskew polynomial ring for which* A *is commutative,* α *is injective and let* $\rho \in \mathbb{K}^*$. *If* A *is* α*-simple and* (A, α, v, ρ) *is singular then* $A[w]$ *is* α*-simple.*

PROOF. Let I be a non-zero α-ideal of $A[w]$. Suppose that $I \cap A = 0$ and let d be the least positive integer such that I contains an element of degree d in w. Let $\tau(I) =$

$$\{a \in A : aw^d + a_{d-1}w^{d-1} + \cdots + a_0 \in I \text{ for some } a_{d-1}, \ldots, a_0 \in A\}.$$

Then, as $A[w]$ is commutative, $\tau(I)$ is a non-zero ideal of A. Recall, from Note 3.4, that $\alpha(w) = \rho w + \alpha(v)$. Hence, for an element f of the form $aw^d + a_{d-1}w^{d-1} + \cdots + a_1 w + a_0$, the coefficient of w^d in $\alpha(f)$ is $\rho^d\alpha(a)$. As I is an α-ideal of $A[w]$, it follows that $\tau(I)$ is an α-ideal of A. By the α-simplicity of A, we may assume that I contains an element of the form $f = w^d + a_{d-1}w^{d-1} + \cdots + a_1 w + a_0$, where each $a_i \in A$. Then $\alpha(f) = (\rho w + v)^d + \alpha(a_{d-1})(\rho w + v)^{d-1} + \ldots$ in which the coefficients of w^d and w^{d-1} are ρ^d and $d\rho^{d-1}v + \rho^{d-1}\alpha(a_{d-1})$ respectively. As $\alpha(f) \in I$ and $\deg(\alpha(f)) = d$, we must have $\alpha(f) = \rho^d f$ and $\rho^d a_{d-1} = \rho^{d-1}(dv + \alpha(a_{d-1}))$, whence $v = \alpha(u) - \rho u$ where $u = -\frac{1}{d}a_{d-1} \in A$. This contradicts our hypotheses so $I \cap A \neq 0$ and, by the α-simplicity of A, $I \cap A = A$ and $I = A[w]$. Thus, $A[w]$ is α-simple. □

THEOREM 3.8. *Suppose that* $\operatorname{char} K = 0$ *and let* $R = R(A, \alpha, v, \rho)$ *be an ambiskew polynomial ring such that* A *is commutative noetherian. Then the ambiskew polynomial* $R = R(A, \alpha, v, \rho)$ *is simple if and only if the following conditions hold:*

(i) A *is* α*-simple;*
(ii) α *is injective;*
(iii) $\rho \neq 0$*;*
(iv) (A, α, v, ρ) *is singular;*
(v) v_n *is a unit in* A *for all* $n \geq 1$.

PROOF. By [**12**, Theorem 3.10] the result is true when θ is an automorphism. For this, note that $\rho \neq 0$ in [**12**] and the change of notation discussed in Remark 2.4 has the consequence that the element $v^{(m)}$ in [**12**, Theorem 3.10] would here be $\alpha^{m-1}(v_m)$, which is 0 precisely when $v_m = 0$. For the remainder of the proof assume that α is not an automorphism.

Suppose that R is simple. Then (i)-(v) hold by Proposition 3.5, bearing in mind that, as A is commutative, $\beta = \alpha$ and each $Av_nA = v_nA$.

Conversely, suppose that (i)-(v) hold, let I be a non-zero ideal of R and let $J = I \cap A[w]$. As α is not an automorphism, $J \neq 0$, by Lemma 3.6.

For $i \geq 0$, let $J_i = \{j \in A[w] : jx^iy^i \in J\}$, which, as $A[w]$ is commutative, is an ideal of $A[w]$.

Let $i \geq 0$ and let $j \in J_i$. By Lemma 3.3(iii), $jx^{i+1}y^{i+1} = jx^iwy^i = jx^iy^iw_{i+1} \in J$ so $J_i \subseteq J_{i+1}$ and $J_i \neq 0$ for $i \geq 0$. Also, $\alpha(j)x^{i+1}y^{i+1} = xjx^iy^{i+1} \in I \cap A[w] = J$ so $\alpha(J_i) \subseteq J_{i+1}$. As $A[w]$ is noetherian, there exists $t \geq 1$ such that $J_t = J_s$ for all $s \geq t$. Then J_t is a non-zero α-ideal of S so, by Lemma 3.7, $J_t = A[w]$ and $x^ty^t \in J \subset I$.

Let $p \geq 0$ be minimal such that $x^t y^p \in I$. Thus, $0 \leq p \leq t$. Suppose that $p > 0$. By Lemma 3.3(v),

$$x^t y^{p-1} v_p = x^t(\rho^p x y^p - y^p x) = \rho^p x^{t+1} y^p - x^t y^p x \in I$$

so, as v_p is a unit in A, $x^t y^{p-1} \in I$, contradicting the minimality of p. Hence, $p = 0$ and $x^t \in I$. Now let $b \geq 0$ be minimal such that $x^b \in I$. Thus, $0 \leq b \leq t$. If $b > 0$ then, by Lemma 3.3(v), $v_b x^{b-1} = x^b y - \rho^p y x^b \in I$ so, as v_b is a unit in A, $x^{b-1} \in I$, contradicting the minimality of b. Thus, $b = 0$, $I = R$ and R is simple. □

EXAMPLE 3.9. Let $A = \mathbb{C}(t)$ be the field of rational functions over $\mathbb{C}$ in one indeterminate t and let $m, n \in \mathbb{N}$ with $n \geq 2$ and $m \leq n$. Let α be the $\mathbb{C}$-endomorphism of A such that $\alpha(t) = t^n$, let $v = t^m$ and let $\rho \in \mathbb{C}^*$. As a ring extension of $\mathbb{C}(t)$, $R = R(\mathbb{C}(t), \alpha, t^m, \rho)$ is generated by x and y subject to the relations

$$xy - \rho yx = t^m, \quad ty = yt^n, \quad xt = t^n x.$$

Clearly α is not an automorphism and A is α-simple and a commutative noetherian domain. Suppose that (A, α, v, ρ) is conformal with splitting element u. Write $u = f(t)/g(t)$, where $f(t), g(t) \in \mathbb{C}[t]$, $g(t) \neq 0$ and $f(t)$ and $g(t)$ have no common zero in $\mathbb{C}$. We can assume that $g(t)$ is monic. Then

$$t^m g(t) g(t^n) = g(t) f(t^n) - \rho g(t^n) f(t). \tag{1}$$

Suppose first that $\deg g(t) = 0$ so that, by monicity, $g(t) = 1$. Then $t^m = f(t^n) - \rho f(t)$. As $m \leq n$, it must be the case that $m = n$ and $f(t) = at + b$ for some $a, b \in \mathbb{C}$, with $a \neq 0$. Comparing coefficients of t, $0 = -\rho a$, giving a contradiction. So $g(t)$ is not constant and must have at least one zero in $\mathbb{C}$.

Let z be a zero in $\mathbb{C}$ of $g(t)$ and let w be an nth root of z in $\mathbb{C}$. Then $g(w^n) = 0$ so, by (1), $g(w)f(w^n) = 0$ and, as $f(t)$ and $g(t)$ have no common zero, $f(w^n) \neq 0$. Thus, $g(w) = 0$. Repeating the argument, for $j \geq 1$, all the n^j-th roots of z are zeros of $g(t)$. Given any non-zero complex number z there is a sequence $\{z_i\}$, with distinct terms, in which $z_1 = z$ and $z_{i+1}^n = z_i$ for $i \geq 1$. As $g(t)$ has only finitely many zeros, it follows that $z = 0$. Hence $g(z) = z^d$, for some $d \geq 1$ and

$$t^{m+(n+1)d} = t^d f(t^n) - \rho t^{dn} f(t).$$

As 0 cannot be a common zero of $f(t)$ and $g(t)$, $f(t)$ has non-zero constant term, s, say. Comparing coefficients of t^d then gives $0 = s$, which is false. So (A, α, v, ρ) is singular.

Also, each $v_i = \rho^{i-1} t^m + \rho^{i-2} t^{mn} + \rho^{i-3} t^{mn^2} + \cdots + t^{mn^{i-1}} \neq 0$. By Theorem 3.8, R is simple.

REMARK 3.10. In contrast to the above example, if D is a field with endomorphisms σ and τ that are not both automorphisms then a diskew polynomial ring $E = D[x, y; \sigma, \tau, b, \rho]$ cannot be simple.

To see this, observe that, as D is a field, σ and τ are both injective. Also, if $b \neq 0$ then $\sigma\tau = \mathrm{id}_D$, σ is surjective and, hence, bijective, and $\tau = \sigma^{-1}$. So we can assume that $b = 0$. In this case the defining relations are $xy = \rho yx$ and, for all $d \in D$, $xd = \sigma(d)x$ and $yd = \tau(d)y$. Therefore Ex is an ideal of E. It is clearly non-zero and proper so E is not simple.

The simplicity criterion for diskew polynomial rings given by [**2**, Theorem 1.3] has the hypotheses that ρ is a unit and $\tau\sigma$ is surjective so is not immediately applicable to make the deduction above. That deduction in itself does not tell us that the simple ambiskew polynomial ring R in Example 3.9 is not a diskew polynomial ring over any ring. The next result gives that conclusion.

PROPOSITION 3.11. *Let A be a field extension of $\mathbb{K}$ and let α be an injective, but not surjective, endomorphism of A. Let $v, \rho \in A$. For any ring D, the ambiskew polynomial ring $R = R(A, \alpha, v, \rho)$ is not an iterated left skew polynomial ring of the form $E = D[u; \tau, \delta][w; \sigma, \partial]$ where $\sigma(D) \subseteq D$. In particular, R is not a diskew polynomial ring over any ring D.*

PROOF. Let R and E be as stated and suppose that there is an isomorphism $\theta : R \to E$. As R is a domain, τ and σ must be injective and D is a domain. The isomorphism θ restricts to an isomorphism between the groups $U(R)$ and $U(E)$. But $U(R) = A\backslash\{0\}$ so $U(E) = \theta(A\backslash\{0\})$. As τ and σ are injective and D is a domain, $U(E) = U(D)$. Hence, $\theta(A) \subseteq D$. Let $B = \theta(A)$. As $\tau(U(D) \subseteq U(D)$ and $\sigma(U(D) \subseteq U(D)$, $\sigma(B) \subseteq B$ and $\tau(B) \subseteq B$. Let $f = \theta^{-1}(u)$ and $g = \theta^{-1}(w)$. Now, $uB = \tau(B)u \subseteq Bu$ and $wB = \sigma(B)w \subseteq Bw$ so $fA \subseteq Af$ and $gA \subseteq Ag$. Express f in its unique form as a sum of monomials $\sum y^i a_{ij} x^j$. Then, for $a, b \in A$, $fa = \sum y^i a_{ij} \alpha^j(a) x^j$ and $bf = \sum y^i \alpha^i(a) a_{ij} x^j$. As A is a field, it follows, as $fA \subseteq Af$, that if $a_{ij} \neq 0$ then $\alpha^j(A) \subseteq \alpha^i(A)$. But the descending chain

$$A \supset \alpha(A) \supset \alpha^2(A) \supset \cdots \supset \alpha^j(A) \supset \ldots$$

is strict so if $a_{ij} \neq 0$ then $j \geq i$. Thus, in the $\mathbb{Z}$-grading on R, $f \in \sum_{d \geq 0} R_d$. Similarly $g \in \sum_{d \geq 0} R_d$ so, as $A = R_0$ and f, g and A generate R, $R = \sum_{d \geq 0} R_d$. As $y \in R_{-1}$, this is a contradiction so the supposed isomorphism θ does not exist. □

EXAMPLE 3.12. For an example of a simple ambiskew polynomial in which A is not a field and α is not bijective, let $A = \mathbb{C}(t)[z]$ be the polynomial ring in z over the rational function field $\mathbb{K}(t)$. With n and m as in Example 3.9, but with $m \geq 2$, let α be the $\mathbb{C}$-endomorphism of A such that $\alpha(z) = z + t$ and $\alpha(t) = t^n$, and let $v = t^m$. Let $\rho \in \mathbb{C}\backslash\{0, 1\}$. Certainly A is a commutative domain and α is not an automorphism of finite order. Also, the elements v_i are as in Example 3.9 and therefore non-zero. As a ring extension of $\mathbb{C}(t)$, $R = R(\mathbb{C}(t)[z], \alpha, t^m, \rho)$ is generated by z, x and y subject to the relations

$$tz = zt,\ xy - \rho yx = t^m,\ ty = yt^n,\ xt = t^n x,\ zy = y(t + z),\ xz = (t + z)x.$$

To see that A is α-simple, suppose that there is a non-zero proper α-ideal I of A. As A is a principal ideal domain, $I = fA$ for some non-zero monic polynomial $p = z^d + f_{d-1}(t)z^{d-1} + \cdots + f_0(t)$ in $\mathbb{C}(t)[z]$ with degree $d \geq 1$. Then $\alpha(p)$ is also monic of degree d and must equal p. Write $f_{d-1} = h(t)/g(t)$ where $h(t), g(t) \in \mathbb{C}[t]$ have no common zeroes and $g(t) \neq 0$. Comparing coefficients of z^{d-1} in the equation $\alpha(p) = p$ gives $dt + \alpha(f_{d-1}(t)) = f_{d-1}(t)$ so

$$dtg(t^n)g(t) + h(t^n)g(t) = h(t)g(t^n).$$

Let $b = \deg(h(t))$ and $c = \deg(g(t))$. The terms on the left hand side of (3.12) have degrees $1 + (n+1)c$ and $nb + c$ and the term on the right hand side has degree $b + nc$. If $b \leq c$ then $nb + c \leq (n+1)c < 1 + (n+1)c$ and $b + nc \leq (n+1)c < 1 + (n+1)c$. Therefore $b > c$ whence $nb + c > b + nc$ which implies that $nb + c = 1 + (n+1)c$ and $nb = nc + 1$. But 1 cannot be a integer multiple of n so we have a contradiction. Thus, A is α-simple.

Suppose that R is conformal with splitting element $u = g_e(t)z^e + g_{e-1}(t)z^{e-1} + \cdots + g_0(t)$ in $\mathbb{C}(t)[z]$, with degree e. Thus, $\alpha(u) - \rho u = t^m$. From Example 3.9, we know that $v \notin \mathbb{C}[t]$ so $e > 0$. Comparing coefficients of z^e in the equation $\alpha(u) - \rho u = t^m$ gives $g_e(t^m) - \rho g_e(t) = 0$. Writing $g_e(t) = r(t)/s(t)$, where $r(t), s(t) \in \mathbb{C}[t]$ and $s(t) \neq 0$, we see that $r(t^m)s(t) = \rho r(t)s(t^m)$. From this it follows, as $m \geq 2$, that $r(t)$ and $s(t)$ must have the same degree. But the leading coefficients of $r(t^m)s(t)$ and $r(t)s(t^m)$ must then be equal so, as $\rho \neq 1$, $r(t^m)s(t) \neq \rho r(t)s(t^m)$. This contradiction shows that R is singular. By Theorem 3.8, $R(A, \alpha, t^m, \rho)$ is simple.

REMARK 3.13. In Examples 3.9 and 3.12, it is easy to amend the choice of v to make R conformal and, hence, not simple. Take any $u \in A$, for example $u = t$, and set $v = \alpha(u) - \rho u$.

It would be interesting to know whether there is an example where A is an affine commutative $\mathbb{C}$-algebra, $R(A, \alpha, v, \rho)$ is simple and α is not bijective. Below we give two naive candidates which fail on the grounds that A is not α-simple.

EXAMPLES 3.14. (i) Let $A = \mathbb{C}[t]$ and let $\alpha(t) = t^2 + 1$. Let ω be a primitive cube root of 1. Then $\alpha(t^2 - \omega) = (t^2 - \omega)(t^2 - \omega^2 + 1)$ so A is not α-simple.

(ii) Let $A = \mathbb{C}[t^{\pm 1}]$ and let $\alpha(t) = t^n$ for some $n \geq 2$. Then $\alpha(t^2 - 1) = t^{2n} - 1 = (t^2 - 1)(t^{2(n-1)} + \cdots + t^2 + 1)$ so A is not α-simple.

4. Primality

In this section, we consider primality of ambiskew polynomial rings with a view to generalizing results, from [**11**], on primality of non-noetherian downup algebras. Most of the results will be for general base rings A but they will be applied to determine which ambiskew polynomial rings of the form $R(\mathbb{K}[t], \alpha, v, \rho)$ are prime. If $\rho \neq 0$ and α is injective then $R(\mathbb{K}[t], \alpha, v, \rho)$ is a domain so primality is only an issue if $\rho = 0$ or if σ is not injective. We shall see, perhaps surprisingly, that, generically, $R(\mathbb{K}[t], \alpha, v, \rho)$ is prime. The precise result, specifying the exceptional cases, is Corollary 4.5.

PROPOSITION 4.1. *If α is not injective and $v_m = 0$ for some $i \geq 1$ then R is not prime.*

PROOF. Let I be as in the proof of Proposition 3.5(ii). Thus, I is a non-zero ideal of R and its elements are finite sums of elements of the form $y^i a x^j$, where $a \in \ker(\alpha)$ and $i, j \in \mathbb{N}_0$. Suppose that $v_m = 0$ for some $m \geq 1$. By Lemma 3.3, $xy^m = \rho^m y x^m + v_m x^{m-1} = \rho^m y x^m$ so, for $j \geq 0$, $x^j y^m = \rho^{mj} y^m x^j$ and, for $a \in \ker(\alpha)$ and $i \geq 0$, $y^i a x^j y^m = \rho^{mj} y^i a y^m x^j = \rho^{mj} y^{i+m} \alpha^m(a) x^j = 0$. Thus, $y^m \in \operatorname{rann}(I)$ whence $\operatorname{rann}(I) \neq 0$ and R cannot be prime. □

PROPOSITION 4.2. *Let $R = R(A, \alpha, v, \rho)$ be an ambiskew polynomial ring. Suppose that for all $j \geq 1$, v_j is regular and that $\ker(\alpha)$ contains a regular element h of A. Then $I \cap A \neq 0$ for every non-zero ideal I of R.*

PROOF. Let $S = [\alpha; y]A$. Thus, $R = S[x; \alpha, \delta]$ with $\alpha(y) = \rho y$, $\delta(y) = v$ and $\delta(A) = 0$. Also, $xh = 0 = hy$.

Let I be a non-zero ideal of R. For $0 \neq a = \sum_{j=0}^{n} f_j x^j \in R$, where each $f_j \in S$, let the *codegree* of a in x be the least j such that $f_j \neq 0$. Choose $0 \neq a = \sum_{j=m}^{n} f_j x^j \in I$ to be of minimal codegree m in x among non-zero elements of I

and suppose that $m > 0$. Then $ay \in I$ and, by Lemma 3.3(v),

$$ay = \sum_{j=m}^{n} f_j(\rho^j yx^j + v_j x^{j-1}).$$

The coefficient of x^{m-1} in ay is $f_m v_m$ which, by the regularity of v_m, is non-zero, contradicting the minimality of m. Thus, $m = 0$, $a = \sum_{j=0}^{n} f_j x^j$ and $f_0 \neq 0$. Then $xh = \beta(h)x = 0$ and $ah = f_0 h$. By the regularity of h, $0 \neq f_0 h = ah \in I \cap S$. Thus, $I \cap S \neq 0$.

For $0 \neq b = \sum_{j=0}^{k} y^j b_j \in S$, where each $b_j \in A$, let the *codegree* of b in y be the least ℓ such that $b_\ell \neq 0$. Choose a non-zero element $b = \sum_{j=\ell}^{n} y^j b_j$ in $I \cap S$ with minimal codegree ℓ in y among non-zero elements of $I \cap S$ and suppose that $\ell > 0$. Then, by Lemma 3.3(v) and as $xh = 0$,

$$xbh = \sum_{j=\ell}^{n} (\rho^j y^j xh + y^{j-1} v_j b_j h) = \sum_{j=\ell}^{n} y^{j-1} v_j b_j h \in I \cap S.$$

The coefficient of $y^{\ell-1}$ in xbh is $v_\ell b_\ell h$ which is non-zero, by the regularity of v_ℓ and h. This contradicts the minimality of ℓ so $\ell = 0$. But then $hb = hb_0 \neq 0$, as $hy = 0$ and h is regular in A, so $I \cap A \neq 0$. □

COROLLARY 4.3. *Let $R = R(A, \alpha, v, \rho)$ where A is a prime $\mathbb{K}$-algebra. Suppose that for all $j \geq 0$, v_j is regular and that* $\ker(\alpha)$ *contains a regular element of A. Then R is prime.*

PROOF. If I and J are non-zero ideals of R such that $IJ = 0$ then, by Proposition 4.2, $I \cap A$ and $J \cap A$ are non-zero ideals of A such that $(I \cap A)(J \cap A) = 0$. □

We now turn our attention to the situation where α is injective and $\rho = 0$.

PROPOSITION 4.4. *Let $R = R(A, \alpha, v, \rho)$ be an ambiskew polynomial ring where A is a domain, α is injective and $\rho = 0$. Then R is prime if and only if $v \neq 0$.*

PROOF. We first note that, for $j \geq 1$, $v_j = \alpha^{j-1}(v)$ so, as α is injective, $v \neq 0$ if and only if $v_j \neq 0$ for all $j \geq 1$. Also, $\beta = \alpha\gamma$ is injective as α is injective and γ is bijective.

Suppose that $v = 0$, in which case γ can be any automorphism of A that commutes with α. The defining relations for R as an extension of A are $xy = 0$ and, for all $a \in A$, $ay = y\alpha(a)$ and $xa = \alpha(a)x$. In this case, Rx is an ideal of R and $(Rx)y = 0$ so R is not prime.

Suppose now that $v \neq 0$. Write $S = [\alpha; y]A$, so that $R = S[x; \beta, \delta]$, where $\delta(y) = v$ and β is extended to S by setting $\beta(y) = 0$. Let I and J be non-zero ideals of R and let m be the minimal degree, in x, of a non-zero element of I. Suppose that $m = 1$ and let $a = s_1 x + s_0$ for some $s_1, s_0 \in S$ with $s_1 \neq 0$. Then $ay = s_1 v + s_0 y \in S$ so, by the minimality of m, $ay = 0$. Finally, suppose that $m > 1$ and let $a = s_m x^m + \cdots + s_1 x + s_0$ for some $s_m, \ldots, s_1, s_0 \in S$ with $s_m \neq 0$. Then

$$ay = s_m x^{m-1} v + \cdots + s_1 v + s_0 y = s_m \alpha^{m-1}(v) x^{m-1} + \cdots + s_1 v + s_0 y,$$

contradicting the minimality of m as $\alpha^{m-1}(v) \neq 0$. So either $m = 0$ or $m = 1$ and $ay = 0$ for any element a of I with degree 1 in x. By symmetry, viewing R as $[\delta, \alpha; y](A[x; \beta]) = [\delta, \alpha; y]S'$, the minimal degree n in y of a non-zero element of J

is either 0 or 1 and, in the latter case $xb = 0$ for any element b of I with degree 1 in y.

There are four combinations to consider. Firstly suppose that $m = n = 0$ and let $0 \neq y^k a_k + \cdots + ya_1 + a_0 = s(y) \in I \cap S$ and $0 \neq b_k x^k + \cdots + b_1 x + b_0 = s'(x) \in J$, where each $a_j, b_j \in A$ and $a_k, b_\ell \neq 0$. Then $s(y)s'(x)$ is the sum of the products $y^i a_i b_j x^j$ of which the only term of the form $y^k c x^\ell$, $c \in A$, is $y^k a_k b_\ell x^\ell$, which is non-zero. Hence, $IJ \neq 0$ in this case.

The second possibility is that I contains an element of the form $r = s_1(y)x + s_0$, where $s_1, s_0 \in S$, $s_1 \neq 0$ and $ry = 0$, and J contains a non-zero element $s'(x) = b_k x^k + \cdots + b_1 x + b_0$ as in the first case. Write $s_1(y) = y^w c_w + \cdots + yc_1 + c_0$ and $s_0(y) = y^v d_v + \cdots + yd_1 + d_0$ where each $c_j, d_j \in A$ and $c_w \neq 0$. The product $rs'(x)$ is the sum of the terms $y^j c_j \beta(b_i) x^{i+1}$ and $y^j d_j b_i x^i$. The only term of the form $y^w a x^{k+1}$ is $y^w c_w x^{k+1}$ which is non-zero so $IJ \neq 0$ in this case. By symmetry, $IJ \neq 0$ if $I \cap S \neq 0$ and J contains an element of the form $r' = ys_1'(x) + s_0'$, where $s_1'(x), s_0'(x) \in S'$, $s_1'(x) \neq 0$ and $xr' = 0$.

The final possibility is that I contains an element of the form $r = s_1(y)x + s_0(y)$, where $s_1(y), s_0(y) \in S$, $s_1 \neq 0$ and $ry = 0$, and J contains an element of the form $r' = ys_1'(x)x + s_0'(x)$, where $s_1'(x), s_0'(x) \in S'$, $s_1'(x) \neq 0$ and $xr' = 0$. Note that if $s_0'(x) = 0$ then $0 = xys_1'(x)x = vs_1'(x)x$, which is impossible, so $s_0'(x) \neq 0$. In this case $rr' = r(ys_1'(x)x + s_0'(x)) = rs_0'(0)$. The same method as in the second possibility shows that $rr' \neq 0$, whence $IJ \neq 0$. Thus, $IJ \neq 0$ in all cases and R is prime. □

COROLLARY 4.5. *Let $R = R(\mathbb{K}[t], \alpha, v(t), \rho)$ be an ambiskew polynomial ring over the polynomial ring $\mathbb{K}[t]$. Then R is not prime if and only if one of the following holds:*

(i) $v(t) = 0$ *and* $\rho = 0$;
(ii) $\alpha(t) = \eta$ *for some* $\eta \in \mathbb{K}$ *and* $v(\eta) = 0 = \rho$;
(iii) $\alpha(t) = \eta$ *for some* $\eta \in \mathbb{K}$, $v(t) = \mu \in \mathbb{K}$, *and* $\mu = 0$ *or* $\rho^\ell = -1$ *for some* $\ell \geq 1$.

PROOF. If α is injective and $\rho \neq 0$ then R is a domain, and hence prime, by two applications of [**13**, Theorem 1.2.9(i)]. If α is injective and $\rho = 0$ then, by Proposition 4.4, R is prime if and only if $v(t) \neq 0$. Thus, if α is injective then R is not prime if and only if (i) holds.

Now assume that α is not injective. As $\mathbb{K}[t]$ is a domain, $\ker(\alpha)$ is a non-zero prime ideal of $\mathbb{K}[t]$, so $t - \eta \in \ker(\alpha)$ and $\alpha(t) = \eta$ for some $\eta \in \mathbb{K}$. By Corollary 4.3 and Proposition 4.1, R is prime if and only if $v_i \neq 0$ for all $i \geq 1$. Let $\mu = \alpha(v(t)) = v(\eta) \in \mathbb{K}$. From Notation 3.2, we see that $v_1 = v(t)$ and $v_j = \rho^{j-1}v(t) + \mu$ for all $j \geq 2$.

If (i) holds (in the non-injective case) then (ii) holds, if (ii) holds then $v_2 = 0$, and if (iii) holds then $v_{\ell+1} = 0$ so each of (i), (ii) and (iii) imply that R is not prime.

Conversely, suppose that R is not prime. If $\rho = 0$ then $v_1 = v(t)$ and $v_j = \mu$ for all $j \geq 2$ so, as R is not prime, $\mu = 0$ and (ii) holds. So we can assume that $\rho \neq 0$. If $\deg v(t) \geq 1$ then $v_i \neq 0$ for all $i \geq 1$ so, as R is not prime, $\deg v(t) < 1$ and $v(t) = \mu$. In this case, $v_1 = \mu$ and, for $j \geq 2$, $v_j = (\rho^{j-1} + 1)\mu$ so, as R is not prime, (iii) holds. □

EXAMPLES 4.6. By [**11**, Theorem 7.1], the non-noetherian down-up algebras of Benkart and Roby [**3**] are the ambiskew polynomial rings $R = R(\mathbb{K}[t], \alpha, t+\mu, \rho)$ where $\alpha(t) = 0$. By Corollary 4.5 or by [**11**, Remark 7.9 and Theorem 7.10], such an algebra is prime if and only if μ and ρ are not both 0.

Cassidy and Shelton [**4**] study the generalized down-up algebras $L(f, r, s, \gamma)$, where $f \in \mathbb{K}[t]$ and $r, s, \gamma \in \mathbb{K}$. As an ambiskew polynomial algebra, $L(f, r, s, \gamma) = R(\mathbb{K}[t], \alpha, -f, s)$ where $\alpha(t) = rt - \gamma$. These algebras are noetherian domains if and only if $rs \neq 0$. By Corollary 4.5, those that are not prime are those where (i) $f = 0$, $s = 0$, (ii) $r = s = 0$, $f(-\gamma) = 0$, or (iii) $r = 0$, $f = \mu$ for some $\mu \in \mathbb{K}$ and $\mu = 0$ or $s^\ell = -1$ for some $\ell \geq 1$.

5. Hyperbolic algebras

This section is concerned with the simplicity of certain quotients of conformal ambiskew polynomial algebras. The choice of name is as explained in the Introduction.

DEFINITION 5.1. With A, α, γ and β be as in Definition 2.1, let u be a γ-normal element of A such that $\alpha(u)$ is also γ-normal, let $\rho = 1$ and let $v = \alpha(u) - u$, which is γ-normal. Then $R(A, \alpha, v, 1)$ is conformal, with splitting element u and Casimir element $z = xy - \alpha(u) = yx - u$. The *hyperbolic algebra* $T = T(A, \alpha, u)$ is the algebra R/zR.

REMARK 5.2. If A is commutative, so that $\gamma = \mathrm{id}\,|_A$, the condition in Definition 5.1 that $\alpha(u)$ is also γ-normal holds trivially. For an example of a γ-normal element u for which $\alpha(u)$ is not γ-normal, let $q \in \mathbb{K}\backslash\{0, 1\}$, take A to be the $\mathbb{K}$-algebra generated by $\{t_i, s_i : i \leq 0\}$ subject to the relations $s_i s_j = s_j s_i$, $t_i t_j = t_j t_i$ and $s_i t_j = \lambda_{ij} t_j s_i$, where $\lambda_{ij} = q$ if $j = i + 1$ and $\lambda_{ij} = 1$ if $j \neq i + 1$, and take α to be the $\mathbb{K}$-endomorphism of A such that $\alpha(s_i) = s_{i+1}$ and $\alpha(t_i) = t_{i+1}$ for $i \geq 0$. If $u = t_0$ then $u \in Z(A)$, in other words, u is $\mathrm{id}\,|_A$-normal, but $\alpha(u) = t_1$ does not commute with s_0.

Let $T = T(A, \alpha, u)$ be a hyperbolic algebra defined as in Definition 5.1. There is a $\mathbb{K}$-homomorphism $\theta : A \to T$ given by $a \mapsto a + zR$. By the right and left universal mapping properties for skew polynomial rings, θ extends to $\mathbb{K}$-homomorphisms $\theta_y : [\alpha; y]A \to T$, with $\theta_y(y) = y + zR$, and $\theta_x : A[x; \beta] \to T$, with $\theta_x(y) = x + zR$.

Let $X = x + zR$, let $Y = y + zR$ and for $a \in A$, let $\overline{a} = a + zR$. Then T is generated by $\theta(A)$, X and Y subject to the relations

$$XY = \overline{\alpha(u)}, YX = \overline{u}, \text{ and, } \forall a \in A,\ \overline{a}Y = Y\overline{\alpha(a)} \text{ and } X\overline{a} = \overline{\beta(a)}X.$$

Note that $X^2Y^2 = X\overline{\alpha(u)}Y = \overline{\alpha^2(u)}XY = \overline{\alpha^2(u)}\ \overline{\alpha(u)}$ and, by induction, for $j \geq 1$,

$$X^jY^j = \overline{\alpha^j(u)} \ldots \overline{\alpha^2(u)}\ \overline{\alpha(u)}. \tag{2}$$

On the other hand, $Y^2X^2 = Y\overline{u}X$ and whether this simplifies to an element of A depends on whether $u \in \alpha(A)$.

REMARK 5.3. In the ambiskew polynomial ring $R(A, \alpha, v, 1)$, we have taken $\rho = 1$ and the symmetry of R discussed in Remark 2.7 extends to T. Formally, the opposite ring $T(A, \alpha, u)^{\mathrm{op}}$ is isomorphic to $T(A^{\mathrm{op}}, \beta, u)$ under an isomorphism that interchanges X and Y.

PROPOSITION 5.4. *Let $T = T(A, \alpha, u)$ be a hyperbolic algebra for which α is injective. Then the homomorphisms $\theta : A \to T$, $\theta_y : [\alpha; y]A \to T$ and $\theta_x : A[x; \beta] \to T$ are injective.*

PROOF. Let $f_n, f_{n-1}, \ldots, f_0 \in [y; \alpha]A$, with $f_n \neq 0$, and let $r = f_n x^n + f_{n-1}x^{n-1} + \cdots + f_0$, an element of R of degree n in x. Then zr has degree $n+1$ in x, with leading coefficient $y\alpha(f_n) \neq 0$. Hence, $zR \cap [y; \alpha]A = 0$ and so θ_y is injective, θ is injective and, by symmetry, θ_x is injective. □

Proposition 5.4 need not hold if α is not injective. In particular if $a \in \ker(\alpha)$ then, in T, $\overline{au} = \overline{a}YX = Y\overline{\alpha(a)}X = 0$. After one example to illustrate this phenomenon, we shall assume that α is injective.

EXAMPLE 5.5. Let $A = \mathbb{K}[t_1, t_2, \ldots]$ be the polynomial algebra over $\mathbb{K}$ in countably many indeterminates t_i, $i \geq 1$. Let α be the $\mathbb{K}$-endomorphism of A such that $\alpha(t_1) = 0$ and, for $i > 1$, $\alpha(t_i) = t_{i-1}$. Let $u = 1$. Then, in T, $\overline{t_1} = \overline{t_1}YX = Y\overline{\alpha(t_1)}X = 0$ and, for $i > 1$, $\overline{t_i} = \overline{t_i}YX = Y\overline{t_{i-1}}X$. By induction, $\overline{t_i} = 0$ for all i and T is the Laurent polynomial ring $\mathbb{K}[X^{\pm 1}]$, with $Y = X^{-1}$.

From now on, we assume that α is injective, identify A with the subalgebra $\theta(A)$ of $T(A, \alpha, u)$ and drop the overlining, writing each $\overline{a}$ as a. Thus, $T(A, \alpha, u)$ is generated by A, X and Y subject to the relations

$$XY = \alpha(u), YX = u, \text{ and, } \forall a \in A,\ aY = Y\alpha(a) \text{ and } Xa = \beta(a)X.$$

EXAMPLE 5.6. Let α be an injective endomorphism of the $\mathbb{K}$-algebra A and let $T = T(A, \alpha, 1)$. Thus, $XY = 1 = YX$ and, for all $a \in A$, $aY = Y\alpha(a)$ and $Xa = \alpha(a)X$. Elements are finite sums of elements of the form $X^{-i}aX^j$, $a \in A$, $i, j \in \mathbb{N}_0$. Here T is, in the terminology of [**7**], the skew Laurent polynomial ring in X over A. It is the left localization of $A[X; \alpha]$ at the left Ore set $\{X^i\}$ and the right localization of $[Y; \alpha]A$ at the right Ore set $\{Y^i\}$. It may be regarded as the prototype for the more general algebras T considered here.

When α is an automorphism, the algebras $T = T(A, \alpha, u)$ in Definition 5.1, are the algebras studied independently by Bavula [**1**], under the name *generalized Weyl algebra*, Rosenberg [**14**], under the name *hyperbolic algebra*, and the author [**9, 10**]. The name *generalized Weyl algebra* is well established and it might seem appropriate to use the same name here. However, that name has been used by Bavula [**2**] for a different generalization, see Definition 5.7 below, to the case of non-bijective endomorphisms, so here we shall use the name *hyperbolic algebra*.

DEFINITION 5.7. [**2**] Let D be a ring with two endomorphisms σ and τ and an element a such that $\tau\sigma(a) = a$ and, for all $d \in D$, $ad = \tau\sigma(d)a$ and $\sigma(a)d = \sigma\tau(d)\sigma(a)$. The *generalized Weyl algebra* $D(\sigma, \tau, a)$ is then generated by D and indeterminates x and y with defining relations $yx = a$, $xy = \sigma(a)$ and, for all $d \in D$, $xd = \sigma(d)x$ and $yd = \tau(d)y$.

REMARK 5.8. In Definition 5.7, if $a = 1$ then the conditions on σ and τ imply that they form a pair of inverse automorphisms. So, in contrast to Example 5.6, a generalized Weyl algebra $D(\sigma, \tau, 1)$, with $XY = 1 = YX$, cannot be constructed unless σ and τ are inverse automorphisms.

REMARK 5.9. In the case where σ and τ are commuting automorphisms of a $\mathbb{K}$-algebra A, the construction in Definition 5.7 is covered by [**12**, Section 5].

The simplicity criterion [**12**, Theorem 5.4] is generalized in [**2**, Theorems 1.1 and 3.2]. An example in which $a \notin Z(A)$ and $D(\sigma, \tau, a)$ is simple is given by [**12**, Example 6.2], which is presented in [**12**] as a simple ambiskew polynomial ring but can be presented as a simple generalized Weyl algebra, over a different base ring, by [**12**, Remark 5.5].

LEMMA 5.10. *Let γ be an automorphism of a $\mathbb{K}$-algebra A, let α be a $\mathbb{K}$-endomorphism of A such that $\alpha\gamma = \gamma\alpha$ and let u be a γ-normal element of A. Let $\beta = \alpha\gamma$. Then $\beta(u) = \alpha(u)$ and $\alpha^i(u)\alpha^j(u) = \alpha^j(u)\alpha^i(u)$ for all $i, j \geq 0$.*

PROOF. For the first part, $\beta(u) = \alpha\gamma(u) = \alpha(u)$. For $d \geq 1$, $\gamma(\alpha^d(u)) = \alpha^d(\gamma(u)) = \alpha^d(u)$. Hence, $u\alpha^d(u) = \alpha^d(u)u$. Applying appropriate powers of α gives $\alpha^i(u)\alpha^j(u) = \alpha^j(u)\alpha^i(u)$ for all $i, j \geq 0$. □

DEFINITION 5.11. We shall refer to elements of the form Y^iaX^j, where $i, j \in \mathbb{N}_0$ and $a \in A$, as *monomials*. The *dual degree* of Y^iaX^j will be the ordered pair (i, j) and the degree will be the degree in the $\mathbb{Z}$-grading, that is $j - i$. Elements of T are finite sums of monomials but such a sum need not be unique. If $i > 0$, $j > 0$ and $a \in \alpha(A)$, with $a = \alpha(b)$, then $Y^iaX^j = Y^i\alpha(b)X^j = Y^{i-1}bYX^j = Y^{i-1}buX^{j-1}$, alternatively $Y^iaX^j = Y^i\alpha(b)X^j = Y^i\beta\gamma^{-1}(b)X^j = Y^iX\gamma^{-1}(b)X^{j-1} = Y^{i-1}u\gamma^{-1}(b)X^{j-1} = Y^{i-1}buX^{j-1}$. This reduction leaves the degree unchanged and changes the dual degree (i, j) to $(i-1, j-1)$. We shall refer to a monomial Y^iaX^j as *irreducible* if $i = 0$ or $j = 0$ or if $i, j > 0$ and $a \notin \alpha(A)$.

LEMMA 5.12. *Suppose that α is injective. Then every element of $T(A, \alpha, u)$ is a unique sum of irreducible monomials Y^iaX^j with distinct dual degrees (i, j).*

PROOF. Firstly we show that every element is a sum of irreducible polynomials of distinct dual degrees. Given $t \in T$, first express t as the sum of its homogeneous components t_d and express each t_d as a sum of monomials. Let $d \geq 0$ be such that $t_d \neq 0$ and let $t_d = a_0X^d + Ya_1X^{d+1} + \cdots + Y^na_nX^{n+d}$, with $n \neq 0$. If not all the monomials in this sum are irreducible, choose the largest j for which $Y^ja_jX^{n+j}$ is not irreducible. Then $j > 0$ and $Y^ja_jX^{n+j} = Y^{j-1}dX^{n+j-1}$ for some $d \in A$. Replace $Y^{j-1}a_{j-1}X^{n+j-1} + Y^ja_jX^{n+j}$ in the sum by $Y^{j-1}(a_{j-1} + d)X^{n+j-1}$. Repeating the argument, the procedure must terminate with a sum of irreducible monomials of distinct dual degrees. A similar argument applies if $d < 0$. Thus, every element of T is a sum of irreducible monomials of distinct dual degrees.

Suppose that, for some element of $T(A, \alpha, u)$, this sum is not unique. Then 0 is a sum $\sum Y^ia_{ij}X^j$ of irreducible monomials of distinct dual degrees. Recall that $T(A, \alpha, u)$ is the quotient of $R = R(A, \alpha, \alpha(u) - u, 1)$ by the ideal zR where $z = xy - \alpha(u) = yx - u$. Then, in R, the ideal zR contains the sum $\sum y^ia_{ij}x^j$ of monomials each of which has $i = 0$ or $j = 0$ or $a_{ij} \notin \alpha(A)$. Also, the pairs (i, j) are distinct. Let $f \in R$ be of degree d in x, with $f = f_dx^d + \cdots + f_1x + f_0$, where each $f_j \in [\alpha; y]A$, and let e be the degree of f_d in y. Thus, $f = y^ebx^d + s$, where $b \in A$ and $s \in R$ is a sum of monomials of the form y^max^n where $a \in A$ and $n < d$ or $n = d$ and $m < e$. It is readily checked that, in Notation 3.2, $v_i = \alpha^i(u) - u$ for $i \geq 1$. Using Lemma 3.3(v),

$$\begin{aligned} zy^ebx^d &= (yx - u)y^ebx^d \\ &= yxy^ebx^d - uy^ebx^d \\ &= y(y^ex + y^{e-1}(\alpha^e(u) - u))bx^d - y^e\alpha^e(u)bx^d \end{aligned}$$

$$\begin{aligned} &= y^{e+1}xbx^d + y^e(\alpha^e(u) - u)bx^d - y^e\alpha^e(u)bx^d \\ &= y^{e+1}\alpha(b)x^{d+1} - y^eubx^d. \end{aligned}$$

By the choice of d and e and corresponding calculations for other monomials, we see that $zf = y^{e+1}\alpha(b)x^{d+1} + s'$, where s' is a sum of monomials of the form $y^m ax^n$ where $a \in A$ and $n < d+1$ or $n = d+1$ and $m < e+1$. Thus, zf cannot be the above sum $\sum y^i a_{ij} x^j$. This completes the proof. □

The formula for the product $Y^iaX^jY^mbX^n$ of two monomials depends on the relative values of j and m. For $\ell \geq 1$, let

$$c_\ell = \alpha^\ell(u) \dots \alpha^2(u)\alpha(u) \in A.$$

By (2), if $m \geq j$ then

$$Y^iaX^jY^mbX^n = Y^iac_jY^{m-j}bX^n = Y^{i+m-j}\alpha^{m-j}(ac_j)bY^n,$$

and if $j > m$,

$$Y^iaX^jY^mbX^n = Y^iaX^{j-m}c_mbX^n = Y^ia\beta^{j-m}(c_mb)X^{n+j-m}.$$

As z is homogeneous in R of degree 0, T inherits the $\mathbb{Z}$-grading from R, with $\deg Y = -1$, $\deg X = 1$ and $\deg Y^iaX^j = j - i$ if $0 \neq a \in A$. For $d \geq 0$, the component T_d consists of finite sums of elements of the form Y^iaX^{i+d} and T_{-d} consists of finite sums of elements of the form $Y^{i+d}aX^i$.

We now aim to determine necessary conditions for a hyperbolic algebra $T(A, \alpha, u)$ to be simple. We first address issues of regularity for Y and X.

LEMMA 5.13. *In a hyperbolic algebra $T = T(A, \alpha, u)$, Y is left regular in T if and only if for all $b \in A\backslash\{0\}$ and all $d \geq 1$, $YbX^d \neq 0$. Also, X is right regular in T if and only if for all $b \in A\backslash\{0\}$ and all $d \geq 1$, $Y^dbX \neq 0$.*

PROOF. The 'only if' part is clear. For the 'if' part, let $t \in T\backslash\{0\}$ be such that $Yt = 0$. As T is $\mathbb{Z}$-graded, we may assume that t is homogeneous of degree d, say. If $d \leq 0$, Ym is irreducible for every irreducible monomial of degree d so $Yt \neq 0$, by Lemma 5.12. So we may assume that $d > 0$. If m is an irreducible monomial of the form Y^iaX^{i+d}, $i \geq 1$, $Ym = Y^{i+1}aX^{i+d}$ is irreducible. All other irreducible monomials are those of the form bX^d, for which YbX^d is either irreducible or, if $b = \alpha(c) \in \alpha(A)$, reduces to cuX^{d-1} which is either irreducible or zero. So, by Lemma 5.12, if $Yt = 0$ then $t = bX^d$ for some $b \in A\backslash\{0\}$. Hence, Y is left regular in T if for all $b \in A\backslash\{0\}$ and all $d \geq 1$, $YbX^d \neq 0$.

By symmetry, X is right regular in T if and only if for all $b \in A\backslash\{0\}$ and all $d \geq 1$, $YbX^d \neq 0$. □

LEMMA 5.14. *Let $T = T(A, \alpha, u)$ be a hyperbolic algebra. If Y and X are regular in T then, for $d \geq 0$, $\alpha^d(u)$ is regular in T, and in A.*

PROOF. Suppose that X and Y are regular in T. As $YX = u$, u is regular in T. Also, it follows from (2) and Lemma 5.10, that $\alpha^d(u)$ is regular in T for all $d \geq 1$. As A embeds in T, $\alpha^d(u)$ is regular in A for all $d \geq 0$. □

LEMMA 5.15. *Let $T = T(A, \alpha, u)$ be a hyperbolic algebra. If T is simple then Y and X are regular in T and $\alpha^d(u)$ is regular in T, and in A, for all $d \geq 0$.*

PROOF. Suppose that T is simple.

Let $I = \cup_{d\geq 1} \operatorname{lann}(Y^d)$. By Lemma 5.12, $1 \notin I$. If $t \in I$ with $tY^d = 0$ then $tY(Y^d) = 0$, $tXY^{d+1} = t\alpha(u)Y^d = tY^d\alpha^{d+1}(u) = 0$ and, for $a \in A$, $taY^d = tY^d\alpha^d(a) = 0$. It follows that I is an ideal of T so, as $I \neq T$, $I = 0$. Hence, Y is right regular. By symmetry, X is left regular.

For left regularity of Y, we apply Lemma 5.13. Let $b \in A$ and $d > 0$ be such that $YbX^d = 0$. Then $0 = XYbX^d = \alpha(u)bX^d$ and, by Proposition 5.4, $\alpha(u)b = 0$. Hence, $XYb = 0$ and, by the left regularity of X and Proposition 5.4, $b = 0$. By Lemma 5.13, Y is left regular and, hence, regular. By symmetry, X is regular.

By Lemma 5.14, for $d \geq 0$, $\alpha^d(u)$ is regular in T, and in A. □

The following technical definition is aimed at determining when Y and X are regular in a hyperbolic algebra $T(A, \alpha, u)$.

DEFINITION 5.16. Let $d \geq 1$, and, for $0 \leq i \leq d$, define $v_{d,i}$ recursively by setting $v_{d,0} = 1$ and, for $0 \leq e \leq d-1$, $v_{d,i+1} = v_{d,i}\alpha^{d-i}(u)$. By Lemma 5.10,

$$v_{d,i} = \alpha^d(u)\alpha^{d-1}(u)\ldots\ldots\alpha^{d-i+1}(u) = \alpha^{d-i+1}(u)\ldots\alpha^{d-1}(u)\ldots\alpha^d(u).$$

We shall say that the triple (A, α, u) is *d-compliant* if, for $a \in A$,

$$av_{d,d} \in \alpha^d(A) + \alpha^{d-1}(A)v_{d,1} + \alpha^{d-2}(A)v_{d,2} + \cdots + \alpha(A)v_{d,d-1} \Rightarrow a \in \alpha(A)$$

and, for symmetry,

$$v_{d,d}a \in \alpha^d(A) + v_{d,1}\alpha^{d-1}(A) + v_{d,2}\alpha^{d-2}(A) + \cdots + v_{d,d-1}\alpha(A) \Rightarrow a \in \alpha(A).$$

If u is regular in A and (A, α, u) is *d-compliant* for all $d \geq 1$, we shall say that (A, α, u) is *compliant*.

Note that it follows from the γ-normality of u that u is right regular in A if and only if u is left regular in A. If (A, α, u) is compliant then $\alpha^i(u)$ is regular for all $i \geq 0$. This can be checked by induction on A. If $a\alpha^i(u) = 0$ and $\alpha^j(u)$ is regular for $1 \leq j \leq i-1$ then, by Lemma 5.10, $a\alpha(u)\alpha^2(u)\ldots\alpha^i(u) = 0 \in \alpha(A)$, so $a = \alpha(b)$ for some $b \in A$, $\alpha(b)\alpha(u)\alpha^2(u)\ldots\alpha^i(u) = 0$, $bu\alpha(u)\ldots\alpha^{i-1}(u) = 0$ (by the injectivity of α), $b = 0$ and $a = 0$.

REMARK 5.17. It may be helpful to note that, for $d = 1$, the implications in Definition 5.16 are

$$a\alpha(u) \in \alpha(A) \Rightarrow a \in \alpha(A) \text{ and } \alpha(u)a \in \alpha(A) \Rightarrow a \in \alpha(A).$$

As $\alpha(u)$ is γ-normal, these are equivalent. In Example 5.20 below, we exhibit an example in which these fail and how this failure leads to non-regularity of Y and X in the hyperbolic algebra $T(A, \alpha, u)$. Note, also, that if (A, α, u) is $(d+1)$-compliant then it is d-compliant. To see this, note that, by Lemma 5.10, $v_{d,i}\alpha^{d+1}(u) = v_{d+1,i+1}$ so that if $av_{d,d} \in \alpha^d(A) + \alpha^{d-1}(A)v_{d,1} + \alpha^{d-2}(A)v_{d,2} + \cdots + \alpha^2(A)v_{d,d-2} + \alpha(A)v_{d,d-1}$ then $av_{d+1,d+1} \in \alpha^d(A)v_{d+1,1} + \alpha^{d-1}(A)v_{d+1,2} + \cdots + \alpha^2(A)v_{d+1,d-1} + \alpha(A)v_{d+1,d}$.

REMARK 5.18. If A is a field extension of $\mathbb{K}$ and $u \neq 0$ then (A, α, u) is easily seen to be compliant, on dividing through by $v_{d,d}$. More generally, for arbitrary A, if u is a unit then so too is $v_{d,d}$ and (A, α, u) is easily seen to be compliant. Another situation in which (A, α, u) is easily seen to be compliant is when α is an automorphism, in which case $\alpha(A) = A$.

EXAMPLE 5.19. Let A be the polynomial algebra $\mathbb{K}[t]$, let α be the $\mathbb{K}$-endomorphism of A such that $\alpha(t) = t^2$ and let $u = t$. For $d \geq 1$, $v_{d,d} = t^{2^{d+1}-2}$ so if $av_{d,d} \in \mathbb{K}[t^2] = \alpha(A)$ then $a \in \mathbb{K}[t^2]$. As $\alpha^i(A)v_{d,d-i} \subset \mathbb{K}[t^2]$ for $0 \leq i \leq d$, (A, α, u) is compliant.

In the next example, (A, α, u) is not compliant and we shall see explicitly that Y is not regular in $T(A, \alpha, u)$.

EXAMPLE 5.20. Let A be the polynomial algebra $\mathbb{K}[s, t]$, let α be the $\mathbb{K}$-endomorphism of A such that $\alpha(t) = t$ and $\alpha(s) = st$ let $u = t = \alpha(u)$. Take $d = 1$ and let $a = s$. Then $a\alpha(u) = st = \alpha(s) \in \alpha(A)$ but $a \notin \alpha(A)$ so (A, α, u) is not 1-compliant. In the hyperbolic algebra $T(A, \alpha, u)$, $YsX - s \neq 0$, because $s \notin \alpha(A)$, but $(YsX - s)Y = Ys\alpha(u) - Y\alpha(s) = Yst - Yst = 0$ so Y is not regular in T. Also, $X(YsX - s) = 0$ so X is not regular in T.

If we replace the polynomial algebra $\mathbb{K}[s, t]$ here by the Laurent polynomial algebra $\mathbb{K}[s^{\pm 1}, t^{\pm 1}]$ then α becomes an automorphism, with $\alpha(st^{-1}) = s$, and $YsX - s$ becomes 0. This shows that if B is an algebra with an injective endomorphism α and a subalgebra A, such that $\alpha(A) \subseteq A$, and $u \in Z(A)$ then $T(A, \alpha, u)$ need not embed in $T(B, \alpha, u)$.

LEMMA 5.21. *Let $T = T(A, \alpha, u)$ be a hyperbolic algebra. Then X and Y are regular in T if and only if (A, α, u) is compliant.*

PROOF. Let $d \geq 1$ and let $a_0, a_1, \ldots, a_d \in A$. Then, by (2) for $1 \leq j \leq d$,

$$(3)\quad \begin{aligned}(Y^d a_d X^d + Y^{d-1}a_{d-1}X^{d-1} \cdots + Ya_1X + a_0)Y^d \\ = Y^d(a_d v_{d,d} + \alpha(a_{d-1})v_{d,d-1} + \ldots \alpha^{d-1}(a_1)v_{d,1} + \alpha^d(a_0)).\end{aligned}$$

Suppose that X and Y are regular in T. Then X^jY^j is regular and, by (2) and Lemma 5.10, $\alpha^d(u)$ is regular in A for all $d \geq 1$. Let $a \in A$ be such that

$$av_{d,d} = \alpha^d(a_0) + \alpha^{d-1}(a_1)v_{d,1} + \alpha^{d-2}(a_2)v_{d,2} + \cdots + \alpha(a_{d-1})v_{d,d-1},$$

and let $a_d = -a$. By (3),

$$(Y^d a_d X^d + Y^{d-1}a_{d-1}X^{d-1} \cdots + Ya_1X + a_0)Y^d = 0$$

so, by the regularity of Y,

$$Y^d a_d X^d + Y^{d-1}a_{d-1}X^{d-1} \cdots + Ya_1X + a_0 = 0.$$

Suppose that $a \notin \alpha(A)$. Then the monomial $Y^d a_d X^d$ is irreducible and $Y^{d-1}a_{d-1}X^{d-1} \cdots + Ya_1X + a_0$ can be expressed as a sum of irreducible monomials of the form $Y^e b_e X^e$, $d - 1 \geq e \geq 1$, or of the form b_0. By Lemma 5.12, $Y^d a_d X^d + Y^{d-1}a_{d-1}X^{d-1} \cdots + Ya_1X + a_0 \neq 0$, contradicting the regularity of Y^d and that of Y. Thus, (A, α, u) is compliant.

Conversely, suppose that (A, α, u) is compliant. By the symmetry of the definition of compliance, it is enough to show that Y is regular. For left regularity, we apply Lemma 5.13. Let $b \in A \backslash \{0\}$, let $d \geq 1$ and suppose that $YbX^d = 0$. Then $0 = XYbX^d = \alpha(u)bX^d$ so, by Proposition 5.4, $\alpha(u)b = 0$. As (A, α, u) is compliant, $\alpha(u)$ is regular so $b = 0$. By Lemma 5.13, Y is left regular in T.

For right regularity, suppose that there exists $0 \neq t \in T$ such that $tY = 0$ but $t \neq 0$. As T is $\mathbb{Z}$-graded and Y is homogeneous, we can assume that t is homogeneous of degree r, say. If $r > 0$ then $Y^rt \neq 0$, by the left regularity of Y, and $Y^rtY = 0$ so, replacing t by Y^rt, we can assume that $r = 0$. If $r < 0$

and $s = -r$ then $t = Y^s w$, for some $w \in T$ of degree 0. In this situation $w \neq 0$, otherwise $t = 0$, and $Y^s wY = 0$ so, by the left regularity of Y, $wY = 0$, and, replacing t by w, we can again assume that $r = 0$. Thus t has the form $Y^d a_d X^d + Y^{d-1} a_{d-1} X^{d-1} \cdots + Y a_1 X + a_0$ for some $d \geq 0$ and some $a_0, a_1, \dots a_d \in A$ such that $a_d \neq 0$ and, for $1 \leq i \leq d$, $Y^i a_i X^i$ is irreducible or zero. If $d = 0$ then, as α is injective, $t = 0$ so we can assume that $d \geq 1$. As $tY = 0$, $tY^d = 0$, so by (3) and the left regularity of Y,

$$a_d v_{d,d} + \alpha(a_{d-1}) v_{d,d-1} + \cdots + \alpha^{d-1}(a_1) v_{d,1} + \alpha^d(a_0) = 0.$$

As $Y^d a_d X^d$ is irreducible, $a_d \notin \alpha(A)$, contradicting the compliance of (A, α, u). Hence, Y is regular. By symmetry, X is regular. □

LEMMA 5.22. *If a hyperbolic algebra $T(A, \alpha, u)$ is simple then $Au + A\alpha^m(u)A = A$ for all $m \geq 1$.*

PROOF. Let $T = T(A, \alpha, u)$ and suppose that T is simple. Suppose that $m \geq 1$ is such that the ideal $Au + A\alpha^m(u)A$ is proper and let M be a maximal ideal of A containing $Au + A\alpha^m(u)A$. For the purposes of this proof, call a monomial $Y^i aX^j$ *special* if $j \geq 1$ or $i \geq m$ or $a \in M$. Let J be the set of all finite sums in T of special monomials. Note that if $i \geq 1$, $j \geq 1$ and $a = \alpha(b)$ then $Y^i aX^j = Y^{i-1} buX^{j-1}$, which, as $bu \in M$, is special. So when an element of J is expressed as a unique sum of irreducible monomials, those monomials are again special. As 1 is not a special monomial, $1 \notin J$.

We claim that J is a left ideal of T. To see this, first note that if $Y^i aX^j$ is special then $Y(Y^i aX^j) = Y^{i+1} aX^j$ is special and, for all $b \in A$, $b(Y^i aX^j) = Y^i \alpha^i(b) aX^j$ is special. Thus $YJ + AJ \subseteq J$. It remains to show that $XJ \subseteq J$. If $a \in M$ or $j \geq 1$ or $i \geq m + 1$ then $X(Y^i aX^j) = \alpha(u) Y^{i-1} aX^j = Y^{i-1} \alpha^i(u) aX^j$ is special and this is also the case when $i = m$ because $\alpha^m(u) \in M$. Clearly $Y^m T \subseteq J$ so $0 \neq Y^m \in \operatorname{ann}(R/J)$. But $1 \notin J$ so the ideal $\operatorname{ann}(R/J)$ is proper and non-zero, contradicting the simplicity of T. □

LEMMA 5.23. *If a hyperbolic algebra $T(A, \alpha, u)$ is simple then A is weakly $\{\alpha, \beta\}$-simple.*

PROOF. Let $T = T(A, \alpha, u)$. Let I be a non-zero proper strong $\{\alpha, \beta\}$-ideal of A. Let J be the set of all finite sums of monomials of the form $Y^i aX^j$ where $a \in I$. Then J is clearly closed under addition, left multiplication by Y, right multiplication by X and right and left multiplication by b for all $b \in A$. If $i > 0$ then $XY^i aX^j = \alpha(u) Y^{i-1} aX^j = Y^{i-1} \alpha^i(u) aX^j$ and if $i = 0$ then $XY^i aX^j = \beta(a) X^{j+1}$. If $j > 0$ then $Y^i aX^j Y = Y^i aX^{j-1} \alpha(u) = Y^i a \beta^{j-1} \alpha(u) X^j$ and if $j = 0$ then $Y^i aX^j Y = Y^{i+1} \alpha(a)$. As I is an $\{\alpha, \beta\}$-ideal of A, it follows that J is closed under left multiplication by X and right multiplication by Y and is an ideal of T. If a monomial $Y^i cX^j \in J$ is not irreducible then $c = \alpha(d) \in I$ so $d \in I$. But then $Y^i cX^j$ reduces to $Y^{i-1} duX^{j-1}$ and $du \in I$. Thus, when an element of J is reduced to a sum of irreducible monomials, those monomials are again in J. By Lemma 5.12, $1 \notin J$, contradicting the simplicity of T. □

COROLLARY 5.24. *Suppose that A satisfies the ascending chain condition (a.c.c.) on ideals. If a hyperbolic algebra $T(A, \alpha, u)$ is simple then A is $\{\alpha, \beta\}$-simple.*

PROOF. Suppose that $T(A, \alpha, u)$ is simple and let I be a non-zero $\{\alpha, \beta\}$-ideal of A. Thus $\alpha(I) \subseteq I$, $\beta(I) \subseteq I$ and $I \subseteq \alpha^{-1}(I) \cap \beta^{-1}(I)$. This gives rise to the ascending chain

$$I \subseteq \alpha^{-1}(I) \subseteq \alpha^{-2}(I) \subseteq \cdots \subseteq \alpha^{-j}(I) \subseteq \dots.$$

As α and β commute and $\beta(I) \subseteq I$, these ideals are β-ideals. By a.c.c. on ideals, the chain must terminate with, say $\alpha^{-j}(I) = \alpha^{-(j+1)}(I)$. Let $J = \alpha^{-j}(I)$. If $a \in \alpha^{-1}(J)$ then $\alpha(a) \in \alpha^{-j}(I)$ so $a \in \alpha^{-(j+1)}(I) = J$ so $\alpha^{-1}(J) \subseteq J$. Conversely, if $a \in J = \alpha^{-(j+1)}(I)$ then $\alpha^{j+1}(a) \in I$ whence $\alpha(a) \in J = \alpha^{-j}(I)$. So $0 \neq J = \alpha^{-1}(J)$. As α and β commute and $\alpha^{-1}(J) = (J)$, $\alpha^{-1}(\beta^{-i}(J)) = \beta^{-i}(J)$ for each i. The chain

$$I \subseteq \beta^{-1}(J) \subseteq \beta^{-2}(J) \subseteq \cdots \subseteq \beta^{-j}(J) \subseteq \dots$$

then terminates in a non-zero proper strong $\{\alpha, \beta\}$-ideal J'. It follows from Lemma 5.23 that $J' = A$. But $J' = \beta^{-k}(J)$ for some k, $1 \in J' = \beta^{-k}(J)$ and $1 = \alpha^k(1) \in J = \alpha^{-j}(I)$. Hence, $1 = \alpha^j(1) \in I$ so $I = A$ and A is $\{\alpha, \beta\}$-simple. □

The following simplicity criterion generalizes the $\mathbb{K}$-algebra case of [**10**, Theorem 6.1].

PROPOSITION 5.25. *Let $T = T(A, \alpha, u)$ be a hyperbolic algebra for which A is commutative noetherian and α is injective. Then T is simple if and only if the following conditions hold:*

(i) *(A, α, u) is compliant;*
(ii) *A is α-simple;*
(iii) *$uA + \alpha^n(u)A = A$ for all $n \geq 1$.*

PROOF. By [**10**, Theorem 6.1] the result is true when α is an automorphism because, in that case, (i) always holds, see Remark 5.18. So we may assume that α is not an automorphism.

Suppose that T is simple. Then Y and X are regular in T, by Lemma 5.15, so (i) holds by Lemma 5.21. As A is commutative, $\beta = \alpha$ and, as A is noetherian, (ii) holds by Corollary 5.24. Finally, A being commutative, (iii) holds by Lemma 5.22.

Conversely, suppose that (i), (ii) and (iii) hold. By Lemma 5.21 and (i), X and Y are regular in T. Also, by Lemma 5.14, $\alpha^i(u)$ is regular in both A and T for $i \geq 0$. Let $0 \neq I$ be an ideal of T and let $J = I \cap A$. Let $S = [\alpha; Y]A$ and recall, from Proposition 5.4, that S is a subalgebra of T. Let $0 \neq b = \sum Y^i a_{ij} X^j \in J$ and let $u = \max\{j : a_{ij} \neq 0 \text{ for some } i\}$. Then $bY^u \in I \cap S$ by (2). As Y is regular in T, $bY^u \neq 0$ and, hence, $I \cap S \neq 0$. Now let d be the minimal degree (in Y) of a non-zero element of $I \cap S$ and let $s = Y^d a_d + \cdots + Y a_1 + a_0 \in I \cap S$. If $d = 0$ then $a_0 \neq 0$. Let $d > 0$ and suppose that $a_0 = 0$. Then, by the regularity of $\alpha^d(u)$,

$$\begin{aligned} Xs &= XY^d a_d + \cdots + XY^2 a_2 + XY a_1 \\ &= \alpha(u) Y^{d-1} a_d + \cdots + \alpha(u) Y a_2 + \alpha(u) a_1 \\ &= Y^{d-1} \alpha^d(u) a_d + \cdots + Y \alpha^2(u) a_2 + \alpha(u) a_1 \in I \cap S, \end{aligned}$$

contradicting the minimality of d as $\alpha^d(u) a_d \neq 0$. Thus $a_0 \neq 0$.

For all $a \in A$,

$$\begin{aligned} & as - sa \\ = \; & Y^d \alpha^d(a) a_d + \cdots + Y \alpha(a) a_1 + a a_0 - (Y^d a_d a + \cdots + Y a_1 a + a_0 a) \\ = \; & Y^d (\alpha^d(a) a_d - a_d a) + \cdots + Y(\alpha(a) a_1 - a_1 a) \end{aligned}$$

As $as - sa \in I$, with degree $\leq d$ and zero constant term, it follows from the above that $as - sa = 0$ and, as A is commutative, that $(\alpha^i(a) - a)a_i = 0$ for $1 \leq i \leq d$. Let $1 \leq i \leq d$ and let I_i be the ideal of A generated by the set $\{\alpha^i(a) - a : a \in A\}$. Then I_i is an α-ideal of A and, as α is not an automorphism, $I_i \neq 0$. By α-simplicity, $I_i = A$ so, as $I_i a = 0$, $a_i = 0$ if $1 \leq i \leq d$. Hence, $d = 0$ and $J \neq 0$.

For $n \geq 0$, let $J_n = \{a \in A : \alpha^n(u) \dots \alpha^2(u)\alpha(u)a \in J\}$. Then

$$J = J_0 \subseteq J_1 \subseteq J_2 \dots \subseteq J_n \subseteq \dots$$

is an ascending chain of non-zero ideals of A and, as A is noetherian, there exists t such that $J_t = J_{t+1}$.

Let $j \in J$. Then

$$\alpha(u)\alpha(j) = XY\alpha(j) = XjY \in J.$$

Let $a \in J_t$ and let $j = \alpha^t(u) \dots \alpha^2(u)\alpha(u)a \in J$. Then

$$(4) \quad \alpha^{t+1}(u)\alpha^t(u) \dots \alpha^2(u)\alpha(u)\alpha(a)$$
$$= \alpha(u)\alpha^{t+1}(u)\alpha^t(u) \dots \alpha^2(u)\alpha(a) = \alpha(u)\alpha(j) \in J$$

so $\alpha(a) \in J_{t+1} = J_t$. Thus J_t is a non-zero α-ideal of A and, by α-simplicity, $J_t = A$. Therefore $1 \in J_t$, $\alpha^t(u) \dots \alpha^2(u)\alpha(u) \in J$ and, by (2), $X^tY^t \in J$.

Let $b \geq 0$ be minimal such that $\alpha^b(u) \dots \alpha^2(u)\alpha(u) \in J$, equivalently $X^bY^b \in J$. Suppose that $b > 0$ and let $a = \alpha^{b-1}(u) \dots \alpha(u)$. Then

$$au^2 = auYX = Y\alpha(a)\alpha(u)X = Y\alpha^b(u) \dots \alpha^2(u)\alpha(u)X \in J.$$

Hence, $\alpha^{b-1}(u) \dots \alpha(u)(u^2A + \alpha^b(u)A) \subseteq J$. But $u^2A + \alpha^b(u)A = A$, as any maximal ideal of A containing $u^2A + \alpha^b(u)A$ would contain $uA + \alpha^b(u)A = A$. It follows that $\alpha^{b-1}(u) \dots \alpha(u) \in J$, contradicting the minimality of b. Hence, $b = 0$, $1 \in J$, $J = A$ and $I = T$. Thus T is simple. □

EXAMPLE 5.26. Let α be an injective $\mathbb{K}$-endomorphism of a commutative noetherian $\mathbb{K}$-algebra A such that A is α-simple. Then $(A, \alpha, 1)$ is compliant, by Example 5.18, and $uA + \alpha^n(u)A = A$ for $n \geq 1$. By Theorem 5.25, the skew Laurent polynomial ring $T(A, \alpha, 1)$ is simple.

EXAMPLE 5.27. Let $A = \mathbb{K}(t)[z]$ and α be as in Example 3.12 with $m = n = 2$. Thus $\alpha(t) = t^2$, $\alpha(z) = z + t$ and A is α-simple. Let $w = \alpha(z) = z + t$. Then $\alpha(w) = w + t^2$, $A = \mathbb{K}(t)[w]$ and $\alpha(A) = \mathbb{K}(t^2)[w]$. Let $u = z$. Then, for $j \geq 1$, $\alpha^j(u) = w + t_j$ where $t_1 = 0$ and $t_j = t^2 + t^4 + \dots + t^{2^{j-1}}$ if $j \geq 2$. Thus $t_j \in \alpha(A)$ for $j \geq 1$. Let $j \geq 1$ and let $a \in A$ be such that $a\alpha^j(u) \in \alpha(A)$. Writing a in the form $f_n(t)w^n + f_{n-1}(t)w^{n-1} + \dots + f_0(t)$, where each $f_i(t) \in \mathbb{K}(t)$, we see that

$$f_n(t)w^{n+1} + (f_{n-1}(t) + t_j f_n(t))w^n + \dots + f_0(t)t_j \in \mathbb{K}(t^2)[w]$$

so $f_n(t) \in \mathbb{K}(t^2)$. Also, $f_{n-1}(t) + t_j f_n(t) \in \mathbb{K}(t^2)$ so, as $t_j \in \mathbb{K}(t^2) \subset \alpha(A)$, $f_{n-1}(t) \in \mathbb{K}(t^2)$. Repeating the argument, we conclude that $f_i(t) \in \mathbb{K}(t^2)$ for $0 \leq i \leq n$ and, hence, that $a \in \alpha(A)$. Thus, for $j \geq 1$,

$$(5) \qquad a\alpha^j(u) \in \alpha(A) \Rightarrow a \in \alpha(A).$$

Now let $d \geq 1$ and suppose that $av_{d,d} \in \alpha^d(A) + \alpha^{d-1}(A)v_{d,1} + \alpha^{d-2}(A)v_{d,2} + \dots + \alpha^2(A)v_{d,d-2} + \alpha(A)v_{d,d-1}$. Each $v_{d,d-1} \in \alpha(a)$ so $av_{d,d} = a\alpha^d(u)\alpha^{d-1}(u) \dots \alpha(u) \in \alpha(A)$ and, by repeated application of (5), we see that $a \in \alpha(A)$. As u is regular in A and A is commutative, it follows that (A, α, u) is compliant.

For $n \geq 1$, $\alpha^n(u) - u = w + t_n - z = t + t_n$ which is a unit in A so $uA + \alpha^n(u)A = A$. By Theorem 5.25, $T(A, \alpha, u)$ is simple.

References

[1] V. V. Bavula, *Generalized Weyl algebras and their representations* (Russian), Algebra i Analiz **4** (1992), no. 1, 75–97; English transl., St. Petersburg Math. J. **4** (1993), no. 1, 71–92. MR1171955

[2] V. V. Bavula, *Generalized Weyl algebras and diskew polynomial rings*, arXiv:1612.08941.

[3] G. Benkart and T. Roby, *Down-up algebras*, J. Algebra **209** (1998), no. 1, 305–344, DOI 10.1006/jabr.1998.7511. MR1652138

[4] T. Cassidy and B. Shelton, *Basic properties of generalized down-up algebras*, J. Algebra **279** (2004), no. 1, 402–421, DOI 10.1016/j.jalgebra.2004.05.009. MR2078408

[5] P. M. Cohn, *Free rings and their relations*, 2nd ed., London Mathematical Society Monographs, vol. 19, Academic Press, Inc. [Harcourt Brace Jovanovich, Publishers], London, 1985. MR800091

[6] K. R. Goodearl and R. B. Warfield Jr., *An introduction to noncommutative Noetherian rings*, 2nd ed., London Mathematical Society Student Texts, vol. 61, Cambridge University Press, Cambridge, 2004. MR2080008

[7] D. A. Jordan, *Bijective extensions of injective ring endomorphisms*, J. London Math. Soc. (2) **25** (1982), no. 3, 435–448, DOI 10.1112/jlms/s2-25.3.435. MR657500

[8] D. A. Jordan, *Iterated skew polynomial rings and quantum groups*, J. Algebra **156** (1993), no. 1, 194–218, DOI 10.1006/jabr.1993.1070. MR1213792

[9] D. A. Jordan, *Krull and global dimension of certain iterated skew polynomial rings*, Abelian groups and noncommutative rings, Contemp. Math., vol. 130, Amer. Math. Soc., Providence, RI, 1992, pp. 201–213, DOI 10.1090/conm/130/1176120. MR1176120

[10] D. A. Jordan, *Primitivity in skew Laurent polynomial rings and related rings*, Math. Z. **213** (1993), no. 3, 353–371, DOI 10.1007/BF03025725. MR1227487

[11] D. A. Jordan, *Down-up algebras and ambiskew polynomial rings*, J. Algebra **228** (2000), no. 1, 311–346, DOI 10.1006/jabr.1999.8264. MR1760967

[12] D. A. Jordan and I. E. Wells, *Simple ambiskew polynomial rings*, J. Algebra **382** (2013), 46–70, DOI 10.1016/j.jalgebra.2013.01.033. MR3034473

[13] J. C. McConnell and J. C. Robson, *Noncommutative Noetherian rings*, Pure and Applied Mathematics (New York), John Wiley & Sons, Ltd., Chichester, 1987. With the cooperation of L. W. Small; A Wiley-Interscience Publication. MR934572

[14] A. L. Rosenberg, *Noncommutative algebraic geometry and representations of quantized algebras*, Mathematics and its Applications, vol. 330, Kluwer Academic Publishers Group, Dordrecht, 1995. MR1347919

School of Mathematics and Statistics, University of Sheffield, Hicks Building, Sheffield S3 7RH, United Kingdom

Email address: d.a.jordan@sheffield.ac.uk

Contemporary Mathematics
Volume **727**, 2019
https://doi.org/10.1090/conm/727/14635

Generalized extending modules via exchange and clean properties

Yeliz Kara and Adnan Tercan

ABSTRACT. A module is called extending or CS provided that every submodule is essential in a direct summand. There have been investigations of several generalized extending conditions in the literature. In this survey paper, we focus on the exchange property on some generalized extending modules. It is an open problem whether extending condition with exchange property implies cleanness or not. We mention some observations on the former problem to construct a counterexample as well as new type of open problems which are replaced by extending with generalized extending conditions.

1. Introduction

All rings are associative with unity and modules are unital right R-modules. Let M be an R-module. Recall from [**16**] or [**22**] that the following module properties are essentially based on direct summands.

C_1 : Every submodule N of M is essential in a direct summand of M.

C_2 : Whenever a submodule N of M is isomorphic to a direct summand of M, then N is itself a direct summand of M.

C_3 : If K and L are direct summands of M such that $K \cap L = 0$, then $K \oplus L$ is a direct summand of M.

Note that a module M is called *extending* or CS if M has C_1 condition. An extending module M with (C_3) C_2 condition is called a *(quasi)-continuous*. It is well known that every continuous module is quasi-continuous.

One of the origin of extending concept is the work of von Neumann in the 1930's concerning continuous geometries and their realization as lattices of principal left ideals of regular rings. Utumi developed the study of von Neumann as the C_1 condition in his work on continuous and self-injective rings in 1960's. The importance of the former class of modules is due to the fact that it is a certain generalization of both injective and semisimple modules. In recent studies, it has been a common interest to investigate extending property with respect to the special classes of submodules. Recall that a submodule N of M is called *projection invariant* [**8**] if $e(N) \subseteq N$ for all $e^2 = e \in \mathrm{End}(M_R)$. To this end, many authors

2010 *Mathematics Subject Classification.* Primary 16D10, 16D50; Secondary 16S50.

Key words and phrases. Extending modules, continuous modules, C_{11}-modules, PI-extending modules, exchange property, clean rings.

The first author is the corresponding author.

have been studying several generalizations of the extending modules including the following classes:

(1) M is called a *C_{11}-module* [**20**] if every submodule of M has a complement which is a direct summand of M;

(2) M is called a *C_{12}-module* [**20**] if every submodule of M can be essentially embedded in a direct summand of M;

(3) M is called *PI-extending* [**2**] if every projection invariant submodule is essential in a direct summand of M.

It is shown that every CS-module is a C_{11}-module and every C_{11}-module is a C_{12}-module [**20**, Proposition 3.2]. Additionally, every C_{11}-module is a PI-extending module by [**2**, Proposition 3.7]. The direct sum and direct summand properties for the foregoing module classes are studied in [**2**, **13**, **20**]. In contrast to extending modules, the aforementioned classes of modules are closed under direct sums [**2**, Corollary 4.11], [**13**, Theorem 3.1], [**20**, Theorem 2.4]. However it is provided that C_{11}, C_{12} and PI conditions are not inherited by direct summands [**21**, Example 4], [**2**, Example 5.5].

Crawley and Jónnson introduced exchange property in [**6**]. Recall that a module M over a ring R is said to have *(finite) exchange property* if for any (finite) index set I, whenever $M \oplus N = \bigoplus_{i \in I} A_i$ for modules N and A_i, then $M \oplus N = M \oplus (\bigoplus_{i \in I} B_i)$ for submodules $B_i \leq A_i$. Note that R is an exchange ring if and only if R_R has exchange property. In [**24**], Warfield proved that a module M has the finite exchange property if and only if $\mathrm{End}(M_R)$ is an exchange ring.

It is presented that every quasi-injective and continuous modules have exchange property in [**16**, Theorem 1.21, Theorem 3.24]. In contrast to continuous modules, quasi-continuous modules, in general, do not have the exchange property [**16**, Examples 3.26]. In [**17**, **19**], the authors showed that for a quasi-continuous module, the finite exchange property is equivalent to the arbitrary exchange property.

Crawley and Jónnson inquired whether the finite exchange property always implies full exchange property for an arbitrary module was raised in [**6**]. Zimmermann-Huisgen and Zimmermann [**25**] answered this question in a positive way. They proved that the implication holds when the module has an indecomposable decomposition. Therefore, they conclude the following interesting corollary:

Corollary [**25**, Corollary 6] *Suppose $M = \bigoplus_{j \in J} M_j$ where each M_j is indecomposable. Then the following conditions are equivalent:*

(i) M has the exchange property,

(ii) M has the finite exchange property,

(iii) All the summands M_j have local endomorphism rings and the family $(M_j)_{j \in J}$ is semi-T-nilpotent.

In [**23**], Yu justified a partial answer to the former problem in [**6**]. It is proved that if a module M has finite exchange property and the idempotent elements of $\mathrm{End}(M_R)$ are all central, then M has the countable exchange property [**23**, Theorem 1.1]. Furthermore, the author also mentioned a result which examines to the answer of an open problem in [**16**] whether the finite exchange property implies the full exchange property for quasi-continuous modules. He spell out a favorable answer as follows:

Theorem [**23**, Theorem 2.1] *Let R be a ring with acc on essential right ideals. Then for every quasi-continuous module M_R, the finite exchange property implies the full exchange property.*

In this survey, we deal with the exchange property on some generalized extending modules. We inspect the classes of modules, which are more general than quasi-continuous, with exchange property. Hence, we obtain some affirmative results whether the finite exchange property implies the full exchange property for the classes of C_{11}-modules, C_{12}-modules as well as PI-extending modules. We simplify our results with corollaries for some special cases. Furthermore, we mention some results on a *clean* ring R in which every element of R can be written as a sum of a unit and an idempotent [**18**]. Clean rings with exchange property are studied in [**5**]. One of the main result on clean modules (i.e., modules whose endomorphism rings are clean) states that every continuous module enjoys with the clean property [**5**, Theorem 3.9]. An open problem is pointed out that whether CS condition with exchange property implies cleannes or not [**5**, Question 4.9]. Thereupon, we interrogate some observations on the former open problem to construct a counterexample. We also deal with the forenamed question for the class of C_{11}-modules. Note that this study is not a comprehensive listing of results and applications of exchange property on generalized extending modules. Good source for literature about extending modules and clean rings with exchange property are [**3, 5, 7, 15–18, 23–25**].

Throughout this paper, if M is an R-module and $X \subseteq M$, then $X \leq M$, $X \leq_e M$, $X \leq_d M$ and $S = \text{End}(M_R)$ denote X is a submodule of M, X is an essential submodule of M, X is a direct summand of M and the endomorphism ring of M_R, respectively. Recall that a ring R is called *Abelian* if every idempotent of R is central. A family $(N_i)_{i\in I}$ of independent submodules of a module M is said to be a *local (direct) summand* if for any finite subset $F \subseteq I$, $\bigoplus_{i\in F} N_i$ is a direct summand of M [**7**]. Other terminology and notation can be found in [**1, 3, 7, 16, 22**].

2. Generalized extending modules with exchange property

In this section, we elaborate the exchange property on the classes of C_{11}-modules, C_{12}-modules and PI-extending modules. We obtain the indecomposable decompositions for the former classes of modules under some additional conditions such as conditional direct summands properties C_2 and C_3. For the sake of completeness, we give the proof of some main results.

LEMMA 2.1. [**13**, Lemma 2.3] *Let R be a ring and M an R-module such that R satisfies acc on right ideals of the form $r(m)$ where $m \in M$. If M_R satisfies*

(i) C_{11} and C_3

or

(ii) C_{12} and C_2

then M has an indecomposable decomposition.

PROOF. The proof follows from [**20**, Lemma 4.6] and [**16**, Theorem 2.17]. □

The next example shows that the converse of Lemma 2.1 is not true, in general.

EXAMPLE 2.2. Let $\mathbb{R}$ be the real field and n any odd integer with $n \geq 3$, and S the polynomial ring $\mathbb{R}[x_1, ..., x_n]$ with indeterminates $x_1, ..., x_n$ over $\mathbb{R}$. Let R be the ring S/Ss where $s = x_1^2 + ... + x_n^2 - 1$. Then the free R-module $M = \bigoplus_{i=1}^n R$ has an indecomposable decomposition. However M satisfies neither C_2 nor C_3 conditions.

PROOF. It is clear that R is a commutative Noetherian domain. Then $M = \bigoplus_{i=1}^{n} R$ has an indecomposable decomposition by [**13**, Lemma 2.1]. Moreover M satisfies both C_{11} and C_{12} properties by [**21**, Example 4] and [**13**, Theorem 3.9]. Recall from [**21**, Example 4] that M contains an indecomposable direct summand K_R which is not C_{11}-module. Therefore M satisfies neither C_2 nor C_3 conditions by [**20**, Theorem 4.3]. □

The following result explains when a module with C_{11} property has an indecomposable decomposition.

PROPOSITION 2.3. [**13**, Proposition 2.4] *The following statements are equivalent for a nonsingular C_{11}-module M which satisfies C_3.*

(i) M has an indecomposable decomposition.

(ii) Every finitely generated submodule of M has finite uniform dimension.

(iii) Every cyclic submodule of M has finite uniform dimension.

(iv) R satisfies acc on right ideals of the form $r(m)$ where $m \in M$.

PROOF. $(i) \Rightarrow (ii)$ There exist an index set I and indecomposable submodules M_i $(i \in I)$ of M such that $M = \bigoplus_{i \in I} M_i$. Since M is C_{11}-module with C_3, M_i is also C_{11} by [**20**, Theorem 4.3]. It follows that M_i $(i \in I)$ is uniform. If L is a finitely generated submodule of M then $L \subseteq \bigoplus_{i \in J} M_i$ for some finite subset J of I and hence L has finite uniform dimension.

$(ii) \Rightarrow (iii)$ Clear.

$(iii) \Rightarrow (iv)$ Let $m \in M$. Suppose that $r(m)$ is essential in a right ideal A of R. Let $a \in A$. There exists an essential right ideal E of R such that $aE \subseteq r(m)$. It follows that $maE = 0$ and hence $ma = 0$ so $a \in r(m)$. Thus $r(m) = A$. Hence $r(m)$ is a complement in R-module R for each $m \in M$. Moreover $R/r(m) \cong mR$ which gives that the R-module $R/r(m)$ has finite uniform dimension. Now (iv) follows by [**7**, Theorem 5.10].

$(iv) \Rightarrow (i)$ By [**20**, Theorem 4.7] and [**16**, Theorem 2.17]. □

Recall from [**25**, Corollary 6] the finite exchange property is equivalent to the full exchange property for a module with indecomposable decomposition. Combining Proposition 2.3 together with [**25**, Corollary 6], we arrive at the following results.

THEOREM 2.4. [**13**, Theorem 2.5] *Let M_R be a nonsingular module such that mR has finite uniform dimension for each $m \in M$. If M_R satisfies C_{11} and C_3 then the finite exchange property implies the full exchange property.*

COROLLARY 2.5. [**13**, Corollary 2.6] *Let R be a ring with finite uniform dimension and M a nonsingular right R-module which satisfies C_{11} and C_3 properties. Then the finite exchange property implies the full exchange property.*

Recall from [**23**, Theorem 2.1], over a ring with acc on essential right ideals, the finite exchange property implies the full exchange property for quasi-continuous modules. In this trend, we get the following result for the classes of C_{11} and C_{12} modules under conditional direct summands properties.

THEOREM 2.6. [**13**, Theorem 2.7] *Let R be a ring with acc on right ideals of* *the form $r(m)$ where $m \in M$ and M a right R-module. If M_R satisfies*

a) C_{11} and C_3

or

b) C_{12} and C_2

then the finite exchange property implies full exchange property.

PROOF. By Lemma 2.1, M has an indecomposable decomposition. Then the result follows from [**25**, Corollary 6]. □

Observe that one can ask whether the assumptions of Theorem 2.6 imply the quasi-continuity of the module or not? However the next example answers this question in a negative way.

EXAMPLE 2.7. [**13**, Example 2.8] Let M denote the $\mathbb{Z}$-module $(\mathbb{Z}/\mathbb{Z}p) \oplus \mathbb{Q}$ for any prime p. Then M satisfies C_{11} and C_2, so does C_{12} and C_3 by [**20**, Proposition 3.2] and [**16**, Proposition 2.2]. Since $M_{\mathbb{Z}}$ is not a CS module, it is not a quasi-continuous module (see, [**20**, Example 4.2]).

Applying Theorem 2.6 to locally Noetherian modules, we have the ensuing corollary.

COROLLARY 2.8. [**13**, Corollary 2.9] *Let M be a locally Noetherian module with C_{11} and C_3. Then the finite exchange property implies the full exchange property.*

Now, we are concerned with the exchange property on the class of PI-extending modules. Before giving our main results, we clarify that a PI-extending module with the full exchange property and C_3 condition (or C_2) need not to have Abelian endomorphism ring. To this end, the following examples demonstrate this fact.

EXAMPLE 2.9. [**13**, Example 2.10] Let $R = \mathbb{Z}_{(p)}$ be the localization of integers $\mathbb{Z}$ at a prime p. Put $M_R = \mathbb{Z}_{(p)} \oplus \mathbb{Z}(p^\infty)$. Then M_R is a PI-extending module with C_3 which does not satisfy C_2. Moreover M has the full exchange property whose endomorphism ring has noncentral idempotents.

PROOF. Observe that M_R is quasi-continuous by [**23**, Example 2.13]. Hence M_R is PI- extending with C_3 by [**2**, Proposition 3.7]. Now, let us consider $f \in End(M_R)$ such that $f(x,y) = (px,y)$ where $(x,y) \in M$. It is easy to see that f is a monomorphism and $f(M) = p\mathbb{Z}_{(p)} \oplus \mathbb{Z}(p^\infty)$ which is essential submodule of M. By [**16**, Lemma 3.14], M_R does not satisfy C_2 condition. For the last part, see [**23**, Example 2.13]. □

EXAMPLE 2.10. [**13**, Example 2.11] Let $M_2(R)$ be the matrix ring as in [**4**, Example 2.7]. Let $T = M_2(R)$. Since T_T is a C_{11}-module then T_T is a PI-extending module by [**2**, Proposition 3.7]. Note that T_T has the full exchange property by [**16**, Theorem 3.24]. Moreover, it can be easily seen that the endomorphism ring of T is not Abelian.

We obtain an indecomposable decomposition of PI-extending modules under some suitable conditions. Hence this fact leads us to apply our result to get exchange property for PI-extending modules.

THEOREM 2.11. [**14**, Theorem 2.8] *Let R be a ring and M an R-module such that R satisfies acc on right annihilators of the form $r(m)$, where $m \in M$. If M is a PI-extending module which satisfies C_3 and S is Abelian, then M has an indecomposable decomposition.*

PROOF. Let $\{X_\lambda \mid \lambda \in I\}$ be an independent family of submodules of M and $X = \bigoplus_{\lambda \in I} X_\lambda$ a local summand of M. Define the canonical projection $\pi_k : X \to \bigoplus_{k \in I, k \neq \lambda} X_k$. Then $f(X) = f(\bigoplus_{\lambda \in I} X_\lambda) = \bigoplus_{\lambda \in I} f(X_\lambda) = \bigoplus_{\lambda \in I} f(ker\pi_\lambda)$ where $f^2 = f \in S$. Since S is Abelian, $f(ker\pi_\lambda) \subseteq ker\pi_\lambda$. Then $f(X) = \bigoplus_{\lambda \in I} f(ker\pi_\lambda) \subseteq \bigoplus_{\lambda \in I} ker\pi_\lambda = X$. Thus X is a projection invariant submodule of M. By [**2**, Corollary 3.2], there exists a direct summand K of M such that $X \cap K = 0$ and $X \oplus K$ is essential in M. Now, consider $X \oplus K$. For any finite subset $A \subseteq I$, $Y = \bigoplus_{\lambda \in A} X_\lambda$ is a direct summand of M. Since M has C_3, $Y \oplus K$ is a direct summand of M. Hence $X \oplus K$ is a local summand of M. By [**20**, Lemma 4.5], $X \oplus K$ is a complement in M. However $X \oplus K$ is an essential submodule of M, then $M = X \oplus K$. So, [**16**, Theorem 2.17] yields that M has an indecomposable decomposition. □

As a result of Theroem 2.11 and [**25**, Corollary 6], we obtain the next result which is related to exchange property for PI-extending modules.

PROPOSITION 2.12. [**14**, Proposition 2.11] *Let R be a ring and M an R-module such that R satisfies acc on $r(m)$ where $m \in M$. If M_R is PI-extending which satisfies C_3 and S is Abelian, then the finite exchange property implies full exchange property.*

On using Theorem 2.11 we get the following useful characterization for PI-extending modules.

THEOREM 2.13. [**14**, Theorem 2.13] *The following statements are equivalent for a nonsingular PI-extending module M which satisfies C_3 and S is Abelian,*

(i) *M has an indecomposable decomposition,*
(ii) *Every finitely generated submodule of M has finite uniform dimension,*
(iii) *Every cyclic submodule of M has finite uniform dimension,*
(iv) *R satisfies acc on $r(m)$ where $m \in M$.*

PROOF. $(i) \Rightarrow (ii)$. There exist an index set I and indecomposable submodules M_i $(i \in I)$ of M such that $M = \bigoplus_{i \in I} M_i$. By [**14**, Proposition 3.4], M_i is also PI-extending. It follows that M_i is uniform. If L is a finitely generated submodule of M, then $L \subseteq \bigoplus_{i \in J} M_i$ for some finite subset J of I and hence L has finite uniform dimension.

$(ii) \Rightarrow (iii)$. It is obvious.

$(iii) \Rightarrow (iv)$. Let $m \in M$. Suppose that $r(m)$ is essential in a right ideal A of R. Let $a \in A$. Then there exists an essential right ideal E of R_R such that $aE \subseteq r(m)$. It follows that $maE = 0$ and hence $ma = 0$ so $a \in r(m)$. Thus $r(m) = A$. Therefore $r(m)$ is a complement in R-module R for each $m \in M$. Moreover $R/r(m) \cong mR$, gives that the R-module $R/r(m)$ has finite uniform dimension. Now (iv) follows by [**7**, Theorem 5.10].

$(iv) \Rightarrow (i)$. It follows from Theorem 2.11. □

We refer Theorem 2.13 to get the following corollaries in some special cases as applications for the former class of modules.

COROLLARY 2.14. [**14**, Corollary 2.14] *Let M be a nonsingular module such that mR has finite uniform dimension for each $m \in M$. If M is PI-extending which satisfies C_3 and S is Abelian then the finite exchange property implies full exchange property.*

COROLLARY 2.15. [**14**, Corollary 2.15] *Let R be a ring with finite uniform dimension and M a nonsingular PI-extending module which satisfies C_3 and S is Abelian. Then the finite exchange property implies full exchange property.*

3. Generalized extending conditions with clean property

It is well known that a ring R is *clean* [**18**] if every element of R is a sum of a unit and an idempotent. In [**18**], Nicholson proved that every clean ring is an exchange ring and a ring with central idempotents is clean if and only if it is an exchange ring. Furthermore, a module M is called *clean* [**5**] if the endomorphism ring of M is clean. In [**5**], the authors proved that the endomorphism ring of a continuous module is clean [**5**, Theorem 3.9]. As a consequence of this result, every quasi-injective module is clean and any right self-injective ring is clean [**5**, Corollaries 3.11 and 3.12]. Moreover, [**5**, Theorem 4.3] makes it clear that a quasi-continuous module M_R is clean if and only if M_R has the (finite) exchange property. Since CS-modules are more general than quasi-continuous modules, it is of interest to investigate which CS-modules are clean.

Recall that an open problem was posed in [**5**, Question 4.9] which namely states that whether CS condition with finite exchange property implies cleanness or not. Recall further that in the same paper, it is examined that the foregoing question for some other kind of subclasses of CS-modules as well as the quasi-continuous modules. Accordingly, the question has positive answer if either R is a right Noetherian ring or M_R is a square-free module [**5**, Theorem 4.10].

We start by mentioning some background results. Let R and S be the rings and M an (R,S)-bimodule. Then $T = \begin{bmatrix} R & M \\ 0 & S \end{bmatrix}$ will denote the generalized upper triangular matrix ring.

LEMMA 3.1. [**12**, Proposition 2.1] *T is an exchange ring if and only if R and S are exchange rings.*

LEMMA 3.2. [**22**, Theorem 3.91] *Assume ${}_RM$ is faithful. Then T is a right nonsingular right CS-ring if and only if*

(i) for every complement K in M_S, there exists an idempotent $e \in R$ such that $K = eM$,

(ii) S is a right nonsingular right CS-ring,

(iii) the right S-module M is nonsingular injective.

LEMMA 3.3. [**9**, Proposition 6.3] *T is a clean ring if and only if R and S are clean.*

Refering to [**5**, Question 4.9] there seems to be very few known examples of modules with finite exchange property which is not clean. Note that the former construction is based on Bergman's example [**11**]. Actually, let K be a field with $char(K) \neq 2$ and $A = K[[x]]$ be the formal power series ring. Let Q be the field of fractions of A and $R = \{r \in \mathrm{End}(A_K)$: there exist $q \in Q$ and $n > 0$ such that

$r(a) = qa$ for all $a \in x^n A\}$. Amongst other several interesting properties, R has nonzero socle I and R/I is a field [**15**, Example 1].

Now, we intend to construct a generalized upper triangular matrix ring T which will answer [**5**, Question 4.9] in a negative way which means that there exists a CS-ring with finite exchange property which need not to be a clean ring. Observe that it is essential to have all of the conditions in Lemma 3.2 to show that T is right nonsingular and right CS-ring. Consequently, the followings are our observations on the related open problem. Note that the following constructions are also based on Bergman's example.

Observation 1. Let $T = \begin{bmatrix} R & A \\ 0 & K \end{bmatrix}$. Since R is a regular ring and K is a field, then R and K are exchange rings. Thus T is an exchange ring by Lemma 3.1. It follows that T_T has the finite exchange property. However T is not clean. To see this, assume the contrary. Then R is a clean ring by Lemma 3.3 which gives a contradiction with [**15**, Example 1]. We conclude that T is an exchange ring which is not clean. On the other hand, to investigate the CS property of T, it is enough to see that T satisfies all of the conditions of Lemma 3.2. Since R is a subring of $End(A_K)$, it is easy to see that A is faithful left R-module. Since A_K is a vector space, it can be easily seen that A_K is nonsingular injective. Moreover K_K is a nonsingular CS-module, as K is a field. Hence T satisfies conditions (ii) and (iii) of Lemma 3.2. However it is hard to see that whether condition (i) of Lemma 3.2 satisfies or not.

Observation 2. Let $T = \begin{bmatrix} R & R/I \\ 0 & R/I \end{bmatrix}$ where I is the socle of R. Note that R has nonzero socle and R/I is a field. Thus T is an exchange ring which is not clean by the similar reasons in Observation 1. Recall that the left annihilator of the left R-module R/I is I which is nonzero. Since R/I is field, all of the conditions of Lemma 3.2 are satisfied apart from the faithfulness of the left R-module ${}_R(R/I)$.

In [**5**], it is investigated that which CS-modules are clean. Since every CS-module satisfies C_{11} property, we wonder whether C_{11} modules are clean or vice versa. Thereby, we have examples clarifying that being clean and having C_{11} property are incomparable. For instance, let $R = T_2(\mathbb{Z})$ be the upper triangular matrix ring over the integers. Since R is the direct sum of uniform submodules, R is a C_{11}-module by [**22**, Corollary 4.28]. However R is not clean by [**10**, Corollary 3]. Now, let us think the ring $R = \begin{bmatrix} \mathbb{Z}_4 & 2\mathbb{Z}_4 \\ 0 & \mathbb{Z}_4 \end{bmatrix}$. Then R is a clean ring by Lemma 3.3. Nevertheless, R_R is not a C_{11}-module by [**22**, Example 4.90]. Furthermore,it is natural to determine which C_{11}-modules are clean. Consequently, we have the next question:

Question. Is a C_{11}-module M_R necessarily clean if M_R has C_2 property?

While we are not able to answer this question in general, we represent an affirmative answer to the former question in the following results. Incidentally, it is clear that C_{11} condition could be replaced by any generalized extending condition in the aforementioned question.

PROPOSITION 3.4. *If M_R is a C_{11}-module with C_2 property satisfying any of the following condition, then M_R is clean.*

(i) $R_R = M_R$ and R is Abelian.

(ii) M_R is a cyclic module and R is commutative.

(iii) M_R is a multiplication module and R is commutative.

PROOF. It is clear from [**22**, Corollary 4.69] that M is a CS-module. Then the proof follows from [**5**, Theorem 3.9]. □

As C_2 implies C_3, we consider whether a C_{11}-module with C_3 is clean or not. To this end, the following example eliminates this situation.

EXAMPLE 3.5. Let $R = \mathbb{Z}[x]$ be the polynomial ring over $\mathbb{Z}$. Then R is not clean by [**10**, Example 2]. Since R is a domain, it is easy to see that R_R is a C_{11}-module with C_3 condition.

Now, we focus on when a C_{11}-module with C_3 condition satisfies the clean property. Thus we obtain the next result:

PROPOSITION 3.6. *Let R be a ring with acc on right ideals of the form $r(m)$ where $m \in M$ and M a right R-module. If M_R has C_{11} and C_3 conditions with the finite exchange property, then M_R is clean.*

PROOF. As a consequence of Lemma 2.1 and Theorem 2.6, M_R has an indecomposable decomposition with exchange property. Then the result follows from [**5**, Theorem 4.7]. □

COROLLARY 3.7. *Let M be the $\mathbb{Z}$-module with finite exchange property. If $M_{\mathbb{Z}}$ is a C_{11}-module with C_3, then M is a clean $\mathbb{Z}$-module.*

References

[1] F. W. Anderson and K. R. Fuller, *Rings and categories of modules*, 2nd ed., Graduate Texts in Mathematics, vol. 13, Springer-Verlag, New York, 1992. MR1245487

[2] G. F. Birkenmeier, A. Tercan, and C. C. Yücel, *The extending condition relative to sets of submodules*, Comm. Algebra **42** (2014), no. 2, 764–778, DOI 10.1080/00927872.2012.723084. MR3169601

[3] G. F. Birkenmeier, J. K. Park, and S. T. Rizvi, *Extensions of rings and modules*, Birkhäuser/Springer, New York, 2013. MR3099829

[4] G. F. Birkenmeier and A. Tercan, *When some complement of a submodule is a summand*, Comm. Algebra **35** (2007), no. 2, 597–611, DOI 10.1080/00927870601074830. MR2294619

[5] V. P. Camillo, D. Khurana, T. Y. Lam, W. K. Nicholson, and Y. Zhou, *Continuous modules are clean*, J. Algebra **304** (2006), no. 1, 94–111, DOI 10.1016/j.jalgebra.2006.06.032. MR2255822

[6] P. Crawley and B. Jónsson, *Refinements for infinite direct decompositions of algebraic systems*, Pacific J. Math. **14** (1964), 797–855. MR0169806

[7] Nguyen Viet Dung, Dinh Van Huynh, P. F. Smith, and R. Wisbauer, *Extending modules*, Pitman Research Notes in Mathematics Series, vol. 313, Longman Scientific & Technical, Harlow; copublished in the United States with John Wiley & Sons, Inc., New York, 1994. With the collaboration of John Clark and N. Vanaja. MR1312366

[8] L. Fuchs, *Infinite abelian groups. Vol. I*, Pure and Applied Mathematics, Vol. 36, Academic Press, New York-London, 1970. MR0255673

[9] A. Haghany and K. Varadarajan, *Study of formal triangular matrix rings*, Comm. Algebra **27** (1999), no. 11, 5507–5525, DOI 10.1080/00927879908826770. MR1713049

[10] J. Han and W. K. Nicholson, *Extensions of clean rings*, Comm. Algebra **29** (2001), no. 6, 2589–2595, DOI 10.1081/AGB-100002409. MR1845131

[11] D. Handelman, *Perspectivity and cancellation in regular rings*, J. Algebra **48** (1977), no. 1, 1–16, DOI 10.1016/0021-8693(77)90289-7. MR0447329

[12] C. Y. Hong, N. K. Kim, and Y. Lee, *Exchange rings and their extensions*, J. Pure Appl. Algebra **179** (2003), no. 1-2, 117–126, DOI 10.1016/S0022-4049(02)00299-2. MR1958378

[13] Y. Kara and A. Tercan, *Modules whose certain submodules are essentially embedded in direct summands*, Rocky Mountain J. Math. **46** (2016), no. 2, 519–532, DOI 10.1216/RMJ-2016-46-2-519. MR3529081

[14] Y. Kara, A. Tercan, and R. Yaşar, *PI-extending modules via nontrivial complex bundles and abelian endomorphism rings*, Bull. Iranian Math. Soc. **43** (2017), no. 1, 121–129. MR3622359

[15] T.-K. Lee, Z. Yi, and Y. Zhou, *An example of Bergman's and the extension problem for clean rings*, Comm. Algebra **36** (2008), no. 4, 1413–1418, DOI 10.1080/00927870701866374. MR2406593

[16] S. H. Mohamed and B. J. Müller, *Continuous and discrete modules*, London Mathematical Society Lecture Note Series, vol. 147, Cambridge University Press, Cambridge, 1990. MR1084376

[17] S. H. Mohamed and B. J. Müller, *On the exchange property for quasi-continuous modules*, Abelian groups and modules (Padova, 1994), Math. Appl., vol. 343, Kluwer Acad. Publ., Dordrecht, 1995, pp. 367–372. MR1378212

[18] W. K. Nicholson, *Lifting idempotents and exchange rings*, Trans. Amer. Math. Soc. **229** (1977), 269–278, DOI 10.2307/1998510. MR0439876

[19] K. Oshiro and S. T. Rizvi, *The exchange property of quasi-continuous modules with the finite exchange property*, Osaka J. Math. **33** (1996), no. 1, 217–234. MR1381627

[20] P. F. Smith and A. Tercan, *Generalizations of CS-modules*, Comm. Algebra **21** (1993), no. 6, 1809–1847, DOI 10.1080/00927879308824655. MR1215548

[21] P. F. Smith and A. Tercan, *Direct summands of modules which satisfy* (C_{11}), Algebra Colloq. **11** (2004), no. 2, 231–237. MR2058772

[22] A. Tercan and C. C. Yücel, *Module theory, extending modules and generalizations*, Frontiers in Mathematics, Birkhäuser/Springer, [Cham], 2016. With the cooperation of Patrick F. Smith. MR3468915

[23] H.-P. Yu, *On modules for which the finite exchange property implies the countable exchange property*, Comm. Algebra **22** (1994), no. 10, 3887–3901, DOI 10.1080/00927879408825055. MR1280098

[24] R. B. Warfield Jr., *Exchange rings and decompositions of modules*, Math. Ann. **199** (1972), 31–36, DOI 10.1007/BF01419573. MR0332893

[25] B. Zimmermann-Huisgen and W. Zimmermann, *Classes of modules with the exchange property*, J. Algebra **88** (1984), no. 2, 416–434, DOI 10.1016/0021-8693(84)90075-9. MR747525

Department of Mathematics, Uludağ University, Bursa, 16059, Turkey
Email address: yelizkara@uludag.edu.tr

Department of Mathematics, Hacettepe University, Ankara, 06532, Turkey
Email address: tercan@hacettepe.edu.tr

Contemporary Mathematics
Volume **727**, 2019
https://doi.org/10.1090/conm/727/14636

A new approach to dualize retractable modules

Derya Keskin Tütüncü and Rachid Tribak

ABSTRACT. Let R be a ring and let M be an R-module with $S = \text{End}_R(M)$. Then M is called *d-retractable* if for $U \leq V \leq M$, $l_S(U)/l_S(V) \ll {}_SS/l_S(V)$ implies that $V/U \ll M/U$. We prove that every nonzero d-retractable module is coretractable but the converse is not true, in general. It is shown that the R-module R_R is d-retractable if and only if R is a semisimple ring. We show that every quasi-injective d-retractable module with essential socle is semisimple. The class of rings R for which every finitely cogenerated R-module is d-retractable, is shown to be exactly that of the right V-rings. Some relevant counterexamples are indicated.

1. Introduction

Throughout this paper, all rings are associative with identity and R denotes such a ring. All modules are unitary right R-modules unless otherwise stated. For an R-module M, we write $\text{Soc}(M)$, $\text{Rad}(M)$ and $E(M)$ for the socle, the radical and the injective hull of M, respectively. The notation $N \subseteq M$ means that N is a subset of M and $N \leq M$ means that N is a submodule of M. Let N be a submodule of a module M. Recall that N is said to be *small* or *superfluous in* M, written $N \ll M$, if $M \neq N + K$ for every proper submodule K of M.

For an R-module M with $S = \text{End}_R(M)$, let $I_S(U) = \{s \in S \mid s(M) \subseteq U\}$ for $U \subseteq M$ and $S_M(H) = HM = \sum_{h \in H} h(M)$ for $H \subseteq S$. Dually, for any $K \subseteq M$ and $I \subseteq S$, we will denote the left annihilator of K in S by $l_S(K) = \{s \in S \mid s(K) = 0\}$ and the right annihilator of I in M by $r_M(I) = \cap\{\text{Ker} g \mid g \in I\}$. An R-module M is called *retractable* if $\text{Hom}_R(M, U) \neq 0$ for any nonzero submodule U of M. For a background on retractable modules we refer the reader to [**6**], [**8**] and [**10**]. It is shown in [**10**, Proposition 2.1] that a retractable R-module M with $S = \text{End}_R(M)$ can be characterized by the following equivalent conditions:

(i) M is a retractable module.

(ii) For any submodule U of M, $S_M I_S(U)$ is essential in U.

(iii) For $U \leq V \leq M$, if $I_S(U)$ is essential in $I_S(V)$ then U is essential in V.

The notion of retractable modules was dualized by Amini-Ershad-Sharif in 2009 [**2**]. An R-module M is said to be *coretractable* if M satisfies the following condition:

(A1) $\text{Hom}_R(M/K, M) \neq 0$ for any proper submodule K of M.

2010 *Mathematics Subject Classification.* Primary 16D10, 16D80.

Key words and phrases. Coretractable modules, retractable modules, (weak) d-retractable modules.

Equivalently, for any proper submodule K of M, there exists a nonzero homomorphism $f : M \to M$ with $f(K) = 0$. It is clear that this approach was obtained by dualizing property (i).

Dualizing property (ii), we obtain the following approach:

(A2) For any proper submodule K of M, $r_M(l_S(K))/K \ll M/K$.

It was shown in [**2**, Lemma 4.1] that the definitions (A1) and (A2) are equivalent. More properties of coretractable modules are established in [**1**], [**9**], [**14**] and [**18**]. We should point out that coretractability is quite useful in connecting the properties of an R-module M with the properties of its endomorphism ring $\mathrm{End}_R(M)$ (see [**14**]).

In this paper, we will dualize property (iii) of the above characterization of retractable modules. We get the following property:

(A3) For $U \leq V \leq M$, $l_S(U)/l_S(V) \ll {}_SS/l_S(V)$ implies that $V/U \ll M/U$.

An R-module M is said to be *d-retractable* if it satisfies the condition (A3). It is shown that every d-retractable module is coretractable. Examples are provided to show that the concept of a d-retractable module is distinct from that of a coretractable module. We show that every semisimple module is d-retractable. It would be desirable to construct a d-retractable module which is not semisimple, but we have not been able to do this. It is also shown that for a ring R, the R-module R_R is d-retractable if and only if R is a semisimple ring. We prove that every quasi-injective d-retractable module with essential socle is semisimple. It is shown that every fully invariant direct summand of a d-retractable module inherits the property. We characterize the class of rings R for which every finitely cogenerated R-module is d-retractable as that of the right V-rings.

2. Results

DEFINITION 2.1. Let M be an R-module with $S = \mathrm{End}_R(M)$. Then M is said to be *d-retractable* if for $U \leq V \leq M$, $l_S(U)/l_S(V) \ll {}_SS/l_S(V)$ implies that $V/U \ll M/U$.

PROPOSITION 2.2. *Every semisimple module is d-retractable.*

PROOF. Let M be a semisimple module with $S = \mathrm{End}_R(M)$. Let U and V be two submodules of M such that $U \subseteq V$ and $l_S(U)/l_S(V) \ll {}_SS/l_S(V)$. As M is semisimple, there exists a submodule U' of M such that $U \oplus U' = M$. By [**2**, Lemma 4.9], we have $S = l_S(U \cap U') = l_S(U) + l_S(U')$. Moreover, $l_S(U) \cap l_S(U') = l_S(M) = 0$. Thus $S = l_S(U) \oplus l_S(U')$. Since $l_S(U)/l_S(V) \ll {}_SS/l_S(V)$, we have $l_S(U) = l_S(V)$. Thus $r_M(l_S(U)) = r_M(l_S(V))$. From [**2**, Lemma 4.1], it follows that $U = V$. Hence $V/U \ll M/U$. This shows that M is a d-retractable module. □

PROPOSITION 2.3. *Let M be a coretractable module. If $S = \mathrm{End}_R(M)$ is a left V-ring, then M is d-retractable.*

PROOF. Let $U \leq V \leq M$ such that $l_S(U)/l_S(V) \ll {}_SS/l_S(V)$. As S is a left V-ring, we have $\mathrm{Rad}({}_SS/l_S(V)) = 0$. Therefore $l_S(U) = l_S(V)$. Let X be a submodule of M such that $U \subseteq X$ and $V/U + X/U = M/U$. Then $V + X = M$ and hence $l_S(V) \cap l_S(X) = 0$. Since $l_S(U) = l_S(V)$, we have $l_S(U) \cap l_S(X) = 0$. But $l_S(X) \subseteq l_S(U)$. Then $l_S(X) = 0$. So $X = M$ as M is coretractable. It follows that $V/U \ll M/U$. □

Combining Proposition 2.3 and [**7**, Corollary 1.23], we get the following corollary. This result gives an idea about the construction of a d-retractable module which is not semisimple. Unfortunately, we have not been able to find an example of such a module.

COROLLARY 2.4. *Let M be a coretractable module such that* $\mathrm{End}_R(M)$ *is a left V-ring which is not regular or not right self-injective. Then M is a d-retractable module which is not semisimple.*

Next, we compare the two notions of coretractable modules and d-retractable modules.

LEMMA 2.5. *Let M be a nonzero d-retractable module with $S = \mathrm{End}_R(M)$. Then for every proper submodule U of M, $l_S(U)$ is not small in ${}_SS$.*

PROOF. Suppose that M has a proper submodule U such that $l_S(U) \ll {}_SS$. Taking $V = M$, we have $U \leq V \leq M$ and $l_S(U)/l_S(V) \ll {}_SS/l_S(V)$ since $l_S(V) = 0$. So $V/U = M/U \ll M/U$. This implies that $U = M$, a contradiction. □

PROPOSITION 2.6. *Every nonzero d-retractable module is coretractable.*

PROOF. Let M be a nonzero d-retractable module with $S = \mathrm{End}_R(M)$. From Lemma 2.5, it follows that $l_S(U) \neq 0$ for every proper submodule U of M. So M is coretractable. □

PROPOSITION 2.7. *Let M be a nonzero module such that $S = \mathrm{End}_R(M)$ is a local ring. Then M is d-retractable if and only if M is a simple module.*

PROOF. To prove the necessity, let N be a nonzero submodule of M. Then $l_S(N)$ is a proper left ideal of S. Since S is local $l_S(N) \ll {}_SS$ and so $l_S(N)/l_S(M) \ll {}_SS/l_S(M)$. Since M is d-retractable, it follows that $M/N \ll M/N$. Hence $M = N$. Therefore M is a simple module. The converse follows from Proposition 2.2. □

Next, we present several examples of coretractable modules which are not d-retractable.

EXAMPLE 2.8. Let p be a prime integer and let $n \geq 2$ be a positive integer. It is well known that the $\mathbb{Z}$-modules $M_1 = \mathbb{Z}(p^\infty)$ and $M_2 = \mathbb{Z}/p^n\mathbb{Z}$ are coretractable. However, since $\mathrm{End}_{\mathbb{Z}}(M_i)$ is a local ring for each $i = 1, 2$, it follows that none of M_1 and M_2 is a d-retractable module by Proposition 2.7.

Before presenting other examples of coretractable modules which are not d-retractable, we introduce the following notion which is a generalization of the concept of d-retractable modules.

DEFINITION 2.9. Let M be an R-module with $S = \mathrm{End}_R(M)$. Then M is said to be *weak d-retractable* if for $U \leq V \lneqq M$, $l_S(U)/l_S(V) \ll {}_SS/l_S(V)$ implies that $V/U \ll M/U$.

Recall that a nonzero module M is called *hollow* if every proper submodule is small in M. If a module M has a proper submodule which contains all other proper submodules, then M is called a *local* module. It is obvious that every local module is a hollow cyclic module.

Clearly, every hollow module is weak d-retractable and every d-retractable module is weak d-retractable. In the next example, we exhibit a weak d-retractable module which is not d-retractable.

EXAMPLE 2.10. Let R be a discrete valuation ring and let $M = R_R$. Since M is a hollow module, M is weak d-retractable. However, if M is coretractable, then $l_R(I) \neq 0$ for every nonzero proper ideal I of R by [**2**, Theorem 2.14]. This contradicts the fact that R is a domain. Therefore M is not coretractable. From Proposition 2.6, it follows that M is not d-retractable either.

The next remark extends Example 2.10.

REMARK 2.11. Any nonzero hollow quasi-projective module M which is not simple is a weak d-retractable module which is not d-retractable. Indeed, by [**5**, 4.28(4)], $\mathrm{End}_R(M)$ is a local ring. Therefore M cannot be d-retractable by Proposition 2.7. Of course, M is a weak d-retractable module.

PROPOSITION 2.12. *Let M be a weak d-retractable module with $S = \mathrm{End}_R(M)$. Assume that M contains a proper submodule U such that $l_S(U) = 0$. Then M/U is a hollow module.*

PROOF. Let V be a proper submodule of M with $U \subseteq V$. Since $l_S(V) \subseteq l_S(U)$, we have $l_S(V) = 0$. Hence $l_S(U)/l_S(V) \ll {}_SS/l_S(V)$. By assumption, we get $V/U \ll M/U$. Therefore M/U is a hollow module. □

COROLLARY 2.13. *Let M be a nonzero module with $S = \mathrm{End}_R(M)$. Assume that $l_S(U) = 0$ for every nonzero proper submodule U of M. The following conditions are equivalent:*

(i) *M is a weak d-retractable module;*

(ii) *M is a hollow module.*

PROOF. (i) $\Rightarrow$ (ii) Let U be a nonzero proper submodule of M. By Proposition 2.12, M/U is a hollow module. Thus M/U is an indecomposable module. From [**17**, 41.4(1)], it follows that M is a hollow module.

(ii) $\Rightarrow$ (i) This is immediate. □

PROPOSITION 2.14. *Let M be a nonzero module such that $S = \mathrm{End}_R(M)$ is a local ring. Then M is weak d-retractable if and only if M is a hollow module.*

PROOF. We first note that M is an indecomposable module since the ring S is local. Assume that M is a weak d-retractable module. Let V be a proper submodule of M and let X be a submodule of M such that $M = V + X$. If $V \cap X = 0$, then $M = V \oplus X$. Therefore, $V = 0$ and $X = M$. Now, if $V \cap X \neq 0$, then $l_S(V \cap X)$ is a proper left ideal of S. Hence, $l_S(V \cap X)$ is small in ${}_SS$. Thus, $l_S(V \cap X)/l_S(V) \ll {}_SS/l_S(V)$ (see [**5**, 2.2(5)]). Since M is weak d-retractable, we have $V/V \cap X \ll M/V \cap X$. So, $(V+X)/X \ll M/X$ by [**5**, 2.2(5)]. As $V+X = M$, we have $X = M$. It follows that $V \ll M$. Therefore, M is hollow. The converse is clear. □

It is well known that every cogenerator is a coretractable module. The following example shows that a cogenerator is not, in general, a weak d-retractable module.

EXAMPLE 2.15. Let R be a commutative artinian local ring which is not quasi-Frobenius. Let m be the maximal ideal of R and let $M = E(R/m)$ be the injective hull of R/m. Clearly, M is a cogenerator. So, M is coretractable. On the other hand, assume that M is weak d-retractable. Since M is an indecomposable injective module, $S = \mathrm{End}_R(M)$ is a local ring (see [**15**, Proposition 3.12]). By Proposition 2.14, M is hollow. Since R is artinian, we have $\mathrm{Rad}(M) \ll M$ by [**4**, Corollary

15.21]. Hence, M is a local module. But $ann_R(M) = 0$ by [**15**, p. 47, Corollary 2]. Then $M \cong R_R$. Since M is injective, R_R is injective. Therefore R is quasi-Frobenius, a contradiction. This shows that M is not (weak) d-retractable.

For a particular example, we can take $R = k[x^2, x^3]/(x^4)$, where k is a field. Indeed, the unique maximal ideal J of R is generated by x^2 and x^3 (mod x^4). So $J^2 = 0$ and hence $\mathrm{Soc}(R) = J$ is not simple. It is clear that R is a commutative artinian local ring. Hence R is not quasi-Frobenius by [**11**, Theorem 15.27].

REMARK 2.16. Example 2.15 shows that even if R is a ring such that every R-module is coretractable, there may exist an R-module M which is not weak d-retractable. In fact, since the ring R is a commutative perfect ring, every R-module is coretractable by [**2**, Theorem 3.14].

The next result provides a sufficient condition for a coretractable module to be weak d-retractable.

PROPOSITION 2.17. *Let M be a nonzero coretractable R-module with $S = \mathrm{End}_R(M)$. Assume that ${}_SS$ is a uniform left S-module. Then M is weak d-retractable.*

PROOF. By [**2**, Corollary 4.6], M is a hollow R-module. Therefore M is weak d-retractable. □

REMARK 2.18. From Proposition 2.6 and Examples 2.8, 2.10 and 2.15, we obtain the following diagram:

$$
\begin{array}{ccc}
 & \mathrm{d-retractable} & \\
\swarrow\not\nearrow & & \searrow\not\nwarrow \\
\mathrm{coretractable} & \not\rightleftarrows & \mathrm{weak\ d-retractable}
\end{array}
$$

Recall that an R-module M is called *semi-injective* if for any $f \in S = \mathrm{End}_R(M)$,

$$Sf = l_S(\mathrm{Ker} f) = l_S(r_M(f)).$$

It is well known that every quasi-injective module is semi-injective.

PROPOSITION 2.19. *Let M be a semi-injective coretractable module with $S = \mathrm{End}_R(M)$. Then the following conditions are equivalent:*

(i) *M is d-retractable;*

(ii) *For all submodules U and V with $U \leq V \leq M$, $l_S(U)/l_S(V) \ll {}_SS/l_S(V)$ implies that $l_S(V)$ is essential in $l_S(U)$.*

PROOF. This follows from definition of d-retractability and [**3**, Theorem 2.2]. □

Recall that a ring R is said to be *right Kasch* if every simple right R-module can be embedded into R_R (see [**11**] or [**13**]). It is shown in [**2**, Theorem 2.14] that a ring R is right Kasch if and only if the module R_R is a coretractable R-module. Next, we will characterize rings R for which the R-module R_R is a d-retractable module. We need the following lemma.

LEMMA 2.20. *The following conditions are equivalent for a ring R:*

(i) *R_R is a (weak) d-retractable R-module;*

(ii) *For $U \leq V \leq R_R$ (with $V \neq R$), $l_R(U)/l_R(V) \ll {}_RR/l_R(V)$ implies that $V/U \ll R/U$.*

PROOF. (i) $\Leftrightarrow$ (ii) Let $S = \mathrm{End}_R(R_R)$. Consider the ring isomorphism $\psi : R \to S$ defined by $\psi(a) = \varphi_a \in S$ with $\varphi_a(x) = ax$ for all $x \in R$. Note that

$$\psi(l_R(I)) = l_S(\psi(I)) \text{ and } \psi^{-1}(l_S(I)) = l_R(I) \text{ for every right ideal } I \text{ of } R.$$

A trivial verification shows that for $U \leq V \leq R_R$,

$$l_R(U)/l_R(V) \ll {}_RR/l_R(V) \text{ if and only if } l_S(U)/l_S(V) \ll {}_SS/l_S(V).$$

This proves the lemma. □

THEOREM 2.21. *The following statements are equivalent for a ring R:*
(i) *The R-module R_R is d-retractable;*
(ii) *$l_R(K)$ is not small in ${}_RR$ for any maximal right ideal K of R;*
(iii) *$l_R(I)$ is not small in ${}_RR$ for any proper right ideal I of R;*
(iv) *R is a right Kasch ring such that $l_R(\mathrm{Soc}(R_R))$ is not small in ${}_RR$ if $\mathrm{Soc}(R_R) \neq R$;*
(v) *R is a semisimple ring.*

PROOF. (i) $\Rightarrow$ (ii) This follows by the same method as in Lemma 2.5 and using Lemma 2.20.

(ii) $\Rightarrow$ (iii) Let I be a proper right ideal of R. Then $I \subseteq K$ for some maximal right ideal K of R. Note that $l_R(K) \subseteq l_R(I)$. Since $l_R(K)$ is not small in ${}_RR$, it follows that $l_R(I)$ is not small in ${}_RR$.

(iii) $\Rightarrow$ (iv) This follows from [**2**, Theorem 2.14].

(iv) $\Rightarrow$ (v) Assume that R is not semisimple. Then $\mathrm{Soc}(R_R) \neq R$. By hypothesis, $L = l_R(\mathrm{Soc}(R_R))$ is not small in ${}_RR$. On the other hand, since R is a right Kasch ring, $\mathrm{Soc}(R_R)$ contains a copy of every simple right R-module. As $L\mathrm{Soc}(R_R) = 0$, it follows that for every maximal right ideal K of R we have $L \subseteq K$. This implies that $L \subseteq \mathrm{Rad}(R_R) = \mathrm{Rad}({}_RR)$. So $L \ll {}_RR$, a contradiction.

(v) $\Rightarrow$ (i) This follows from Proposition 2.2. □

REMARK 2.22. (i) From Theorem 2.21 and [**2**, Theorem 2.14], it follows easily that for any right Kasch ring which is not semisimple, R_R is a coretractable module which is not d-retractable. For particular examples of such rings see for example, [**2**, Example 3.3] and [**11**, Examples 8.29(6) and Exercise 13, p. 285].

(ii) Applying Theorem 2.21, we see that for any local ring R which is not semisimple, the R-module R_R is weak d-retractable but is not d-retractable.

QUESTION 2.23. From Lemma 2.5 and Theorem 2.21, we may conjecture that for an R-module M with $S = \mathrm{End}_R(M)$, the following conditions

(i) M is d-retractable,

(ii) For every proper submodule U of M, $l_S(U)$ is not small in ${}_SS$,

may possibly be equivalent. This question is still unanswered. Note that if M is a coretractable module with $\mathrm{Rad}(S) = 0$, then M satisfies the condition (ii).

PROPOSITION 2.24. *The following conditions are equivalent for a quasi-injective R-module M with essential socle (e.g., M is finitely cogenerated):*
(i) *M is d-retractable;*
(ii) *M is semisimple.*

PROOF. Let $S = \mathrm{End}_R(M)$.

(i) $\Rightarrow$ (ii) By [**17**, 22.1(5)], $\mathrm{Rad}(S) = l_S(\mathrm{Soc}(M))$. Assume that M is not semisimple. Then $\mathrm{Soc}(M) \neq M$. Hence $l_S(\mathrm{Soc}(M))$ is not small in ${}_SS$ by Lemma 2.5. That is, $\mathrm{Rad}(S)$ is not small in ${}_SS$, a contradiction.

(ii) $\Rightarrow$ (i) By Proposition 2.2. □

COROLLARY 2.25. *The following conditions are equivalent for a ring R:*
(i) *Every finitely cogenerated R-module is d-retractable;*
(ii) *$E(R/K)$ is d-retractable for every maximal right ideal K of R;*
(iii) *R is a right V-ring.*

PROOF. (i) $\Rightarrow$ (ii) This is immediate.

(ii) $\Rightarrow$ (iii) Let L be a simple R-module. Then $L \cong R/K$ for some maximal right ideal K of R. By hypothesis, $E(L) \cong E(R/K)$ is d-retractable. So $E(L)$ is semisimple by Proposition 2.24. This implies that $L = E(L)$ is an injective R-module. So R is a right V-ring.

(iii) $\Rightarrow$ (i) This follows from Proposition 2.2 and the fact that every finitely cogenerated module over a right V-ring is semisimple. □

PROPOSITION 2.26. *The following statements are equivalent for a right noetherian ring R:*
(i) *Every injective R-module is d-retractable;*
(ii) *Every indecomposable injective R-module is d-retractable;*
(iii) *R is a semisimple ring.*

PROOF. (i) $\Rightarrow$ (ii) This is obvious.

(ii) $\Rightarrow$ (iii) Let M be an indecomposable injective R-module. By [**15**, Proposition 3.12], $\mathrm{End}_R(M)$ is a local ring. So M is a simple R-module by Proposition 2.7. Now let E be an injective R-module. Then E is a direct sum of indecomposable injective R-modules (see [**15**, Theorem 4.4]). Thus E is a semisimple module. This clearly forces that R is a semisimple ring since $R \leq E(R)$.

(iii) $\Rightarrow$ (i) By Proposition 2.2. □

Recall that a submodule N of a module M is called *fully invariant* if $f(N) \subseteq N$ for every endomorphism f of M. Let M be an R-module with $S = \mathrm{End}_R(M)$. For $\mathcal{H} \subseteq S$ and $U \subseteq M$, let $l_{\mathcal{H}}(U) = \{f \in \mathcal{H} \mid f(U) = 0\}$. The next result shows that the class of (weak) d-retractable modules is closed under fully invariant direct summands.

PROPOSITION 2.27. *Let $M = M_1 \oplus M_2$ be a weak d-retractable module with $M_2 \neq 0$. If M_1 is fully invariant in M, then M_1 is a d-retractable module.*

PROOF. Assume that M_1 is fully invariant in M. Then $\mathrm{Hom}_R(M_1, M_2) = 0$. Let $S_1 = \mathrm{End}_R(M_1)$, $S_2 = \mathrm{End}_R(M_2)$ and $S = \mathrm{End}_R(M)$. We write $\mathcal{H}_1$ for the set of endomorphisms f of M with $f(M_2) = 0$ and $\mathcal{H}_2$ for the set of endomorphisms f of M with $f(M_1) = 0$. As $\mathrm{Hom}_R(M_1, M_2) = 0$, we have $\mathcal{H}_1 = \{f \in \mathrm{Hom}_R(M, M_1) \mid f(M_2) = 0\}$. Note that $\mathcal{H}_1$ and $\mathcal{H}_2$ are left ideals of S such that $S = \mathcal{H}_1 \oplus \mathcal{H}_2$. Let us show that M_1 is d-retractable. Let $U_1 \leq V_1 \leq M_1$ such that $l_{S_1}(U_1)/l_{S_1}(V_1) \ll {}_{S_1}S_1/l_{S_1}(V_1)$. Note that $l_S(U_1) = l_{\mathcal{H}_1}(U_1) + \mathcal{H}_2$ and $l_S(V_1) = l_{\mathcal{H}_1}(V_1) + \mathcal{H}_2$. If $f_1 \in S_1$ and $f_2 \in S_2$, we define $f = f_1 \oplus f_2 \in S$ by $f(x_1 + x_2) = f_1(x_1) + f_2(x_2)$ for every $(x_1, x_2) \in M_1 \times M_2$. Since $\mathrm{Hom}_R(M_1, M_2) = 0$, S_1 can be regarded as a left S-module in the obvious way. Consider the isomorphism of left S-modules

$\psi_1 : S_1 \to \mathcal{H}_1$ defined by $\psi(f_1) = f_1 \oplus 0$ for every $f_1 \in S_1$. Using the mapping ψ_1, it is easy to check that $l_{\mathcal{H}_1}(U_1)/l_{\mathcal{H}_1}(V_1) \ll {}_S\mathcal{H}_1/l_{\mathcal{H}_1}(V_1)$. This yields

$$(l_{\mathcal{H}_1}(U_1) + \mathcal{H}_2)/(l_{\mathcal{H}_1}(V_1) + \mathcal{H}_2) \ll {}_SS/(l_{\mathcal{H}_1}(V_1) + \mathcal{H}_2).$$

That is $l_S(U_1)/l_S(V_1) \ll {}_SS/l_S(V_1)$. Since M is weak d-retractable and $V_1 \neq M$, we have $V_1/U_1 \ll M/U_1$. But $M/U_1 = (M_1/U_1)\oplus((M_2+U_1)/U_1)$. Hence $V_1/U_1 \ll M_1/U_1$ by [**12**, Lemma 4.2(2)]. This completes the proof. □

A module M is called a *duo* module if every submodule of M is fully invariant in M.

PROPOSITION 2.28. *Let $M = M_1 \oplus M_2$ be a duo module. If M_1 and M_2 are d-retractable, then M is also d-retractable.*

PROOF. We follow the notation of the proof of Proposition 2.27. Assume that M_1 and M_2 are d-retractable. Let $U \leq V \leq M$ such that $l_S(U)/l_S(V) \ll {}_SS/l_S(V)$. Since M is a duo module, we have $U = (U\cap M_1)\oplus(U\cap M_2)$ and $V = (V\cap M_1)\oplus(V\cap M_2)$. It follows easily that $l_S(U) = l_{\mathcal{H}_1}(U\cap M_1)\oplus l_{\mathcal{H}_2}(U\cap M_2)$ and $l_S(V) = l_{\mathcal{H}_1}(V\cap M_1)\oplus l_{\mathcal{H}_2}(V\cap M_2)$. Let us prove that $l_{\mathcal{H}_1}(U\cap M_1)/l_{\mathcal{H}_1}(V\cap M_1) \ll \mathcal{H}_1/l_{\mathcal{H}_1}(V\cap M_1)$. Let X be a left ideal of S such that $l_{\mathcal{H}_1}(V\cap M_1) \subseteq X \subseteq \mathcal{H}_1$ and $l_{\mathcal{H}_1}(U\cap M_1)+X = \mathcal{H}_1$. Then $l_{\mathcal{H}_1}(U\cap M_1)+l_{\mathcal{H}_2}(U\cap M_2)+X+\mathcal{H}_2 = S$. Since $l_S(U)/l_S(V) \ll {}_SS/l_S(V)$, we have $l_{\mathcal{H}_1}(V \cap M_1) + l_{\mathcal{H}_2}(V \cap M_2) + X + \mathcal{H}_2 = S$. Hence $X \oplus \mathcal{H}_2 = \mathcal{H}_1 \oplus \mathcal{H}_2$. So $X = \mathcal{H}_1$. We thus get $l_{\mathcal{H}_1}(U \cap M_1)/l_{\mathcal{H}_1}(V \cap M_1) \ll \mathcal{H}_1/l_{\mathcal{H}_1}(V \cap M_1)$. Using the mapping ψ_1, we deduce that $l_{S_1}(U \cap M_1)/l_{S_1}(V \cap M_1) \ll S_1/l_{S_1}(V \cap M_1)$. In the same manner we can see that $l_{S_2}(U \cap M_2)/l_{S_2}(V \cap M_2) \ll S_2/l_{S_2}(V \cap M_2)$. By hypothesis, $(V \cap M_1)/(U \cap M_1) \ll M_1/(U \cap M_1)$ and $(V \cap M_2)/(U \cap M_2) \ll M_2/(U \cap M_2)$. A trivial verification shows that $V/U \ll M/U$. □

REMARK 2.29. Let P be the set of prime numbers. It is shown in [**2**, Example 2.9] that although the $\mathbb{Z}$-modules $\{\mathbb{Z}/p\mathbb{Z} \mid p \in P\}$ are coretractable, the $\mathbb{Z}$-module $M = \prod_{p\in P} \mathbb{Z}/p\mathbb{Z}$ is not coretractable. The same example shows also that an infinite direct product of d-retractable modules is not, in general, d-retractable.

A module M is said to *have* (C_3) if for every direct summands M_1 and M_2 of M with $M_1\cap M_2 = 0$, $M_1 \oplus M_2$ is a direct summand of M. Dually, a module M is said to *have* (D_3) if for every direct summands M_1 and M_2 of M with $M_1 + M_2 = M$, $M_1 \cap M_2$ is a direct summand of M (see [**12**]).

LEMMA 2.30. *Let M be a module with $S = \mathrm{End}_R(M)$. If ${}_SS$ has (C_3), then M has (D_3).*

PROOF. Let A and B be two direct summands of M such that $M = A + B$. Then there exist idempotents $e_1, e_2 \in S$ such that $e_1(M) = A$ and $e_2(M) = B$. Therefore $l_S(A) = l_S(e_1(M)) = S(1 - e_1)$ and $l_S(B) = l_S(e_2(M)) = S(1 - e_2)$ are direct summands of ${}_SS$. Moreover, we have $l_S(M) = l_S(A) \cap l_S(B) = 0$. Since ${}_SS$ has (C_3), $l_S(A) \oplus l_S(B)$ is a direct summand of ${}_SS$. So, $l_S(A) \oplus l_S(B) = Sf$ for some idempotent f in S. It is easy to check that $r_Ml_S(A)\cap r_Ml_S(B) = r_M(l_S(A)\oplus l_S(B)) = r_M(Sf) = (1 - f)(M)$, $r_Ml_S(A) = A$ and $r_Ml_S(B) = B$. It follows that $A \cap B = (1 - f)(M)$ is a direct summand of M. □

PROPOSITION 2.31. *Let M be a coretractable module with $S = \mathrm{End}_R(M)$ such that M has (D_3). Then ${}_SS$ has (C_3).*

PROOF. Let Se and Sf be two direct summands of ${}_SS$ with $Se \cap Sf = 0$, where e and f are idempotents in S. Note that

$$l_S((1-e)(M) + (1-f)(M)) = l_S((1-e)(M)) \cap l_S((1-f)(M)) = Se \cap Sf = 0.$$

Since M is coretractable, it follows that $(1-e)(M) + (1-f)(M) = M$. As M has (D_3), there exists an idempotent $g \in S$ such that $(1-e)(M) \cap (1-f)(M) = g(M)$. Note that $l_S((1-e)(M)) \cap l_S((1-f)(M)) = 0$. Applying [**2**, Lemma 4.9], we get

$$\begin{aligned} &l_S((1-e)(M) \cap (1-f)(M)) \\ &\quad = l_S((1-e)(M)) \oplus l_S((1-f)(M)) = Se \oplus Sf = l_S(g(M)) = S(1-g). \end{aligned}$$

It follows that $Se \oplus Sf$ is a direct summand of ${}_SS$. Therefore ${}_SS$ has (C_3). □

COROLLARY 2.32. *Let M be a d-retractable module with $S = \mathrm{End}_R(M)$ such that M has (D_3). Then ${}_SS$ has (C_3).*

PROOF. By Propositions 2.6 and 2.31. □

For a submodule U of a module M, $W(U)$ will denote the set of maximal submodules of M containing U.

LEMMA 2.33. *Let M be a finitely generated R-module. For $U \leq V \leq M$, we have*

$$V/U \ll M/U \Leftrightarrow W(U) = W(V).$$

PROOF. ($\Rightarrow$) Let $U \leq V \leq M$ and let K be a maximal submodule of M with $U \subseteq K$. Since $V/U \ll M/U$, $V/U \subseteq \mathrm{Rad}(M/U)$. Hence $V/U \subseteq K/U$. So $V \subseteq K$. Therefore $W(U) \subseteq W(V)$. Clearly, we have $W(V) \subseteq W(U)$. We thus get $W(U) = W(V)$.

($\Leftarrow$) Assume that $W(U) = W(V)$. Suppose that there exists a proper submodule T of M such that $U \subseteq T$ and $M/U = V/U + T/U$. Since M is finitely generated, there exists a maximal submodule K of M with $T \subseteq K$. Hence $U \subseteq K$. By hypothesis, $V \subseteq K$. This contradicts the fact that $M = V + T$. Therefore $V/U \ll M/U$. □

PROPOSITION 2.34. *The following conditions are equivalent for a finitely generated module M with $S = \mathrm{End}_R(M)$:*

(i) *M is (weak) d-retractable;*

(ii) *For $U \leq V \leq M$ (with $V \neq M$), $W(l_S(U)) = W(l_S(V))$ implies that $W(U) = W(V)$.*

PROOF. This is a direct consequence of Lemma 2.33. □

Combining Lemmas 2.20 and 2.33, we obtain the following result.

PROPOSITION 2.35. *The following conditions are equivalent for a ring R:*

(i) *R_R is a (weak) d-retractable R-module;*

(ii) *For $U \leq V \leq R_R$ (with $V \neq R$), $W(l_R(U)) = W(l_R(V))$ implies that $W(U) = W(V)$.*

COROLLARY 2.36. *Let M be a finitely generated (weak) d-retractable module with $S = \mathrm{End}_R(M)$. Then:*

(i) *For $U \leq K \leq M$ such that K is a maximal submodule of M, $l_S(U) = l_S(K)$ implies that M/U is a local module.*

(ii) *For $U \leq V \leq M$ (with $V \neq M$), $l_S(U) = l_S(V)$ implies that $W(U) = W(V)$.*

PROOF. (i) Let K be a maximal submodule of M and let U be a submodule of M such that $U \subseteq K$ and $l_S(U) = l_S(K)$. Since M is weak d-retractable, we have $W(U) = W(K)$ by Proposition 2.34. This implies that $W(U) = \{K\}$. It is clear that every proper submodule of M/U is contained in K/U. This implies that every proper submodule of M/U is small in M/U. So M/U is a hollow module. But M/U is finitely generated. Then M/U is a local module.

(ii) This follows easily from Proposition 2.34. □

An R-module M is called *small* if $M \ll L$ for some R-module L. It is well known that an R-module M is a small module if and only if M is a small submodule of its injective hull $E(M)$. Following [**16**], a module M is said to be *noncosingular* if $\cap\{N \leq M \mid M/N \text{ is a small module}\} = M$.

THEOREM 2.37. *Let M be a nonzero noncosingular module with $S = \mathrm{End}_R(M)$. Assume that for every proper left ideal I of S, $r_M(I)$ is not small in M. Then the following conditions are equivalent:*

(i) *M is a d-retractable module;*

(ii) *For every submodules U and V of M with $U \leq V \leq M$,*

$$l_S(U)/l_S(V) \ll {}_SS/l_S(V) \text{ if and only if } V/U \ll M/U.$$

PROOF. (i) $\Rightarrow$ (ii) Let $0 \neq U \leq V \leq M$. If $l_S(U)/l_S(V) \ll {}_SS/l_S(V)$, then $V/U \ll M/U$ since M is d-retractable. Now assume that $V/U \ll M/U$. Let I be a left ideal of S such that $l_S(V) \subseteq I$ and $l_S(U)/l_S(V)+I/l_S(V) = S/l_S(V)$. Then $S = l_S(U)+I$ and $r_Ml_S(U) \cap r_M(I) = 0$. We claim that $(r_M(I) \oplus r_Ml_S(U))/r_Ml_S(U) \ll M/r_Ml_S(U)$. Indeed, let X be a submodule of M such that $r_Ml_S(U) \subseteq X$ and $(r_M(I) \oplus r_Ml_S(U)) + X = M$. So $r_M(I) + X = M$. Since $l_S(V) \subseteq I$, we have $r_M(I) \subseteq r_Ml_S(V)$. It follows that $r_Ml_S(V)+X = M$. Note that M is coretractable by Proposition 2.6. Hence $r_Ml_S(V)/V \ll M/V$ by [**2**, Lemma 4.1]. Therefore $V + X = M$. But $V/U \ll M/U$. Then $U + X = M$. But $U \subseteq r_Ml_S(U) \subseteq X$. So $X = M$, as required. Since $r_M(I) \cong (r_M(I) \oplus r_Ml_S(U))/r_Ml_S(U)$, it follows that $r_M(I)$ is a small module. By [**16**, Lemma 2.3(1)], we get $r_M(I) \ll M$. By hypothesis, we have $I = S$.

(ii) $\Rightarrow$ (i) This is clear. □

COROLLARY 2.38. *Let M be a finitely generated noncosingular module with $S = \mathrm{End}_R(M)$. Assume that for every proper left ideal I of S, $r_M(I)$ is not small in M. Then the following conditions are equivalent:*

(i) *M is a d-retractable module;*

(ii) *For $U \leq V \leq M$, $W(l_S(U)) = W(l_S(V)) \Leftrightarrow W(U) = W(V)$.*

PROOF. By Lemma 2.33 and Theorem 2.37. □

References

[1] A. N. Abyzov and A. A. Tuganbaev, *Retractable and coretractable modules* (Russian, with English and Russian summaries), Fundam. Prikl. Mat. **19** (2014), no. 2, 5–20; English transl., J. Math. Sci. (N.Y.) **213** (2016), no. 2, 132–142. MR3431912

[2] B. Amini, M. Ershad, and H. Sharif, *Coretractable modules*, J. Aust. Math. Soc. **86** (2009), no. 3, 289–304, DOI 10.1017/S1446788708000360. MR2529325

[3] T. Amouzegar, *Endomorphism rings of semi-injective coretractable modules*, Malays. J. Math. Sci. **9** (2015), no. 2, 317–323. MR3350187

[4] F. W. Anderson and K. R. Fuller, *Rings and categories of modules*, Springer-Verlag, New York-Heidelberg, 1974. Graduate Texts in Mathematics, Vol. 13. MR0417223

[5] J. Clark, C. Lomp, N. Vanaja, and R. Wisbauer, *Lifting modules*, Frontiers in Mathematics, Birkhäuser Verlag, Basel, 2006. Supplements and projectivity in module theory. MR2253001

[6] Ş. Ecevit and M. T. Koşan, *On rings all of whose modules are retractable*, Arch. Math. (Brno) **45** (2009), no. 1, 71–74. MR2591662

[7] K. R. Goodearl, *von Neumann regular rings*, Monographs and Studies in Mathematics, vol. 4, Pitman (Advanced Publishing Program), Boston, Mass.-London, 1979. MR533669

[8] A. Haghany, O. A. S. Karamzadeh, and M. R. Vedadi, *Rings with all finitely generated modules retractable* (English, with English and Persian summaries), Bull. Iranian Math. Soc. **35** (2009), no. 2, 37–45, 270. MR2642924

[9] D. Keskin Tütüncü and B. Kaleboğaz, *On coretractable modules*, Hokkaido Math. J. **44** (2015), no. 1, 91–99, DOI 10.14492/hokmj/1470052355. MR3532102

[10] S. M. Khuri, *Nonsingular retractable modules and their endomorphism rings*, Bull. Austral. Math. Soc. **43** (1991), no. 1, 63–71, DOI 10.1017/S000497270002877X. MR1086718

[11] T. Y. Lam, *Lectures on modules and rings*, Graduate Texts in Mathematics, vol. 189, Springer-Verlag, New York, 1999. MR1653294

[12] S. H. Mohamed and B. J. Müller, *Continuous and discrete modules*, London Mathematical Society Lecture Note Series, vol. 147, Cambridge University Press, Cambridge, 1990. MR1084376

[13] W. K. Nicholson and M. F. Yousif, *Quasi-Frobenius rings*, Cambridge Tracts in Mathematics, vol. 158, Cambridge University Press, Cambridge, 2003. MR2003785

[14] N. O. Ertaş, D. K. Tütüncü, and R. Tribak, *A Variation of Coretractable Modules*, Bull. Malays. Math. Sci. Soc. **41** (2018), no. 3, 1275–1291, DOI 10.1007/s40840-016-0390-7. MR3818417

[15] D. W. Sharpe and P. Vámos, *Injective modules*, Cambridge University Press, London-New York, 1972. Cambridge Tracts in Mathematics and Mathematical Physics, No. 62. MR0360706

[16] Y. Talebi and N. Vanaja, *The torsion theory cogenerated by M-small modules*, Comm. Algebra **30** (2002), no. 3, 1449–1460, DOI 10.1081/AGB-120004881. MR1892609

[17] R. Wisbauer, *Foundations of module and ring theory*, Revised and translated from the 1988 German edition, Algebra, Logic and Applications, vol. 3, Gordon and Breach Science Publishers, Philadelphia, PA, 1991. A handbook for study and research. MR1144522

[18] J. Žemlička, *Completely coretractable rings*, Bull. Iranian Math. Soc. **39** (2013), no. 3, 523–528. MR3095342

Department of Mathematics, Hacettepe University, 06800 Beytepe, Ankara, Turkey
Email address: keskin@hacettepe.edu.tr

Centre Régional des Métiers de L'Education et de la Formation (CRMEF)-Tanger, Avenue My Abdelaziz, Souani, B.P.:3117, Tangier, Morocco
Email address: tribak12@yahoo.com

Contemporary Mathematics
Volume **727**, 2019
https://doi.org/10.1090/conm/727/14637

Essentially ADS modules and rings

M. Tamer Koşan, Truong Cong Quynh, and Jan Žemlička

Abstract. This paper introduces the notion of essentially ADS (e-ADS) modules. Basic structural properties and examples of e-ADS modules are presented. In particular, it is proved that (1) The class of all e-ADS modules properly contains all ADS as well as automorphism invariant modules. e-ADS modules serves also as a tool for characterization of various classes of rings. It is shown that: (2) R is a QF-ring if and only if every projective right R-module is e-ADS; (3) R is a semisimple Artinian ring if and only if every e-ADS module is injective. The final part of this paper describes properties of e-ADS rings, which allow to prove a criterion of e-ADS modules for non-singular rings: (4) Let R be a right non-singular ring and Q be its the right maximal ring of quotients. Then R is a right e-ADS ring if and only if either $eQ \ncong (1-e)Q$ for any idempotent $e \in R$ or $R \cong \mathbb{M}_2(A)$ for a suitable right automorphism invariant ring A.

1. Introduction

The absolute direct summand (ADS) property for modules was introduced by Fuchs in [**6**] and recently was intensively studied by Alahmadi, Jain and Leroy in [**1**]. Recall that a right module M over a ring R is said to be ADS if for every decomposition $M = S \oplus T$ and every complement T' of S, we have $M = S \oplus T'$.

In recent works [**5, 8, 12**], the notion of automorphism invariant modules was shown to be and important tool for finding correspondences between various concept of injectivity. A module M is called automorphism invariant if it is invariant under automorphisms of its injective hull, equivalently if every isomorphism between two essential submodules of M extends to an automorphism of M [**8**]. Quasi-injective modules are automorphism invariant. Assume that an R-module M has a decomposition $M = S \oplus T$ such that T' is a complement of S, $T' \cap T = 0$ and $S \cap (T' \oplus T) \leq^e S$. It is easy to see (cf. Lemma 2.4) that $E(S) \cong E(T)$, where E denotes the injective hull. In light of this observation, we define essentially ADS-modules (shortly e-ADS), as an R-module M such that for every decomposition $M = S \oplus T$ of M and every complement T' of S with $T' \cap T = 0$ and $S \cap (T' \oplus T) \leq^e S$, we have $M = S \oplus T'$. This definition naturally generalizes both notions mentioned above. Furthermore, recall that when a module M is quasi-continuous, for each decomposition $M = A \oplus B$, A and B are relatively injective. This property of modules

2010 *Mathematics Subject Classification.* Primary 16D40, 16E50.
Key words and phrases. Essentially ADS-modules and ring, automorphism invariant modules and ring.

is known to be equivalent to ADS modules ([**1**, Lemma 3.1]) and automorphism invariant modules ([**8**, Theorem 5]). Since an R-module M is e-ADS if and only if for each decomposition $M = A \oplus B$, A and B are relatively automorphism invariant (see Lemma 2.8), e-ADS modules arise as a generalization of quasi-continuous modules, ADS modules as well as automorphism invariant modules. The goal of this paper is to present a list of significant structural properties of e-ADS modules and to exhibit relations with other notions. We show that e-ADS as well as automorphism invariant or ADS modules have a description in the language of the lattice theory (Lemmas 2.1, 2.15, 4.4). Important for further study is the division of the class of e-ADS modules into trivial and non-trivial case (cf. Theorem 2.9). Moreover it is proved in Theorem 2.9 that, if $E(A) \ncong E(B)$ for each decomposition $M = A \oplus B$, then M is e-ADS. On the other hand if M is an e-ADS module with a decomposition $M = A \oplus B$ such that $E(A) \cong E(B)$, then $A \cong B$ and the modules A and B are automorphism invariant. This result is key to our work and is used to characterize many well-known classes of modules in terms of e-ADS modules. For example, we show in Theorem 2.18 that for an e-ADS module M with a decomposition $M = A \oplus B$ such that $E(A) \cong E(B)$, M satisfies the exchange property if and only if $End(M)$ is semiregular.

The final part of the article is devoted to rings which are e-ADS as right modules over themselves. By applying elementary lattice theoretical tools on rings induced by idempotents we characterize when non-singular rings are e-ADS. Based on the key observation that a non-trivial e-ADS ring is isomorphic to a 2×2 full matrix ring over an automorphism invariant ring (Lemma 4.9) we prove a characterization of non-singular e-ADS rings. Namely, a non-singular ring is e-ADS if and only if it is either trivial e-ADS or it is a product of a self-injective ring and a matrix ring $M_2(S)$ over an automorphism invariant ring S with many central idempotents (Theorem 4.11).

Throughout this article, unless otherwise stated, all rings have unity and all modules are unital. For a submodule N of M, we use $N \leq M$ ($N < M$) to mean that N is a submodule of M (respectively, proper submodule), and we write $N \leq^e M$ to indicate that N is an essential submodule of M. For any term not defined here the reader is referred to [**2**], [**4**] and [**9**].

2. e-ADS modules

Let M and N be two modules. The module M is called automorphism N-invariant if for any essential submodule A of N, any essential monomorphism from A to M can be extended to some $g \in Hom(N, M)$ ([**12**]).

We note that M is automorphism invariant if M is automorphism M-invariant by [**8**, Theorem 2].

LEMMA 2.1. *Let M and N be modules and $X = M \oplus N$. The following conditions are equivalent:*

(1) *M is automorphism N-invariant.*
(2) *For any complement K of M in X with $K \cap N = 0$ and $M \cap (K \oplus N) \leq^e M$, the module X has a decomposition $X = M \oplus K$.*

PROOF. Consider the natural projections $\pi_M : X \to M$ and $\pi_N : X \to N$. Note that $\pi_M(K) = M \cap (K + N)$ for each submodule K of X.

(1) $\Rightarrow$ (2) Let K be a complement of M in X with $K \cap N = 0$ and $\pi_M(K) \leq^e M$. Clearly, $M \oplus K = M \oplus \pi_N(K)$ so that $\pi_N(K)$ is essential in N. Consider the homomorphism $\theta : \pi_N(K) \to \pi_M(K)$ defined by $\theta(n) = m$ whenever $k = m + n \in K$ for $k \in K, m \in M, n \in N$. It is easy to see that θ is an isomorphism ($K \cap N = 0$ by the assumption). Since M is automorphism N-invariant, the homomorphism θ can be extended to some $g : N \to M$. Set $T := \{n + g(n) | n \in N\}$. Clearly, $M \oplus T = X$ and T contains K essentially by modularity. Since K is a complement, we obtain $T = K$.

(2) $\Rightarrow$ (1) Let A be an essential submodule of N and $f : A \to M$ be an essential monomorphism. Set $H := \{a - f(a)| : a \in A\}$. Clearly, $H \cap N = 0$, $H \cap M = 0$ and $\pi_M(H) = f(A)$ is essential in M. Then $M \oplus H = M \oplus \pi_N(H) = M \oplus A$, which is essential in X. Let K be a complement of M in X containing H. Then $H \leq^e K$. Hence $K \cap N = 0$ because $H \cap N = 0$. Moreover, $\pi_M(H) \leq \pi_M(K)$ which implies that $\pi_M(K) \leq^e M$. By the assumption, we have $M \oplus K = X$. Now let $\pi : M \oplus K \to M$ be the projection. Then writing an element $a \in A$ in the form $a = a - f(a) + f(a)$, the restriction of π to N is the desired extension of f. □

LEMMA 2.2 ([**12**, Theorem 2.2]). *The following are equivalent for modules M and N:*

(1) *M is automorphism N-invariant.*
(2) *$\alpha(N) \leq M$ for every isomorphism $\alpha : E(N) \to E(M)$.*

As an immediate consequence of Lemmas 2.1 and 2.2 , we obtain the following observation.

COROLLARY 2.3. *If M and N are relatively automorphism invariant modules and $E(M) \cong E(N)$, then $M \cong N$*

LEMMA 2.4. *Let M be a module with a decomposition $M = S \oplus T$. If T' is a complement of S with $T' \cap T = 0$ and $S \cap (T' \oplus T) \leq^e S$, then $T \oplus T' \leq^e M$ and $E(S) \cong E(T)$.*

PROOF. Note that $S \oplus T' \leq^e M$ because T' is a complement of S. Since

$$T \oplus [S \cap (T' \oplus T)] \subseteq T \oplus T' \text{ and } T \oplus [S \cap (T' \oplus T)] \leq^e T \oplus S = M,$$

we get $T \oplus T' \leq^e M$. Moreover, the injective hulls $E(S)$, $E(T)$ and $E(T')$ can be taken as submodules of the injective hull $E(M)$ such that $S \leq^e E(S)$, $T \leq^e E(T)$ and $T' \leq^e E(T')$. Since $S \cap T' = 0 = T \cap T'$, it is easy to see that

$$E(S) \cap E(T') = 0 = E(T) \cap E(T').$$

On the other hand,

$$E(S) + E(T') = E(M) = E(T) + E(T')$$

because both $E(S) + E(T') = E(S) \oplus E(T')$ and $E(T) + E(T') = E(T) \oplus E(T')$ are injective submodules of $E(M)$, and both $S \oplus T'$ and $T \oplus T'$ are essential in $E(M)$. Thus

$$\begin{aligned} E(T) \cong (E(T) + E(T'))/E(T') &= E(M)/E(T') \\ &= (E(S) + E(T'))/E(T') \cong E(S). \end{aligned}$$

□

In light of Lemma 2.4, we call M an essentially ADS-module, shortly e-ADS, if for every decomposition $M = S \oplus T$ of M and every complement T' of S with $T' \cap T = 0$ and $S \cap (T' \oplus T) \leq^e S$, we have $M = S \oplus T'$.

Clearly, ADS-modules are e-ADS. The following examples show that the converse is not true in general and that the class of e-ADS modules is not closed under taking direct summands, respectively.

EXAMPLE 2.5. Let T be a torsion abelian group which is not divisible and put $M := \mathbb{Z} \oplus T$. Then every decomposition $M = A \oplus B$ contains a subgroup which is isomorphic to $\mathbb{Z}$ while the second is torsion, hence $E(A) \ncong E(B)$. By Lemma 2.4, there exists no a decomposition satisfying the hypothesis of the definition of e-ADS exists. So that the conditions for having an e-ADS modules are vacuously satisfied and the module is e-ADS. Hence it is an e-ADS abelian group.

On the other hand, since T is not divisible, we obtain T is not $\mathbb{Z}$-injective and so M is not ADS by [**1**, Lemma 3.1].

EXAMPLE 2.6. Put $M := \mathbb{Z} \oplus \mathbb{Z}_p \oplus \mathbb{Z}_{p^2}$ for some prime number p. Then M is e-ADS by Example 2.5. Since $\mathbb{Z}_p$ is not automorphism $\mathbb{Z}_{p^2}$-invariant, we obtain that $Z_p \oplus \mathbb{Z}_{p^2}$ is not e-ADS by Lemma 2.8.

Let us mention the following equivalent conditions for a module to be e-ADS.

THEOREM 2.7. *The following conditions are equivalent for a module M:*

(1) *M is e-ADS.*
(2) *For every decomposition $M = S \oplus T$, if T' is a complement of S in M and T is a complement of T' in M, then $M = S \oplus T'$.*

PROOF. (1) $\Rightarrow$ (2) Suppose that $M = S \oplus T$ is a decomposition of M, T' is complement of S and T is a complement of T' in M. Then $S \cap (T' \oplus T) \leq^e S$ since $T \oplus T' \leq^e M$. By (1), we have $M = S \oplus T'$.

(2) $\Rightarrow$ (1) Let $M = S \oplus T$ of M and $S \cap (T' \oplus T) \leq^e S$ for a complement T' of S with $T' \cap T = 0$. By Lemma 2.4, $T \oplus T' \leq^e M$. Since T is a direct summand of M, we get T is a complement of T' in M. By (2), we have $M = S \oplus T'$. □

In [**1**, Lemma 3.1], it is shown that an R-module M is ADS if and only if for each decomposition $M = A \oplus B$, A and B are mutually injective.

LEMMA 2.8. *An R-module M is e-ADS if and only if for each decomposition $M = A \oplus B$, A and B are relatively automorphism invariant.*

PROOF. This is clear from Lemma 2.1. □

The following characterization proves to be quite useful.

THEOREM 2.9. *Let M be an R-module.*

(1) *If $E(A) \ncong E(B)$ for each decomposition $M = A \oplus B$, then M is e-ADS.*
(2) *If M is an e-ADS module with a decomposition $M = A \oplus B$ such that $E(A) \cong E(B)$, then $A \cong B$ and the modules A and B are automorphism invariant.*

PROOF. (1) This follows from Lemmas 2.2 and 2.8.

(2) By Lemma 2.8 and Corollary 2.3, we have $A \cong B$. Thus A is automorphism A-invariant, i.e. automorphism invariant. □

In the following observation, we continue to obtain equivalent conditions for a module to be e-ADS.

THEOREM 2.10. *The following conditions are equivalent for a module M:*

(1) *M is e-ADS,*
(2) *Assume that M has a decomposition $M = A \oplus B$. For any isomorphism $f \in Hom(E(B), E(A))$, the module M has a decomposition $M = A \oplus X$, where $X = \{b + f(b) |\ b \in B, f(b) \in A\}$.*
(3) *For every decomposition $M = A \oplus B$ such that $E(A) \cong E(B)$, the module $A \cong B$ is automorphism invariant.*
(4) *Either $E(A) \ncong E(B)$ for every decomposition $M = A \oplus B$ or there exists an automorphism invariant module X for which $M \cong X \oplus X$ and for every two decompositions $X = P_1 \oplus Q_1 = P_2 \oplus Q_2$ with $E(P_1) \oplus E(P_2) \cong E(Q_1) \oplus E(Q_2)$ we have $(P_1 \oplus P_2) \cong (Q_1 \oplus Q_2)$ is automorphism invariant.*

PROOF. (1) $\Rightarrow$ (2) We show that $X = \{b + f(b) | b \in B, f(b) \in A\}$ is a complement of A in M. Notice that $A \cap X = 0$, $X \cap B = 0$ and $A \cap (X \oplus B) \leq^e A$. Let L be a submodule of M such that $L \cap A = 0$ and $X \leq L$. Consider the natural projections π_A and π_B of M onto A and B, respectively.
Claim: $\pi_A(x) = f\pi_B(x)$ for all $x \in L$: Assume that there exists $x \in L$ such that $(\pi_A - f\pi_B)(x) \neq 0$. Since $A \leq^e E(A)$, there exists $r \in R$ such that $0 \neq (\pi_A - f\pi_B)(xr) \in A$. As $xr \in L$ and $\pi_B(xr) + f\pi_B(xr) \in X \subseteq L$, we have

$$\pi_A(xr) - f\pi_B(xr) = xr - (\pi_B(xr) + f\pi_B(xr)) \in A \cap L = 0,$$

a contradiction. Thus $\pi_A(x) = f\pi_B(x)$ for all $x \in L$.

For $x \in L$, we have

$$x = \pi_A(x) + \pi_B(x) = f(\pi_B(x)) + \pi_B(x) \in X,$$

which implies $L \subseteq X$.

(2) $\Rightarrow$ (3) If the module M has a decomposition $M = A \oplus B$ for an isomorphism $f \in Hom(E(B), E(A))$, we obtain $M = A \oplus X$ with $X = \{b + f(b) |\ b \in B, f(b) \in A\}$. Clearly, $f(B) \leq A$ and hence A is automorphism B-invariant by Lemma 2.2. Symmetrically, $f(A) \leq B$ and so A is automorphism and $A \cong B$.

(3) $\Rightarrow$ (1) This is a direct consequence of Lemma 2.8.

(1) $\Rightarrow$ (4) This follows from Theorem 2.9(2).

(4) $\Rightarrow$ (3) If $E(A) \ncong E(B)$ for each decomposition $M = A \oplus B$, there is nothing to prove. Assume that M has a decomposition $M = X_1 \oplus X_2$ for submodules X_1 and X_2 of M such that $X \cong X_1 \cong X_2$. We suppose furthermore that M has an another decomposition $M = A \oplus B$ such that $E(A) \cong E(B)$. By [**3**, Theorem 3], both the modules X_1, X_2 and M satisfy the exchange property. Thus there exist submodules $P_1 \subseteq A$, $Q_1 \subseteq B$ such that $M = X_1 \oplus P_1 \oplus Q_1$. Note that $X_2 \cong M/X_1 \cong P_1 \oplus Q_1$ is automorphism invariant, hence there exist submodules $P_2 \subseteq A$, $Q_2 \subseteq B$ such that $M = P_1 \oplus Q_1 \oplus P_2 \oplus Q_2$. Clearly, as $P_1 \oplus P_2 \subseteq A$ and $Q_1 \oplus Q_2 \subseteq B$, we get $A = P_1 \oplus P_2$ and $B = Q_1 \oplus Q_2$, hence

$$E(P_1) \oplus E(P_2) \cong E(P_1 \oplus P_2) \cong E(A)$$

and

$$E(Q_1) \oplus E(Q_2) \cong E(Q_1 \oplus Q_2) \cong E(B).$$

Now, since $E(A) \cong E(B)$, the hypothesis of (4) implies that $A \cong B$ is automorphism invariant. □

For modules M and N, N is said to be M-injective if every homomorphism from each submodule of M to N extends to a homomorphism from M to N, and M and N are called relatively injective if N is M-injective and M is N-injective. The module M is called quasi-injective if M is M-injective. It is well-known that a module is quasi-injective if and only if it is invariant under automorphisms and idempotent endomorphisms of its injective hull.

In [**8**], Lee and Zhou discussed when an automorphism invariant module is quasi-injective or injective and they obtained the following observation.

LEMMA 2.11 ([**8**, Theorem 5]). *If $M \oplus N$ is automorphism invariant, then M and N are relatively injective.*

Combining Lemmas 2.8 and 2.11, we have

COROLLARY 2.12. *Every automorphism invariant module is e-ADS.*

The following example shows that the converse of Corollary 2.12 is not true in general.

EXAMPLE 2.13. Take any continuous module M which is not quasi-injective (e.g. if R is the ring of all sequences of real numbers that are eventually rational, then R_R is continuous but not quasi-injective), then clearly M is ADS (and hence e-ADS) but not automorphism invariant.

We recall Example 2.6. It also shows that e-ADS modules are not closed with respect to general direct summands. On the other hand, Corollary 2.12 and Theorem 2.9 prove that the class of all e-ADS modules is closed under taking some important cases of direct summands. We can then show:

COROLLARY 2.14. *Let M be an e-ADS module. If M has a decomposition $M = A \oplus B$ such that $E(A) \cong E(B)$, then A is e-ADS.*

In view of the claim of Theorem 2.9, we say that a module M is trivial e-ADS if it has no a decomposition $M = A \oplus B$ such that $E(A) \cong E(B)$.

The following observation shows that the trivial e-ADS modules can be described using lattices of their submodules.

PROPOSITION 2.15. *Let M be a module. Then M is trivial e-ADS if and only if for every decomposition $M = A \oplus B$ no complement of A is a complement of B.*

PROOF. Suppose that the module M has a decomposition $M = A \oplus B$ such that $E(A) \cong E(B)$. The isomorphism $\varphi : E(B) \cong E(A)$ implies that the restriction of φ on $C = \varphi^{-1}(A) \cap N$ forms an essential monomorphism $\psi : C \to A$. Put $H := \{c - \varphi(c)| : a \in C\}$. Now if we follow the same way as in the proof of (2) $\Rightarrow$ (1) of Lemma 2.1, we have fixed a complement K of B containing H. Since $K \cap B = 0$ and $A \cap (K + B) \leq^e A$, we obtain that K is complement of B.

Conversely, suppose that M has a decomposition $M = A \oplus B$ and K is simultaneously complement of A and B. Then

$$E(M) = E(A) \oplus E(K) = E(B) \oplus E(K),$$

hence $E(A) \cong E(B)$ (here we notice that all injective hulls are considered as submodules of $E(A)$). □

Now, we provide several useful necessary conditions of trivial e-ADS modules.

LEMMA 2.16. *Let M be a nonzero module. If every idempotent of* $\text{End}(M)$ *can be extended to a central idempotent of* $\text{End}(E(M))$, *then M is trivial e-ADS.*

PROOF. Suppose that M has a decomposition $M = A \oplus B$ and consider an idempotent $e \in \text{End}(M)$ defined by the rule $e(a+b) = a$ for all $a \in A$, $b \in B$. By the hypothesis, there exists a central idempotent $\tilde{e} \in \text{End}(E(M))$ satisfying $\tilde{e}(m) = e(m)$ for each $m \in M$. Now, assume that we have an isomorphism $i : E(A) \to E(B)$ and extend it to an endomorphism $j \in \text{End}(E(M))$ such that $j(a+b) = i(a)$ for all $a \in E(A)$ and $b \in E(B)$. Since $A \neq 0 \neq B$ by the hypothesis and i is an isomorphism, $i(A) \cap B$ is essential in $E(B)$, hence there exists nonzero element $a \in A$ for which $0 \neq i(a) \in B$. As $\tilde{e}$ is central, i.e. $\tilde{e}j = j\tilde{e}$, we have

$$0 \neq i(a) = j(a) = je(a) = j\tilde{e}(a) = \tilde{e}j(a) = ei(a) = 0,$$

a contradiction. □

Since every idempotent endomorphism of a module M can be extended to an idempotent endomorphism of $E(M)$ we obtain the following consequence:

COROLLARY 2.17. *If M is a nonzero module such that every idempotent of* $\text{End}(E(M))$ *is central, then M is trivial e-ADS.*

A right R-module M is said to satisfy *the exchange property* if for every right R-module A and any two direct sum decompositions $A = M_1 \oplus N = \oplus_{i\in I} A_i$ with $M_1 \cong M$, there exist submodules B_i of A_i such that $A = M_1 \oplus (\oplus_{i\in I} B_i)$.

A ring R is called *semiregular* if, for every $a \in R$, there exists $b \in R$ such that $bab = b$ and and $a - aba \in J(R)$ ([**10**]).

THEOREM 2.18. *Let M be a non trivial e-ADS module. Then*

(1) *M satisfies the exchange property.*
(2) *End(M) is semiregular.*

PROOF. (1) By Theorem 2.9(2), we obtain $M \cong A \oplus A$ where A is automorphism invariant. Moreover, A satisfies the exchange property by [**3**, Theorem 3]. Hence M satisfies the exchange property because the class of modules satisfying the exchange property is closed under taking finite direct sums.
(2) It follows from Theorem 2.9(2), [**3**, Proposition 1] and [**11**, Theorem 29]. □

Recall an easy observation about central idempotents.

LEMMA 2.19. *Let A and B be direct summands of a module M and f a central idempotent of* $\text{End}(M)$. *If $A \cong B$, then $f(A) \cong f(B)$.*

PROOF. Let $\varphi : A \to B$ be an isomorphism and consider the natural projection $\pi_A : M \to A$ and the natural embedding $\nu_B : B \to M$. Put $h = \nu_B\varphi\pi_A \in \text{End}(M)$. Since f is a central idempotent we get $h = fhf \oplus (1-f)h(1-f)$, hence fhf induces an isomorphisms between $f(A)$ and $f(B)$. □

Note that direct sums of two e-ADS modules need not be e-ADS (as it can be illustrated, e.g. by the direct sum of two trivial e-ADS modules $\mathbb{Z}_2$ and $\mathbb{Z}_4$). The following theorem shows some kind of restrictive closure property of e-ADS modules.

THEOREM 2.20. *Let M be a trivial e-ADS and N a nontrivial e-ADS module. If* $\text{Hom}(E(M), E(N)) = 0 = \text{Hom}(E(N), E(M))$, *then $M \oplus N$ is trivial e-ADS.*

PROOF. Let $X = M \oplus N$ and assume that there exists a decomposition $X = A \oplus B$ such that $E(A)$ and $E(B)$ are isomorphic. Note that we may suppose all modules and their injective hulls as submodules of $E(X)$.

Since N satisfies exchange property by Theorem 2.18, there exist submodules $C \subseteq A$ and $D \subseteq B$ such that $X = N \oplus C \oplus D$. Obviously, $M \cong X/N \cong C \oplus D$. Thus $E(M) \cong E(C) \oplus E(D)$ where $E(C)$ and $E(D)$ are considered as submodules of $E(A)$ and $E(B)$, respectively. Note that there are injective submodules $E_A \subseteq E(A)$ and $E_B \subseteq E(B)$ for which $E_A \oplus E(C) = E(A)$ and $E_B \oplus E(D) = E(B)$. Now it is easy to see that $E(N) \cong E_A \oplus E_B$. By the hypothesis, we get $\mathrm{End}(E(X)) \cong \mathrm{End}(E(M)) \times \mathrm{End}(E(N))$, hence there exists a central idempotent $f \in \mathrm{End}(E(X))$ for which $f(E(X)) = E(M)$ and $(1 - f)(E(X)) = E(N)$. By Lemma 2.19, we obtain that $f(E(A)) \cong f(E(B))$. As $f(E(A)) = E(C)$ and $f(E(B)) = E(D)$, a contradiction. □

3. Classes of e-ADS modules and some ring conditions

Let $\sigma[M]$ denote the Wisbauer category of a module M, i.e. the full category of R-Mod consisting of submodules of quotients of direct sums of copies of M (see [**14**]).

THEOREM 3.1. *The following conditions are equivalent for a module M:*

(1) *M is semisimple.*
(2) *Every module in $\sigma[M]$ is e-ADS.*
(3) *Every finitely generated module in $\sigma[M]$ is e-ADS.*
(4) *Every 4-generated module in $\sigma[M]$ is e-ADS.*

PROOF. (1) $\Rightarrow$ (2) $\Rightarrow$ (3) $\Rightarrow$ (4) are clear.

(4) $\Rightarrow$ (1) Let $N \in \sigma[M]$ be a cyclic module and $x \in M$. Then

$$(N \oplus xR) \oplus (N \oplus xR)$$

is a 4-generated module in $\sigma[M]$ and hence is e-ADS by the hypothesis. By Lemma 2.8, $N \oplus xR$ is automorphism $N \oplus xR$-invariant and N is xR-injective by Lemma 2.11. By [**9**, Theorem 1.4], N is M-injective. Thus M is semisimple by [**4**, Corollary 7.14]. □

Theorem 3.1 gives immediately the following.

COROLLARY 3.2. *A ring R is semisimple Artinian if and only if every 4-generated R-module is e-ADS.*

The following observation gives an another characterization of e-ADS modules in the category $\sigma[M]$.

THEOREM 3.3. *The following conditions are equivalent for a module M:*

(1) *M is semisimple.*
(2) *The direct sum of every two e-ADS modules in $\sigma[M]$ is e-ADS.*
(3) *Every e-ADS module in $\sigma[M]$ is M-injective.*
(4) *The direct sum of any family of e-ADS modules in $\sigma[M]$ is e-ADS.*

PROOF. (1) $\Rightarrow$ (4) $\Rightarrow$ (2) They are obvious.

(2) $\Rightarrow$ (3) Let N be an e-ADS module. By our assumption, $(N \oplus E_M(N)) \oplus (N \oplus E_M(N))$ is e-ADS. Then $N \oplus E_M(N)$ is automorphism invariant. Hence N is $E_M(N)$-injective by Lemma 2.11. It follows that N is M-injective.

(3) $\Rightarrow$ (1) We consider a family $\{S_i | i \in \mathbb{N}\} (\subset \sigma[M])$ of simple right R-modules. It follows that $\oplus_{i\in\mathbb{N}} S_i$ is semisimple and so is e-ADS. By (3), $\oplus_{i\in\mathbb{N}} S_i$ is M-injective. Therefore $\oplus_{i\in\mathbb{N}} S_i$ is a direct summand of $\oplus_{i\in\mathbb{N}} E_M(S_i)$. But $\oplus_{i\in\mathbb{N}} S_i$ is essential in $\oplus_{i\in\mathbb{N}} E_M(S_i)$ and then $\oplus_{i\in\mathbb{N}} S_i = \oplus_{i\in\mathbb{N}} E_i$ is M-injective. Thus M is locally Noetherian. We can write $E_M(M) = \oplus_{i\in I} K_i$ for some indecomposable right R-modules K_i in $\sigma[M]$ by [**14**, 27.4]. We have that every K_i is M-injective and obtain that every K_i is uniform. For each $i \in I$, let $0 \neq x \in K_i$. Since K_i is uniform, xR is uniform as well, hence xR is e-ADS. Then xR is M-injective by (3). It follows that xR is a direct summand of K_i and we have $xR = K_i$. Thus K_i is simple for all $i \in I$. That means $E_M(M)$ is semisimple. Thus M is semisimple. □

COROLLARY 3.4. *The following conditions are equivalent for a ring R:*

(1) *R is semisimple Artinian.*
(2) *The direct sum of every two e-ADS modules is e-ADS.*
(3) *Every e-ADS module is injective.*
(4) *The direct sum of any family of e-ADS modules is e-ADS.*

We note that if $M \oplus E(M)$ is e-ADS for an R-module M, then $M \cong E(M)$ by Theorem 2.9 and so M is injective.

THEOREM 3.5. *The following conditions are equivalent for a ring R:*

(1) *R is right Noetherian.*
(2) *The direct sum of injective right R-modules is e-ADS.*
(3) *For any injective right R-module X, $X^{(\mathbb{N})}$ is e-ADS.*

PROOF. (1) $\Rightarrow$ (2) $\Rightarrow$ (3) They are obvious.

(3) $\Rightarrow$ (1) Let X be an injective module. Clearly, $X \oplus E(R_R)$ is also injective. Let $M = X \oplus E(R_R)$. Since $4 \cdot |\mathbb{N}| = |\mathbb{N}|$, we obtain that $(M^{(\mathbb{N})})^{(4)} \cong M^{(\mathbb{N})}$. By (3), $M^{(\mathbb{N})} \oplus M^{(\mathbb{N})}$ is automorphism invariant. It follows that $M^{(\mathbb{N})}$ is quasi-injective. On the other hand, $X^{(\mathbb{N})}$ is isomorphic to a direct summand of $M^{(\mathbb{N})}$. It implies that $X^{(\mathbb{N})}$ is $E(R_R)$-injective and so $X^{(\mathbb{N})}$ is injective. Hence R is right Noetherian. □

A ring R is called a right *V-ring* if every simple right R-module is injective.

THEOREM 3.6. *The following conditions are equivalent for a ring R:*

(1) *R is a right V-ring,*
(2) *$S \oplus E(S)$ is e-ADS for every simple right R-module S.*

PROOF. (1) $\Rightarrow$ (2) This is obvious.

(2) $\Rightarrow$ (1) Assume that $S \oplus E(S)$ is e-ADS for every simple right R-module S. Let S be a simple right R-module. By the hypothesis, $S \oplus E(S)$ is e-ADS. Then, by Theorem 2.9(1), $S \cong E(S)$, and so S is injective. □

THEOREM 3.7. *The following conditions are equivalent for a ring R:*

(1) *R is a QF-ring.*
(2) *Every projective right R-module is e-ADS.*
(3) *Every essential extension of any free right R-module is e-ADS.*

PROOF. (1) $\Rightarrow$ (2) and (1) $\Rightarrow$ (3) are obvious.

(2) $\Rightarrow$ (1) Let I be a non-empty set. Clearly $(R^{(I)})^4$ is also a projective module. By (2), $R^{(I)} \oplus R^{(I)}$ is automorphism invariant. It follows that $R^{(I)}$ is quasi-injective. Therefore $R^{(I)}$ is injective. Thus R is Σ-injective and so R is a QF-ring.

(3) $\Rightarrow$ (1) Let F be a free right R-module. Then $F \oplus E(F)$ is an essential extension of a free right module F^2. By (3), $F \oplus E(F)$ is e-ADS, hence F is injective. Now we have proved that every projective right R-module is injective. Thus R is QF by the Faith-Walker theorem. $\square$

4. The Structure of e-ADS rings

We say that a ring R is right e-ADS if it is an e-ADS module over itself. A right e-ADS ring R is called trivial if R_R is trivial e-ADS, i.e. the module R_R does not have a decomposition $R_R = A \oplus B$ such that $E(A) \cong E(B)$. Otherwise R is said to be a nontrivial e-ADS ring.

Let R be a ring, e be an idempotent of R, $S := eRe$ and $n \in \mathbb{N}$. Denote by $\mathcal{L}(eR^n)$ the lattice of all submodules of the projective R-module eR^n, and $\mathcal{L}(S^n)$ the lattice of all submodules of the free module S^n. Define two mappings

$$\Phi : \mathcal{L}(eR^n) \to \mathcal{L}(S^n)$$

and

$$\Psi : \mathcal{L}(S^n) \to \mathcal{L}(eR^n)$$

by the rules

$$\Phi(I) = Ie, \qquad \Psi(J) = JR$$

for arbitrary $I \in \mathcal{L}(eR^n)$ and $J \in \mathcal{L}(S^n)$.

Lemma 4.1. *Φ and Ψ are well-defined monotonic mappings. Moreover, Φ is a lattice homomorphism and Ψ is compatible with the operation $+$.*

Proof. Straightforward from the above notation. $\square$

Note that the inclusion $\Psi(J_1 \cap J_2) \subseteq \Psi(J_1) \cap \Psi(J_2)$ holds generally for arbitrary $J_1, J_2 \in \mathcal{L}(S^n)$ but the following example shows that the reverse need not be true.

Example 4.2. Let $R = \{(a_{ij}) \in M_{3\times3}(\mathbb{Q}) |\ a_{31} = a_{32} = 0\}$ be a subring of matrix ring $M_{3\times3}(\mathbb{Q})$. Put $e := \begin{pmatrix} 1 & 0 & 0 \\ 0 & 1 & 0 \\ 0 & 0 & 0 \end{pmatrix}$, $f := \begin{pmatrix} 1 & 0 & 0 \\ 1 & 0 & 0 \\ 0 & 0 & 0 \end{pmatrix}$, $g := \begin{pmatrix} 1 & 0 & 0 \\ 2 & 0 & 0 \\ 0 & 0 & 0 \end{pmatrix}$, $S := eRe$, $J_1 := fS$, and $J_2 := gS$. Then it is easy to see that

$$J_1 \cap J_2 = 0$$

and

$$J_1R \cap J_2R = \{ \begin{pmatrix} 0 & 0 & u \\ 0 & 0 & v \\ 0 & 0 & 0 \end{pmatrix} \mid u, v \in \mathbb{Q}\}.$$

Thus $(J_1 \cap J_2)R \neq J_1R \cap J_2R$.

Lemma 4.3. *Let R be a ring and $e \in R$ be an idempotent such that $ReR = R$. Then Φ and Ψ are mutually inverse lattice isomorphisms.*

Proof. Let $S := ReR$. Since both Φ and Ψ are monotonic, it is enough to show that $\Phi\Psi$ and $\Psi\Phi$ are identity mappings on $\mathcal{L}(S)$ and $\mathcal{L}(eR)$, respectively. Let $I \in \mathcal{L}(eR)$ and $J \in \mathcal{L}(S)$. Since $ReR = R$, we get

$$\Psi\Phi(I) = IeR = IReR = IR = I.$$

On the other hand $S = eRe$ and $J = Je$ imply that

$$\Phi\Psi(J) = JRe = JeRe = JS = J.$$

□

Recall that essentiality of modules can be expressed as a condition of lattices of submodules:

LEMMA 4.4. *Let $A \subseteq B$ are submodules of a module M. Then $A \leq^e B$ if and only if there exists no submodule $C \subseteq B$ such that $A \cap C = 0$.*

PROOF. This is well known. □

The following general consequence is a special case of [**15**, Theorem 1.2] for the lattice isomorphism from Lemma 4.3.

COROLLARY 4.5. *Let R and S be rings, M an R-module, N an S-module and K, L submodules of M. Suppose that $\phi : \mathcal{L}(M_R) \to \mathcal{L}(N_S)$ is an isomorphism of lattices of all submodules of M and N. Then K is a complement of L if and only if $\phi(K)$ is a complement of $\phi(L)$.*

Lemmas 4.4, 2.1, 2.15 and Corollary 4.5 show that e-ADS, trivial e-ADS and relative automorphism invariant are lattice conditions. Thus the assertions of the following theorem hold true because lattices of all submodules of M and N are isomorphic.

THEOREM 4.6. *Let R and S be rings, M an R-module and N an S-module. Assume $\phi : \mathcal{L}(M_R) \to \mathcal{L}(N_S)$ is an isomorphism of lattices.*

(1) *M is (trivial) e-ADS if and only if N is a (trivial) e-ADS.*
(2) *If $M = A \oplus B$, then $N = \phi(A) \oplus \phi(B)$ and A is B-automorphism invariant if and only if $\phi(A)$ is $\phi(B)$-automorphism invariant.*

Let $n \in \mathbb{N}$ and e be an idempotent of a ring R such that $ReR = R$. Recall that $L(eR^n_R)$ and $L(S^n_S)$ are isomorphic lattices by Lemma 4.3 for every $n \in \mathbb{N}$, where $S = eRe$.

THEOREM 4.7. *Let R be a ring, $n \in \mathbb{N}$ and $e \in R$ be an idempotent such that $ReR = R$.*

(1) *eR^n_R is a (trivial) e-ADS module if and only if $eR^n e$ is (trivial) e-ADS as a right eRe-module.*
(2) *Let $eR^n = A \oplus B$. Then A is B-automorphism invariant if and only if Ae is Be-automorphism invariant.*
(3) *eR is automorphism invariant if and only if S_S is automorphism invariant, where $S = eRe$.*

PROOF. (1) and (2) follow immediately from Theorem 4.6.
(3) It suffices to apply (2) for the decomposition $eR^2 = eR \oplus eR$. □

The next observation shows that the class of e-ADS rings is closed under taking finite products.

PROPOSITION 4.8. *If R_1 and R_2 are e-ADS rings, then $R_1 \times R_2$ is e-ADS as well.*

Proof. Put $R := R_1 \times R_2$ and let e_i be orthogonal central idempotents such that $R_i = Re_i$ for $i = 1, 2$. It is easy to see that $e_1 + e_2 = 1$, $E(R) = E(R_1) \oplus E(R_2)$ and $E(R_i) = E(R)e_i$ for $i = 1, 2$. Suppose that $R = A \oplus B$ is a module decomposition, $C \leq^e A$, $D \leq^e B$ and $f : C \to D$ is an isomorphism. Then $f_i : Ce_i \to De_i$ defined by $f_i(r) = re_i$ is an isomorphism for each $i = 1, 2$. We note that $Ce_i \leq^e Ae_i$ and $De_i \leq^e Be_i$ for each $i = 1, 2$. By the hypothesis, there exist extensions $g_i : Ae_i \to Be_i$ of f_i. Clearly, $g = g_1 \oplus g_2 : A \to B$ extends f. □

We denote the set of all $n \times n$ matrices over a ring R by $M_n(R)$.

Lemma 4.9. *If R is a non-trivial e-ADS ring, then there exists a right automorphism invariant ring S such that $R \cong M_2(S)$.*

Proof. Since R is a non-trivial e-ADS ring, there exists an idempotent $e \in R$ for which $E(eR) \cong E((1-e)R)$. Thus $eR \cong (1-e)R$ is automorphism invariant by Theorem 2.9. Put $S := eRe$. Then

$$R \cong \mathrm{End}(eR \oplus eR) \cong M_2(S)$$

and S is a right automorphism invariant ring by Theorem 4.7(3). □

Let R be a ring. Recall that R is said to be right non-singular if its right singular ideal $Z(R) = \{r \in R : rI = 0 \text{ for some essential right ideal } I \text{ of } R\}$ is zero, and R is called normal if if moreover its idempotents are central. Note that every abelian regular ring or every product of rings without non-trivial idempotents can serve as elementary examples of normal rings.

Proposition 4.10. *Let R be a right non-singular normal automorphism invariant ring. Then*

(1) *R is trivial e-ADS,*
(2) *$M_2(R)$ is non-trivial e-ADS.*

Proof. Denote by Q the maximal right ring of quotients R. Obviously $eQ = E(eR)$ for every idempotent e.

(1) As every central idempotent of R is a central idempotent of Q, the assertion follows from Lemma 2.16.

(2) By Theorem 4.7 it is enough to prove that $M = R \oplus R$ is a non-trivial e-ADS module. Clearly, M cannot be trivial. So it suffices to prove Theorem 2.10(4). Suppose $R = e_iR \oplus f_iR$ for every $i = 1, 2$, where (e_i, f_i) is a pair of orthogonal idempotents such that $e_1Q \oplus e_2Q \cong f_1Q \oplus f_2Q$. We claim that $A := e_1R \oplus e_2R \cong B := f_1R \oplus f_2R$ (and that A is automorphism invariant).

Since R is a normal ring, i.e., all idempotents e_i, f_i of R, are central for each $i = 1, 2$, we have

$$e_iQ = e_ie_jQ \oplus e_if_jQ$$

$$f_iQ = f_ie_jQ \oplus f_if_jQ$$

for $i \neq j$. Hence $Q = e_1e_2Q \times e_1f_2Q \times f_1e_2Q \times f_1f_2Q$, where there is no nonzero homomorphism between two distinct components. Thus

$$E(A) = e_1Q + e_2Q \cong (e_1e_2Q)^{(2)} \oplus e_1f_2Q \oplus e_2f_1Q$$

and

$$E(B) = f_1Q + f_2Q \cong (f_1f_2Q)^{(2)} \oplus e_1f_2Q \oplus e_2f_1Q.$$

We have observed that $\mathrm{Hom}(e_1e_2Q, E(B)) = 0$ as well as $\mathrm{Hom}(e_1e_2Q, E(B)) = 0$ which implies that $e_1e_2 = 0 = f_1f_2$. Hence

$$E(A) \cong e_1f_2Q \oplus e_2f_1Q \cong E(B)$$

and so

$$A \cong e_1f_2R \oplus e_2f_1R \cong B.$$

Finally, since $e_1f_2R \oplus e_2f_1R$ is isomorphic to a direct summand of R which is automorphism invariant, we obtain that A is automorphism invariant by [**8**, Lemma 4]. □

We finish the section with the following criterion.

THEOREM 4.11. *Let R be a right non-singular ring and Q be its the maximal right ring of quotients. Then the following is equivalent:*

(1) *R is right e-ADS,*
(2) *Either $eQ \ncong (1-e)Q$ for any idempotent $e \in R$ or $R \cong M_2(S)$ for a suitable right automorphism invariant ring S,*
(3) *Either $eQ \ncong (1-e)Q$ for any idempotent $e \in R$ or $R \cong T \times M_2(S)$ for a suitable self-injective ring T and a normal right automorphism invariant ring S.*

PROOF. (1) $\Rightarrow$ (2) If R is a right trivial e-ADS ring, then $Q \cong E(R)$ has no a decomposition $Q = A \oplus B$ with a isomorphic summand, which implies that $eQ \ncong (1-e)Q$ for any idempotent $e \in R$.

If R is a non-trivial e-ADS ring, then there exists a right automorphism invariant ring S such that $R \cong M_2(S)$ by Lemma 4.9.

(2) $\Rightarrow$ (3) Assume $R \cong M_2(S_0)$ for a right automorphism invariant ring S_0. Clearly, S_0 is, moreover, non-singular, hence there exists a right selfinjective ring S_1 and a normal right automorphism invariant ring S such that $S_0 \cong S_1 \times S$ by [**5**, Theorem 7]. Now it is easy to see that

$$M_2(S_0) \cong M_2(S_1) \times M_2(S)$$

and $T = M_2(S_0)$ is self-injective by [**7**, Corollary 9.3].

(3) $\Rightarrow$ (1) We remark that the first condition implies that R is a trivial e-ADS ring. Suppose that $R \cong T \times M_2(S)$ where T is a self-injective ring and S is a normal right automorphism invariant ring. Note that T is an e-ADS ring and $M_2(S)$ is e-ADS by Lemma 4.10. So, R is right e-ADS by Lemma 4.8. □

COROLLARY 4.12. *Every simple non-trivial right e-ADS ring is necessarily self-injective.*

PROOF. It follows from Theorem 4.11 and [**5**, Corollary 10]. □

References

[1] A. Alahmadi, S. K. Jain, and A. Leroy, *ADS modules*, J. Algebra **352** (2012), 215–222, DOI 10.1016/j.jalgebra.2011.10.035. MR2862183

[2] F. W. Anderson and K. R. Fuller, *Rings and categories of modules*, Springer-Verlag, New York-Heidelberg, 1974. Graduate Texts in Mathematics, Vol. 13. MR0417223

[3] P. A. Guil Asensio and A. K. Srivastava, *Automorphism-invariant modules satisfy the exchange property*, J. Algebra **388** (2013), 101–106, DOI 10.1016/j.jalgebra.2013.05.003. MR3061680

[4] N. V. Dung, D. V. Huynh, P. F. Smith, and R. Wisbauer, *Extending modules*, Pitman Research Notes in Mathematics Series, vol. 313, Longman Scientific & Technical, Harlow; copublished in the United States with John Wiley & Sons, Inc., New York, 1994. With the collaboration of John Clark and N. Vanaja. MR1312366

[5] N. Er, S. Singh, and A. K. Srivastava, *Rings and modules which are stable under automorphisms of their injective hulls*, J. Algebra **379** (2013), 223–229, DOI 10.1016/j.jalgebra.2013.01.021. MR3019253

[6] L. Fuchs, *Infinite abelian groups. Vol. I*, Pure and Applied Mathematics, Vol. 36, Academic Press, New York-London, 1970. MR0255673

[7] K. R. Goodearl, *von Neumann regular rings*, Monographs and Studies in Mathematics, vol. 4, Pitman (Advanced Publishing Program), Boston, Mass.-London, 1979. MR533669

[8] T.-K. Lee and Y. Zhou, *Modules which are invariant under automorphisms of their injective hulls*, J. Algebra Appl. **12** (2013), no. 2, 1250159, 9, DOI 10.1142/S0219498812501599. MR3005608

[9] S. H. Mohamed and B. J. Müller, *Continuous and discrete modules*, London Mathematical Society Lecture Note Series, vol. 147, Cambridge University Press, Cambridge, 1990. MR1084376

[10] W. K. Nicholson, *Semiregular modules and rings*, Canad. J. Math. **28** (1976), no. 5, 1105–1120, DOI 10.4153/CJM-1976-109-2. MR0422343

[11] W. K. Nicholson and Y. Zhou, *Semiregular morphisms*, Comm. Algebra **34** (2006), no. 1, 219–233, DOI 10.1080/00927870500346214. MR2194360

[12] T. C. Quynh and M. T. Koşan, *On ADS modules and rings*, Comm. Algebra **42** (2014), no. 8, 3541–3551, DOI 10.1080/00927872.2013.788185. MR3196060

[13] T. C. Quynh and M. T. Koşan, *On automorphism-invariant modules*, J. Algebra Appl. **14** (2015), no. 5, 1550074, 11, DOI 10.1142/S0219498815500747. MR3323335

[14] R. Wisbauer, *Foundations of module and ring theory*, Revised and translated from the 1988 German edition, Algebra, Logic and Applications, vol. 3, Gordon and Breach Science Publishers, Philadelphia, PA, 1991. A handbook for study and research. MR1144522

[15] J. M. Zelmanowitz, *Correspondences of closed submodules*, Proc. Amer. Math. Soc. **124** (1996), no. 10, 2955–2960, DOI 10.1090/S0002-9939-96-03469-7. MR1342050

Department of Mathematics, Faculty of Sciences, Gazi University, 06560 Ankara, Turkey

Email address: mtamerkosan@gazi.edu.tr

Department of Mathematics, Danang University, 459 Ton Duc Thang, DaNang city, Vietnam

Email address: tcquynh@live.com

Department of Algebra, Charles University in Prague, Faculty of Mathematics and Physics Sokolovská 83, 186 75 Praha 8, Czech Republic

Email address: zemlicka@karlin.mff.cuni.cz

Contemporary Mathematics
Volume **727**, 2019
https://doi.org/10.1090/conm/727/14638

A perspective on amalgamated rings via symmetricity

Handan Kose, Burcu Ungor, Yosum Kurtulmaz, and Abdullah Harmanci

ABSTRACT. In this paper, we deal with some versions of reversibility and symmetricity on amalgamated rings along an ideal.

1. Introduction

In [**3**], if R is a local Cohen-Macaulay ring, it is given necessary and sufficient conditions for $R \bowtie I$ to be a Gorenstein ring, where I is an ideal of R. The amalgamated duplication $R \bowtie E$ of a commutative ring R along an R-module E (that it is assumed to be an ideal in some overring of R and so E is an R-submodule of the total ring of fractions K of R), introduced by D'Anna and Fontana [**4**], is a subring $R \bowtie E = \{(r, r + e) \mid r \in R \text{ and } e \in E\}$ of $R \times K$ (endowed with the usual componentwise operations).

Let A and B be arbitrary rings with unity and I be a proper ideal of B and $f : A \to B$ be a ring homomorphism. We consider the following subring of $A \times B$:

$$A \bowtie^f I := \{(a, f(a) + x) \mid a \in A, x \in I\}$$

called *the amalgamation of A with B along I with respect to f*. This ring is a generalization of the amalgamated duplication of a commutative ring along an ideal which is introduced and studied by D'Anna and Fontana in [**4, 5**].

In [**5**], the authors studied characterizations for $A \bowtie^f I$ to be a Noetherian ring, an integral domain, a reduced ring and they characterized those distinguished pullbacks that can be expressed as an amalgamation provided the rings are commutative. Moreover, in [**4**], they investigated also the prime spectrum, chain properties and Krull dimension of $A \bowtie^f I$. This paper aims at studying the transfer of the notion of reversible rings, weakly reversible rings, symmetric rings, weak symmetric rings, generalized weakly symmetric rings and weakly symmetric rings to the amalgamation of rings along ideals. In addition to the Introduction, it is divided sections and each section is devoted to a notion related to the versions of reversibility or symmetricity.

2010 *Mathematics Subject Classification.* Primary 16N40, 16N80, 16U99.

Key words and phrases. Reversible ring, weakly reversible ring, symmetric ring, weak symmetric ring, GWS ring, weakly symmetric ring, amalgamated ring along an ideal.

The first author was supported by Ahi Evran University Scientific Research Projects Coordination Unit. Project Number: FEF.A4.16.001.

In the sequel, nil(A) will denote the set of all nilpotent elements of a ring A, also by $\mathbb{Z}$ and $\mathbb{Z}_n$, we denote the integers and the ring of integers modulo n for a positive integer n.

2. Reversibility of amalgamated rings

A ring R is called *reversible* if for any a, $b \in R$, $ab = 0$ implies $ba = 0$. In [**2**], Cohn studied reversible rings. Further studies are done in [**1**], [**7**], [**9**] and [**12**]. In this section, we investigate the conditions on the reversibility of the rings of the form $A \bowtie^f I$. We start with some examples to illustrate the definition.

EXAMPLE 2.1. Let $A = \mathbb{Z}_2$ and $B = \begin{bmatrix} \mathbb{Z}_2 & \mathbb{Z}_2 \\ 0 & \mathbb{Z}_2 \end{bmatrix}$ be the rings and $I = \begin{bmatrix} 0 & \mathbb{Z}_2 \\ 0 & 0 \end{bmatrix}$ the ideal of B and $f : A \to B$ defined by $f(a) = \begin{bmatrix} a & 0 \\ 0 & a \end{bmatrix}$ where $a \in \mathbb{Z}_2$. Then A is a reversible ring, B is not reversible. For $\begin{bmatrix} 0 & 0 \\ 0 & 1 \end{bmatrix}, \begin{bmatrix} 1 & 1 \\ 0 & 0 \end{bmatrix} \in B$, $\begin{bmatrix} 0 & 0 \\ 0 & 1 \end{bmatrix} \begin{bmatrix} 1 & 1 \\ 0 & 0 \end{bmatrix} = 0$ but $\begin{bmatrix} 1 & 1 \\ 0 & 0 \end{bmatrix} \begin{bmatrix} 0 & 0 \\ 0 & 1 \end{bmatrix} \neq 0$. Also $f(A) + I$ is reversible, and

$$A \bowtie^f I = \left\{ (0, \begin{bmatrix} 0 & 0 \\ 0 & 0 \end{bmatrix}), (0, \begin{bmatrix} 0 & 1 \\ 0 & 0 \end{bmatrix}), (1, \begin{bmatrix} 1 & 0 \\ 0 & 1 \end{bmatrix}), (1, \begin{bmatrix} 1 & 1 \\ 0 & 1 \end{bmatrix}) \right\}$$

is reversible.

THEOREM 2.2. *Let A and B be a pair of rings, $f : A \to B$ a ring homomorphism and I a proper ideal of B. Then the following hold.*

(1) *If $A \bowtie^f I$ is reversible, then A is reversible.*
(2) *If A and $f(A) + I$ are reversible, so is $A \bowtie^f I$.*
(3) *Assume that f is injective. If $f(A) + I$ is reversible, then so are A and $A \bowtie^f I$.*
(4) *Let $f^{-1}(I) = \{0\}$.*
    (a) *If B is reversible, then $A \bowtie^f I$ is reversible.*
    (b) *If $f(A) + I$ is reversible, then $A \bowtie^f I$ is reversible.*

PROOF. (1) Let a, $b \in A$ with $ab = 0$. Then $(a, f(a))(b, f(b)) = 0$ in $A \bowtie^f I$. By hypothesis, $(b, f(b))(a, f(a)) = 0$. So $ba = 0$.
(2) Let $(a, f(a)+x)$, $(b, f(b)+y) \in A \bowtie^f I$ with $(a, f(a)+x)(b, f(b)+y) = 0$. Then $ab = 0$ and $(f(a)+x)(f(b)+y) = 0$. By hypothesis, $ba = 0$ and $(f(b)+y)(f(a)+x) = 0$. It follows that $A \bowtie^f I$ is reversible.
(3) We first show that A is reversible. Let a, $b \in A$ with $ab = 0$. Then $f(a) + 0, f(b) + 0 \in f(A) + I$ and $(f(a) + 0)(f(b) + 0) = f(ab) = 0$. By hypothesis, $(f(b)+0)(f(a)+0) = 0 = f(ba)$. By the injectivity of f, $ba = 0$. So A is reversible. By (2), $A \bowtie^f I$ is reversible.
(4) (a) Note that $f(A) + I$ is reversible as a subring of the reversible ring B. It is clear that $f(A) + I$ is isomorphic to $A \bowtie^f I$ by the homomorphism α defined by $\alpha(a, f(a) + x) = f(a) + x$ where $(a, f(a) + x) \in A \bowtie^f I$.
(b) Clear by the property that reversibility is preserved under isomorphism. $\square$

EXAMPLE 2.3. Let $A = \mathbb{Z}_2$ and $B = \begin{bmatrix} \mathbb{Z}_2 & 0 & \mathbb{Z}_2 \\ 0 & \mathbb{Z}_2 & 0 \\ 0 & 0 & \mathbb{Z}_2 \end{bmatrix}$ be the rings and $I = \begin{bmatrix} 0 & 0 & 0 \\ 0 & \mathbb{Z}_2 & 0 \\ 0 & 0 & 0 \end{bmatrix}$ the ideal of B and $f : A \to B$ defined by $f(a) = \begin{bmatrix} a & 0 & 0 \\ 0 & a & 0 \\ 0 & 0 & a \end{bmatrix}$. Then A is reversible and B is not reversible. Also f is injective and

$$f(A) + I = \left\{ \begin{bmatrix} 0 & 0 & 0 \\ 0 & 0 & 0 \\ 0 & 0 & 0 \end{bmatrix}, \begin{bmatrix} 0 & 0 & 0 \\ 0 & 1 & 0 \\ 0 & 0 & 0 \end{bmatrix}, \begin{bmatrix} 1 & 0 & 0 \\ 0 & 1 & 0 \\ 0 & 0 & 1 \end{bmatrix}, \begin{bmatrix} 1 & 0 & 0 \\ 0 & 0 & 0 \\ 0 & 0 & 1 \end{bmatrix} \right\}$$

is reversible. By Theorem 2.2 (3), $A \bowtie^f I$ is a reversible ring.

The following example shows that the reverse implication of (1) in Theorem 2.2 need not be true in general.

EXAMPLE 2.4. Let $A = \mathbb{Z}_2$ and $B = \begin{bmatrix} \mathbb{Z}_2 & \mathbb{Z}_2 \\ 0 & \mathbb{Z}_2 \end{bmatrix}$ be the rings and $I = \begin{bmatrix} 0 & \mathbb{Z}_2 \\ 0 & \mathbb{Z}_2 \end{bmatrix}$ the ideal of B and $f : A \to B$ defined by $f(a) = \begin{bmatrix} a & 0 \\ 0 & a \end{bmatrix}$ where $a \in \mathbb{Z}_2$. Let $u = (0, \begin{bmatrix} 0 & 1 \\ 0 & 0 \end{bmatrix})$, $v = (1, \begin{bmatrix} 1 & 1 \\ 0 & 0 \end{bmatrix}) \in A \bowtie^f I$. Then $uv = 0$ and $vu \neq 0$. Hence $A \bowtie^f I$ is not reversible. Similarly, let $x = \begin{bmatrix} 0 & 1 \\ 0 & 1 \end{bmatrix}$, $y = \begin{bmatrix} 1 & 0 \\ 0 & 0 \end{bmatrix} \in f(A) + I$. Then $xy = 0$ and $yx \neq 0$. Hence $f(A) + I$ is not reversible.

The following lemma is clear from definitions.

LEMMA 2.5. *Let A and B be a pair of rings, $f : A \to B$ a ring homomorphism and I a proper ideal of B. Then the following hold.*

(1) *If A and B are reversible rings, then $A \bowtie^f I$ is reversible.*
(2) *Let $B = A$, $f = id_A$ and $I = (0)$. In this situation, $A \bowtie^f I$ is reversible if and only if A and B are reversible.*
(3) *If $A \bowtie^f I$ is reversible, then every subring is reversible, in particular, the ideals $(0) \bowtie^f I$, $A \bowtie^f (0)$ and $I_1 \bowtie^f I$ are reversible where I_1 is any ideal of A.*

DEFINITION 2.6. We recall that, if $\alpha : A \to C$, $\beta : B \to C$ are ring homomorphisms, the subring $D := \alpha \times_C \beta := \{(a, b) \in A \times B \mid \alpha(a) = \beta(b)\}$ of $A \times B$ is called the *pullback* (or *fiber product*) of α and β (see [**5**]).

THEOREM 2.7. *Let A and B be rings and I an ideal of B and $f : A \to B$ a homomorphism. Let $u : A \to A \times \big((f(A)+I)/I\big)$ defined by $u(a) = (a, f(a)+I)$ and $v : A \times (f(A) + I) \to A \times \big((f(A) + I)/I\big)$ defined by $v(a, f(b) + x) = (a, f(b) + I)$, where $a, b \in A$ and $x \in I$. Then $A \bowtie^f I$ is the pullback of the maps u and v.*

PROOF. Consider the following diagram with $\alpha(a, f(a)+t) = a$ and $\beta(a, f(a)+t) = (a, f(a)+t) \in A \times (f(A)+I)$ where $(a, f(a)+t) \in A \bowtie^f I$.

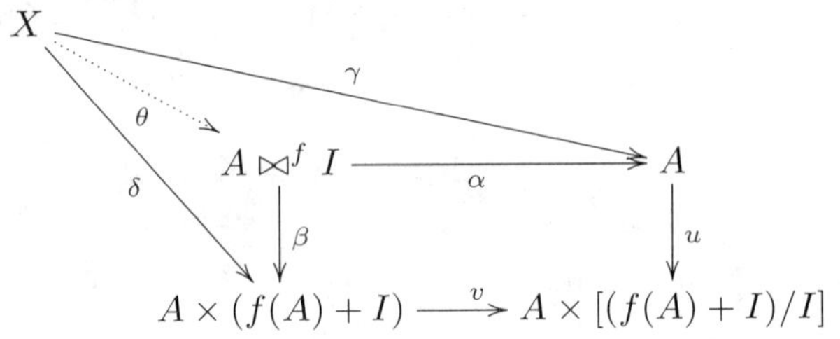

Then $u\alpha = v\beta$. For if $(a, f(a)+t) \in A \bowtie^f I$, then $u\alpha(a, f(a)+t) = u(a) = (a, f(a)+I)$, $v\beta(a, f(a)+t) = (a, f(a)+I)$. Let X be any ring and γ and δ ring homomorphisms such that $v\delta = u\gamma$. For any $x \in X$, set $\gamma(x) = a \in A$. Then $u(a) = (a, f(a)+I)$ and $\delta(x) = (b, f(c)+i)$ for some b, $c \in A$ and $i \in I$. Hence $v\delta = u\gamma$ implies $a = b$ and $f(c) - f(a) \in I$. Define $\theta(x) = (a, f(a)+i)$. Then $\beta\theta = \delta$ and $\alpha\theta = \gamma$. On the other hand, θ is unique, for if ν satisfies $\alpha\nu = \gamma$ and $\beta\nu = \delta$, then $\nu = \theta$ since β is a monomorphism. This completes the proof. □

PROPOSITION 2.8. *With the notation of Definition 2.6, we have:*

(1) *If A is reversible and β is injective, then D is reversible.*
(2) *If at least one of the following conditions holds*
  (a) *A is reversible and $nil(B) \cap Ker\beta = \{0\}$,*
  (b) *B is reversible and $nil(A) \cap Ker\alpha = \{0\}$,*

  then D is reversible.

PROOF. (1) Assume that A is reversible. Let (a, b) and (c, d) be in D with $(a, b)(c, d) = 0$, then $ac = 0$, $bd = 0$. We have $ca = 0$ and $\alpha(ca) = \alpha(c)\alpha(a) = \beta(d)\beta(b) = \beta(db) = 0$, then $db = 0$ since β is injective. So D is reversible.
(2) By the symmetry of conditions (a) and (b), it is enough to show that condition (a) holds. Let (a, b) and (c, d) be in D with $(a, b)(c, d) = 0$. Then $ac = 0$ and $bd = 0$. Since A is reversible, we have $ca = 0$. Also, $\beta(db) = \beta(d)\beta(b) = \alpha(c)\alpha(a) = 0$, so $db \in Ker\beta$. Also $(db)^2 = dbdb = 0$, this implies that $db \in \text{nil}(B) \cap Ker\beta = \{0\}$. Therefore D is reversible. □

3. Weakly reversibility of amalgamated rings

Reversible rings are generalized by Liang-Gang in [**11**], that is, a ring R is called *weakly reversible* if for all $a, b, r \in R$ such that $ab = 0$, $Rbra$ is a nil left ideal of R (equivalently, $braR$ is a nil right ideal of R). Weakly reversible rings are also studied in [**7**]. Reversible rings are weakly reversible. There are weakly reversible rings that are not reversible as the following example shows.

EXAMPLE 3.1. Let $A = \mathbb{Z}_2$ and $B = \begin{bmatrix} \mathbb{Z}_2 & \mathbb{Z}_2 \\ 0 & \mathbb{Z}_2 \end{bmatrix}$ be the rings and $I = \begin{bmatrix} 0 & \mathbb{Z}_2 \\ 0 & \mathbb{Z}_2 \end{bmatrix}$ the ideal of B and $f : A \to B$ defined by $f(a) = \begin{bmatrix} a & 0 \\ 0 & a \end{bmatrix}$ where $a \in \mathbb{Z}_2$. Then f is injective, $f(A) \cap I = \{0\}$ and A is weakly reversible. By [**11**, Proposition 2.3], B is weakly reversible. Note that $A \bowtie^f I$ has the following elements
$(0, \begin{bmatrix} 0 & 0 \\ 0 & 0 \end{bmatrix}), (0, \begin{bmatrix} 0 & 1 \\ 0 & 0 \end{bmatrix}), (0, \begin{bmatrix} 0 & 0 \\ 0 & 1 \end{bmatrix}), (0, \begin{bmatrix} 0 & 1 \\ 0 & 1 \end{bmatrix}), (1, \begin{bmatrix} 1 & 0 \\ 0 & 1 \end{bmatrix}), (1, \begin{bmatrix} 1 & 1 \\ 0 & 1 \end{bmatrix}),$

$(1, \begin{bmatrix} 1 & 0 \\ 0 & 0 \end{bmatrix}), (1, \begin{bmatrix} 1 & 1 \\ 0 & 0 \end{bmatrix})$. It is easy to check that $A \bowtie^f I$ is weakly reversible. Now $(0, \begin{bmatrix} 0 & 0 \\ 0 & 1 \end{bmatrix})(1, \begin{bmatrix} 1 & 1 \\ 0 & 0 \end{bmatrix}) = 0$ in $A \bowtie^f I$, but $(1, \begin{bmatrix} 1 & 1 \\ 0 & 0 \end{bmatrix})(0, \begin{bmatrix} 0 & 0 \\ 0 & 1 \end{bmatrix}) \neq 0$. So $A \bowtie^f I$ is not reversible.

THEOREM 3.2. *Let A and B be rings, $f : A \to B$ a ring homomorphism and I a proper ideal of B. Then the following hold.*

(1) *Assume that f is injective. If $A \bowtie^f I$ is weakly reversible and $f(A) \cap I = \{0\}$, then both A and $f(A) + I$ are weakly reversible.*
(2) *If A and $f(A) + I$ are weakly reversible, so is $A \bowtie^f I$.*
(3) *Assume that f is injective. If $f(A) + I$ is weakly reversible, then so are A and $A \bowtie^f I$.*
(4) *Let $f^{-1}(I) = \{0\}$.*
    (a) *If B is weakly reversible, then $A \bowtie^f I$ is weakly reversible.*
    (b) *If $f(A) + I$ is weakly reversible, then $A \bowtie^f I$ is weakly reversible.*

PROOF. (1) Let $a, b \in A$ with $ab = 0$. Then $(a, f(a) + 0)(b, f(b) + 0) = 0$ in $A \bowtie^f I$. By hypothesis, $(A \bowtie^f I)(b, f(b)+0)(c, f(c)+z)(a, f(a)+0)$ is a nil left ideal of $A \bowtie^f I$ for all $(c, f(c) + z) \in A \bowtie^f I$. Since operations are componentwise, $Abca$ is a nil left ideal of A, and so A is weakly reversible. Let $f(a)+x$, $f(b)+y \in f(A)+I$ with $(f(a) + x)(f(b) + y) = 0$. Then $(a, f(a) + x)(b, f(b) + y) = 0$. By hypothesis, $(A \bowtie^f I)(b, f(b) + y)(c, f(c) + z)(a, f(a) + x)$ is a nil left ideal of $A \bowtie^f I$. Then $(f(A) + I)(f(b) + y)(f(c) + z)(f(a) + x)$ is a nil left ideal of $f(A) + I$.
(2) Let $(a, f(a)+x), (b, f(b)+y) \in A \bowtie^f I$ with $(a, f(a)+x)(b, f(b)+y) = 0$. Hence $ab = 0$ and $(f(a)+x)(f(b)+y) = 0$. Since A and $f(A)+I$ are weakly reversible, we have $Abca$ is a nil left ideal of A for all $c \in A$ and $(f(A)+I)(f(b)+y)(f(c)+z)(f(a)+x)$ is a nil left ideal of $f(A)+I$ for all $f(c)+z \in f(A)+I$. Thus there exist $m > 0$ and $n > 0$ such that $(rbca)^m = 0$ for all $r \in A$ and $[(f(r)+s)(f(b)+y)(f(c)+z)(f(a)+x)]^n = 0$. For $k = max\{n, m\}$, $[(r, f(r)+s)(b, f(b)+y)(c, f(c)+z)(a, f(a)+x)]^k = 0$. We have $(A \bowtie^f I)(b, f(b) + y)(c, f(c) + z)(a, f(a) + x)$ is a nil left ideal of $A \bowtie^f I$ for all $(c, f(c) + z) \in A \bowtie^f I$.
(3) It is enough to show that A is weakly reversible. Let $a, b \in A$ with $ab = 0$. Since $f(A) + I$ is weakly reversible, $(f(A) + I)f(b)(f(c) + x)f(a)$ is a nil left ideal of $f(A)+I$. That is, $(f(r)+s)f(b)(f(c)+x)f(a)$ is a nilpotent element of $f(A)+I$ for all $r, c \in A$ and $s, x \in I$. In particular, for $s = x = 0$ we have $f(r)f(b)f(c)f(a)$ is nilpotent. So there exists $m > 0$ such that $f((rbca)^m) = 0$. By hypothesis, $(rbca)^m = 0$ in A. So A is weakly reversible. By (2), $A \bowtie^f I$ is also weakly reversible.
(4)(a) Let $a, b \in A$ with $ab = 0$. Then $f(a)f(b) = 0$ in B. Since the subring $f(A)$ of B is weakly reversible, we have $f(A)f(b)f(c)f(a)$ is a nil left ideal of $f(A)$ for all $c \in A$. So there exists $m > 0$ such that $f((Abca)^m) = 0 \in I$. Thus $(Abca)^m \subseteq f^{-1}(I) = \{0\}$, we have $(Abca)^m = 0$. That is, $Abca$ is a nilpotent left ideal, hence it is a nil left ideal of A. So A is weakly reversible. By (2), we have that $A \bowtie^f I$ is weakly reversible.
(b) It is clear from (a). □

There are rings A and B with an injective homomorphism $f : A \to B$ and an ideal I of B with $f(A) \cap I = 0$ but $A \bowtie^f I$ is not weakly reversible.

EXAMPLE 3.3. Let $A = \mathbb{Z}_2$ and $X = \begin{bmatrix} \mathbb{Z}_2 & \mathbb{Z}_2 \\ \mathbb{Z}_2 & \mathbb{Z}_2 \end{bmatrix}$ be the rings and $Y = \begin{bmatrix} \mathbb{Z}_2 & \mathbb{Z}_2 \\ 0 & \mathbb{Z}_2 \end{bmatrix}$ and $B = \begin{bmatrix} X & 0 \\ 0 & Y \end{bmatrix}$ the ring direct sum of X and Y. Let $I = \begin{bmatrix} X & 0 \\ 0 & 0 \end{bmatrix}$ be the ideal of B and $f(a) = a(e_{11} + e_{22} + e_{33} + e_{44})$ for $a \in A$ where e_{ij} denotes the 4×4 matrix units in B for each i, j. Let $U = (1, e_{11} + e_{12} + e_{33} + e_{44})$, $V = (0, e_{12} + e_{22})$ and $T = (0, e_{12} + e_{21})$. Then $UV = 0$. But $VTU = (0, e_{11} + e_{12} + e_{21} + e_{22})$ is a nilpotent element. Then f is injective and $f(A) \cap I = \{0\}$. On the other hand, $(A \bowtie^f I)VTU$ is not a nil left ideal. For if $Z = (0, e_{11})$, then $ZVTU = (0, e_{11} + e_{12})$ and $(ZVTU)^2 = ZVTU$. Hence $A \bowtie^f I$ is not weakly reversible.

4. Symmetricity of amalgamated rings

Symmetric rings are defined by Lambek in [**10**]. A ring R is called *symmetric* if $abc = 0$ implies $acb = 0$ for a, b, $c \in R$. In [**6**], this concept is extended to the central symmetric ring, that is, if $abc = 0$ implies acb is central in R for a, b, $c \in R$. Clearly the symmetric property of rings are preserved under isomorphisms and under subrings. In this section we study necessary and sufficient conditions for $A \bowtie^f I$ to be symmetric.

THEOREM 4.1. *Let A and B be a pair of rings, $f : A \to B$ a ring homomorphism and I a proper ideal of B. Then the following hold.*

(1) *If $A \bowtie^f I$ is symmetric, then so is A.*
(2) *If A and $f(A) + I$ are symmetric, then so is $A \bowtie^f I$.*
(3) *Assume that $I \cap S \neq \emptyset$ where S is the set of regular central elements of B. Then $A \bowtie^f I$ is a symmetric ring if and only if $f(A) + I$ and A are symmetric rings.*
(4) *Assume that f is injective and $f(A)+I$ is a symmetric ring. Then $A \bowtie^f I$ is a symmetric ring.*
(5) *Assume that $f^{-1}(I) = \{0\}$. If $f(A) + I$ is a symmetric ring, then $A \bowtie^f I$ is a symmetric ring.*

PROOF. (1) Assume that $A \bowtie^f I$ is symmetric. Let a, b, $c \in A$ with $abc = 0$. Then $(a, f(a))(b, f(b))(c, f(c)) = 0$ in $A \bowtie^f I$. By assumption, $(a, f(a))(c, f(c))(b, f(b)) = 0$ from which we have $acb = 0$. So A is symmetric.
(2) Assume that A and $f(A) + I$ are symmetric. Let $(a, f(a) + x)$, $(b, f(b) + y)$, $(c, f(c) + z) \in A \bowtie^f I$ with $(a, f(a) + x)(b, f(b) + y)(c, f(c) + z) = 0$. Then $abc = 0$ and $(f(a) + x)(f(b) + y)(f(c) + z) = 0$. By assumption, $acb = 0$ and $(f(a) + x)(f(c) + z)(f(b) + y) = 0$. Hence $(a, f(a) + x)(c, f(c) + z)(b, f(b) + y) = 0$. So $A \bowtie^f I$ is symmetric.
(3) It is enough to show that if $A \bowtie^f I$ is symmetric, then $f(A) + I$ is symmetric. For if $s \in S \cap I$ and $f(a) + x$, $f(b) + y$, $f(c) + z \in f(A) + I$ such that $(f(a) + x)(f(b) + y)(f(c) + z) = 0$, then $(0, s(f(a) + x))(0, s(f(b) + y))(0, s(f(c) + z)) = 0$. By hypothesis, $(0, s(f(a) + x))(0, s(f(c) + z))(0, s(f(b) + y)) = 0$. Hence $s^3((f(a) + x)(f(c) + z)(f(b) + y)) = 0$. By the regularity of s, we have $(f(a) + x)(f(c) + z)(f(b) + y) = 0$.
(4) Assume that f is injective and $f(A) + I$ is symmetric. Let $a, b, c \in A$ with $abc = 0$. Hence $f(a)f(b)f(c) = 0$ in $f(A) + I$. Since $f(A) + I$ is symmetric, $f(a)f(c)f(b) = 0$. Then $f(acb) = 0$. By hypothesis, $acb = 0$ in A. So A is a symmetric ring. By (2), we have that $A \bowtie^f I$ is symmetric.

(5) Assume that $f^{-1}(I) = \{0\}$ and $f(A)+I$ is a symmetric ring. It is enough to show that A is symmetric. Let $abc = 0$ for $a, b, c \in A$. Then $f(a)f(b)f(c) = 0$ in $f(A)+I$. Since the subring $f(A)$ of $f(A) + I$ is symmetric, we have $f(a)f(c)f(b) = 0 \in I$. Thus $acb \in f^{-1}(I) = \{0\}$. Hence $acb = 0$. That is, A is symmetric. By (2), $A \bowtie^f I$ is symmetric. □

EXAMPLE 4.2. Let x, y and z be indeterminates and consider the set

$$R = \{a_0 + a_1x + a_2y + a_3z | a_0, a_1, a_2, a_3 \in \mathbb{Z}\}$$

with componentwise addition and defining multiplication

$$(a_0 + a_1x + a_2y + a_3z)(b_0 + b_1x + b_2y + b_3z) =$$
$$a_0b_0 + (a_0b_1 + a_1b_0)x + (a_0b_2 + a_2b_0)y + (a_0b_3 + a_3b_0 + a_1b_2)z.$$

Then R is a ring with identity. From this multiplication all products are zero except that $xy = z$ and that 1 acts as an identity. Let $A = B = R$. For every ring homomorphism $f\colon A \to B$ and every proper ideal I of B, $A \bowtie^f I$ is not symmetric because A is not symmetric [**6**, Example 2.2].

5. Weak symmetricity of amalgamated rings

In this section we study weak symmetric rings. In [**13**], Ouyang and Chen discussed weak symmetric rings. A ring R is called *weak symmetric* if $abc \in \text{nil}(R)$ implies $acb \in \text{nil}(R)$ for all a, b, $c \in R$. In [**13**], it is proved that all symmetric rings are weak symmetric. Clearly the weak symmetric property of rings are preserved under isomorphisms and under subrings. In this section we study necessary and sufficient conditions for $A \bowtie^f I$ to be weak symmetric.

THEOREM 5.1. *Let A and B be a pair of rings, $f : A \to B$ a ring homomorphism and I a proper ideal of B. Then the following hold.*

(1) *If $A \bowtie^f I$ is weak symmetric, then so is A.*
(2) *If A and $f(A) + I$ are weak symmetric, then so is $A \bowtie^f I$.*
(3) *Assume that $I \cap S \neq \emptyset$ where S is the set of regular central elements of B. Then $A \bowtie^f I$ is a weak symmetric ring if and only if $f(A) + I$ and A are weak symmetric rings.*
(4) *Assume that $f^{-1}(I) \subseteq$ nil(A). If $f(A) + I$ is a weak symmetric ring, then $A \bowtie^f I$ is a weak symmetric ring.*
(5) *If $f(A) + I$ is a weak symmetric ring and f is injective, then A and I are weak symmetric.*

PROOF. (1) Assume that $A \bowtie^f I$ is weak symmetric. Let a, b, $c \in A$ such that abc is nilpotent. Then $(a, f(a))(b, f(b))(c, f(c)) = (abc, f(abc))$ is nilpotent in $A \bowtie^f I$. By assumption $(a, f(a))(c, f(c))(b, f(b)) = (acb, f(acb))$ is nilpotent. So acb is nilpotent. It follows that A is weak symmetric.
(2) Assume that A and $f(A) + I$ are weak symmetric. Let $(a, f(a) + x)(b, f(b) + y)(c, f(c)+z)$ be nilpotent in $A \bowtie^f I$. Then $abc \in A$ and $(f(a)+x)(f(b)+y)(f(c)+z) \in f(A)+I$ are nilpotent. By assumption, acb and $(f(a)+x)(f(c)+z)(f(b)+y)$ are nilpotent respectively, in A and $f(A)+I$. Hence $(a, f(a)+x)(c, f(c)+z)(b, f(b)+y)$ is nilpotent in $A \bowtie^f I$ and so $A \bowtie^f I$ is weak symmetric.
(3) For the necessity: By (1), A is weak symmetric. To prove $f(A) + I$ is weak symmetric, let $f(a)+x$, $f(b)+y$, $f(c)+z \in f(A)+I$ with $(f(a)+x)(f(b)+y)(f(c)+z)$

nilpotent and $s \in I \cap S$. Then

$$(0, f(0) + s(f(a) + x))(0, f(0) + s(f(b) + y))(0, f(0) + s(f(c) + z))$$

is nilpotent in $A \bowtie^f I$. By hypothesis,

$$(0, f(0) + s(f(a) + x))(0, f(0) + s(f(c) + z))(0, f(0) + s(f(b) + y))$$

is nilpotent in $A \bowtie^f I$. Then $s^3(f(a) + x)(f(c) + z)(f(b) + y)$ is nilpotent and $(f(a)+x)(f(c)+z)(f(b)+y)$ is nilpotent since s is central regular. Hence $f(A)+I$ is weak symmetric.
The sufficiency is clear from (2).
(4) Assume that $f^{-1}(I) \subseteq \text{nil}(A)$. It is sufficient to prove that A is weak symmetric. Let $a, b, c \in A$ with $abc \in \text{nil}(A)$. So there exists $m > 0$ such that $(abc)^m = 0$. Hence $f(a)f(b)f(c)$ is a nilpotent element of $f(A)+I$. Since $f(A)+I$ is weak symmetric, $f(a)f(c)f(b)$ is nilpotent. So there exists $n > 0$ such that $f(acb)^n = 0 \in I$. Thus $(acb)^n \in f^{-1}(I)$. By assumption $(acb)^n \in \text{nil}(A)$. So acb is a nilpotent element of A. By (2), we have that $A \bowtie^f I$ is weak symmetric.
(5) Let $a, b, c \in A$ with abc nilpotent. So $f(a)f(b)f(c)$ is nilpotent in $f(A)+I$. Since $f(A)+I$ is weak symmetric, $f(a)f(c)f(b)$ is nilpotent. So there exists $m > 0$ such that $(f(a)f(c)f(b))^m = 0 = f((acb)^m)$. By hypothesis $(acb)^m = 0$. We have that A is weak symmetric. Since the weak symmetric property of rings are preserved under subrings, I is weak symmetric. This completes the proof. □

EXAMPLE 5.2. Let R be a domain and $T_3(R)$ the ring of 3×3 upper triangular matrices. Consider the ideal $I = \begin{bmatrix} R & R & R \\ 0 & 0 & 0 \\ 0 & 0 & 0 \end{bmatrix}$ of $T_3(R)$. Let $A = T_3(R)$ and $B = T_3(R)/I$ and $f\colon A \to B$ be the natural projection. By [**13**, Example 2.2.], A is weak symmetric. Since B is isomorphic to the ring of 2×2 upper triangular matrices, B is also weak symmetric by [**13**, Proposition 2.3]. Being $f(A) + I = B$ and Theorem 5.1(2) imply $A \bowtie^f I$ is weak symmetric.

6. Generalized weakly symmetricity of amalgamated rings

In [**14**], generalized weakly symmetric rings (or GWS, for short) are discussed. A ring R is called *GWS* if $abc = 0$ implies bac is nilpotent for a, b, $c \in R$. Clearly, $abc = 0$ implies bac is nilpotent for a, b, $c \in R$ if and only if $abc = 0$ implies acb is nilpotent for a, b, $c \in R$ if and only if $abc = 0$ implies $Racrb$ is a nil left ideal of R for any $r \in R$ if and only if $abc = 0$ implies $brcaR$ is a nil right ideal of R for any $r \in R$. The class of GWS rings is a proper generalization of that of central symmetric rings.

THEOREM 6.1. *Let A and B be a pair of rings, $f : A \to B$ a ring homomorphism and I a proper ideal of B. Then the following hold.*

(1) *If $A \bowtie^f I$ is GWS, then A is GWS.*
(2) *If A and $f(A) + I$ are GWS, so is $A \bowtie^f I$.*
(3) *Assume that f is injective. If $f(A)+I$ is GWS, then so are A and $A \bowtie^f I$.*
(4) *Let $f^{-1}(I) = \{0\}$.*
    (a) *If B is GWS, then $A \bowtie^f I$ is GWS.*
    (b) *If $f(A) + I$ is GWS, then $A \bowtie^f I$ is GWS.*

PROOF. (1) Assume that $A \bowtie^f I$ is GWS. Let a, b, $c \in A$ with $abc = 0$. Then $(a, f(a) + 0)(b, f(b) + 0)(c, f(c) + 0) = 0$. By assumption, $(a, f(a) + 0)(c, f(c) + 0)(b, f(b) + 0)$ is nilpotent. Then acb is nilpotent.
(2) Suppose that A and $f(A)+I$ are GWS and $(a, f(a)+x)(b, f(b)+y)(c, f(c)+z) = 0$ for some $(a, f(a) + x)$, $(b, f(b) + y)$, $(c, f(c) + z) \in A \bowtie^f I$. Then $abc = 0$ and $(f(a) + x)(f(b) + y)(f(c) + z) = 0$. By supposition, acb and $(f(a) + x)(f(c) + z)(f(b)+y)$ are nilpotent with $(acb)^n = 0$ and $\big((f(a)+x)(f(c)+z)(f(b)+y)\big)^t = 0$ for some positive integers n and t. Let $m = max\{n, t\}$. Then $\big((a, f(a)+x)(c, f(c)+z)(b, f(b) + y)\big)^m = 0$. Hence $(a, f(a) + x)(c, f(c) + z)(b, f(b) + y)$ is nilpotent, and so $A \bowtie^f I$ is GWS.
(3) Let a, b, $c \in A$ with $abc = 0$. Then $(f(a) + 0)(f(b) + 0)(f(c) + 0) = 0$. By hypothesis, $(f(a) + 0)(f(c) + 0)(f(b) + 0)$ is nilpotent and so acb is nilpotent since f is injective. Then A is GWS. By (2), $A \bowtie^f I$ is GWS.
(4) Let $f^{-1}(I) = \{0\}$. (a) Assume that B is GWS, then $f(A) + I$ is GWS since GWS property is preserved under subrings. Let a, b, $c \in A$ with $abc = 0$. Then $f(a)f(b)f(c) = f(abc)$ is zero. Hence $f(A) + I$ being GWS implies that $f(acb)$ is nilpotent. So there exists $n > 0$ such that $f(acb)^n = 0 \in I$. Thus $(acb)^n \in f^{-1}(I)$. Therefore $(acb)^n = 0$ and so A is GWS. By (2), $A \bowtie^f I$ is GWS.
(b) It is clear from (4)(a). □

EXAMPLE 6.2. Let F be a field and consider the ring $R = \begin{bmatrix} F & F & F \\ 0 & F & F \\ 0 & 0 & F \end{bmatrix}$ and the ideal $I = \begin{bmatrix} 0 & 0 & F \\ 0 & 0 & F \\ 0 & 0 & F \end{bmatrix}$. Let $A = R$ and $B = R/I$ and f be the natural projection $\pi\colon A \to B$. By [**14**, Theorem 2.7] A is GWS. Since B is isomorphic to the ring of 2×2 upper triangular matrices over F, B is GWS. Note that $f(A)+I = B$. Then $A \bowtie^f I$ is GWS by Theorem 6.1(2).

7. Weakly symmetricity of amalgamated rings

The present authors discussed weakly symmetric rings in [**8**]. A ring R is *weakly symmetric* if for all a, b, c, $r \in R$, abc is nilpotent implies $Racrb$ is a nil left ideal of R (equivalently, $acrbR$ is a nil right ideal of R). Every symmetric ring is reversible, similarly every weakly symmetric ring is weakly reversible. In fact, if R is a weakly symmetric ring and $ab = 0$, then $1ab = 0$ and so $R1bra$ is a nil left ideal for all $r \in R$. Hence $Rbra$ is a nil left ideal of R.

THEOREM 7.1. *Let A and B be rings, $f : A \to B$ a ring homomorphism and I a proper ideal of B. Then the following hold.*

(1) *If $A \bowtie^f I$ is weakly symmetric, then A is weakly symmetric.*
(2) *If A and $f(A) + I$ are weakly symmetric, so is $A \bowtie^f I$.*
(3) *Assume that f is injective. If $f(A) + I$ is weakly symmetric, then so are A and $A \bowtie^f I$.*
(4) *Let $f^{-1}(I) = \{0\}$.*
    (a) *If B is weakly symmetric, then $A \bowtie^f I$ is weakly symmetric.*
    (b) *If $f(A) + I$ is weakly symmetric, then $A \bowtie^f I$ is weakly symmetric.*

PROOF. (1) Assume that $A \bowtie^f I$ is weakly symmetric. Let a, b, $c \in R$. If abc is nilpotent, then $(a, f(a)+0)(b, f(b)+0)(c, f(c)+0)$ is nilpotent in $A \bowtie^f I$. So $(A \bowtie^f I)(a, f(a)+0)(c, f(c)+0)(x, f(x)+t)(b, f(b)+0)$ is a nil left ideal of $A \bowtie^f I$ for all $(x, f(x)+t) \in A \bowtie^f I$. Then $Aacxb$ is a nil left ideal for all $x \in A$.
(2) Suppose that A and $f(A)+I$ are weakly symmetric. Let $(a, f(a)+x)$, $(b, f(b)+y)$, $(c, f(c)+z) \in A \bowtie^f I$. If $(a, f(a)+x)(b, f(b)+y)(c, f(c)+z)$ is nilpotent, then abc and $(f(a)+x)(f(b)+y)(f(c)+z)$ are nilpotent in A and $f(A)+I$, respectively. Hence $Aacxb$ is a nil left ideal of A for all $x \in A$ and $(f(A)+I)(f(a)+x)(f(c)+z)(f(d)+v)(f(b)+y)$ is a nil left ideal of $f(A)+I$ for all $f(d)+v \in f(A)+I$. Thus $(A \bowtie^f I)(a, f(a)+x)(c, f(c)+z)(d, f(d)+v)(b, f(b)+y)$ is a nil left ideal of $A \bowtie^f I$.
(3) Assume that f is injective and $f(A)+I$ is weakly symmetric. Let a, b, $c \in A$. If abc is nilpotent, then $(f(a)+0)(f(b)+0)(f(c)+0)$ is nilpotent in $f(A)+I$. By assumption, $(f(A)+I)(f(a)+0)(f(c)+0)(f(d)+z)(f(b)+0)$ is a nil left ideal of $f(A)+I$ for all $f(d)+z \in f(A)+I$. It implies that $f(Aacdb)$ is a nil subset of $f(A)+I$. Since f is injective, $Aacdb$ is a nil left ideal of A for all $d \in A$. Since A and $f(A)+I$ are weakly symmetric, by (2), $A \bowtie^f I$ is weakly symmetric.
(4) (a) Suppose that $f^{-1}(I) = \{0\}$ and B is weakly symmetric. Let $(a, f(a)+x)$, $(b, f(b)+y)$, $(c, f(c)+z) \in A \bowtie^f I$. If $(a, f(a)+x)(b, f(b)+y)(c, f(c)+z)$ is nilpotent, then abc and $(f(a)+x)(f(b)+y)(f(c)+z)$ are nilpotent in A and $f(A)+I$, respectively. The ring B being weakly symmetric, $Bf(a)f(c)xf(b)$ is a nil left ideal of B for all $x \in B$. For any t, $d \in A$, $f(tacdb)$ is nilpotent, say $f((tacdb)^n) = 0$ for some positive integer n. Then $(tacdb)^n \in f^{-1}(I)$. Hence $(tacdb)^n = 0$. Thus $Aacdb$ is a nil left ideal of A. On the other hand, any subring of a weakly symmetric ring is weakly symmetric. Therefore $f(A)+I$ is weakly symmetric. By (2), $A \bowtie^f I$ is weakly symmetric.
(b) Suppose that $f^{-1}(I) = \{0\}$ and $f(A)+I$ is weakly symmetric. Let $(a, f(a)+x)$, $(b, f(b)+y)$, $(c, f(c)+z) \in A \bowtie^f I$. Assume that $(a, f(a)+x)(b, f(b)+y)(c, f(c)+z)$ is nilpotent in $A \bowtie^f I$. Then abc, $f(a)f(b)f(c)$ and $(f(a)+x)(f(b)+y)(f(c)+z)$ are nilpotent. Hence $f(A)f(a)f(c)f(d)f(b)$ is a nil left ideal and $(f(A)+I)(f(a)+x)(f(c)+z)(f(d)+t)(f(b)+y)$ is a nil left ideal of $f(A)+I$ for all $f(d)+t \in f(A)+I$. It follows that $(A \bowtie^f I)(a, f(a)+x)(c, f(c)+z)(d, f(d)+t)(b, f(b)+y)$ is a nil left ideal of $A \bowtie^f I$ for all $(d, f(d)+t) \in A \bowtie^f I$. □

EXAMPLE 7.2. Let F be a field and consider the ring $R = \begin{bmatrix} F & F \\ 0 & F \end{bmatrix}$ and the ideal $I = \begin{bmatrix} F & F \\ 0 & 0 \end{bmatrix}$. Let $A = B = R$ and $f\colon A \to B$ be a homomorphism defined by $f\left(\begin{bmatrix} a & b \\ 0 & c \end{bmatrix}\right) = \begin{bmatrix} a & -b \\ 0 & c \end{bmatrix}$. Note that $f(A)+I = B$. It is easy to check that A is weakly symmetric. Therefore $A \bowtie^f I$ is weakly symmetric by Theorem 7.1(2).

Acknowledgments. The authors would like to thank the referee for his/her helpful suggestions to improve the presentation of this paper.

References

[1] D. D. Anderson and V. Camillo, *Semigroups and rings whose zero products commute*, Comm. Algebra **27** (1999), no. 6, 2847–2852, DOI 10.1080/00927879908826596. MR1687281

[2] P. M. Cohn, *Reversible rings*, Bull. London Math. Soc. **31** (1999), no. 6, 641–648, DOI 10.1112/S0024609399006116. MR1711020

[3] M. D'Anna, *A construction of Gorenstein rings*, J. Algebra **306** (2006), no. 2, 507–519, DOI 10.1016/j.jalgebra.2005.12.023. MR2271349

[4] M. D'Anna and M. Fontana, *An amalgamated duplication of a ring along an ideal: the basic properties*, J. Algebra Appl. **6** (2007), no. 3, 443–459, DOI 10.1142/S0219498807002326. MR2337762

[5] M. D'Anna, C. A. Finocchiaro, and M. Fontana, *Amalgamated algebras along an ideal*, Commutative algebra and its applications, Walter de Gruyter, Berlin, 2009, pp. 155–172. MR2606283

[6] G. Kafkas, B. Ungor, S. Halicioglu, and A. Harmanci, *Generalized symmetric rings*, Algebra Discrete Math. **12** (2011), no. 2, 72–84. MR2952903

[7] H. Kose, B. Ungor, S. Halicioglu, and A. Harmanci, *A generalization of reversible rings*, Iran. J. Sci. Technol. Trans. A Sci. **38** (2014), no. 1, 43–48. MR3288572

[8] B. Ungor, H. Kose, Y. Kurtulmaz, and A. Harmanci, *A nil approach to symmetricity of rings*, Indian J. Math. **60** (2018), no. 2, 337–357. MR3837768

[9] J. Krempa and D. Niewieczerzał, *Rings in which annihilators are ideals and their application to semi-group rings* (English, with Russian summary), Bull. Acad. Polon. Sci. Sér. Sci. Math. Astronom. Phys. **25** (1977), no. 9, 851–856. MR0506445

[10] J. Lambek, *On the representation of modules by sheaves of factor modules*, Canad. Math. Bull. **14** (1971), 359–368, DOI 10.4153/CMB-1971-065-1. MR0313324

[11] Z. Liang and Y. Gang, *On weakly reversible rings*, Acta Math. Univ. Comenian. (N.S.) **76** (2007), no. 2, 189–192. MR2385031

[12] G. Marks, *Reversible and symmetric rings*, J. Pure Appl. Algebra **174** (2002), no. 3, 311–318, DOI 10.1016/S0022-4049(02)00070-1. MR1929410

[13] L. Ouyang and H. Chen, *On weak symmetric rings*, Comm. Algebra **38** (2010), no. 2, 697–713, DOI 10.1080/00927870902828702. MR2598907

[14] J. Wei, *Generalized weakly symmetric rings*, J. Pure Appl. Algebra **218** (2014), no. 9, 1594–1603, DOI 10.1016/j.jpaa.2013.12.011. MR3188858

Department of Mathematics, Ahi Evran University, 40100 Kirsehir, Turkey
Email address: handan.kose@ahievran.edu.tr

Department of Mathematics, Ankara University, 06560 Ankara, Turkey
Email address: bungor@science.ankara.edu.tr

Department of Mathematics, Bilkent University, 06800 Ankara, Turkey
Email address: yosum@fen.bilkent.edu.tr

Department of Mathematics, Hacettepe University, 06800 Ankara, Turkey
Email address: harmanci@hacettepe.edu.tr

Contemporary Mathematics
Volume **727**, 2019
https://doi.org/10.1090/conm/727/14639

Jacobson pairs and Bott-Duffin decompositions in rings

T. Y. Lam and Pace P. Nielsen

Abstract. We define the Bott-Duffin decompositions of elements in a ring, which generalize the strongly clean decompositions, and prove that the Bott-Duffin decompositions of $1 - ab$ are in a natural bijection with those of $1 - ba$. This bijection respects a number of well known additive decompositions of elements in a ring. For instance, the result implies that $1 - ab$ is strongly clean (respectively, strongly nil-clean, Drazin invertible, quasipolar, or pseudopolar) if and only if so is $1 - ba$. Examples and further applications are given.

Introduction

Given a ring R and two elements $a, b \in R$, we say that $(1 - ab, 1 - ba)$ is a *Jacobson pair*. The reason for this terminology is the following result, attributed to Jacobson.

Jacobson's Lemma. *If $\alpha = 1 - ab$ is a unit with inverse $s \in R$, then $\beta = 1 - ba$ is a unit with inverse $t = 1 + bsa \in R$.*

The equation $t = 1 + bsa$ is sometimes referred to as the "Desert Island Formula", in tribute to a vintage remark of Kaplansky duly quoted in the introductory section of [**21**]. There are many generalizations of this magic lemma. For instance, if (α, β) is a Jacobson pair and α is Drazin invertible, then β is also Drazin invertible [**7, 24**]. We say accordingly that *Jacobson's Lemma holds* for Drazin invertible elements. In [**3**] a formula for the Drazin inverse of β was found in terms of the Drazin inverse of α, and in [**21**] further expressions for the Drazin inverse were proven. Each of these formulas can be thought of as generalizations of the Desert Island Formula.

To give another example, again suppose (α, β) is a Jacobson pair. If α is (von Neumann) regular, then so is β; i.e. Jacobson's Lemma holds for regular elements. Furthermore, if s is an inner inverse for the element α, then $t = 1 + bsa$ is an inner inverse for β. It turns out that while the usual Desert Island Formula works, nevertheless it can still be improved and this was done by the authors in the recent paper [**20**].

2010 *Mathematics Subject Classification.* Primary 16D70; Secondary 16P70, 16U60, 16U80.
Key words and phrases. Jacobson pair, Fitting decomposition, Bott-Duffin decomposition, quasi-nilpotent, strongly clean elements, Drazin inverse and Bott-Duffin inverse.

This paper is devoted to generalizing Jacobson's Lemma to a very large class of examples. We recapture the fact that Jacobson's Lemma holds for Drazin invertible elements, but extend this to (uniquely) strongly clean elements and strongly nil-clean elements, quasipolar and pseudopolar elements, among other examples introduced in §1. This is accomplished using the notion of a Bott-Duffin decomposition, named for the pioneering work done in [**2**], and to be formally defined in §2. The main theorem of this paper is Theorem 2.5, where we prove a natural bijection between the Bott-Duffin decompositions of any Jacobson pair. The remainder of the paper is devoted to exploring what other properties this natural one-to-one correspondence respects, and multiple examples are given which demonstrate the great generality that the Bott-Duffin decompositions provide.

As usual, given a ring R we write $\mathrm{U}(R)$ for the group of units, $\mathrm{idem}(R)$ for the set of idempotents, $J(R)$ for the Jacobson radical, and $\mathrm{nil}(R)$ for the set of nilpotent elements.

1. Generalized ABAB-decompositions

In this section we give a quick introduction to the notions of strongly clean, polar, quasipolar, and pseudopolar elements of rings. Many of these ring-theoretic properties were originally introduced in the context of linear algebra and matrix factorizations. Thus, in this section we emphasize a module-theoretic interpretation for certain additive decompositions of elements in rings.

This interpretation generalizes an observation which had its birth in the work of Nicholson [**23**] and further expansions by Diesl in [**8**] and [**9**, Lemma 2.8]. This connects certain additive element-wise decompositions of an endomorphism with corresponding module direct sum decompositions. Throughout, M_k will denote a right k-module, where k is some ring, and we will write endomorphisms on the left of M. (The ring k will not play an important role in what follows, and may be safely ignored. We also note in passing that every ring R can be viewed as the endomorphism ring $\mathrm{End}\,(R_R)$.) Given an endomorphism $\varphi \in \mathrm{End}\,(M_k)$ and a submodule $N \subseteq M$, we write $\varphi\restriction_N$ to denote the restriction of φ to the submodule N. In what follows, we will always restrict maps to φ-invariant submodules; that is $\varphi(N) \subseteq N$.

For any ring R and any subset $S \subset R$, the multiplicative *centralizer* (or commutant) of S in R is

$$C(S) := \{r \in R : rs = sr \text{ for all } s \in S\}.$$

We write $C^2(S)$ for the *double-centralizer* $C(C(S))$, and when $S = \{s\}$ is a single element, we write $C(s)$ rather than $C(\{s\})$.

PROPOSITION 1.1. *Fix an element $\alpha \in R = \mathrm{End}(M_k)$. Given $x \in R$, there exists an additive decomposition $\alpha = e + x$, where $e \in \mathrm{idem}(R) \cap C(\alpha)$, if and only if there exists a direct sum decomposition diagram*

$$\begin{array}{ccccc} M & = & A & \oplus & B \\ & & {\scriptstyle \alpha\restriction_A = x\restriction_A}\big\downarrow & & \big\downarrow{\scriptstyle (1-\alpha)\restriction_B = -x\restriction_B} \\ M & = & A & \oplus & B. \end{array}$$

PROOF. ($\Rightarrow$): Assume $\alpha = e + x$ for some idempotent $e^2 = e \in R$ which commutes with α. Since $x = \alpha - e$, we see that e commutes with x as well. We find that $\alpha(1-e) = (e+x)(1-e) = x(1-e)$ and so $\alpha \upharpoonright_{(1-e)M} = x \upharpoonright_{(1-e)M}$. Similarly, we find that $(1-\alpha)e = (1-e-x)e = -xe$ so $(1-\alpha) \upharpoonright_{eM} = -x \upharpoonright_{eM}$. Setting $A := (1-e)M$ and $B := eM$, we have that these are α-invariant direct sum complements, which yields the needed module decomposition.

($\Leftarrow$): Suppose we are given a diagram as in the statement of the proposition. Let $e \in R$ be the (idempotent) projection to B with kernel A. Since both A and B are α-invariant, we see that e commutes with α. One checks directly that α and $e+x$ behave the same on both A and B, proving the equality $\alpha = e+x$. □

This proposition motivates the following definition.

Definition 1.2 **(ABAB-Decomposition)**. Given an endomorphism $\alpha \in R = \mathrm{End}\,(M_k)$ and another endomorphism $e \in \mathrm{idem}(R) \cap C(\alpha)$, we define $x := \alpha - e$ and say that the diagram

$$\begin{array}{ccccc} M & = & A & \oplus & B \\ & & {\scriptstyle \alpha\upharpoonright_A = x\upharpoonright_A}\big\downarrow & & \big\downarrow{\scriptstyle (1-\alpha)\upharpoonright_B = -x\upharpoonright_B} \\ M & = & A & \oplus & B \end{array}$$

is the *ABAB-decomposition* corresponding to e.

When studying additive decompositions $\alpha = e + x$ (with $e^2 = e$ and $ex = xe$), one is naturally led to posit additional conditions on the element x. We recall a few common examples found in the literature. Throughout these examples we set $R := \mathrm{End}\,(M_k)$.

Example 1.3. **(Strongly clean elements).** Following [**23**], if $\alpha = e + x$ for some $e \in \mathrm{idem}(R) \cap C(\alpha)$, and additionally $x \in \mathrm{U}(R)$, we say this is a *strongly clean decomposition* for α, or that α is *strongly clean*. These elements arise in the study of exchange rings. Strongly clean decompositions for α are in one-to-one correspondence with ABAB-decomposition of the form

$$\begin{array}{ccccc} M & = & A & \oplus & B \\ & & {\scriptstyle \alpha\upharpoonright_A = \mathrm{unit}}\big\downarrow & & \big\downarrow{\scriptstyle (1-\alpha)\upharpoonright_B = \mathrm{unit}} \\ M & = & A & \oplus & B \end{array}$$

noting that $(1-\alpha)\upharpoonright_B = -x \upharpoonright_B$ is a unit (i.e. an automorphism of B) if and only if $-(1-\alpha)\upharpoonright_B = x \upharpoonright_B$ is a unit.

Example 1.4. **(Strongly nil-clean elements).** Following [**9**], an element $\alpha \in R$ is *strongly nil-clean* if $\alpha = f + t$ for some $f \in \mathrm{idem}(R) \cap C(\alpha)$ and some $t \in \mathrm{nil}(R)$. Setting $e := 1 - f$ and $x := (1 - 2e + t)$, we see that $\alpha = e + x$, which for the purposes of this paper will be called a *strongly nil-clean complementary decomposition* for α. It is easy to compute that $-xe = (1-t)e$ and $x(1-e) = (1+t)(1-e)$. Recall that a ring element is said to be "unipotent" when it is of the form "1+nilpotent". Hence, the corresponding ABAB-decomposition looks like

$$\begin{array}{ccccc} M & = & A & \oplus & B \\ & & {\scriptstyle \alpha\upharpoonright_A = \mathrm{unipotent}}\big\downarrow & & \big\downarrow{\scriptstyle (1-\alpha)\upharpoonright_B = \mathrm{unipotent}} \\ M & = & A & \oplus & B. \end{array}$$

Conversely, an ABAB-decomposition of this form corresponds to a strongly nil-clean complementary decomposition. If such a decomposition exists for α, then it is necessarily unique by [**14**, Theorem 3] or [**9**, Corollary 3.8].

EXAMPLE 1.5. **(Drazin invertible elements).** Again suppose $\alpha = e + x$ where $e \in \operatorname{idem}(R) \cap C(\alpha)$ and $x \in \mathrm{U}(R)$, as in the case of strongly clean elements. If additionally $\alpha e \in \operatorname{nil}(R)$, then we say that α is *Drazin invertible.* (Such elements are also called *strongly π-regular* or *polar* in the literature.[1]) Equivalently, $-xe$ is unipotent in the corner ring eRe. In this case, such an equation $\alpha = e + x$ is *unique.* We'll call it the *Fitting decomposition* of α (in the ring R), and e is called the *spectral idempotent* of α. From these remarks, we see that $\alpha \in R = \operatorname{End}(M_k)$ is Drazin invertible if and only if there is an ABAB-decomposition of the form

$$\begin{array}{ccccc} M & = & A & \oplus & B \\ & & \alpha\restriction_A = \text{unit} \big\downarrow & & \big\downarrow (1-\alpha)\restriction_B = \text{unipotent} \\ M & = & A & \oplus & B. \end{array}$$

In fact, this is just the usual description of α being a Fitting endomorphism on the module M_k, and the equation $M = A \oplus B$ above is *also* called the Fitting decomposition of M (with respect to α).

We can generalize the previous example by replacing nilpotence with a more general property. Following Harte [**13**], an element $q \in R$ is called *quasi-nilpotent* if $1 - qr \in \mathrm{U}(R)$ for every $r \in C(q)$, and we let $\operatorname{qnil}(R)$ denote the set of all quasi-nilpotent elements of R. It is apparent that $\operatorname{nil}(R) \cup J(R) \subseteq \operatorname{qnil}(R)$. (The quasi-nilpotents are exactly the *topologically nil* elements of Banach algebras.) Just as an element $x \in R$ is unipotent when $x = 1 + t$ for some $t \in \operatorname{nil}(R)$, we say that x is *quasi-unipotent* if $x = 1 + q$ for some $q \in \operatorname{qnil}(R)$.

EXAMPLE 1.6. **(Quasipolar elements).** Generalizing Example 1.5, if $\alpha = e + x$ for some $e \in \operatorname{idem}(R) \cap C(\alpha)$ with $x \in \mathrm{U}(R)$ and $\alpha e \in \operatorname{qnil}(R)$, then we call this a *quasi-Fitting decomposition* (or "generalized Fitting decomposition" modelling the terminology of [**6**]). Such a decomposition corresponds to an ABAB-decomposition of the form

$$\begin{array}{ccccc} M & = & A & \oplus & B \\ & & \alpha\restriction_A = \text{unit} \big\downarrow & & \big\downarrow (1-\alpha)\restriction_B = \text{quasi-unipotent} \\ M & = & A & \oplus & B. \end{array}$$

If we strengthen the assumption that $e \in C(\alpha)$ to $e \in C^2(\alpha)$, then the element α is called *quasipolar.*[2] For any given $\alpha \in R$, there is at most one quasipolar representation, and hence at most one ABAB-decomposition as above with $e \in C^2(\alpha)$.

---

[1]When defining polar elements, some authors instead of writing $\alpha = e + x$, use $\alpha + e = x$ (still with $e \in \operatorname{idem}(R) \cap C(\alpha)$, $x \in \mathrm{U}(R)$, and $\alpha e \in \operatorname{nil}(R)$). Essentially this has the effect of replacing α with $-\alpha$ and x with $-x$.

[2]Here we are following the terminology in [**19**]. Many authors have previously used $\alpha + e = x$, rather than $\alpha = e + x$, to define quasipolarity. The reader should keep in mind that our renormalization will have payouts later in the element-wise bijections we construct.

EXAMPLE 1.7. **(Pseudopolar elements).** In the previous example, if we replace the condition "$\alpha e \in \mathrm{qnil}(R)$" with the weaker condition "there is some integer $n \geq 1$ such that $(\alpha e)^n \in J(R)$" then we call the corresponding decompositions *pseudo-Fitting*.

Further, replacing $e \in C(\alpha)$ with $e \in C^2(\alpha)$, the element α is called *pseudopolar*, and there is at most one such decomposition [**26**]. (Again, the comments in footnotes 1 and 2 apply.)

The ABAB-decompositions above make it clear we have the following containments:

$$\{\text{strongly nil-clean complementary decompositions}\} \subseteq \{\text{Fitting decompositions}\}$$
$$\subseteq \{\text{quasi-Fitting decompositions}\} \subseteq \{\text{pseudo-Fitting decompositions}\}$$
$$\subseteq \{\text{strongly clean decompositions}\}.$$

We will give one more (large) class of examples, and their connection to ABAB-decompositions.

DEFINITION 1.8 ($\mathcal{P}$-**Clean**). Let $\mathcal{P}$ be a property of elements in rings. Following [**9**], we say that $\alpha \in R$ is *strongly $\mathcal{P}$-clean* if $\alpha = e + x$ where $e \in \mathrm{idem}(R) \cap C(x)$, and $x \in R$ has $\mathcal{P}$. In this case we say that $\alpha = e + x$ is a *strongly $\mathcal{P}$-clean decomposition*, and that e is a *strongly $\mathcal{P}$-cleansing* idempotent for α.

DEFINITION 1.9 (**Strong Corner Property**). Let $\mathcal{P}$ be a property of elements in rings. We say that $\mathcal{P}$ is a *strong corner property* if for any $x \in R$ and any idempotent $e \in C(x)$, x has $\mathcal{P}$ in R if and only if exe and $(1-e)x(1-e)$ have $\mathcal{P}$ in eRe and $(1-e)R(1-e)$, respectively.

For instance, the property of being a unit, nilpotent, unipotent, or idempotent are all strong corner properties. We will see later (in Proposition 3.3) that so is being a quasi-nilpotent element.

PROPOSITION 1.10. *Let $\mathcal{P}$ be a strong corner property, and let $\alpha \in R$. The set of strongly $\mathcal{P}$-clean decompositions $\alpha = e + x$ is in natural bijection with the set of ABAB-decompositions of the form*

$$\begin{array}{ccccc} M & = & A & \oplus & B \\ & & {\scriptstyle \alpha\restriction_A \text{ has } \mathcal{P}} \big\downarrow & & \big\downarrow {\scriptstyle -(1-\alpha)\restriction_B \text{ has } \mathcal{P}} \\ M & = & A & \oplus & B. \end{array}$$

PROOF. This follows by the proof of [**9**, Lemma 2.8], *mutatis mutandis*. □

2. Bott-Duffin Decompositions

In the theory of generalized inverses in rings, there is one very useful kind of invertibility that is known as Bott-Duffin invertibility. For a given idempotent e in a ring R, an element $x \in R$ is said to be *Bott-Duffin invertible relative to e* if $exe \in \mathrm{U}(eRe)$. In this case, the inverse of exe in the corner ring eRe is known as the *Bott-Duffin inverse* of x (relative to e). Making use of these definitions, we now introduce the main object of study in this section, which is a special class of additive decompositions of elements in a ring based on the notion of Bott-Duffin invertibility.

DEFINITION 2.1 (**Bott-Duffin Decomposition**). Let R be a ring and $\alpha \in R$. If $\alpha = e + x$ where $e \in \text{idem}(R) \cap C(x)$ and x is Bott-Duffin invertible relative to e, we call this a *Bott-Duffin decomposition* of α. We will say that e is the *spectral idempotent* of the Bott-Duffin decomposition $\alpha = e + x$.

Note that since $ex = xe$ in Definition 2.1, the condition for x to be Bott-Duffin invertible relative to e simply boils down to $xe \in \mathrm{U}(eRe)$. Thus, we'll be using the element-wise notion of Bott-Duffin invertibility only in a rather easy special case. In this case, the Bott-Duffin inverse of x relative to e (if it exists) is simply the unique element $v \in eRe$ such that $v(ex) = (ex)v = e$, or equivalently, just $vx = xv = e$. The example below illustrates how in the simplest cases Bott-Duffin decompositions can behave in rings.

EXAMPLE 2.2. Every element $\alpha \in R$ has at least one Bott-Duffin decomposition, namely $\alpha = 0 + \alpha$, but in general there may exist many other decompositions. The element $0 \in R$ has a Bott-Duffin decomposition $0 = e + (-e)$ for every idempotent $e \in R$. On the other hand, the element $1 \in R$ has only the trivial Bott-Duffin decomposition $1 = 0 + 1$. Any strongly clean decomposition $\alpha = e + x$ is clearly a Bott-Duffin decomposition (since $x \in \mathrm{U}(R) \cap C(e)$ has Bott-Duffin inverse $x^{-1}e$ relative to e). In particular, any idempotent $\varepsilon \in R$ has a Bott-Duffin decomposition $\varepsilon = (1 - \varepsilon) + (2\varepsilon - 1)$ (with spectral idempotent $1 - \varepsilon$).

We have also a close relationship between the Bott-Duffin decompositions of an endomorphism and certain ABAB-decompositions, as in the following result which is the special case of Proposition 1.1 with the added requirement that $xe \in \mathrm{U}(eRe)$.

PROPOSITION 2.3. *Given $\alpha \in R = \mathrm{End}(M_k)$, the Bott-Duffin decompositions $\alpha = e + x$ are in one-to-one correspondence with diagrams*

$$\begin{array}{ccccc} M & = & A & \oplus & B \\ & & {\scriptstyle \alpha\restriction_A}\Big\downarrow & & \Big\downarrow{\scriptstyle (1-\alpha)\restriction_B = \text{unit}} \\ M & = & A & \oplus & B. \end{array}$$

Note that in this diagram, the left vertical arrow labelled $\alpha \restriction_A$ is intended to mean just that the summand A is α-invariant, with no other conditions attached. This is due to the fact that, if A is taken to be $\ker(e)$, then $\alpha \restriction_A = x \restriction_A$, which is *not* subject to any requirements in Definition 2.1. We note also that the right vertical arrow labelled $(1-\alpha)\restriction_B$ is only required to be an automorphism of B, with again no other conditions attached. Therefore, the decomposition examples given in (1.3)–(1.7) in §1 enable us to draw the following useful conclusions. (In retrospect, of course, it was these conclusions that have provided us the main motivation for introducing the notion of Bott-Duffin decompositions in Definition 2.1.)

COROLLARY 2.4. *Strongly clean, Fitting, quasi-Fitting, pseudo-Fitting, strongly unipotent-clean, and strongly nil-clean complementary decompositions are Bott-Duffin decompositions.*

We next prove the first main theorem of this paper, which connects the Bott-Duffin decompositions for the two elements in a Jacobson pair (α, β). The main point of this theorem is that it introduces a procedure whereby a Bott-Duffin decomposition of α will give rise to a "corresponding" Bott-Duffin decomposition of β. Furthermore, when the same procedure is repeated on the latter Bott-Duffin decomposition, one gets back the original Bott-Duffin decomposition of α.

THEOREM 2.5. *Let $a, b \in R$, $\alpha := 1 - ab$, and $\beta := 1 - ba$. Suppose that $\alpha = e + x$ is a Bott-Duffin decomposition in R, and let v denote the Bott-Duffin inverse of x relative to e.*

(A) *The element $f := -bva$ is an idempotent that is isomorphic to e. (Equivalently, $1-e$ forms a Jacobson pair with $1-f$.) Also, we have $ReR = RfR$.*
(B) *If $y := \beta - f$, then (x, y) is a Jacobson pair. The commutation relations $\alpha a = a\beta$ and $b\alpha = \beta b$ are inherited by e and f; that is, $ea = af$, $be = fb$, and also, $xa = ay$, $bx = yb$.*
(C) *The equation $\beta = f + y$ is a Bott-Duffin decomposition, where y has Bott-Duffin inverse $-bv^2a$ relative to f; that is, $(fy)^{-1} = -bv^2a$ in the corner ring fRf.*
(D) *We have $e = -a\,(fy)^{-1}b$. Thus, if we apply the same procedure starting with the Bott-Duffin decomposition $\beta = f + y$ (and switching the roles of a and b), we will get back $\alpha = e + x$. In particular, our procedure gives a bijective correspondence between the Bott-Duffin decompositions of α and those of β.*
(E) *$(1 - e + xe,\ 1 - f + yf)$ is a Jacobson pair in R, and so is $(1 - e - xe,\ 1 - f - yf)$.*
(F) *For any integer $n \geq 0$ and any ideal $J \subseteq R$, $(\alpha e)^n \in J$ iff $(\beta f)^n \in J$.*
(G) *For any ideal $J \subseteq R$, $\overline{\alpha e}$ is quasi-nilpotent in R/J iff $\overline{\beta f}$ is quasi-nilpotent in R/J.*
(H) (Double-Centralizer Property) *$e \in C^2(\alpha)$ iff $f \in C^2(\beta)$.*

PROOF. (A) Left multiplying $ab = 1 - e - x$ by $v = ve$ gives $vab = -vx = -e$, and similarly, $abv = -e$. Thus,

$$f^2 = (-bva)(-bva) = b(vab)va = -beva = -bva = f.$$

By [**18**, Theorem 21.20], the idempotent $f = (-bv)\,a$ is isomorphic to the idempotent $e = a\,(-bv)$. The equation $f = -beva$ shows that $RfR \subseteq ReR$. After proving (D) below, we can likewise conclude that $ReR \subseteq RfR$, so we have in fact $ReR = RfR$.

(B) We have $ea = -abva = af$, and similarly, $be = fb$. Combining these with $\alpha a = a\beta$ and $b\alpha = \beta b$, we get $xa = ay$ and $bx = yb$. Clearly, $x = \alpha - e = 1 - a\,(b - bv)$ forms a Jacobson pair with $1 - (b - bv)\,a = 1 - ba + bva = \beta - f = y$.

(C) To begin with, $f\beta = f - fba = f - bea = f - baf = \beta f$. This implies that $fy = yf$. The rest of the proof of part (C) will follow if we can show that $-bv^2a \in fRf$ and that it provides an inverse for fy in fRf. The fact that $-bv^2a \in fRf$ follows from $f\,(bv^2a) = bev^2a = bv^2a$ and a similar equation $(bv^2a)f = bv^2a$. Left multiplying the former by $-y$ (which commutes with f) gives

$$(fy)\,(-bv^2a) = -ybv^2a = -bxv^2a = -bva = f,$$

and a similar computation gives $(-bv^2a)\,(fy) = f$.

(D) Using (C) and the equations $e = -abv = -vab$, we have

$$-a\,(fy)^{-1}b = -a\,(-bv^2a)\,b = (abv)\,(vab) = (-e)^2 = e.$$

Noting that $(fy)^{-1}$ is the Bott-Duffin inverse of y relative f, the rest of (D) clearly follows.

(E) Since $1-e+xe = 1+vab-xvab = 1+(va-xva)\,b$, it forms a Jacobson pair with

$$1+b\,(va-xva) = 1-f-bxva = 1-f-ybva = 1-f+yf.$$

Clearly, the same calculation works for the pair $(1-e-xe, 1-f-yf)$.

(F) It suffices to prove the "only if" statement, so assume that $(\alpha e)^n \in J$. By repeated use of the commutation rule $\beta b = b\alpha$, we have

$$(\beta f)^n = \beta^n f = \beta^n(-bva) = -b\alpha^n (ev)\,a \in R\,(\alpha e)^n R \subseteq J.$$

(G) After replacing R by R/J, we may assume that $J=0$. Proceeding as in (F), we note that $\beta f = \beta\,(-bva) = -b\,(\alpha va)$, while $-(\alpha va)\,b = \alpha\,(-vab) = \alpha e$. Thus, we see that (G) follows from the fact that, for any two elements $s,t \in R$, st is quasi-nilpotent iff ts is quasi-nilpotent; see [**7**]. (We will actually re-prove this fact as a part of Proposition 3.3 below.)

(H) This is possibly the most tricky part in this theorem. Assuming that $e \in C^2(\alpha)$, consider any element $r \in C(\beta)$. *Our goal is to show that* $r \in C(f)$. To begin with, note that $r \in C(ba)$, so we have an equation $rba = bar$. Using this, we get $ab\,(arb) = a\,(rba)\,b = arb\,(ab)$. Therefore, $arb \in C(ab) = C(\alpha)$, which implies that $arb \in C(e)$ and hence also $arb \in C(x)$. This implies that arb also commutes with the Bott-Duffin inverse of x relative to e, that is, $arb \in C(v)$, either by a direct computation or appealing to Drazin's commutation theorem in [**11**, Theorem 2.3]. Now consider the element bv^2arba, which we'll compute in two different ways. First,

$$bv^2a\,(rba) = bv^2a\,(bar) = bv\,(-e)\,ar = -bvar = fr.$$

Next, commuting arb twice with v, we get

$$bv^2(arb)\,a = b\,(arb)\,v^2a = (rba)\,bv^2a = -rbeva = -rbva = rf.$$

These computations show that $r \in C(f)$, as desired. The converse follows by symmetry. □

REMARK 2.6. The significance of the fact proved in (E) is as follows. Since $xe \in \mathrm{U}(eRe)$, we have $1-e+xe \in \mathrm{U}(R)$. Therefore, (E) implies that $1-f+yf \in \mathrm{U}(R)$. From this (and the fact that $yf = fy$), it is *also* easy to show that $fy \in \mathrm{U}(fRf)$. Moreover, using the Desert Island Formula for the Jacobson pairs in (E), it is possible to compute *explicitly* $(fy)^{-1}$. In fact, in our work, this was how we first arrived at the formula $(fy)^{-1} = -bv^2a$ in (C). But just as in the case of the Desert Island Formula, once this "answer" is known, it is very easy to verify by an explicit calculation that the answer does work. Since this "method of proof" is a bit easier to write down, we have for convenience chosen to follow it in the proof of (C) above, instead of first digressing to prove the fact (E).

REMARK 2.7. A few words are perhaps in order to explain why we required that $ex = xe$ in Definition 2.1. If we define $\alpha = e+x$ to be a "general Bott-Duffin decomposition" by assuming only $e^2 = e$ and that x is Bott-Duffin invertible relative to e, then we can no longer expect any one-to-one correspondence between the "general Bott-Duffin decompositions" of α and those of β. For an explicit example to illustrate this, let $R = \mathbb{M}_2(S)$ where S is any nonzero ring. Let $a = \begin{pmatrix} 1 & 0 \\ 0 & 0 \end{pmatrix}$ and $b = \begin{pmatrix} 0 & 0 \\ 1 & 0 \end{pmatrix}$ in R. Then $ab = 0$, and $\alpha := I_2 - ab = I_2$ is easily seen to have only one "general Bott-Duffin decomposition" (namely, $0 + I_2$). However, $\beta := I_2 - ba = \begin{pmatrix} 1 & 0 \\ -1 & 1 \end{pmatrix} \in \mathrm{U}(R)$ has (at least) the following family of "general Bott-Duffin decompositions":

$$\beta = 0 + \beta = \begin{pmatrix} 0 & s \\ 0 & 1 \end{pmatrix} + \begin{pmatrix} 1 & -s \\ -1 & 0 \end{pmatrix} \quad \text{(for any } s \in \mathrm{U}(S)).$$

Here, the last matrix is Bott-Duffin invertible relative to $\begin{pmatrix} 0 & s \\ 0 & 1 \end{pmatrix}$, with Bott-Duffin inverse $\begin{pmatrix} 0 & -1 \\ 0 & -s^{-1} \end{pmatrix}$. (In particular, β could have as many "general Bott-Duffin decompositions" as we want, by suitably choosing S.) Nevertheless, according to Theorem 2.5, the trivial decomposition $\beta = 0 + \beta$ is the *only* Bott-Duffin decomposition of β (in the sense of Definition 2.1). We note, incidentally, that all decompositions of β listed above are clean decompositions.

We close this section with a result on the case where $ab = ba$. In this case, $\alpha = \beta$, so it would be of interest to see which Bott-Duffin decompositions $\alpha = e + x$ correspond to themselves under the one-to-one correspondence given in Theorem 2.5. The answer to this question is given as follows.

PROPOSITION 2.8. *Let $a, b \in R$ be such that $ab = ba$, and let $\alpha := 1 - ab = e + x$ be a Bott-Duffin decomposition, with a corresponding decomposition $\beta := 1 - ba = f + y$ as in* Theorem 2.5. *Then $e = f$ iff $e \in C(a)$, iff $e \in C(b)$.*

PROOF. First assume that $e = f$. By Theorem 2.5(B), we have $ea = af = ae$ and $be = fb = eb$, so $e \in C(a) \cap C(b)$. Conversely, assume that $e \in C(a)$. Since $b \in C(a)$, we have $x = 1 - ab - e \in C(a)$ too. As in the proof of Theorem 2.5, let $v \in eRe$ be the Bott-Duffin inverse of x relative to e. Then Drazin's commutation theorem [**11**, (2.3)] shows (as before) that $v \in C(a)$. Recalling that $e = -abv$ (from the proof of Theorem 2.5(A)), we have $f := -bva = -bav = -abv = e$. If we assume, instead, that $e \in C(b)$, a similar calculation shows that $v \in C(b)$, so again $f = -bva = -vba = -vab = e$. (We note that, a priori, it was not at all clear that $e \in C(a)$ would imply that $e \in C(b)$, or vice versa.) □

EXAMPLE 2.9. Instead of working with a Jacobson pair $(1 - ab, 1 - ba)$, one may also work directly with the pair (ab, ba) too. In this case, however, the Bott-Duffin decompositions of ab may no longer be in one-to-one correspondence with those of ba. For instance, if $a, b \in R$ are such that $ab = 1 \neq ba$, then ab has only one Bott-Duffin decomposition $ab = 0 + 1$. On the other hand, ba has at least two Bott-Duffin decompositions; namely, $ba = 0 + ba = (1 - ba) + (2ba - 1)$.

3. Preserving More Properties

The bijection we constructed in §2 does more than simply preserve the underlying Bott-Duffin structure. Indeed, as we will see shortly, the following very general definition will lead to a large class of ring theoretic decomposition properties that are also preserved by the mapping constructed in Theorem 2.5.

DEFINITION 3.1 (**Strong Corner Jacobson Property**). Let $\mathcal{P}$ be a condition on elements of rings. We say that $\mathcal{P}$ is a *strong corner Jacobson property* if the following three conditions hold.

(i) $\mathcal{P}$ is a "Jacobson property", in the sense that, if (α, β) is a Jacobson pair in R and α has $\mathcal{P}$ in R, then β has $\mathcal{P}$ in R.
(ii) The property $\mathcal{P}$ is a strong corner property, in the sense of Definition 1.9.
(iii) The element $1 \in R$ always has $\mathcal{P}$ in R.

EXAMPLE 3.2. Being a unit is a strong corner Jacobson property by an application of Jacobson's Lemma. Somewhat easier is the fact that *unipotence* is also a strong corner Jacobson property. Indeed, if $1 - xy$ is unipotent, then xy is nilpotent, and so is yx (with the index of nilpotence changing by at most one). This means that $1 - yx$ is indeed unipotent. (The remaining properties among (i), (ii) and (iii) above are all easy to check.) At this point, it would be natural to ask whether the same fact holds if unipotence is replaced by *quasi-unipotence.* We'll supply a positive answer to this question below as we have not been able to find a full reference for the following result in the literature.

PROPOSITION 3.3. *Quasi-unipotence is a strong corner Jacobson property.*

PROOF. It turns out that a fairly substantial argument is needed. Again, to check that quasi-unipotence is a Jacobson property rests on checking that

$$(*) \qquad xy \text{ being quasi-nilpotent implies that } yx \text{ is quasi-nilpotent.}$$

This has been done, for instance, in [**7**] (for Banach algebras). For completeness, we'll include a proof here for general rings. To show that yx remains quasi-nilpotent, consider any element $r \in R$ with $r\,(yx) = (yx)\,r$. Then xr^2y commutes with xy since

$$(xr^2y)xy = xr^2(yx)y = x(yx)r^2y = xy(xr^2y).$$

From the quasi-nilpotence of xy, we have $1 - (xr^2y)xy \in \mathrm{U}(R)$. By Jacobson's Lemma,

$$1 - y(xr^2y)x = 1 - (yxr)(ryx) = 1 - (yxr)\,(yxr) = (1 - yxr)(1 + yxr) \in \mathrm{U}(R).$$

Since $1 - yxr$ and $1 + yxr$ commute, this proves that $1 + yxr \in \mathrm{U}(R)$, and hence yx is quasi-nilpotent.

Since (3.1)(iii) trivially holds, it remains only to show that, if $e = e^2 \in R$ and $z \in R$ are such that $ez = ze$, then z is quasi-unipotent in R iff eze is quasi-unipotent in eRe and $e'ze'$ is quasi-unipotent in $e'Re'$, where $e' := 1 - e$. The "only if" part is easy. For the "if" part, assume the stated properties on eze and $e'ze'$, say $eze = e + q$ and $e'ze' = e' + q'$ for some quasi-nilpotent elements $q \in eRe$ and $q' \in e'Re'$. We need to show that $q + q'$ is quasi-nilpotent in R. For the remainder of this proof, we will freely write elements of R in the form of 2×2 matrices via their Peirce decompositions with respect to the idempotent e. Thus, for instance, we write $q + q' = \begin{pmatrix} q & 0 \\ 0 & q' \end{pmatrix}$. Consider any $r = \begin{pmatrix} s & t \\ t' & s' \end{pmatrix} \in R$ that commutes with $q + q'$; that is, $sq = qs$, $s'q' = q's'$, $t'q = q't'$, and $tq' = qt$. We must show that $1 + r(q + q') \in \mathrm{U}(R)$. Write

$$1 + r\,(q + q') = \begin{pmatrix} e + sq & tq' \\ t'q & e' + s'q' \end{pmatrix}. \tag{3.4}$$

Since $sq = qs$ and $q \in eRe$ is quasi-nilpotent, $e + sq$ is a unit in eRe, say with inverse u; and similarly $e' + s'q'$ has inverse v in $e'Re'$. As $e + sq$ commutes with

q, so does u; that is, $uq = qu$. Similarly, we have $vq' = q'v$. Multiplying (3.4) on the left by the unit $\begin{pmatrix} e & 0 \\ -vt'q & e' \end{pmatrix}\begin{pmatrix} u & 0 \\ 0 & v \end{pmatrix}$, it suffices to show that

$$\begin{pmatrix} e & utq' \\ 0 & e' - vt'qutq' \end{pmatrix} \in \mathrm{U}(R). \tag{3.5}$$

Using the four commuting relations $t'q = q't'$, $tq' = qt$, $vq' = q'v$ and $uq = qu$, we have

$$(vt'qut)\,q' = vq't'uqt = q'(vt'qut).$$

As q' is quasi-nilpotent in fRf, this implies that the lower-right corner of the matrix in (3.5) is a unit in fRf, and so the whole matrix is a unit in R, as desired. □

The presentation of the main result in this section depends crucially on the next proposition on the transfer of a strong corner Jacobson property "from $\pm ex$ to $\pm fy$" for a corresponding pair of Bott-Duffin decompositions in the notations of Theorem 2.5.

PROPOSITION 3.6. *Let $\mathcal{P}$ be a strong corner Jacobson property. Let $(\alpha, \beta) = (1 - ab, 1 - ba)$ be a Jacobson pair, and let $\alpha = e + x$ and $\beta = f + y$ be a pair of corresponding Bott-Duffin decompositions under the bijection given in Theorem 2.5.*

(A) *If $ex = e\,(\alpha - 1)$ has $\mathcal{P}$ in eRe, then $fy = f(\beta - 1)$ has $\mathcal{P}$ in fRf.*
(B) *If $-ex = e\,(1 - \alpha)$ has $\mathcal{P}$ in eRe, then $-fy = f(1 - \beta)$ has $\mathcal{P}$ in fRf.*
(C) *If $(1-e)\,x = (1-e)\,\alpha$ has $\mathcal{P}$ in $(1-e)R\,(1-e)$, then $(1-f)\,y = (1-f)\beta$ has $\mathcal{P}$ in $(1-f)R(1-f)$.*

PROOF. To simplify the notations, let $e' = 1 - e$ and $f' = 1 - f$.

(A) Suppose that ex has $\mathcal{P}$ in eRe. The element $e' = 1_{e'Re'}$ has $\mathcal{P}$ in $e'Re'$ by (3.1)(iii), and hence $e' + ex$ has $\mathcal{P}$ in R by (3.1)(ii). By Theorem 2.5(E), $e' + ex$ and $f' + fy$ form a Jacobson pair. Hence $f' + fy$ has $\mathcal{P}$ in R by (3.1)(i), and hence so does fy in fRf by (3.1)(ii).

(B) This case is not directly covered by case (A) since elements satisfying property $\mathcal{P}$ are not necessarily closed under negation. However, the argument used for case (A) still works here, as Theorem 2.5(E) showed that $(e' - ex,\ f' - fy)$ also form a Jacobson pair.

(C) The argument here is again similar to that used for proving (A), but with e and e' playing reversed roles. Assume that $e'x$ has $\mathcal{P}$ in $e'Re'$. As $e = 1_{eRe}$ has $\mathcal{P}$ in eRe by (3.1)(iii), so does $e + e'x$ in R by (3.1)(ii). Noting that $fy = f(f' - ba) = -fba = -bea$ (by Theorem 2.5(B)), we see that

$$e + e'x = e + e'\alpha = \alpha + e - e\alpha = 1 - ab + eab$$

forms a Jacobson pair with

$$1 - ba + bea = \beta - fy = f + y - fy = f + f'y.$$

Thus, $f + f'y$ has $\mathcal{P}$ in R by (3.1)(i), and hence so does $f'y$ in $f'Rf'$ by (3.1)(ii). □

We are now ready to prove the second main result of this paper.

THEOREM 3.7. *The one-to-one correspondence in Theorem 2.5 respects strongly clean, Fitting, quasi-Fitting, pseudo-Fitting, strongly unipotent-clean, and strongly nil-clean complementary decompositions.*

PROOF. First of all, the various kinds of decompositions listed above are all Bott-Duffin decompositions by Corollary 2.4. Furthermore, all of the needed conditions on ex, $-ex$, or $(1-e)\,x$ are strong corner Jacobson properties, as we can see by applying (1.10), (3.2), and (3.3) to Examples (1.3)–(1.7). All desired conclusions now follow from this observation and Proposition 3.6. $\square$

Certainly, the result above covers some recently known cases of properties that are preserved by the passage from $\alpha = 1 - ab$ to $\beta = 1 - ba$. This includes, for instance, the case where $1 - ab$ is Drazin invertible (as proved in [**7**, **21**, **24**]), and more generally, the case where $1 - ab$ is quasi-Drazin invertible (as proved in [**27**]). Strongly nil clean elements were handled recently in [**17**, Theorem 2.9 and Corollary 2.10]. According to Theorem 3.7 above, we now also know that the same result holds for a number of other cases; for instance, if $1 - ab$ is strongly clean, or uniquely strongly clean, then so is $1 - ba$.[3] In a way, it is no longer surprising that such results *do not* apply, for instance, to clean elements since clean decompositions for $\alpha = 1 - ab$ need not be Bott-Duffin decompositions (in the sense of Definition 2.1). Taking the "uniquely strongly clean" case, we conclude this section with the following nontrivial application of Theorem 3.7.

COROLLARY 3.8. *If $ab \in R$ is an idempotent, then $\beta := 1 - ba \in R$ is uniquely strongly clean. It is also strongly regular, with Fitting decomposition $\beta = (ba)^2 + (1 - ba - (ba)^2)$.*

PROOF. Since $\alpha := 1 - ab$ is an idempotent, it has a unique strongly clean decomposition (namely, $\alpha = e + x$ with $e = 1 - \alpha = ab$ and $x = 1 - 2ab$). Applying Theorem 3.7, we see that the same holds for $\beta = 1 - ba$. To work out the strongly clean decomposition for β, we note that x has Bott-Duffin inverse ex^{-1} relative to e. By the proof of Theorem 2.5, the "corresponding" decomposition $\beta = f + y$ is supposed to have

$$f = -b\,((ab)\,x^{-1})\,a = -bab\,(1 - 2ab)\,a = -(ba)^2 + 2b\,(ab)^2 a = (ba)^2.$$

Therefore, $y = \beta - f = 1 - ba - (ba)^2$. (Incidentally, y^{-1} is given by $1 + ba - 3\,(ba)^2$.) Finally, the fact that α is strongly regular implies that β is also strongly regular by [**21**, Corollary 2.7], so $\beta = f + y$ is necessarily the Fitting decomposition of β. $\square$

4. Mock Drazin Inverses for Strongly Clean Decompositions

In this section, we shall look more closely at what happens to the results in §2 and §3 in the special case where the property $\mathcal{P}$ is taken to be "being a unit" (in a ring). In this case, a decomposition $\alpha = e + x$ subject to the requirements $e = e^2$ and $x \in \mathrm{U}(R) \cap C(e)$ would simply be a *strongly clean decomposition* of an element $\alpha \in R$. As we have stated in Example 2.2, such a decomposition is always a Bott-Duffin decomposition. In general, if $\alpha = e + x$ is a Bott-Duffin decomposition, the "complementary equation" $1 - \alpha = (1 - e) + (-x)$ need not be Bott-Duffin decomposition. As an easy exercise, the reader can check that $1 - \alpha = (1 - e) + (-x)$ is *also* a Bott-Duffin decomposition iff $x \in \mathrm{U}(R)$; that is, iff $\alpha = e + x$ is a strongly clean decomposition. The main goal of this section

[3]The fact that $1 - ab$ being strongly clean implies the same for $1 - ba$ was announced by the authors in [**21**] in 2012, but had so far remained unpublished. A (different) proof of this result by Gürgün has now appeared in [**12**].

will be to give more detailed information in this important special case. The extra leverage in this case is that we can introduce the interesting notion of the *mock Drazin inverse* of such a decomposition, as follows.

DEFINITION 4.1 (**Mock Drazin Inverse**). Let $\alpha = e+x$ be any strongly clean decomposition. As $x \in \mathrm{U}(R)$, we define the element $\alpha' := (1-e)\,x^{-1}$ to be the **mock Drazin inverse** of the given decomposition $\alpha = e+x$. (Note that, although we have used here the convenient notation α' for the mock Drazin inverse, it *does* depend on the choice of the decomposition $\alpha = e+x$, and is not determined by the element α alone.)

To motivate the above definition, we recall that, in the classical case where $\alpha \in R$ is *polar* (i.e. Drazin invertible) with (unique) Fitting decomposition $\alpha = e+x$ as defined in Example 1.5, the standard Drazin inverse of α is given by the element $(1-e)\,x^{-1}$; see e.g. [**16**, Theorem 4.2]. In general, an element $\alpha \in R$ *may not* be polar, but it may still have a strongly clean decomposition $\alpha = e+x$. In this case, it would therefore be useful to study the mock Drazin inverse $(1-e)\,x^{-1}$ associated with such a decomposition of α. Fortuitously, it turns out that some of the facts about Drazin invertible elements in rings can be extended to such a more general framework. To begin with, note that

$$\alpha\alpha' = (e+x)\,(1-e)\,x^{-1} = x\,(1-e)\,x^{-1} = 1-e, \tag{4.2}$$

which may be called the *associated idempotent* of the decomposition $\alpha = e+x$. From (4.2), we see easily that $\alpha'\alpha\alpha' = \alpha'$, which is the first axiomatic property of the classical Drazin inverse. Next, for any integer $n \geq 0$, we have

$$\alpha^n - \alpha^{n+1}\alpha' = \alpha^n(1-\alpha\alpha') = \alpha^n e. \tag{4.3}$$

In order that α is polar and that $e+x$ is its Fitting decomposition, the extra condition needed is that $(\alpha e)^n = \alpha^n e = 0$ for some integer $n \geq 1$ (see, e.g. [**16**]), which, by (4.3) above, amounts precisely to the second axiomatic property $\alpha^n = \alpha^{n+1}\alpha'$ for α' to be the classical Drazin inverse of α. The smallest integer $n \geq 1$ for which $(\alpha e)^n = 0$ (that is, the nilpotence index of αe) is known as the *Drazin index* of α. In the case of *quasipolar* or *pseudopolar* elements, we have basically the same characterizations, except that in the quasipolar case we require that αe is quasi-nilpotent, and in the pseudopolar case we require that some power of αe lies in the Jacobson radical $J(R)$. The quasi-Drazin inverse and the pseudo-Drazin inverse of α are still given by the same expression $(1-e)\,x^{-1}$ arising from a (unique) quasi-Fitting or pseudo-Fitting decomposition $\alpha = e+x$; see [**16**], [**27**], and [**26**]. In our more general set-up, however, $\alpha = e+x$ can be *any* strongly clean decomposition of the element α. The point of Definition 4.1 is that a mock Drazin inverse can be defined for such a decomposition, without requiring any property to be satisfied by the element $\alpha e = e\alpha$.

For any Jacobson pair (α, β), we have carried out a study in our earlier paper [**21**] on how the Fitting decompositions (and Drazin inverses) of α and β are related, when α is assumed to be polar. Our next goal now is to prove the third main result of this paper, which is to extend a number of our earlier results in [**21**] to the case of *general* strongly clean decompositions of α and β, as follows.

THEOREM 4.4. *Let $(\alpha, \beta) = (1-ab, 1-ba)$ be a Jacobson pair, and let $\alpha = e+x$ and $\beta = f+y$ be a pair of "corresponding" strongly clean decompositions (in the sense of* Theorem 3.7). *Then the following conclusions hold.*

(A) *$f = -b\,(ex^{-1})\,a$ and $y^{-1} = 1 + b\,(x^{-1} - ex^{-2})\,a$.*
(B) *$\alpha' = 1 + abx^{-1}$, $\beta' = 1 + bay^{-1}$, and (α', β') is a Jacobson pair satisfying the commutation relations $\alpha' a = a\beta'$ and $b\alpha' = \beta' b$.*
(C) *The mock Drazin inverse of β associated with the strongly clean decomposition $\beta = f + y$ is given by $\beta' = 1 - f + b\,\alpha' a$.*
(D) *If $a, b \in \mathrm{U}(R)$ or R is an IC ring (in the sense of [**15**]), the two spectral idempotents e and f are similar.*
(E) *Suppose α is polar, with Fitting decomposition $\alpha = e + x$. Then β is also polar (with the same Drazin index), and the corresponding strongly clean decomposition $\beta = f + y$ is the Fitting decomposition of β. Finally, we have $f = bera$ where $r = (1 - \alpha e)^{-1}$.*
(F) *Without the Drazin index part,* (E) *above holds also with "polar" replaced by "quasipolar" (resp. "pseudopolar") and "Fitting" replaced by "quasi-Fitting" (resp. "pseudo-Fitting").*

PROOF. We shall use freely the notations and formulas in the proof of Theorem 2.5. Here, the Bott-Duffin inverse of x relative to e is simply given by $v := ex^{-1} \in \mathrm{U}(eRe)$.

(A) Recalling Theorem 2.5(A), we have $f = -bva = -b\,(ex^{-1})\,a$. Next, $x = 1 - a\,(b - bv)$ forms a Jacobson pair with $y = 1 - (b - bv)\,a$. Since $x \in \mathrm{U}(R)$, it follows that $y \in \mathrm{U}(R)$, and the Desert Island Formula gives

$$(\dagger) \qquad y^{-1} = 1 + (b - bv)\,x^{-1}a = 1 + b\,(x^{-1} - ex^{-2})\,a.$$

This may be thought of as a higher form of the Desert Island Formula (with an extra "quadratic term"). Indeed, in the case where $e = 0$, we'll have $f = 0$ too since f is isomorphic to e. In this situation, $x = \alpha = 1 - ab$ and $y = \beta = 1 - ba$, in which case (†) reverts to the classical Desert Island Formula.

(B) The equation giving the mock Drazin inverse α' follows from

$$\alpha' := (1 - e)\,x^{-1} = (x + ab)\,x^{-1} = 1 + abx^{-1},$$

and similarly, $\beta' = 1 + bay^{-1}$. Recalling the commutation rule $xa = ay$ from Theorem 2.5(B), we have $ay^{-1} = x^{-1}a$. Thus, $\beta' = 1 + bx^{-1}a$, which forms a Jacobson pair with $1 + abx^{-1} = \alpha'$. Finally, the other commutation rule $ea = af$ from Theorem 2.5(B) gives

$$\alpha' a = (1 - e)\,x^{-1}a = (1 - e)\,ay^{-1} = a\,(1 - f)\,y^{-1} = a\beta',$$

and a similar argument gives $b\alpha' = \beta' b$.

(C) Using (A), (B), and the fact that $ay^{-1} = x^{-1}a$, we have

$$\begin{aligned} \beta' &= 1 + bay^{-1} = 1 + bx^{-1}a = 1 + b\,[e + (1 - e)]\,x^{-1}a \\ &= 1 + bex^{-1}a + b\,(1 - e)\,x^{-1}a = 1 - f + b\,\alpha' a, \end{aligned}$$

(D) If R is an IC ring, the isomorphic idempotents e, f must be similar by [**15**, Corollary 7.6]. For the other case, recall from the proof of Proposition 3.6(C) that $fy = -bea$. If we assume that $a, b \in \mathrm{U}(R)$, then since $y \in \mathrm{U}(R)$ too, the idempotent f is *equivalent* to e. (Two idempotents e_1, e_2 are called equivalent if $e_1 = ue_2w$ for some $u, w \in \mathrm{U}(R)$.) By a theorem of Song and Guo [**25**, Theorem 2], e and f are similar in R. (Note that, in general, e and f need not be similar. For instance, suppose $ab = 1 \neq ba$. Then $\alpha = 1 - ab = 0$ has a unique strongly clean decomposition $1 + (-1)$, with $e = 1$, while the idempotent $\beta = 1 - ba$ has a unique strongly clean decomposition $f + (1 - 2\,ba)$, with $f = ba$ not similar to $e = 1$.)

(E) All conclusions in (E) follow from Theorem 2.5 and Theorem 3.7, except for the fact that $f = -b\,(ex^{-1})\,a$ can now be written in the new form $f = bera$ where $r = (1-\alpha e)^{-1}$. To see this, let $t := \alpha e = e + ex$, which is nilpotent. Since $e\,(1-t) = e - \alpha e = -ex$, we have $ex^{-1} = -e\,(1-t)^{-1} = -er$, so $f = -b\,(ex^{-1})\,a = bera$. (The advantage of this new formula is that it expresses f directly in terms of a, b and the spectral idempotent e of α without explicitly using x or x^{-1}. If α has Drazin index n, then $t^n = 0$ and $r = 1 + t + \cdots + t^{n-1}$, so $f = bera$ can further be written as $f = besa$ where $s = 1 + \alpha + \cdots + \alpha^{n-1}$. We also note that $1 - f = \beta\beta'$, the associated idempotent of β, can be expressed in the form $(1 - bea)^m$ for any $m \geq n$. This fact was fully proved in [**21**, Theorem 2.1].)

(F) The same proof for (E) essentially carries over verbatim. Here, instead of being nilpotent, αe is quasi-nilpotent (resp. has a power lying in $J(R)$), so the inverse $r = (1 - \alpha e)^{-1}$ exists nevertheless. From part (F) and part (G) of Theorem 2.5, it follows that βf remains quasi-nilpotent (resp. has a power lying in $J(R)$). The only other thing to be careful about is that we have to ensure that the double-centralizer property $f \in C^2(\beta)$ holds so that $f + y$ is a quasi-Fitting (resp. pseudo-Fitting) decomposition of β. This follows from Theorem 2.5(H) since we have $e \in C^2(\alpha)$ to begin with. □

In the case where $\alpha = 1 - ab$ is polar, the equation in part (C) relating the mock Drazin inverse of β to that of α boils down to the Drazin inverse formula for Jacobson pairs obtained in [**21**, Theorem 2.1], and the equation $f = bera$ in part (E) recovers [**21**, Theorem 2.4]; see also [**3**, Theorem 3.6].[4] In the case where α is quasipolar (resp. pseudopolar), part (C) and part (F) recover the results of Zhuang, Chen and Cui in [**27**] and Cvetković-Ilić and Harte in [**7**] (resp. those of Gürgün in [**12**]). Our treatment here is, however, much more efficient, in that one single short proof is used to get the results in all three cases. Moreover, we have shown that the first four parts of Theorem 4.4 hold quite generally for any corresponding pair of strongly clean decompositions of α and β.

In the case where *all* elements of a ring R are polar, we can draw another more precise conclusion about Jacobson pairs in R, as follows.

COROLLARY 4.5. *Let R be a strongly π-regular ring. If (α, β) is a Jacobson pair in R, then the spectral idempotents of α and β are similar.*

PROOF. From Ara's Theorem 3 in [**1**], regular elements in R are unit-regular, so R is an IC ring by [**15**, Theorem 1.1]). Therefore, the desired conclusion follows from part (E) of Theorem 4.4. □

One of the most famous results in the classical theory of generalized inverses is "Cline's Formula" stated in Theorem 4.6 below, which was proved in [**5**] seven years after the introduction of the theory of Drazin inverses in [**10**]. It took many more years for Jacobson's Desert Island Formula to be extended to the case of Drazin inverses, but still, no link has ever been provided between the two formulas. In the following, we'll show that Cline's Formula is actually a natural consequence

[4]The proof of [**3**, Theorem 3.6] was harder and more indirect since it involved a lengthy reduction to the case where α is group-invertible.

of the (generalized) Desert Island Formula, by a quick application of the "complementation principle" via the Jacobson map $r \mapsto 1 - r$ for rings. We have to be careful here, since in general this map does not preserve Drazin invertibility (or its quasi- and pseudo-versions). However, it *does* preserve strong cleanness, while the first half of Theorem 4.4 is fully applicable to strongly clean decompositions. With this observation, a judicious application of part (A) and part (E) of Theorem 4.4 yields quickly a new conceptual derivation of "Cline's Formula" for polar (or Drazin invertible) elements in the second part of the theorem below.

THEOREM 4.6. *For any $a, b \in R$, the strongly clean decompositions of ab are in one-to-one correspondence with those of ba. If ab is polar with Drazin index n and Drazin inverse z, then ba is polar with Drazin index $\leq n+1$ and Drazin inverse bz^2a.*

PROOF. The first part of the theorem follows from Theorem 3.7 and the fact that the strongly clean decompositions of ab (resp. ba) are in one-to-one correspondence with those of $1 - ab$ (resp. $1 - ba$). For the second part of the theorem, assume ab is polar, and let $ab = e_0 - x$ be the Fitting decomposition of ab, so $(abe_0)^n = 0$ and ab has Drazin inverse $z = -ex^{-1}$ where $e := 1 - e_0$. Then $\alpha := 1 - ab = e + x$ is a strongly clean decomposition. If $f + y$ is the corresponding strongly clean decomposition of $\beta := 1 - ba$, we have the strongly clean decomposition $ba = (1 - f) + (-y)$. We'll show that $[ba\,(1 - f)]^{n+1} = 0$. Using the commutation rule $af = ea$, we have $ba\,(1 - f) = ba - bea = b\,(1 - e)\,a$. Since $[ab\,(1 - e)]^n = (abe_0)^n = 0$, it follows that $[ba\,(1 - f)]^{n+1} = 0$. This shows that ba is polar, with Drazin index $\leq n + 1$. Finally, the Drazin inverse of ba is given by $f(-y)^{-1}$. Using Theorem 4.4(A) and the commutation rule $fb = be$, we have

$$
\begin{aligned}
f(-y)^{-1} &= -f\,[1 + b\,(x^{-1} - ex^{-2})\,a] = -f - b\,(ex^{-1})\,a + be^2x^{-2}a \\
&= b\,(ex^{-1})^2a = bz^2a,
\end{aligned}
$$

upon recalling that $f = -b\,(ex^{-1})\,a$. □

With the above proof in place for Cline's Formula for polar elements, we can now quickly get its full analogues for both the quasipolar and the pseudopolar elements, as follows.

THEOREM 4.7. *For any $a, b \in R$, if $ab \in R$ is quasipolar with quasi-Drazin inverse z, then ba is quasipolar with quasi-Drazin inverse bz^2a. If ab is pseudopolar instead, a similar result also holds.*

PROOF. We will only handle the quasipolar case, as the pseudopolar case is nearly identical. In the quasipolar case, the proof of Theorem 4.6 can be carried over essentially verbatim. Here, we would start with $ab = e_0 - x$ being the quasi-Fitting decomposition of ab, so $e_0 \in C^2(ab)$ and abe_0 is quasi-nilpotent. Writing $\alpha := 1 - ab = e + x$ as before where $e := 1 - e_0 \in C^2(ab) = C^2(\alpha)$. Using the same notations as in the proof of Theorem 4.6, we have a corresponding strongly clean decomposition $\beta := 1 - ba = f + y$. Then Theorem 2.5(H) gives $f \in C^2(\beta)$; or equivalently, $1 - f \in C^2(ba)$. Arguing as in the proof of Theorem 4.6 (and recalling the statement $(*)$ in the proof of Proposition 3.3), we see that $abe_0 = ab\,(1 - e)$ being quasi-nilpotent implies that $ba\,(1 - f)$ is quasi-nilpotent. From the strongly clean decomposition $ba = (1 - f) + (-y)$ where $1 - f \in C^2(ba)$, it follows that ba is quasipolar. The formula for the quasi-Drazin inverse of ba now follows from the same calculation as in the last part of the proof of Theorem 4.6. □

To complete our references to the literature, we note that Cline's Formula in the quasipolar case was proved by Liao, Chen and Cui in [**22**, Theorem 2.2], and by Gürgün in [**12**, Theorem 2.8]. The corresponding result for the pseudopolar case was first proved by Wang and Chen in [**26**, Theorem 3.6], and later by Gürgün in [**12**, Theorem 2.9].

EXAMPLE 4.8. Revisiting the theme of Proposition 2.8, we'll give here an explicit example in which $ab = ba$ (so $\alpha = \beta$), but the correspondence of strongly clean decompositions for $\alpha = \beta$ constructed in Theorem 2.5 is far from being the identity map. Let $R = \mathbb{M}_2(S)$ where S is a nonzero ring in which $1 = u + w$ for two suitable units u, w, and let $a = \begin{pmatrix} u & -u \\ 0 & u \end{pmatrix}$ and $b = \begin{pmatrix} 1 & 1 \\ 0 & 1 \end{pmatrix}$. Since $ab = ba = u\,I_2$, $\alpha := I_2 - ab = w\,I_2$ forms a Jacobson pair with $\beta = I_2 - ba = \alpha$. For any $s \in S$ such that $su = us$, consider the idempotent matrix $e = \begin{pmatrix} 1 & s \\ 0 & 0 \end{pmatrix}$. For $x := \alpha - e = \begin{pmatrix} -u & -s \\ 0 & w \end{pmatrix} \in \mathrm{U}(R)$, we check easily that $ex = xe$, so we have a strongly clean decomposition $\alpha = e + x$. Since $be = e$ is not equal to $eb = \begin{pmatrix} 1 & 1+s \\ 0 & 0 \end{pmatrix}$, we know from Proposition 2.8 that the decomposition $\alpha = e + x$ does not correspond to itself (in the sense of Theorem 2.5). To compute the "corresponding" strongly clean decomposition $f + y$ for $\beta = \alpha$, recall from Theorem 4.4(A) that f is given by $-bex^{-1}a$. A quick computation shows that $f = \begin{pmatrix} 1 & s-1 \\ 0 & 0 \end{pmatrix} \neq e$, so we have the following two "corresponding" (but distinct) strongly clean decompositions (for any $s \in S$ commuting with u):

$$\alpha = \begin{pmatrix} w & 0 \\ 0 & w \end{pmatrix} = \begin{pmatrix} 1 & s \\ 0 & 0 \end{pmatrix} + \begin{pmatrix} -u & -s \\ 0 & w \end{pmatrix} = \begin{pmatrix} 1 & s-1 \\ 0 & 0 \end{pmatrix} + \begin{pmatrix} -u & 1-s \\ 0 & w \end{pmatrix}.$$

Of course, there are two other (more obvious) strongly clean decompositions for α too; namely, $\alpha = 0 + \alpha$, and $\alpha = I_2 - uI_2$. Under the correspondence constructed in Theorem 2.5, each of these decompositions corresponds to itself. (After all, these decompositions involve only the trivial idempotents, which are central.) Finally, we observe that all of the constructions in this example work already in the ring of upper-triangular matrices $\mathbb{T}_2(S)$. However, the existence of the equation $1 = u + w$ in S did play an essential role, as it is well known (e.g. from [**4**]) that $\mathbb{T}_n(S)$ is a uniquely strongly clean ring for any $n \geq 1$ if S is a Boolean ring.

5. Acknowledgments

The authors would like to thank Alexander Diesl and Thomas Dorsey for many enlightening comments on this paper, as well as the anonymous referee for helpful comments. This work was partially supported by a grant from the Simons Foundation (#315828 to Pace Nielsen). The project was sponsored by the National Security Agency under Grant Number H98230-16-1-0048.

References

[1] Pere Ara, *Strongly π-regular rings have stable range one*, Proc. Amer. Math. Soc. **124** (1996), no. 11, 3293–3298, DOI 10.1090/S0002-9939-96-03473-9. MR1343679

[2] R. Bott and R. J. Duffin, *On the algebra of networks*, Trans. Amer. Math. Soc. **74** (1953), 99–109, DOI 10.2307/1990850. MR0056573

[3] N. Castro-González, C. Mendes-Araújo, and Pedro Patricio, *Generalized inverses of a sum in rings*, Bull. Aust. Math. Soc. **82** (2010), no. 1, 156–164, DOI 10.1017/S0004972710000080. MR2669152

[4] Jianlong Chen, Zhou Wang, and Yiqiang Zhou, *Rings in which elements are uniquely the sum of an idempotent and a unit that commute*, J. Pure Appl. Algebra **213** (2009), no. 2, 215–223, DOI 10.1016/j.jpaa.2008.06.004. MR2467398

[5] R. E. Cline, *An application of representations for the generalized inverse of a matrix*, Tech. Summary Rep. 592, Math. Research Center, U.S. Army, Univ. Wisconsin, Madison (1965).

[6] Jian Cui and Jianlong Chen, *Characterizations of quasipolar rings*, Comm. Algebra **41** (2013), no. 9, 3207–3217, DOI 10.1080/00927872.2012.660670. MR3169449

[7] Dragana Cvetkovic-Ilic and Robin Harte, *On Jacobson's lemma and Drazin invertibility*, Appl. Math. Lett. **23** (2010), no. 4, 417–420, DOI 10.1016/j.aml.2009.11.009. MR2594854

[8] Alexander James Diesl, *Classes of strongly clean rings*, ProQuest LLC, Ann Arbor, MI, 2006. Thesis (Ph.D.)–University of California, Berkeley. MR2709132

[9] Alexander J. Diesl, *Nil clean rings*, J. Algebra **383** (2013), 197–211, DOI 10.1016/j.jalgebra.2013.02.020. MR3037975

[10] M. P. Drazin, *Pseudo-inverses in associative rings and semigroups*, Amer. Math. Monthly **65** (1958), 506–514, DOI 10.2307/2308576. MR0098762

[11] Michael P. Drazin, *Commuting properties of generalized inverses*, Linear Multilinear Algebra **61** (2013), no. 12, 1675–1681, DOI 10.1080/03081087.2012.753593. MR3175394

[12] Orhan Gürgün, *On Cline's formula for some certain elements in a ring*, An. Ştiinţ. Univ. Al. I. Cuza Iaşi. Mat. (N.S.) **62** (2016), no. 2, vol. 1, 403–410. MR3680218

[13] Robin Harte, *On quasinilpotents in rings*, Panamer. Math. J. **1** (1991), 10–16. MR1088863

[14] Yasuyuki Hirano, Hisao Tominaga, and Adil Yaqub, *On rings in which every element is uniquely expressible as a sum of a nilpotent element and a certain potent element*, Math. J. Okayama Univ. **30** (1988), 33–40. MR976729

[15] Dinesh Khurana and T. Y. Lam, *Rings with internal cancellation*, J. Algebra **284** (2005), no. 1, 203–235, DOI 10.1016/j.jalgebra.2004.07.032. MR2115012

[16] J. J. Koliha and Pedro Patricio, *Elements of rings with equal spectral idempotents*, J. Aust. Math. Soc. **72** (2002), no. 1, 137–152, DOI 10.1017/S1446788700003657. MR1868712

[17] Tamer Koşan, Zhou Wang, and Yiqiang Zhou, *Nil-clean and strongly nil-clean rings*, J. Pure Appl. Algebra **220** (2016), no. 2, 633–646, DOI 10.1016/j.jpaa.2015.07.009. MR3399382

[18] T. Y. Lam, *A first course in noncommutative rings*, 2nd ed., Graduate Texts in Mathematics, vol. 131, Springer-Verlag, New York, 2001. MR1838439

[19] T. Y. Lam, *Excursions in Ring Theory, in progress*, 2018.

[20] T. Y. Lam and Pace P. Nielsen, *Inner inverses and inner annihilators in rings*, J. Algebra **397** (2014), 91–110, DOI 10.1016/j.jalgebra.2013.08.023. MR3119217

[21] T. Y. Lam and Pace P. Nielsen, *Jacobson's lemma for Drazin inverses*, Ring theory and its applications, Contemp. Math., vol. 609, Amer. Math. Soc., Providence, RI, 2014, pp. 185–195, DOI 10.1090/conm/609/12127. MR3204360

[22] Yihua Liao, Jianlong Chen, and Jian Cui, *Cline's formula for the generalized Drazin inverse*, Bull. Malays. Math. Sci. Soc. (2) **37** (2014), no. 1, 37–42. MR3151689

[23] W. K. Nicholson, *Strongly clean rings and Fitting's lemma*, Comm. Algebra **27** (1999), no. 8, 3583–3592, DOI 10.1080/00927879908826649. MR1699586

[24] Pedro Patrício and R. E. Hartwig, *The link between regularity and strong-pi-regularity*, J. Aust. Math. Soc. **89** (2010), no. 1, 17–22, DOI 10.1017/S1446788710001448. MR2727061

[25] Guangtian Song and Xuejun Guo, *Diagonability of idempotent matrices over noncommutative rings*, Linear Algebra Appl. **297** (1999), no. 1-3, 1–7, DOI 10.1016/S0024-3795(99)00059-2. MR1723833

[26] Zhou Wang and Jianlong Chen, *Pseudo Drazin inverses in associative rings and Banach algebras*, Linear Algebra Appl. **437** (2012), no. 6, 1332–1345, DOI 10.1016/j.laa.2012.04.039. MR2942354

[27] Guifen Zhuang, Jianlong Chen, and Jian Cui, *Jacobson's lemma for the generalized Drazin inverse*, Linear Algebra Appl. **436** (2012), no. 3, 742–746, DOI 10.1016/j.laa.2011.07.044. MR2854904

Department of Mathematics, University of California, Berkeley, California 94720
Email address: lam@math.berkeley.edu

Department of Mathematics, Brigham Young University, Provo, Utah 84602
Email address: pace@math.byu.edu

Contemporary Mathematics
Volume **727**, 2019
https://doi.org/10.1090/conm/727/14640

Remarks on the Jacobson radical

André Leroy and Jerzy Matczuk

ABSTRACT. The aim of the paper is to investigate the set $\Delta(R) =: \{r \in R \mid r + U(R) \subseteq U(R)\}$ of a ring R. This set is a ring closely related to the Jacobson radical of R. It is shown that $\Delta(R)$ is the largest Jacobson radical subring of R which is closed with respect to multiplication by units of R. The behavior of Δ under ring constructions is studied, some families of rings for which $\Delta(R) = J(R)$ are presented. Methods of constructing rings with $\Delta(R) \neq J(R)$ are also described.

Introduction

It is well-known that if r is an element of the Jacobson radical $J(R)$ of R, then $r + u \in U(R)$ for every $u \in U(R)$, i.e.

$$J(R) \subseteq \Delta(R) =: \{r \in R \mid r + U(R) \subseteq U(R)\}.$$

It was already observed in [**L**, Exercise 4.24] that the above inclusion can be strict and that equality holds in the case of rings of stable range 1 (cf. [**L**, Exercise 20.10B]).

The aim of this paper is to investigate the set $\Delta(R)$ in details. We offer various characterizations of $\Delta(R)$ and then study the behavior of the operator Δ under some standard ring constructions. We also extend the definition and properties of Δ to rings without unity.

We show, in Theorems 3 and 13, that $\Delta(R)$ is the largest Jacobson radical subring of R which is closed with respect to multiplication by all units (quasi-invertible elements) of R, $\Delta(R) = J(T)$, where T is the subring of R generated by units of R, and the equality $\Delta(R) = J(R)$ holds if and only if $\Delta(R)$ is an ideal of R. As a consequence, we obtain that Δ is a closure operator, i.e. $\Delta(\Delta(R)) = \Delta(R)$ and that $\Delta(M_n(R)) = J(M_n(R)) = M_n(J(R))$, for any ring R and $n \geq 2$. Some further instances of rings for which $\Delta(R) = J(R)$ are presented in Theorem 11. Various examples exhibiting the differences in the behavior of Δ and the Jacobson radical are given in Examples 18 (including more instances of rings where the inclusion $J(R) \subseteq \Delta(R)$ is strict).

We also investigate Δ of polynomial rings and corners of rings. It appears (cf. Corollary 17) that for any ring R with 2 invertible, $\Delta(eRe) \subseteq e\Delta(R)e$, where e is denotes an idempotent of R. Moreover, contrary to the Jacobson radical,

2010 *Mathematics Subject Classification.* Primary 16N20.
Key words and phrases. Jacobson radical, quasi-invertible elements.

this inclusion can be strict. We conclude this short note by showing that, for a polynomial ring $R[x]$ over a 2-primal ring R, we have $\Delta(R[x]) = \Delta(R) + J(R[x])$.

In all the text R will stand for a ring usually with identity and $U(R)$ will denote the group of units of R. $J(R)$ and $B(R)$ stand for the Jacobson and the prime radicals of R, respectively.

The description and properties of $\Delta(R)$

We begin with the following lemma which collects basic properties of

$$\Delta(R) = \{r \in R \mid \forall_{u \in U(R)}\ r + u \in U(R)\}.$$

LEMMA 1. *For any ring R, we have:*

1. $\Delta(R) = \{r \in R \mid \forall_{u \in U(R)}\ ru + 1 \in U(R)\} = \{r \in R \mid \forall_{u \in U(R)}\ ur + 1 \in U(R)\}$;
2. *For any $r \in \Delta(R)$ and $u \in U(R)$, $ur, ru \in \Delta(R)$;*
3. *$\Delta(R)$ is a subring of R;*
4. *$\Delta(R)$ is an ideal of R if and only if $\Delta(R) = J(R)$;*
5. *For any rings R_i, $i \in I$, $\Delta(\prod_{i \in I} R_i) = \prod_{i \in I} \Delta(R_i)$.*

PROOF. Notice that, for any $u \in U(R)$ and $r \in R$, $ru^{-1} + 1 \in U(R)$ iff $r + u \in U(R)$ iff $u^{-1}r + 1 \in U(R)$. This yields (1). The statement (2) is an easy consequence of (1).

(3) Let $r, s \in \Delta(R)$. Then $-r + s + U(R) \subseteq -r + U(R) = -r - U(R) \subseteq U(R)$, i.e. Δ is a subgroup of the additive group of R. Then also $rs = r(s+1) - r \in \Delta(R)$, as $r(s+1) \in \Delta(R)$ by (2).

(4) Clearly $J(R) \subseteq \Delta(R)$. Suppose that $\Delta(R)$ is an ideal of R and $r \in R$. Then $rx + 1 \in U(R)$, for any $x \in \Delta(R)$ and $\Delta(R) \subseteq J(R)$ follows, i.e. $\Delta(R) = J(R)$. The reverse implication is clear.

We leave the easy proof of (5) to the reader. □

When e is an idempotent of a ring R, then the element $1 - 2e$ is a unit of R. This observation and Lemma 1(2) give immediately the following corollary.

COROLLARY 2. *For any ring R:*

1. *$\Delta(R)$ is closed with respect to multiplication by nilpotent elements;*
2. *If $2 \in U(R)$, then $\Delta(R)$ is closed with respect to multiplication by idempotents.*

Recall that if R is a ring (not necessarily with 1), then the circle monoid $R_\circ = (R, \circ)$ of R is the set R with the circle operation $\circ$ defined by $x \circ y = x + y - xy$. If R is a unital ring, then the monoid $R_\circ$ is isomorphic to the multiplicative monoid $(R, \cdot)$ of R, the isomorphism is given by assigning to every x of $R_\circ$ the element $1 - x$. In particular $y \in R$ is invertible as an element of the monoid $R_\circ$ (such elements are called quasi-invertible or quasi-regular) if and only if $1 - y$ is invertible as an element of the ring R and the group of units $U(R)$ of R is isomorphic to the group $U_\circ(R)$ of quasi-invertible elements of R. The inverse of y in $R_\circ$ is called the quasi-inverse of y. It is known that $I = J(R)$ is the largest ideal of R such that $U_\circ(I) = I$.

THEOREM 3. *Let R be a unital ring and T be the subring of R generated by $U(R)$. Then:*

(1) $\Delta(R) = J(T)$ *and* $\Delta(S) = \Delta(R)$, *for any subring S of R such that $T \subseteq S$;*
(2) $\Delta(R)$ *is the largest Jacobson radical ring contained in R which is closed with respect to multiplication by units of R.*

PROOF. (1) Notice that the subring T consists of all finite sums of units of R. Thus the statements (2) and (4) of Lemma 1 imply that $\Delta(T)$ is an ideal of T and $\Delta(T) = J(T)$, respectively.

If $r \in \Delta(R)$, then $r + U(R) \subseteq U(R)$. This means that r can be presented as a sum of two units. In particular, $r \in T$ and $\Delta(R) \subseteq T$ follows.

Let S be a subring of R such that $T \subseteq S$. Then $U(S) = U(R)$, so $\Delta(S) = \{r \in S \mid r + U(S) \subseteq U(S)\} = \{r \in S \mid r + U(R) \subseteq U(R)\} = S \cap \Delta(R) = \Delta(R)$, as $\Delta(R) \subseteq T \subseteq S$.

(2) By (1), $\Delta(R)$ is a Jacobson radical subring of R and Lemma 1(2) shows that $\Delta(R)$ is closed by left and right multiplication by units of R.

Now let S be a Jacobson radical ring contained in R which is closed by multiplication by units. If $s \in S$ and $u \in U(R)$, then $su \in S = J(S)$. Thus su is quasi-regular in S and hence $1 + su \in U(R)$. Now, Lemma 1(1) shows that $s \in \Delta(R)$, i.e. $S \subseteq \Delta(R)$, for any Jacobson radical subring of R. This proves (2). □

The characterization of $\Delta(R)$ given in Theorem 3(2) leads immediately to the following corollary.

COROLLARY 4. *Let R be a ring such that every element of R is a sum of units. Then $\Delta(R) = J(R)$.*

The classical theorem of Amitsur states that the Jacobson radical of an F-algebra R over a field F is nil, provided $\dim_F R < |F|$. Applying statement (1) of Theorem 3 we get the following corollary.

COROLLARY 5. *Let R be an algebra over a field F. If $\dim_F R < |F|$, then $\Delta(R)$ is a nil ring.*

One can also apply directly the Amitsur's arguments to prove the above corollary. In particular, when R is an algebra over a field F, Amitsur's argument shows that algebraic elements from $\Delta(R)$ are nilpotent.

For a unital ring R and its, not necessary unital, subring S, $\hat{S}$ will denote the subring of R generated by $S \cup \{1\}$. For further applications we will need the following

PROPOSITION 6. *Let R be a unital ring. Then:*

(1) *Let S be a subring of R such that $U(S) = U(R) \cap S$. Then $\Delta(R) \cap S \subseteq \Delta(S)$.*
(2) $U(\widehat{\Delta(R)}) = U(R) \cap \widehat{\Delta(R)}$;
(3) *Let I be an ideal of R such that $I \subseteq J(R)$. Then $\Delta(R/I) = \Delta(R)/I$.*

PROOF. The statement (1) is an easy consequence of the definition of Δ.

(2) Let us notice that if $r \in \Delta(R)$, then $v = 1+r \in U(R)$ and $v^{-1} = 1-rv^{-1} \in \widehat{\Delta(R)} \cap U(R)$, as by Lemma 1, $-rv^{-1} \in \Delta(R)$.

Let $u = r + k \cdot 1 \in \widehat{\Delta(R)} \cap U(R)$, where $r \in \Delta(R)$ and $k \in \mathbb{Z}$. We claim that $\bar{k} = k \cdot 1 \in U(R)$. We have $u - \bar{k} = r \in \Delta(R)$ and using Lemma 1, we obtain $1 - \bar{k}u^{-1} = (u - \bar{k})u^{-1} = ru^{-1} \in \Delta(R)$. Therefore $\bar{k}u^{-1} = 1 - (1 - \bar{k}u^{-1}) \in U(R)$ and $\bar{k} \in U(R)$ follows. As $\Delta(R)$ is closed by multiplications by units we can apply the first part of the proof to $v = u\bar{k}^{-1} = 1 + r\bar{k}^{-1}$ to obtain $u^{-1}\bar{k} = v^{-1} \in \widehat{\Delta(R)}$, i.e. $u^{-1}\bar{k} = s + \bar{l}$, for some $s \in \Delta(R)$ and $l \in \mathbb{Z}$. This implies, as $s\bar{k}^{-1} \in \Delta(R)$, $u^{-1} = s\bar{k}^{-1} + \bar{k}^{-1}\bar{l} \in \widehat{\Delta(R)}$ and shows that $U(R) \cap \widehat{\Delta(R)} \subseteq U(\widehat{\Delta(R)})$. The reverse inclusion $U(\widehat{\Delta(R)}) \subseteq U(R) \cap \widehat{\Delta(R)}$ is clear and (2) follows.

(3) Let $^-$ denote the canonical epimorphism of R onto R/I. Notice that, as $I \subseteq J(R)$, $U(\bar{R}) = \overline{U(R)}$.

Let $\bar{r} \in \Delta(\bar{R})$ and $u \in U(R)$. Then $\bar{r} + \bar{u} \in U(\bar{R})$ and there are elements $v \in U(R)$ and $j \in I$ such that $r + u = v + j$. Moreover $v + j \in U(R)$, as $I \subseteq J(R)$. This implies that $\Delta(\bar{R}) \subseteq \overline{\Delta(R)}$. Due to equality $U(\bar{R}) = \overline{U(R)}$, the reverse inclusion is clear, i.e. (3) holds. □

As an application of the above proposition we get the following

COROLLARY 7. *For any unital ring R, $\Delta(\widehat{\Delta(R)}) = \Delta(R)$, i.e. Δ is a closure operator.*

PROOF. Notice that $\Delta(R)$ is a Jacobson radical ideal of $T = \widehat{\Delta(R)}$. Hence, $\Delta(R) \subseteq T$.

Since $\Delta(R)$ contains all central nilpotent elements, $T/\Delta(R)$ is either isomorphic to $\mathbb{Z}$ or to $\mathbb{Z}_n := \mathbb{Z}/n\mathbb{Z}$, for some $n > 1$ which is square free. Therefore, by Proposition 6(3) and Corollary 4 we have $\Delta(T)/\Delta(R) = \Delta(T/\Delta(R)) = J(T/\Delta(R)) = 0$. This means that $\Delta(T) = \Delta(R)$, as required. □

Proposition 6(1) applies to $S = Z(R)$ - the center of R. Therefore, as for the Jacobson radical, we have:

COROLLARY 8. $\Delta(R) \cap Z(R) \subseteq \Delta(Z(R))$.

We will see in Example 18(4) that the inclusion from the above corollary can be strict even when $J(R) = J(Z(R)) = 0$.

For a ring R, let $T_n(R)$ denote the ring of all n by n upper triangular matrices over R, $J_n(R)$ the ideal of $T_n(R)$ consisting of all strictly upper triangular matrices and $D_n(R)$ the subring of diagonal matrices. As a direct consequence of Proposition 6(3) we get

COROLLARY 9. *For any ring R:*

(1) $\Delta(T_n(R)) = D_n(\Delta(R)) + J_n(R)$;
(2) $\Delta(R[x]/(x^n)) = \Delta(R)[x]/(x^n)$;
(3) $\Delta(R[[x]]) = \Delta(R)[[x]]$.

COROLLARY 10. *For any ring R, $\Delta(R) = J(R)$ if and only if $\Delta(R/J(R)) = 0$.*

The following theorem indicates a few classes of rings in which $\Delta(R) = J(R)$.

THEOREM 11. *$\Delta(R) = J(R)$ if R is a ring satisfying one of the following conditions:*

(1) *$R/J(R)$ is isomorphic to a product of matrix rings and division rings.*
(2) *R is a semilocal ring.*

(3) *R is a clean ring such that $2 \in U(R)$.*
(4) *R is a UJ-ring, i.e. when $U(R) = 1 + J(R)$.*
(5) *R has stable range 1.*
(6) *$R = FG$ is a group algebra over a field F.*

PROOF. (1). Let R be as in (1). In virtue of Corollary 10 it is enough to show that $\Delta(R/J(R)) = 0$. For doing this, we may assume that $J(R) = 0$, i.e. R is a product of matrix rings and division rings. If R is a matrix ring $M_n(S)$, for some unital ring S and $n \geq 2$ then, by Theorem of M. Henriksen [**H**], every element of R is a sum of three units so by Corollary 4 $\Delta(R) = J(R) = 0$. When S is a division ring, then clearly $\Delta(S) = 0$. Now (1) is a direct consequence of Lemma 1(5).

The statement (2) is a special case of (1).

(3). Suppose R is a clean ring such that $2 \in U(R)$. If $e \in R$ is an idempotent, then $1 - 2e \in U(R)$ and $e = (\frac{1}{2} - \frac{1}{2}(1 - 2e))$ is a sum of two units. This shows that every element of R is a sum of three units, so the thesis is a consequence of Corollary 4.

(4). Suppose $U(R) = 1 + U(R)$. Let us now suppose that the ring R is a UJ-ring. Then if $r \in \Delta(R)$ we have $r + U(R) \subseteq U(R)$ i.e. $r + 1 + J(R) \subseteq 1 + J(R)$. This immediately shows that $r \in J(R)$ and $\Delta(R) = J(R)$ follows.

(5) This statement is Exercise 20.10(b) in the book [**L**]. We repeat the short argument for the convenience of the reader. Let us assume the ring R has stable range 1. We only have to show that if $r \in R$ is such that $r + U(R) \subseteq U(R)$ then $r \in J(R)$. Now, for any $s \in R$ we have, $Rr + R(1 - sr) = R$. Hence, by the stable range 1 condition, we know that there exists $x \in R$ such that $r + x(1 - sr) \in U(R)$. This gives $x(1 - sr) \in r + U(R) \subseteq U(R)$ and shows that $r \in J(R)$.

If $R = FG$ is as in (6), then clearly every element of R is a sum of units, so (6) follows as above. □

It is known that semilocal rings have stable range 1, thus the statement (2) of the above proposition is also a consequence of (5).

Let us record the following easy observation.

LEMMA 12. *Let G be a subgroup of the additive group of a unital ring R. Then G is closed with respect to multiplication by invertible elements if and only if it is closed with respect to multiplication by quasi-invertible elements of R.*

PROOF. Let $r \in R$. As G is an additive group, $rG \subseteq G$ if and only if $(1-r)G \subseteq G$. This observation yields the lemma. □

Therefore, using the above we have:

THEOREM 13. *Let R be a unital ring and G a subgroup of the additive group of R. Then the following conditions are equivalent:*
(1) $G = \Delta(R)$
(2) *G is the largest Jacobson radical subring of R which is closed with respect to multiplication by quasi-invertible elements of R*
(3) *G is the largest additive subgroup of R consisting of quasi-invertible elements and closed with respect to multiplication by quasi-invertible elements of R.*

PROOF. Theorem 3 (2) and Lemma 12 show that $\Delta(R)$ is a Jacobson radical subring of R which is closed by multiplication by quasi-invertible elements. Let G be an additive subgroup consisting of quasi-invertible elements and closed with respect to multiplication by quasi-invertible elements of R. In particular, G is a Jacobson radical non unital subring of R and, by Lemma 12, G is closed by multiplication by units of R. Therefore Theorem 3 (2) gives $G \subseteq \Delta(R)$. This yields the theorem. □

One can modify the definition of Δ to work for rings without unity. Namely, let us set $\Delta_\circ(R) = \{r \in R \mid r + U_\circ(R) \subseteq U_\circ(R)\}$. Then it is clear that if R is a unital ring, then $\Delta_\circ(R) = \Delta(R)$. For any ring R, not necessary containing 1, let R^1 denote the ring obtained from R by adjoining unity with the help of $\mathbb{Z}$. Then, making use of the fact that $U_\circ(\mathbb{Z}) = 0$, it is easy to check that

LEMMA 14. *For any, not necessary unital, ring R, one has* $\Delta_\circ(R) = \Delta_\circ(R^1) = \Delta(R^1)$.

The above lemma says that we can extend the definition of Δ to all, not necessary, unital rings and all statements from Theorem 13 remain equivalent for arbitrary rings. Moreover, if one of the equivalent conditions holds, then $\Delta(\Delta(R)) = \Delta(R)$

A well-known classical result about the Jacobson radical $J(R)$ of a ring R states that $J(eRe) = eJ(R)e$, for any idempotent e of R. We will show that such equality is not satisfied in general for $\Delta(R)$. However the inclusion $e\Delta(R)e \subseteq \Delta(eRe)$ always holds under the mild assumption that $e\Delta(R)e \subseteq \Delta(R)$. We have seen in Corollary 2 that this assumption holds automatically provided $2 \in U(R)$.

PROPOSITION 15. *For any ring R, the following conditions hold:*

(1) *Let $e^2 = e$ be such that $e\Delta(R)e \subseteq \Delta(R)$. Then $e\Delta(R)e \subseteq \Delta(eRe)$.*
(2) *$\Delta(R)$ does not contain nonzero idempotents.*
(3) *$\Delta(R)$ does not contain nonzero unit regular elements.*

PROOF. (1) Let us first remark that if $y \in U(eRe)$, then $y_1 = y + (1-e) \in U(R)$ is such that $y = ey_1e$. Now, let $r \in e\Delta(R)e \subseteq \Delta(R)$, we want to show that for any unit $y \in U(eRe)$ we have that $e - yr \in U(eRe)$. As above, let $y_1 := y + 1 - e \in U(R)$. Since $r \in e\Delta(R)e \subseteq \Delta(R)$, we know that $1 - y_1r \in U(R)$. Hence there exists $b \in R$ such that $b(1 - y_1r) = 1$ and so $e = eb(1 - y_1r)e = eb(e - y_1re)e = eb(e - (y + 1 - e)re = eb(e - yre) + eb(1 - e)re = ebe(e - yre)$, where the last equality comes from the fact that $r \in eRe$. This gives that $e - yre = e - yr$ is left invertible in eRe. Since $1 - y_1r \in U(R)$ we also have $(1 - y_1r)b = 1$ and hence $1 = (1 - (y + 1 - e)r)b = (1 - yr)b$. Multiplying on both sides by e we get $e = e(1 - yr)be = (e - yr)be = (e - yr)ebe$. This shows that ebe is a right and left inverse of $e - yr$, as required.

(2) If $e^2 = e \in \Delta(R)$ then $1 - e = e + (1 - 2e) \in U(R)$, as $1 - 2e$ is a unit. This forces $e = 0$, i.e. (2) holds.

(3) if $a \in \Delta(R)$ is a unit regular element, then there exists an invertible element $u \in U(R)$ such that au is an idempotent. Statement (2) above shows that a must then be zero. □

Proposition 15 gives immediately the following corollary:

COROLLARY 16. *Let R be a unit regular ring, then $\Delta(R) = 0$.*

It is known (cf. [**G**, Proposition 4.12]) that unit regular rings are exactly regular rings with stable range 1. Thus the above corollary is also a consequence of

Theorem 11. We do know whether, for any regular ring R, $\Delta(R)$ has to be equal to 0.

Corollary 2 and Proposition 15(1) yield the following:

COROLLARY 17. *Suppose* $2 \in U(R)$. *Then* $e\Delta(R)e \subseteq \Delta(eRe)$, *for any idempotent* e *of* R.

The following examples show instances where $\Delta(R) \neq J(R)$ and indicate limits for obtained results.

EXAMPLES 18. (1) Let us observe that Theorem 13 implies that if A is a subring of a ring R such that $U(R) = U(A)$, then $J(A) \subseteq \Delta(R)$. In particular taking A to be a commutative domain with $J(A) \neq 0$ and $R = A[x]$, we obtain $0 = J(R) \subset J(A) \subseteq \Delta(R)$. (see also [**L**, Exercise 4.24])

(2) (cf. [**DL**, Example 2.5]) Let $R = \mathbb{F}_2 < x, y > / < x^2 >$. Then $J(R) = 0$ and $U(R) = 1 + \mathbb{F}_2 x + xRx$. In particular $\mathbb{F}_2 x + xRx$ is contained in $\Delta(R)$ but $J(R) = 0$.

(3) Let S be any ring such that $J(S) = 0$ and $\Delta(S) \neq 0$ and let $R = M_2(S)$. Then, by Theorem 11(1), $\Delta(R) = J(R) = 0$. Therefore, if $e = e_{11} \in R$, then $e\Delta(R)e = eJ(R)e = J(eRe) = 0$. and $\Delta(eRe) \simeq \Delta(S) \neq 0$. This shows that the inclusion $e\Delta(R)e \subseteq \Delta(eRe)$ from Proposition 15 can be strict, in general.

(4) Let A be a commutative domain with $J(A) \neq 0$ and $S = A[x]$. Then, by (1), $0 \neq J(A) \subseteq \Delta(S)$ and clearly $J(S) = 0$. $R = M_2(S)$, where A is a commutative local domain. By Theorem 11, $\Delta(R) = J(R) = 0$. Notice that the center $Z = Z(R)$ of $R = M_2(S)$ is isomorphic to S and $U(Z) = U(R) \cap Z$. Therefore $0 = \Delta(R) \cap Z \subseteq \Delta(Z) \simeq J(A) \neq 0$. Thus the inclusion from Corollary 8 can be strict even when $J(R) = 0 = J(Z(R))$.

The Example 18(2) was attributed to Bergman in [**DL**]. Let us mention that this example appeared earlier in [**CLP**, Example 6], where prime rings with commuting nilpotent elements were considered.

Recall that a 2-primal ring is a ring such that the set of all nilpotent elements $N(R)$ of R coincides with the prime radical $B(R)$, i.e. $R/B(R)$ is a reduced ring. We close the paper with the following result.

PROPOSITION 19. *Let* R *be 2-primal ring. Then* $\Delta(R[x]) = \Delta(R) + J(R[x])$.

PROOF. Suppose first that R is a reduced ring. Then $U(R[x]) = U(R)$ (cf. [**KLM**]). Thus, by definition of $\Delta(R[x])$, we have $\Delta(R) \subseteq \Delta(R[x])$. Let $a + a_0 \in \Delta(R[x])$, where $a \in R[x]x$ and $a_0 \in R$. Then, for any $u \in U(R)$, $a + a_0 + u \in U(R)$. This forces $a_0 + u \in U(R)$ and $a = 0$ and $\Delta(R) = \Delta(R[x])$ follows in this case.

Suppose now that R is 2-primal. Clearly $B(R[x]) = B(R)[x] \subseteq J(R[x])$. As R is 2-primal, $R/B(R)$ is reduced, so $J(R[x]) = B(R[x]) = B(R)[x]$. By the first part of the proof applied to $R/B(R)$ and Proposition 6(2), we have

$$\Delta(R) + B(R)[x] = \Delta(R/B(R)[x]) = \Delta(R[x]/J(R[x])) = \Delta(R[x])/J(R[x]).$$

The above yields the desired equality. □

References

[CLP] M. Chebotar, P.-H. Lee, and E. R. Puczyłowski, *On prime rings with commuting nilpotent elements*, Proc. Amer. Math. Soc. **137** (2009), no. 9, 2899–2903, DOI 10.1090/S0002-9939-09-09894-3. MR2506447

[DL] P. V. Danchev and T.-Y. Lam, *Rings with unipotent units*, Publ. Math. Debrecen **88** (2016), no. 3-4, 449–466, DOI 10.5486/PMD.2016.7405. MR3491753

[G] K. R. Goodearl, *von Neumann regular rings*, Monographs and Studies in Mathematics, vol. 4, Pitman (Advanced Publishing Program), Boston, Mass.-London, 1979. MR533669

[H] M. Henriksen, *Two classes of rings generated by their units*, J. Algebra **31** (1974), 182–193, DOI 10.1016/0021-8693(74)90013-1. MR0349745

[KLM] P. Kanwar, A. Leroy, and J. Matczuk, *Clean elements in polynomial rings*, Noncommutative rings and their applications, Contemp. Math., vol. 634, Amer. Math. Soc., Providence, RI, 2015, pp. 197–204, DOI 10.1090/conm/634/12699. MR3307398

[L] T. Y. Lam, *Exercises in classical ring theory*, 2nd ed., Problem Books in Mathematics, Springer-Verlag, New York, 2003. MR2003255

Department of Mathematics, Université d'Artois, Rue Jean Souvraz, 62300 Lens, France

Email address: andre.leroy@univ-artois.fr

Department of Mathematics, University of Warsaw, ul. Banacha 2, 02-097 Warsaw, Poland

Email address: jmatczuk@mimuw.edu.pl

Contemporary Mathematics
Volume **727**, 2019
https://doi.org/10.1090/conm/727/14641

Panov's theorem for weak Hopf algebras

Christian Lomp, Alveri Sant'Ana, and Ricardo Leite dos Santos

Abstract. Panov proved necessary and sufficient conditions to extend the Hopf algebra structure of an algebra R to an Ore extension $R[x;\sigma,\delta]$ with x being a skew-primitive element. In this paper we extend Panov's result to Ore extensions over weak Hopf algebras. As an application we study Ore extensions of connected groupoid algebras.

1. Introduction

In [**9**], Panov found necessary and sufficient conditions for an Ore extension $H = R[x;\sigma,\delta]$ over a Hopf algebra R, to have a structure of a Hopf algebra which extends the Hopf algebra structure of R and such that x is a skew-primitive element, i.e. $\Delta(x) = g\otimes x + x\otimes h$, for some $g,h\in R$. Later in [**4**], Brown et al. extended Panov's result by allowing x to be more general. In this paper we extend Panov's characterisation to Ore extensions of weak Hopf algebras, where by weak Hopf algebras we mean a generalisation of Hopf algebras in the sense of Böhm et al. [**1**] see also [**5, 8**]. A weak Hopf algebra R is a (unital, associative) algebra and a (counital, coassociative) coalgebra, with counit ϵ and comultiplication Δ, satisfying certain conditions. The difficulty for Panov's characterisation to carry over to weak Hopf algebras lies in the fact that the counit ϵ of a weak Hopf algebra R is not multiplicative and that the comultiplication Δ does not map 1 to $1\otimes 1$. Indeed it is well-known that a weak Hopf algebra is a Hopf algebra if and only if $\Delta(1) = 1\otimes 1$ if and only if ϵ is an algebra map (see [**1**, page 391]). Hence if R is a weak Hopf algebra over a field k and x is a primitive element (in the usual sense), i.e. $\Delta(x) = 1\otimes x + x\otimes 1$, then the coassociativity implies

$$\Delta(1)\otimes x + 1\otimes x\otimes 1 + x\otimes 1\otimes 1 = 1\otimes 1\otimes x + 1\otimes x\otimes 1 + x\otimes\Delta(x)$$

If x and 1 are linearly independent, then comparing the tensorands on both sides implies $\Delta(1) = 1\otimes 1$, which shows that R is a Hopf algebra. Note that the counity implies $\epsilon(x) = 0$. Hence x and 1 are only linearly dependent if $x = 0$. Thus we will

The first author is a member of CMUP (UID/MAT/00144/2013), which is funded by FCT with national (MEC) and European structural funds (FEDER), under the partnership agreement PT2020.

This work emerged out of the project "Internacionalização da Pós-Graduação no Rio Grande do Sul", financed by FAPERGS, and is part of the last author's doctoral thesis. The last author was partially supported by Capes, Brazil. Some results of this work were obtained during the last author's visit to the University of Porto. He would like to thank the Mathematics Department of the Faculty of Science of the University of Porto for its hospitality.

not be able of extending Panov's characterisation with the ordinary definition of a primitive element.

Recall that if R is a Hopf algebra, then R^* is an algebra with the convolution product and R is an R^*-bimodule with the left and right action given by $f \cdot a := a_1 f(a_2)$ and $a \cdot f := f(a_1)a_2$ for any $a \in R$ and $f \in R^*$. Let $X(R)$ denote the set of characters of R, i.e. the set of $\chi \in R^*$ that are algebra maps. Following [**3**, I.9.25], the left (resp. right) winding automorphisms of R are the automorphisms τ^l_χ (resp. τ^r_χ) given by $\tau^l_\chi(a) = a \cdot \chi = \chi(a_1)a_2$ (resp. $\tau^r_\chi(a) = \chi \cdot a = a_1\chi(a_2)$), for $a \in R$.

In [**9**, Theorem 1.3], Panov showed that an Ore extension $H = R[x; \sigma, \delta]$ of a Hopf algebra R carries also a Hopf algebra structure which extends the one of R and such that x is skew-primitive with $\Delta(x) = g \otimes x + x \otimes 1$ for some group-like element $g \in R$ if and only if there exists a character $\chi : R \to k$ such that $\sigma = \tau^r_\chi = Ad_g \tau^l_\chi$ and $\Delta(\delta(a)) = ga_1 \otimes \delta(a_2) + \delta(a_1) \otimes a_2$ for all $a \in R$. Here Ad_g is the conjugation by g, i.e. $Ad_g(a) = gag^{-1}$ for all $a \in R$.

Note that while any element g in a Hopf algebra H that satisfy $\Delta(g) = g \otimes g$ is a group-like element, and hence invertible, the identity maps e_x, for an object x of a groupoid $\mathcal{G}$, satisfy $\Delta(e_x) = e_x \otimes e_x$, but are not invertible elements in the groupoid algebra $k[\mathcal{G}]$ unless $\mathcal{G}$ has only one object and hence $k[\mathcal{G}]$ is actually a group algebra and therefore a ordinary Hopf algebra. This means that additional care has to be taken in defining group-like and skew-primitive elements for weak Hopf algebras.

The paper is organised as follows. In Sections 2 and 3 we develop the necessary machinery that will be used throughout the work. In section 4 we present our main results, Theorems 4.2 and 4.4. In the first one we show necessary and sufficient conditions to extend the structure of a weak bialgebra to an Ore extension of it and in the latter we treat the same problem in the case of a weak Hopf algebras. In both results we needed to assume certain additional hypotheses that are trivially satisfied in the case of ordinary Hopf algebras, thus obtaining a generalisation of Panov's result announced above for the context of weak Hopf algebras. In the last section we apply our results to the study of Ore extensions of connected groupoid algebras. Throughout the text, k will denote a field and tensor products are taken over k. Automorphisms and derivations of k-algebras are assumed to be k-linear.

2. Group-like elements in and characters of weak bialgebras

We start this section recalling the concepts of weak bialgebras and weak Hopf algebras. For more details and properties of these structures we refer [**1**, **8**].

DEFINITION 2.1. A weak bialgebra is a k-vector space R with the structures of an associative unital algebra and a coassociative counital coalgebra with comultiplication Δ and counit ϵ such that:

(i) The comultiplication Δ is a multiplicative k-linear map such that

$$(\Delta \otimes id)(\Delta(1)) = (\Delta(1) \otimes 1)(1 \otimes \Delta(1)) = (1 \otimes \Delta(1))(\Delta(1) \otimes 1);$$

(ii) The counit ϵ is a k-linear map which is weak multiplicative in the sense of

$$\epsilon(abc) = \epsilon(ab_1)\epsilon(b_2c) = \epsilon(ab_2)\epsilon(b_1c), \quad \text{for all } a, b, c \in R.$$

We recall the linear endomorphisms $\epsilon_t, \epsilon_s, \epsilon'_t$ and ϵ'_s on a weak bialgebra R:

$$\epsilon_t(a) = \epsilon(1_1 a)1_2, \qquad \epsilon_s(a) = \epsilon(a1_2)1_1, \qquad \epsilon'_t(a) = \epsilon(a1_1)1_2, \qquad \epsilon'_s(a) = \epsilon(1_2 a)1_1,$$

for all $a \in R$, where we are using the Sweedler notation without summation symbol $\Delta(a) = a_1 \otimes a_2$, for any $a \in R$. Moreover, any weak bialgebra R has the two subalgebras

$$R_s := \mathrm{Im}(\epsilon_s) = \{a \in R \mid \Delta(a) = 1_1 \otimes a 1_2\}, \qquad R_t := \mathrm{Im}(\epsilon_t) = \{a \in R \mid \Delta(a) = 1_1 a \otimes 1_2\},$$

which are separable algebras over the base field k (see [**5**, 36.6, 36.8]).

DEFINITION 2.2. A weak Hopf algebra is a weak bialgebra R with multiplication μ and comultiplication Δ, such that there exists a linear map $S : R \to R$, named antipode, that satisfies

$$\epsilon_t = \mu(id \otimes S)\Delta, \quad \epsilon_s = \mu(S \otimes id)\Delta \quad \text{and} \quad S(a_1)a_2S(a_3) = S(a), \text{ for all } a \in R.$$

Thus, if R is a weak Hopf algebra with antipode S, then $\epsilon_t(a) = a_1S(a_2)$ and $\epsilon_s(a) = S(a_1)a_2$ holds for every $a \in R$. Moreover, using the weak counity property 2.1(ii) one easily deduces for any $a, b \in R$:

$$\epsilon(ab) = \epsilon(a\epsilon_t(b)) = \epsilon(a\epsilon_s'(b)) = \epsilon(\epsilon_t'(a)b) = \epsilon(\epsilon_s(a)b) \tag{1}$$

Hence if $\epsilon_t(b) = 1$, then $\epsilon(ab) = \epsilon(a)$ and if $\epsilon_s(a) = 1$ then $\epsilon(ab) = \epsilon(b)$. Using equation (1) and induction one proves:

LEMMA 2.3. *Let R be a weak bialgebra and $g \in R$. Then:*

(i) $\epsilon_t(g) = 1$ *if and only if* $\epsilon(ag^n) = \epsilon(a)$, *for any* $a \in R$ *and* $n \geq 0$;
(ii) $\epsilon_s(g) = 1$ *if and only if* $\epsilon(g^n a) = \epsilon(a)$, *for any* $a \in R$*and* $n \geq 0$.

The same is true if we change ϵ_t by ϵ_s' in (i) and ϵ_s by ϵ_t' in (ii) of Lemma above.

2.1. Weak group-like Elements. As mentioned before, any element g in a Hopf algebra H that satisfy $\Delta(g) = g \otimes g$ is a group-like element, and hence invertible. However this might not be anymore the case for weak Hopf algebras and therefore additional care has to be taken in defining group-like elements for weak Hopf algebras.

DEFINITION 2.4 (cf. [**7**, Definiton 4.1]). An element g of a weak bialgebra is called *weak group-like* if $\Delta(g) = \Delta(1)(g \otimes g)$ and $\Delta(g) = (g \otimes g)\Delta(1)$ hold. A *group-like element* is an invertible weak group-like element.

If R is a weak bialgebra, then we denote the set of weak group-like elements of R by $G_w(R)$. Thus, $G(R) := G_w(R) \cap \mathcal{U}(R)$, is the set of group-like elements of R, where $\mathcal{U}(R)$ denotes the unit group of R. Note that 1 is always a group-like element in any weak bialgebra. Furthermore it is easy to check that $G_w(R)$ is a monoid with the multiplication of R and that $G(R)$ is a subgroup of $\mathcal{U}(R)$.

EXAMPLE 2.5. Let $\mathcal{G} = (\mathcal{G}_0, \mathcal{G}_1)$ be a groupoid with $|\mathcal{G}_0|$ finite. Let $R = k\mathcal{G}_1$ be its groupoid algebra, which is a weak Hopf algebra. Then any element $g \in \mathcal{G}_1$ satisfies $\Delta(g) = g \otimes g$ and as $\Delta(g) = \Delta(1)\Delta(g) = \Delta(g)\Delta(1)$, g is a weak group-like element. A particular groupoid algebra is $R = M_n(k)$, the $n \times n$ matrix ring over a field k. Here $|\mathcal{G}_0| = n$ and for any $i, j \in \mathcal{G}_0$ there exist exactly one morphism E_{ij} from i to j. Clearly E_{ii} are the identity maps and it is easy to check that $E_{ij}E_{st} = \delta_{j,s}E_{it}$ satisfies the ordinary multiplication rule of the elementary units of the $n \times n$ matrix ring $M_n(k)$. Thus we can identify $R = k\mathcal{G}_1$ with $M_n(k)$ such that

$$\Delta(E_{ij}) = E_{ij} \otimes E_{ij}, \qquad \epsilon(E_{ij}) = 1, \qquad \text{and} \qquad S(E_{ij}) = E_{ji}.$$

In particular, the monoid $G_w(R)$ of weak group-like elements of $R = M_n(k)$ is given by

$$G_w(M_n(k)) = \left\{ \sum_{i\in I} E_{i\sigma(i)} \mid I \subseteq \{1,\ldots,n\},\ \sigma : I \to \{1,\ldots,n\} \text{ injective} \right\}.$$

The group-like elements are precisely the permutation matrices, i.e. the elements of the form $g_\sigma = \sum_{i=1}^n E_{i\sigma(i)}$ for some $\sigma \in S_n$. The assignment $\sigma \mapsto g_\sigma$ is an isomorphism of groups between the symmetric group S_n and $G(M_n(k))$.

To see the above equality, we take $\gamma = \sum_{i,j=1}^n \lambda_{ij}E_{ij} \in G_w(R)$. Since $\Delta(\gamma) = \Delta(1)(\gamma\otimes\gamma)$ we have that $\sum_{i,j=1}^n \lambda_{ij}E_{ij}\otimes E_{ij} = \sum_{i,j,t=1}^n \lambda_{ij}\lambda_{it}E_{ij}\otimes E_{it}$. Consequently, we obtain $\lambda_{ij}^2 = \lambda_{ij}$ and $\lambda_{ij}\lambda_{it} = 0$ whenever $t \neq j$, for all $1 \leq i,j,t \leq n$, and it follows that $\lambda_{ij} \in \{0,1\}$. Moreover, if $\lambda_{ij} \neq 0$ then $\lambda_{it} = 0$, for all $t \neq j$. Set $I = \{i \in \{1,\ldots,n\} : \exists j : \lambda_{ij} = 1\}$. Then we can define a map $\sigma : I \to \{1,\ldots,n\}$ by $\sigma(i) = j$ if and only if $\lambda_{ij} = 1$ and it follows that $\gamma = \sum_{i\in I} E_{i\sigma(i)}$. The reverse inclusion is clear.

Using the counity properties we have the following easy Lemma:

LEMMA 2.6. *Let g be a weak group-like element in a weak bialgebra R. Then $g = \epsilon_t(g)g = g\epsilon_s(g)$. If R is a weak Hopf algebra then $\epsilon_t(g) = gS(g)$ and $\epsilon_s(g) = S(g)g$ are idempotents.*

It follows from the last lemma that for any weak group-like element g that has a right inverse $\epsilon_t(g) = \epsilon'_s(g) = 1$ holds, while $\epsilon_s(g) = \epsilon'_t(g) = 1$ holds if g has a left inverse. Consequently, if R is a weak Hopf algebra, then $S(g)$ is a right inverse of a weak group-like element g if and only if $\epsilon_t(g) = 1$ as well as $S(g)$ is a left inverse of g if and only if $\epsilon_s(g) = 1$. The example of the matrix units of $R = M_n(k)$ shows that in general $\epsilon_t(g)$ might be different from 1 for a weak group-like element g.

2.2. Weak Characters. A character χ of a Hopf algebra R is a (unital) algebra homomorphism $\chi : R \to k$. In particular, characters are group-like elements of the finite dual R° of a Hopf algebra. One particular example of a character of a Hopf algebra is its counit $\epsilon \in R^*$. In the case of a weak bialgebra, the counit is not an algebra homomorphism as it is neither unital nor multiplicative. However the notion of a left (respectively right) winding map τ_χ^l (resp τ_χ^r) as mentioned in the introduction (see also [**3**, I.9.25]) makes sense for weak bialgebra.

DEFINITION 2.7. A weak right character of a weak bialgebra R is a linear map $\chi : R \to k$ such that τ_χ^r is a unital algebra homomorphisms. We denote the set of all weak right characters by $X_w^r(R)$.

LEMMA 2.8. *Let R be a weak bialgebra. Then $X_w^r(R)$ is a submonoid of the multiplicative monoid R^*. Moreover, for any $\chi \in X_w^r(R)$:*

1. *$\tau_\chi^r(a) = a$ for any $a \in R_t$ and $\chi \in X_w^r(R)$.*
2. *If χ is invertible in R^* with inverse χ', then τ_χ^r is an automorphism of R and $\chi' \in X_w^r(R)$.*
3. *If τ_χ^r is an automorphism of R, then χ has a left inverse χ' in R^*.*

PROOF. Since R is a coalgebra, $\tau_\epsilon^r = id$, i.e. $\epsilon \in X_w^r(R)$. Let χ and χ' be any elements of R^*. For any $a \in R$ we have

$$\tau_{\chi*\chi'}^r(a) = a_1\chi*\chi'(a_2) = a_1\chi(a_2)\chi'(a_3) = \tau_\chi^r(a_1\chi'(a_2)) = \tau_\chi^r\left(\tau_{\chi'}^r(a)\right). \tag{2}$$

Hence if χ and χ' are in $X_w^r(R)$, then $\tau^r_{\chi*\chi'} = \tau^r_\chi \circ \tau^r_{\chi'}$ is a unital algebra map.

(1) Let $a \in R_t$. Then $\Delta(a) = 1_1 a \otimes 1_2$ (see [**5**, 36.6]). Hence $\tau^r_\chi(a) = 1_1 a \chi(1_2) = \tau^r_\chi(1)a = a$, since τ^r_χ is unital.

(2) This follows from equation (2), because if χ' is a (two-sided) inverse of χ in R^*, then

$$\tau^r_\chi \circ \tau^r_{\chi'} = \tau^r_{\chi*\chi'} = \tau_\epsilon = id = \tau_\epsilon = \tau^r_{\chi'*\chi} = \tau^r_{\chi'} \circ \tau^r_\chi.$$

Hence, $\tau^r_{\chi'}$ is the inverse function of the algebra map τ^r_χ and therefore itself an algebra map, i.e. $\chi' \in X_w^r(R)$.

(3) If τ^r_χ is an automorphism with inverse σ, then $\chi' = \epsilon\sigma$ satisfaz

$$\chi' * \chi(a) = \chi'(a_1)\chi(a_2) = \epsilon\sigma(\tau^r_\chi(a)) = \epsilon(a), \qquad \forall a \in R.$$

Hence χ' is a left inverse of χ in R^*. □

Analogously to the definition of a weak right character, we can define a weak left character as an element $\chi \in R^*$ such that τ^l_χ is a unital algebra homomorphism. The set of such elements shall be denoted by $X^l_w(R)$. It is clear that left and right characters are the same if R is cocommutative. With an analogous proof as in the last lemma we have

LEMMA 2.9. *Let R be a weak bialgebra. Then $X^l_w(R)$ is a submonoid of the multiplicative monoid R^*. Moreover, for any $\chi \in X^l_w(R)$:*

1. *$\tau^l_\chi(a) = a$ for any $a \in R_s$ and $\chi \in X^l_w(R)$.*
2. *If χ is invertible in R^* with inverse χ', then τ^l_χ is an automorphism of R and $\chi' \in X^l_w(R)$.*
3. *If τ^l_χ is an automorphism of R, then χ has a right inverse χ' in R^*.*

Furthermore, we set $X_w(R) = X^l_w(R) \cap X^r_w(R)$, whose elements are called *weak characters* and $X(R) = U(R^*) \cap X_w(R)$, whose elements are called *characters*.

EXAMPLE 2.10. Let $R = M_n(k)$ be the weak Hopf algebra from Example 2.5 and take $\chi \in X_w(R)$. As R is cocommutative, $\sigma := \tau^l_\chi = \tau^r_\chi$ is a unital algebra homomorphism. For each i,j there are elements $\lambda_{ij} := \chi(E_{ij})$. In particular, $\sigma(E_{ij}) = \lambda_{ij}E_{ij}$ and hence $\sum_i E_{ii} = 1 = \sigma(1) = \sum_k \lambda_{ii}E_{ii}$, i.e. $\lambda_{ii} = 1$ for all i. Moreover,

$$\lambda_{il}\lambda_{lj}E_{ij} = \sigma(E_{il})\sigma(E_{lj}) = \sigma(E_{il}E_{lj}) = \sigma(E_{ij}) = \lambda_{ij}E_{ij}.$$

Thus $\lambda_{ij} = \lambda_{il}\lambda_{lj}$ for all i,j,l. The case $j = i$ shows that $1 = \lambda_{ii} = \lambda_{il}\lambda_{li}$, i.e. $\lambda_{il} = \lambda_{li}^{-1}$, for any i,l. Thus $\lambda_{ij} = \lambda_{il}\lambda_{lj} = \lambda_{li}^{-1}\lambda_{lj}$, which holds for any l. Fixing $l = 1$ one can show that $\lambda_{12}, \ldots, \lambda_{1n}$ uniquely determine all such elements λ_{ij} and hence the algebra map σ. Conversely, it is clear that for any $q_1 = 1, q_2, \ldots, q_n \in \mathcal{U}(k) =: k^\times$ one can define a weak character χ by setting $\chi(E_{ij}) := q_i^{-1}q_j$. Note that any such weak character is actually invertible. Hence we have shown that $X_w(R) = X(R) \simeq (k^\times)^{n-1}$ as groups. Furthermore, if $n > 1$, then the characters χ defined like this are not multiplicative, since $\chi(E_{11}E_{22}) = 0$, while $\chi(E_{11})\chi(E_{22}) = 1$.

The following Proposition is important for Panov's theorem.

PROPOSITION 2.11. *Let R be a weak bialgebra and $\sigma : R \to R$ a unital algebra homomorphism.*

1. *$\sigma = \tau^r_\chi$ for some $\chi \in X^r_w(R)$ if and only if $\Delta\sigma = (\mathrm{id} \otimes \sigma)\Delta$.*

(2) $\sigma = \tau^l_\chi$ *for some* $\chi \in X^l_w(R)$ *if and only if* $\Delta\sigma = (\sigma \otimes \mathrm{id})\Delta$.

In both cases $\chi = \epsilon\sigma$.

PROOF. (1) Let $\sigma = \tau^r_\chi$ for some $\chi \in X^r_w(R)$. Then for any $a \in R$:

$$\Delta\sigma(a) = \Delta(a_1)\chi(a_2) = a_1 \otimes a_2\chi(a_3) = a_1 \otimes \tau^r_\chi(a_2) = (\mathrm{id} \otimes \sigma)\Delta(a).$$

On the other hand if σ satisfies $\Delta\sigma = (\mathrm{id} \otimes \sigma)\Delta$, then applying $\mathrm{id} \otimes \epsilon$ and setting $\chi = \epsilon\sigma \in R^*$ yields for all $a \in R$:

$$\sigma(a) = (\mathrm{id} \otimes \epsilon)\Delta(\sigma(a)) = (\mathrm{id} \otimes \chi)\Delta(a) = a_1\chi(a_2) = \tau^r_\chi(a).$$

The proof of (2) is similar to (1). □

COROLLARY 2.12. *Let R be a weak Hopf algebra with antipode S, then for any* $\chi \in X_w(R)$:

(i) $S * \tau^r_\chi = \epsilon_s\tau^r_\chi$ *and* $\tau^l_\chi * S = \epsilon_t\tau^l_\chi$;

(ii) *If* χS *is the inverse of* χ *in* R^* *then* $S = \tau^l_\chi S\tau^r_\chi = \tau^r_\chi S\tau^l_\chi$.

PROOF. (*i*) For any $a \in R$: $S * \tau^r_\chi(a) = S(a_1)a_2\chi(a_3) = \epsilon_s(a_1\chi(a_2)) = \epsilon_s(\tau^r_\chi(a))$. Similarly $\tau^l_\chi(a) * S = \chi(a_1)a_2S(a_3) = \epsilon_t(\chi(a_1)a_2) = \epsilon_t(\tau^l_\chi(a))$.

(*ii*) Suppose that χS is the inverse of χ. We calculate

$$\tau^l_\chi(S(\tau^r_\chi(a))) = S(a_1)\chi(S(a_2))\chi(a_3) = S(a_1)(\chi S * \chi)(a_2) = S(a),$$

for any $a \in R$. Similarly $\tau^r_\chi(S(\tau^l_\chi(a))) = \chi(a_1)\chi(S(a_2))S(a_3) = (\chi * \chi S)(a_1)S(a_2) = S(a)$. □

2.3. Tensor Products. The tensor product of two weak bialgebras (resp. weak Hopf algebras) is again a weak bialgebra (resp. weak Hopf algebra). More precisely if A and B are weak bialgebras with coalgebra structures $(A, \Delta_A, \epsilon_A)$ and $(B, \Delta_B, \epsilon_B)$, then the algebra $A \otimes B$ has the coalgebra structure $\Delta(a \otimes b) = \mathrm{flip}(\Delta_A \otimes \Delta_B)(a \otimes b) = (a_1 \otimes b_1) \otimes (a_2 \otimes b_2)$ and counit $\epsilon(a \otimes b) = \epsilon_A(a)\epsilon_B(b)$. If A and B have antipodes S_A and S_B, respectively, then $S(a \otimes b) = S_A(a) \otimes S_B(b)$ defines an antipode on $A \otimes B$. In particular $G_w(A \otimes B) = G_w(A) \otimes G_w(B)$ and $X_w(A \otimes B) = X_w(A) \otimes X_w(B) \subseteq A^* \otimes B^*$, where for $\chi \in X_w(A)$ and $\chi' \in X_w(B)$ one naturally means by $\chi'' = \chi \otimes \chi'$ the map that sends $a \otimes b$ to $\chi(a)\chi'(b)$.

Note that for any Hopf algebra H, the matrix ring $M_n(H) = M_n(k) \otimes H$ is a weak Hopf algebra that is a Hopf algebra if and only if $n = 1$.

Going back to the example of a groupoid algebra $R = k\mathcal{G}$, Example 2.5, we note that any groupoid can be decomposed in its connected component, i.e. the partition of the vertices $\mathcal{G}_0$ such that between two different parts there does not exist any morphism. Then R decomposes into a direct product of groupoid algebras of its connected components. Furthermore if $\mathcal{G}$ is connected and e_1 is any vertex in $\mathcal{G}_0$, then $G = \mathrm{Aut}(e_1)$, the set of automorphisms of e_1 is a group. Suppose that $\mathcal{G}$ is connected and that $\mathcal{G}_0 = \{e_1, \ldots, e_n\}$ is the set of vertices. For any $i > 1$ fix a morphism α_i from e_1 to e_i. Then any morphism β from e_i to e_j is of the form $\beta = \alpha_i^{-1}g\alpha_j$, where $g = \alpha_i\beta\alpha_j^{-1} \in G$ (where we write the composition of maps from left to right, as in the case of going along the arrows in a diagram). Hence the map $\beta \mapsto gE_{ij}$ defines an algebra automorphism from the groupoid algebra $k\mathcal{G}$ to the matrix ring $M_n(kG)$ over the group algebra kG. Hence we will identify R with $M_n(kG)$ and as such $R = M_n(k) \otimes kG$ is a tensor product of the Hopf algebra kG and the weak Hopf algebra $M_n(k)$. In particular given any group character

$\rho : G \to k^\times$ and elements $q_1 = 1, q_2, \dots, q_n \in k^\times$ (see Example 2.10) we have a character χ of R defined by $\chi(gE_{ij}) = q_i^{-1} q_j \rho(g)$.

3. Skew-primitive elements in and skew-coderivations of weak bialgebras

3.1. Skew-primitive elements. A weak Hopf algebra that contains a primitive element in the usual sense must be a Hopf algebra, as mentioned in the introduction. We propose the following definition.

DEFINITION 3.1. Let H be a weak bialgebra and $g, h \in H$ two weak group-like elements. An element $x \in H$ is called (g,h)-primitive if $\Delta(x) = \Delta(1)(g \otimes x + x \otimes h)$ and $\Delta(x) = (g \otimes x + x \otimes h)\Delta(1)$. An element is called skew-primitive if it is (g,h)-primitive for some weak group-like elements g, h.

REMARK 3.2. Let $R \subseteq H$ be an extension of weak Hopf algebras, $x \in H$ and $g, h \in R$ such that $\Delta(1)(g \otimes x + x \otimes h) = \Delta(x) = (g \otimes x + x \otimes h)\Delta(1)$. Suppose $R \cap Rx = 0 = R \cap xR$. The coassociativity of Δ applied to the first equality shows that

$$\begin{aligned}\Delta^2(1)(g \otimes g \otimes x + g \otimes x \otimes h + x \otimes \Delta(h))\\ = (\Delta \otimes \mathrm{id})\Delta(x) = \Delta^2(1)(\Delta(g) \otimes x + g \otimes x \otimes h + x \otimes h \otimes h).\end{aligned}$$

Hence comparing the coefficient of $1 \otimes 1 \otimes x$ in both expressions leads to $\Delta^2(1)(g \otimes g \otimes 1) = \Delta^2(1)(\Delta(g) \otimes 1)$. Applying $\mathrm{id} \otimes \epsilon$ yields $\Delta(1)(g \otimes g) = \Delta(g)$. Analogously, using the second equality, one concludes $\Delta(g) = (g \otimes g)\Delta(1)$ which shows that g needs to be a weak group-like element. The same is true for h, i.e. x is (g,h)-primitive.

LEMMA 3.3. *Let R be a weak bialgebra and x a (g,h)-primitive element. Then*

$$x = \epsilon_t(g)x + \epsilon_t(x)h = g\epsilon_s(x) + x\epsilon_s(h).$$

PROOF. This follows from the counity condition and the comultiplication of x. □

REMARK 3.4. As a consequence from the last lemma one concludes that $\epsilon_t(x) = 0$ if $\epsilon_t(g) = 1$ and h is left R_t-torsion free, while $\epsilon_s(x) = 0$ if $\epsilon_s(h) = 1$ and g is right R_s-torsion free.

Moreover if $R \subseteq H$ is an extension of weak bialgebras such that $x \in H$, $g, h \in R$ and $R \cap Rx = R \cap xR = 0$, then $x = \epsilon_t(g)x = x\epsilon_s(h)$ and $\epsilon_t(x)h = g\epsilon_s(x) = 0$. Thus if x is left R_t-torsion free, $\epsilon_t(g) = 1$, which would imply that g has a right inverse in case R is a weak Hopf algebra, by Lemma 2.6. Similarly if x is right R_s-torsion free, then $\epsilon_s(h) = 1$, which would imply that h has a left inverse in case of R being a weak Hopf algebra.

EXAMPLE 3.5. Let H be a Hopf algebra and consider the weak Hopf algebra $R = M_n(H)$ for some $n > 0$. The coalgebra structure of R is given by (cf. Subsection 2.3)

$$\Delta(hE_{ij}) = \Delta(h)(E_{ij} \otimes E_{ij}), \quad \epsilon(hE_{ij}) = \epsilon(h), \quad S(hE_{ij}) = S(h)E_{ji}, \forall h \in H, \forall i, j$$

Let $x \in H$ be a primitive element. Then xE_{ij} is a (E_{ij}, E_{ij})-primitive element of R since $\Delta(xE_{ij}) = E_{ij} \otimes xE_{ij} + xE_{ij} \otimes E_{ij}$. Note that if $n > 1$, then neither is E_{ij} invertible, i.e. $\epsilon_t(E_{ij}) \neq 1$, nor is E_{ij} a left R_t-torsion-free element. However $\epsilon(xE_{ij}) = 0$.

3.2. Coderivations. A derivation δ that satisfies Panov's condition is called a coderivation. We briefly discuss basic properties of such maps. Given a coalgebra C, a *coderivation* on C is a map $\delta : C \to C$ such that $\Delta\delta = (\mathrm{id} \otimes \delta + \delta \otimes \mathrm{id})\Delta$ holds (see [**6**, p.44]). For any $\chi \in C^*$ the map $\delta := (\mathrm{id} \otimes \chi - \chi \otimes \mathrm{id})\Delta$ is a coderivation, since

$$\begin{aligned}
\Delta\delta(a) &= \Delta(a_1\chi(a_2) - \chi(a_1)a_2) \\
&= a_1 \otimes a_2\chi(a_3) - \chi(a_1)a_2 \otimes a_3 \\
&= a_1 \otimes a_2\chi(a_3) - a_1 \otimes \chi(a_2)a_3 + a_1\chi(a_2) \otimes a_3 - \chi(a_1)a_2 \otimes a_3 \\
&= a_1 \otimes \delta(a_2) + \delta(a_1) \otimes a_2 = (\mathrm{id} \otimes \delta + \delta \otimes \mathrm{id})\Delta(a).
\end{aligned}$$

Any such coderivation is called *inner*. It follows from [**6**, Theorem 3] that any coderivation of a coseparable coalgebra is inner. Thus $\delta = 0$ is the only coderivation of a coseparable, cocommutative coalgebra, e.g. $C = M_n(k)$.

For an element $a \in R$ of an algebra we denote by $\lambda_a : R \to R$ the right R-linear map given by left multiplication by the element a, i.e. $\lambda_a(x) = ax$.

DEFINITION 3.6. Let R be a weak bialgebra and $g, h \in G_w(R)$. A map $\delta : R \to R$ is called (g,h)-coderivation if $\Delta\delta = (\lambda_g \otimes \delta + \delta \otimes \lambda_h)\Delta$. A skew-coderivation is a (g,h)-coderivation for some $g, h \in G_w(R)$.

LEMMA 3.7. *Let δ be a (g,h)-coderivation of a weak bialgebra R such that $\epsilon_s(g) = \epsilon_s(h) = 1$. Then $\epsilon\delta = 0$.*

PROOF. For $a \in R$ we have by Lemma 2.3

$$\delta(a) = (\epsilon \otimes \mathrm{id})\Delta\delta(a) = (\epsilon \otimes \mathrm{id})(ga_1 \otimes \delta(a_2) + \delta(a_1) \otimes ha_2) = \delta(a) + \epsilon(\delta(a_1))ha_2$$

Hence $\epsilon(\delta(a_1))ha_2 = 0$. Applying ϵ again and using equation (1) leads to $0 = \epsilon(\delta(a_1))\epsilon(ha_2) = \epsilon(\delta(a_1))\epsilon(a_2)) = \epsilon(\delta(a))$, using again Lemma 2.3. □

4. The main results

Before we present necessary and sufficient conditions to extend a weak Hopf algebra structure from an algebra R to an Ore extension $H = R[x;\sigma,\delta]$ we would like to point out that it is possible to apply Panov's original characterisation for Hopf algebras to produce proper weak Hopf algebra structures on Ore extensions.

Let A be any Hopf algebra with automorphism σ, σ-derivation δ and group-like element g. Suppose that the Hopf algebra structure of A can be extended to the Ore extension $A[x;\sigma,\delta]$ such that x is $(g,1)$-primitive, e.g. if this data satisfies Panov's criteria. Then given any weak Hopf algebra R, the algebra $R \otimes A$ has a natural structure of weak Hopf algebra with group like element $\overline{g} = 1 \otimes g$. Moreover, $\overline{\sigma} = id_R \otimes \sigma$ is an automorphism of $R \otimes A$ and $\overline{\delta} = id_R \otimes \delta$ is an $\overline{\sigma}$-derivation of $R \otimes A$. Since $(R \otimes A)[y;\overline{\sigma},\overline{\delta}]$ is isomorphic to $R \otimes A[x;\sigma,\delta]$ with $y \mapsto 1 \otimes x$, the weak Hopf algebra structure of $R \otimes A[x;\sigma,\delta]$ can be lifted to a weak Hopf algebra structure on $(R \otimes A)[y;\overline{\sigma},\overline{\delta}]$ which extends the weak Hopf algebra structure of $R \otimes A$ and would be an ordinary Hopf algebra if and only if R was an ordinary Hopf algebra.

A particular example is the groupoid algebra $H = M_n(kG)$ for some group G, field k and $n > 0$. Since $H = R \otimes A$ with $R = M_n(k)$ a weak Hopf algebra and $A = kG$ an ordinary Hopf algebra, we have that if $kG[x;\sigma,\delta]$ has a Hopf algebra structure extending the one of kG, then $M_n(kG)[y,\overline{\sigma},\overline{\delta}]$ becomes a weak Hopf algebra extending the weak Hopf algebra structure of $M_n(kG)$.

PROPOSITION 4.1 (Necessary conditions). *Let R be a weak bialgebra with automorphism σ, σ-derivation δ and weak group-like element g, such that the weak bialgebra structure on R extends to a weak bialgebra structure on $H = R[x;\sigma,\delta]$ with x an $(g,1)$-primitive element. Then*

(i) $\epsilon_t(g) = 1$;
(ii) δ *is a* $(g,1)$*-coderivation;*
(iii) $\sigma = \tau^l_\chi$, *for some weak left character* $\chi \in X^l_w(R)$, *which has a right inverse in* R^*;
(iv) $\Delta(\sigma(a))(g\otimes 1) = (g\otimes 1)(id\otimes\sigma)\Delta(a)$ *holds for any* $a \in R$.

PROOF. By Remark 3.4, $\epsilon_t(g) = 1$. By equation (1) and Lemma 2.3 we have therefore $\epsilon(Hx) = 0$. For any $a \in R$ we have:

$$\begin{aligned}\Delta(x)\Delta(a) &= (g\otimes x + x\otimes 1)\Delta(a)\\ &= (ga_1\otimes\sigma(a_2))(1\otimes x) + ga_1\otimes\delta(a_2) + (\sigma(a_1)\otimes a_2)(x\otimes 1) + \delta(a_1)\otimes a_2\\ &= (ga_1\otimes\sigma(a_2))(1\otimes x) + (\sigma(a_1)\otimes a_2)(x\otimes 1) + ga_1\otimes\delta(a_2) + \delta(a_1)\otimes a_2\end{aligned}$$

On the other hand using the multiplication rule in H:

$$\begin{aligned}\Delta(x)\Delta(a) &= \Delta(\sigma(a))\Delta(x) + \Delta(\delta(a))\\ &= \Delta(\sigma(a))(g\otimes 1)(1\otimes x) + \Delta(\sigma(a))(x\otimes 1) + \Delta(\delta(a))\end{aligned}$$

Hence comparing the coefficients of $1\otimes x$, $x\otimes 1$ and $1\otimes 1$ we conclude

$$\Delta(\sigma(a))(g\otimes 1) = ga_1\otimes\sigma(a_2) \tag{3}$$

$$\Delta(\sigma(a)) = \sigma(a_1)\otimes a_2 \tag{4}$$

$$\Delta(\delta(a)) = ga_1\otimes\delta(a_2) + \delta(a_1)\otimes a_2. \tag{5}$$

The first of these equations is condition (iv), while the last equation shows that δ is a $(g,1)$-coderivation. By Proposition 2.11, the second equation yields $\sigma = \tau^l_\chi$ for $\chi = \epsilon\circ\sigma \in X^l_w(R)$. Since σ is an automorphism, χ has a right inverse, by Lemma 2.9. □

For an invertible element $g \in R$ the linear automorphism of R that sends an element a to gag^{-1} is called the (left) adjoint map by g and denoted by Ad_g.

THEOREM 4.2 (Extending the weak bialgebra structure). *Let R be a weak bialgebra with automorphism σ, σ-derivation δ and group-like element g with $\epsilon(a\delta(b)) = 0$ for all $a,b \in R$.*

Then the following statements are equivalent:

(a) *The weak bialgebra structure on R extends to a weak bialgebra structure on $H = R[x;\sigma,\delta]$ such that x is a $(g,1)$-primitive element.*
(b) *There exists character χ such that $\sigma = \tau^l_\chi = Ad_g\tau^r_\chi$ and δ is a $(g,1)$-coderivation.*

PROOF. $(a)\Rightarrow(b)$ By Proposition 4.1 δ is a $(g,1)$-coderivation and there exists a weak left character $\chi = \epsilon\sigma$ such that $\sigma = \tau^l_\chi$. Moreover by the same proposition, one has $\Delta(\sigma(a))(g\otimes 1) = (g\otimes 1)(\mathrm{id}\otimes\sigma)\Delta(a)$, for any $a \in R$. Multiplying $g^{-1}\otimes 1$ from the right side yields $\Delta\sigma = (Ad_g\otimes\sigma)\Delta$ and applying $(id\otimes\epsilon)$ to both sides from the right brings us to $\sigma = Ad_g\tau^r_{\epsilon\sigma}$. Since $\epsilon\sigma = \epsilon\tau^l_\chi = \chi$, we have $\sigma = Ad_g\tau^r_\chi$ or equivalently $\tau^r_\chi = Ad_{g^{-1}}\sigma$, which shows that τ^r_χ is also an automorphism and hence $\chi \in X^r_w(R)$ is also a weak right character and χ has a right inverse, i.e. χ is a character.

$(b) \Rightarrow (a)$: Suppose $\sigma = Ad_g\tau_\chi^r$, then for any $a \in R$:

$$\begin{aligned}\Delta(\sigma(a))(g\otimes 1) &= \Delta(ga_1\chi(a_2)g^{-1})(g\otimes 1)\\ &= (ga_1g^{-1})\otimes(ga_2g^{-1})(g\otimes 1)\chi(a_3) = ga_1\otimes\sigma(a_2).\end{aligned}$$

Assuming $\epsilon(a\delta(b)) = 0$ for all $a, b \in R$, we also calculate for all $a \in R$:

$$\begin{aligned}(\dagger)\qquad (g\otimes x + x\otimes 1)\Delta(1)\Delta(a) &= ga_1\otimes\sigma(a_2)x + ga_1\otimes\delta(a_2)\\ &\quad + \sigma(a_1)x\otimes a_2 + \delta(a_1)\otimes a_2\\ &= \Delta(\sigma(a))(g\otimes x) + \Delta(\sigma(a))(x\otimes 1) + \Delta(\delta(a))\end{aligned}$$

where we also used that δ is a $(g,1)$-coderivation, together with Proposition 2.11. In particular for $a = 1$ we obtain $(g\otimes x + x\otimes 1)\Delta(1) = \Delta(1)(g\otimes x + x\otimes 1)$. Moreover, by the universal property of the Ore extension we obtain a (non-unital) algebra homomorphism $\Delta : H \to H\otimes H$ that extends the comultiplication of R and satisfies $\Delta(x) = \Delta(1)(g\otimes x + x\otimes 1)$ since the equation $(\dagger)$ shows that $\Delta(x)\Delta(a) = \Delta(\sigma(a))\Delta(x) + \Delta(\delta(a))$ holds for all $a \in R$. In order to prove the coassociativity of Δ, we show $(\Delta\otimes\mathrm{id})\Delta(x) = (\mathrm{id}\otimes\Delta)\Delta(x)$ since then by the uniqueness part of the universal property of the Ore extension both maps $(\Delta\otimes\mathrm{id})\Delta$ and $(\mathrm{id}\otimes\Delta)\Delta$ are equal. As a shorthand we write $\Delta^2(a) = (\Delta\otimes\mathrm{id})\Delta(a)$ for any $a \in R$. Note that

$$\Delta^2(a)(\Delta(1)\otimes 1) = \Delta^2(a)(1\otimes\Delta(1)) = \Delta^2(a).$$

Hence we calculate:

$$\begin{aligned}(\Delta\otimes\mathrm{id})\Delta(x) &= \Delta^2(1)(\Delta(g)\otimes x + \Delta(x)\otimes 1)\\ &= \Delta^2(1)(g\otimes g\otimes x + g\otimes x\otimes 1 + x\otimes 1\otimes 1)\\ &= \Delta^2(1)(g\otimes\Delta(x)\otimes x + x\otimes\Delta(1))\\ &= (\mathrm{id}\otimes\Delta)\Delta(x).\end{aligned}$$

This shows the coassociativity of Δ.

For the counity, we define a linear map $\epsilon : H \to k$ by $\epsilon\left(\sum_{i=0}^n a_ix^i\right) = \epsilon(a_0)$ for any $a_i \in R$. The map ϵ is well defined since the powers of x form a basis of H over R. Note that for all $a, b \in R$: $\epsilon(axb) = \epsilon(a\sigma(b)x) + \epsilon(a\delta(b)) = 0$, where we use our additional assumption that $\epsilon(a\delta(b)) = 0$ holds for any $a, b \in R$. By induction it is easy to prove $\epsilon(HxH) = 0$.

It is enough to verify the property of the counit for elements of the form ax^n with $n > 0$. A short induction argument shows that for any $n > 0$, there exist elements $C_{i,j}^{(n)} \in R$ such that

$$(6)\qquad (g\otimes x + x\otimes 1)^n = \sum_{i,j=0}^n C_{i,j}^{(n)}x^i\otimes x^j,$$

where $C_{n,0}^{(n)} = 1$ and $C_{i,0}^{(n)} = 0$ for any $i < n$, as well as $C_{0,n}^{(n)} = g^n$ and $C_{0,j}^{(n)} \in \mathrm{span}\{a\delta(b) : a, b \in R\}$ for any $j < n$. Then

$$(\mathrm{id}\otimes\epsilon)\Delta(ax^n) = \sum_{i,j=0}^n a_1C_{i,j}^{(n)}x^i\epsilon(a_2x^j) = \sum_{i=0}^n aC_{i,0}^{(n)}x^i = ax^n,$$

since $C_{i,0}^{(n)} = 0$ for $i < n$ and $C_{n,0}^{(n)} = 1$. Moreover,

$$(\epsilon \otimes \mathrm{id})\Delta(ax^n) = \sum_{i,j=0}^{n} \epsilon(a_1 C_{i,j}^{(n)} x^i) a_2 x^j = \underbrace{\epsilon(a_1 g^n)}_{=\epsilon(a_1)} a_2 x^n + \sum_{j=1}^{n} \underbrace{\epsilon(a_1 C_{0,j}^{(n)})}_{=0} a_2 x^j = ax^n.$$

where we use Lemma 2.3 for $\epsilon(ag^n) = \epsilon(a)$, since $\epsilon_t(g) = 1$ by Lemma 2.6, and $\epsilon(a_1 C_{0,j}^{(n)}) = 0$ as by equation (6), $a_1 C_{0,j}^{(n)} \in \mathrm{span}\{a\delta(b) : a, b \in R\}$ and $\epsilon(a\delta(b)) = 0$, by our additional assumption for $(b) \Rightarrow (a)$. Hence H is coalgebra.

Now, we are going to show the weak counity property. Let $\alpha = \tilde{\alpha}x + a$, $\beta = \tilde{\beta}x + b$ and $\gamma = \tilde{\gamma}x + c \in H$. Using $\epsilon(HxH) = 0$, what we have mentioned earlier, it is easy to check that $\epsilon(\alpha\beta\gamma) = \epsilon(abc)$. Note that $\Delta(\beta) = \Delta(\tilde{\beta})(g \otimes x + x \otimes 1) + \Delta(b)$. Thus

$$\begin{aligned}&\epsilon(\alpha\beta_1)\epsilon(\beta_2\gamma)\\ &= \epsilon(\alpha\tilde{\beta}_1 g)\underbrace{\epsilon(\tilde{\beta}_2 x\gamma)}_{=0} + \underbrace{\epsilon(\alpha\tilde{\beta}_1 x)}_{=0}\epsilon(\tilde{\beta}_2\gamma) + \epsilon(\alpha b_1)\epsilon(b_2\gamma) = \epsilon(\alpha b_1)\epsilon(b_2\gamma) = \epsilon(ab_1)\epsilon(b_2 c)\end{aligned}$$

Similarly $\epsilon(\alpha\beta_2)\epsilon(\beta_1\gamma) = \epsilon(\alpha b_2)\epsilon(b_1\gamma) = \epsilon(ab_2)\epsilon(b_1 c)$. Since R is a weak bialgebra we have

$$\epsilon(\alpha\beta_1)\epsilon(\beta_2\gamma) = \epsilon(ab_1)\epsilon(b_2 c) = \epsilon(abc) = \epsilon(ab_2)\epsilon(b_1 c) = \epsilon(\alpha\beta_2)\epsilon(\beta_1\gamma).$$

This shows that H is a weak bialgebra extending the structure of R. □

REMARK 4.3. The extra condition $\epsilon(a\delta(b)) = 0$, for all $a, b \in R$ in the Theorem, seems to be strange and it is not clear to the authors whether it is really needed. If g a group-like element and hence invertible, then $\epsilon_s(g) = 1$ by Lemma 2.6. Moreover, if δ is a $(g,1)$-coderivation and $\epsilon_s(g) = 1$, then $\epsilon\delta = 0$ by Lemma 3.7. If we had $\delta(R_s) = 0$ and $\sigma = \tau_\chi^l$ for some weak left character $\chi \in X_w^l(R)$, then for all $a, b \in R$:

$$\epsilon(a\delta(b)) = \epsilon(\epsilon_s(a)\delta(b)) = \epsilon(\sigma(\epsilon_s(a))\delta(b)) = \epsilon(\delta(\epsilon_s(a)b)) - \epsilon(\delta(\epsilon_s(a))b) = 0,$$

where we use equation (1) in the first equality, the R_s-linearity of τ_χ^l in the second equality (see Lemma 2.9) and $\delta(R_s) = 0$ and $\epsilon\delta = 0$ in the last one. Clearly $\delta(R_s) = 0$ for an arbitrary Hopf algebra, or for $\delta = 0$, i.e. in case R is a coseparable, cocommutative weak Hopf algebra like a groupoid algebra.

THEOREM 4.4. *Let R be a weak Hopf algebra with antipode S, σ an automorphism of R, δ an σ-derivation of R and g a group-like element. Suppose that $\delta(R_s) = 0$. Then the following conditions are equivalent:*

(a) *The weak Hopf algebra structure of R extends to a weak Hopf algebra structure on $H = R[x; \sigma, \delta]$, such that x is $(g,1)$-primitive element and $S(x) = -S(g)x$.*
(b) *The following statements are satisfied:*
    (i) *$\sigma = \tau_\chi^l = Ad_g \tau_\chi^r$, for some character χ of R;*
    (ii) *δ is a $(g,1)$-coderivation;*
    (iii) *$Ad_g S = \sigma S \sigma$;*
    (iv) *$\delta S \sigma = \lambda_g S \delta$, where λ_g denotes the left multiplication by g.*

PROOF. $(a) \Rightarrow (b)$ By Proposition 4.1, $\sigma = \tau_\chi^l$ for some weak left characters χ and δ is a $(g,1)$-coderivation. Hence by Remark 4.3, $\epsilon(a\delta(b)) = 0$ for all $a, b \in R$. Thus, we can apply Theorem 4.2 to obtain that χ is a character. Under the assumption that $S(x) = -S(g)x$ and using the commutation rule in $H = R[x;\sigma,\delta]$ we calculate for any $a \in R$:

$$\begin{aligned} S(a)S(g)x &= -S(xa) \\ &= -S(\sigma(a)x) - S(\delta(a)) \\ &= S(g)xS(\sigma(a)) - S(\delta(a)) \\ &= S(g)\sigma(S(\sigma(a)))x + [S(g)\delta(S(\sigma(a))) - S(\delta(a))] \end{aligned}$$

Comparing the coefficients of x and 1 leads to the equations $S(a)S(g) = S(g)\sigma(S(\sigma(a)))$ and $S(g)\delta(S(\sigma(a))) = S(\delta(a))$ for all $a \in R$. Hence $Ad_g S = \sigma S \sigma$, since $S(g) = g^{-1}$ by Lemma 2.6, and $\delta S \sigma = \lambda_g S \delta$.

$(b) \Rightarrow (a)$: Let $\sigma = \tau^l_\chi$ for some character χ. Using the assumptions $\delta(R_s) = 0$ and $\epsilon \circ \delta = 0$, we have $\epsilon(a\delta(b)) = 0$ for all $a, b \in R$, by Remark 4.3. Hence by Proposition 4.2 the weak bialgebra structure of R extends to $H = R[x;\sigma,\delta]$ with x being $(g,1)$-primitive and $\epsilon(Hx) = 0$. Since $\epsilon(axb) = \epsilon(a\sigma(b)x) + \epsilon(a\delta(b)) = 0$ for all $a, b \in R$ we also have $\epsilon(HxH) = 0$.

Let $\gamma = -S(g)x$. Using properties (iii) and (iv) we calculate for all $a \in R$:

$$\begin{aligned} S(a)\gamma &= -S(a)S(g)x - S(\delta(a)) + S(\delta(a)) \\ &= -S(g)\sigma(S(\sigma(a)))x - S(g)\delta(S(\sigma(a))) + S(\delta(a)) \\ &= -S(g)(xS(\sigma(a))) + S(\delta(a)) \\ &= \gamma S(\sigma(a)) + S(\delta(a)) \end{aligned}$$

Hence by the universal property of the Ore extension, there exists a unique unital anti-algebra homomorphism $S : H \to H$ that extends S and sends x to $S(x) = \gamma = -S(g)x$. We will verify the antipode axioms of a weak Hopf algebra. It is clear that $a_1 S(a_2) = \epsilon_t(a)$, $S(a_1)a_2 = \epsilon_s(a)$ and $S(a_1)a_2 S(a_3) = S(a)$ holds for all $a \in R$. For any $h \in H$ we have $\epsilon_t(hx) = \epsilon(1_1 hx)1_2 = 0$, since $\epsilon(Hx) = 0$. Therefore

$$(hx)_1 S((hx)_2) = h_1 g S(h_2 x) + h_1 x S(h_2) = -h_1 g S(g) x S(h_2) + h_1 x S(h_2) = 0 = \epsilon_t(hx),$$

as $gS(g) = \epsilon_t(g) = 1$ by assumption (b.i).

Similar, $\epsilon_s(hx) = \epsilon(hx1_1)1_2 = 0$, since $\epsilon(HxH) = 0$. Furthermore note that the elements of R_s commute with x, because $\sigma = \tau^l_\chi$ is R_s-linear by Lemma 2.9 and $\delta(R_s) = 0$ by assumption, i.e. $xa = \sigma(a)x + \delta(a) = ax$ for any $a \in R_s$. Hence using $S(x) = -S(g)x$ and $S(a_1)a_2 \in R_s$ we calculate for any $a \in R$: $S((ax)_1)(ax)_2 = S(a_1 g)a_2 x + S(a_1 x)a_2 = S(g)(\epsilon_s(a)x - x\epsilon_s(a)) = 0$. Suppose that $S(h_1)h_2 = 0$ for all $h = ax^n$ with $a \in R$ and some $n \geq 0$. Then we also calculate: $S((hx)_1)(hx)_2 = S(h_1 g)h_2 x + S(h_1 x)h_2 = S(g)S(h_1)h_2 x + S(x)S(h_1)h_2 = 0$. Hence by induction on n we have that $S(h_1)h_2 = \epsilon_s(h)$ for any $h = ax^n$ with $a \in R$ and $n \geq 0$. In particular we have that $H_s = \epsilon_s(H) = \epsilon_s(R) = R_s$, i.e. all elements of H_s commute with x in H.

Finally we will show

$$S(h_1)h_2 S(h_3) = S(h) \tag{7}$$

for all $h \in H$. Again it is enough to show this only for the monomials $h = ax^n$ with $n \geq 0$. For $n = 0$ and $h = a \in R$ we have $S(h_1)h_2 S(h_3) = S(h)$ since S extends the antipode of R. Suppose that (7) holds for all elements $h = ax^n$ for some $n \geq 0$.

Then

$$\begin{aligned}
S((hx)_1)(hx)_2S((hx)_3)\\
&= S(h_1g)h_2gS(h_3x) + S(h_1g)h_2xS(h_3) + S(h_1x)h_2S(h_3)\\
&= S(g)S(h_1)h_2gS(x)S(h_3) + S(g)S(h_1)h_2xS(h_3) + S(x)S(h_1)h_2S(h_3)\\
&= -S(g)\epsilon_s(h_1)gS(g)xS(h_2) + S(g)\epsilon_s(h_1)xS(h_2) + S(x)\epsilon_s(h_1)S(h_2)\\
&= -S(g)x\epsilon_s(h_1)S(h_2) + S(g)x\epsilon_s(h_1)S(h_2) + S(x)S(h)\\
&= S(x)S(h) = S(hx)
\end{aligned}$$

□

REMARK 4.5. Condition (b.iii) is automatically satisfied in case g is invertible and the inverse of χ is of the form χS, which is always true for an ordinary Hopf algebra. This is so, because in this case, by Corollary 2.12, one has that $S = \tau^r_\chi S \tau^l_\chi$. Since by condition $(b.i)$, $\sigma = \tau^l_\chi = Ad_g \tau^r_\chi$, we have $Ad_g S = \sigma S \sigma$, i.e. $gS(a)S(g) = \sigma(S(\sigma(a)))$, for all $a \in R$.

We present now an example of an Ore extension $R[x;\sigma]$ which extend the structure of a weak Hopf algebra R and that was not obtained by tensoring a Hopf algebra that admits a Hopf-Ore extension (in the sense of Panov) by a weak Hopf algebra.

EXAMPLE 4.6. (cf. [**2**, Example 9]) Let k be a field and N a positive integer not divisible by the characateristic of k. The algebraic quantum torus is defined as the algebra

$$R = k\left\langle U, V, V^{-1} \mid U^N = 1, VU = qUV\right\rangle.$$

where $q \in k$ satisfies $q^N = 1$. It is well known that R is a weak Hopf algebra with structure given by

$$\Delta(U^nV^m) = \frac{1}{N}\sum_{i=1}^{N}(U^{i+n}V^m \otimes U^{-i}V^m), \quad \epsilon(U^nV^m) = \begin{cases} N, & U^n = 1 \\ 0, & U^n \neq 1 \end{cases},$$

and $S(U^nV^m) = V^{-m}U^n$.

Now consider the automorphism $\sigma : R \to R$ given by $\sigma(U) = U$ and $\sigma(V) = UV$. Then $\sigma = \tau^l_\chi = \tau^r_\chi$ for the weak character $\chi \in R^*$ defined by $\chi(U^nV^m) = Nq^{m-1}$ if N divides $n+m$ and zero otherwise. Then, by Theorem 4.4 $R[x;\sigma]$ is a weak Hopf algebra extending the structure of R, with x being a $(1,1)$-primitive element.

5. Ore extensions of connected groupoid algebras

We finish by discussing the Ore extensions of the groupoid algebra $k\mathcal{G}$ of a "connected" groupoid $\mathcal{G}$, by which we mean a groupoid such that any two objects are connected by a morphism. We furthermore assume that the set of objects is finite. Then $R = k\mathcal{G} = M_n(kG)$, for some $n > 0$ and some group G. Panov characterised such Ore extensions for the case $n = 1$ (see [**9**, Proposition 2.2]).

PROPOSITION 5.1. *Let R be a cocommutative weak Hopf algebra, with automorphism σ and σ-derivation δ such that the weak Hopf algebra structure of R extends to one of $H = R[x;\sigma,\delta]$ with x being an $(g,1)$-primitive element, for some group-like element g, such that $S(x) = -S(g)x$, $\delta(R_s) = 0$ and the inverse of χ is χS, then g is central.*

PROOF. Since $\sigma(a) = \tau^l_\chi(a) = Ad_g\tau^r_\chi(a)$ by Theorem 4.4, we have

$$a = \chi(S(a_1))\sigma(a_2) = Ad_g\sigma(\sigma^{-1}(a)) = Ad_g(a).$$

Hence $ag = ga$, for all $a \in R$. □

The Proposition generalises [**9**, Corollary 1.4]. Moreover, the last proposition applies in particular to groupoid algebras $k\mathcal{G}$ of a connected groupoid with finitely many objects, i.e. $R = M_n(kG)$ for some $n > 0$ and group G. Generalising Example 2.5, any weak group-like element is of the form $\gamma = \sum_{i\in I} g_i E_{i,\sigma(i)}$ for some subset $I \subseteq \{1,\ldots,n\}$ and injective map $\sigma : I \to \{1,\ldots,n\}$ and elements $g_i \in G$. Moreover, if γ has a left inverse, then $I = \{1,\ldots,n\}$ and σ has to be a bijection. In particular γ has also a right inverse which is $S(\gamma) = \sum_{i=1}^n g_i^{-1} E_{\sigma(i),i}$. Furthermore the characters $\chi \in X_w(R)$ of $R = M_n(kG)$ are of the form $\chi(gE_{ij}) = p_i^{-1}p_j\rho(g)$, where $\rho \in kG^*$ is a group character and $p_1,\ldots,p_n \in k^\times$ are non-zero scalars. Clearly χS is the inverse of χ.

Let R be a cocommutative weak Hopf algebra, $\chi \in X(R)$ and g a central group-like element of R. Suppose that $\alpha \in R^*$ satisfies

$$\alpha(ab) = \alpha(a)\epsilon(b) + \chi(a)\alpha(b), \qquad \forall a, b \in R. \tag{8}$$

Then τ^l_α satisfies

$$\tau^l_\alpha(ab) = \alpha(a_1b_1)a_2b_2 = \alpha(a_1)a_2\epsilon(b_1)b_2 + \chi(a_1)a_2\alpha(b_1)b_2 = \tau^l_\alpha(a)b + \tau^l_\chi(a)\tau^l_\alpha(b). \tag{9}$$

Define the map $\delta : R \to R$, by $\delta(a) := (1-g)\tau^l_\alpha(a)$, for all $a \in R$. Since $1-g$ is central, we have $\delta(ab) = \delta(a)b + \tau^l_\chi(a)\delta(b)$, i.e. δ is an τ^l_χ-derivation. Furthermore δ is a $(g,1)$-coderivation, because

$$\begin{aligned}
\Delta\delta(a) &= \alpha(a_1)\Delta(1-g)\Delta(a_2)\\
&= \alpha(a_1)((1-g)\otimes 1 + g\otimes(1-g))\Delta(a_2)\\
&= \alpha(a_1)(1-g)a_2\otimes a_3 + ga_1\otimes(1-g)\alpha(a_2)a_3\\
&= \delta(a_1)\otimes a_2 + ga_1\otimes\delta(a_2) = (\delta\otimes \mathrm{id} + \lambda_g\otimes\delta)\Delta(a)
\end{aligned}$$

where we use the cocommutativity in the third equation. Note that $Ad_g = \mathrm{id}$ as g is central and $\tau^l_\chi = Ad_g\tau^r_\chi = \tau^r_\chi$, as R is cocommutative.

THEOREM 5.2. *Let G be any group, $n > 0$, $R = M_n(kG)$, $\chi \in X(R)$ and g a central element of G. Suppose that there exists $\alpha \in R^*$ such that $\alpha(ab) = \alpha(a)\epsilon(b) + \chi(a)\alpha(b)$ holds for any $a, b \in R$ and $\alpha(E_{ii}) = 0$ for any i. Let $\sigma = \tau^l_\chi$ and $\delta(a) := (1-g)\tau^l_\alpha(a)$, for all $a \in R$. Then δ is a σ-derivation and a $(g,1)$-coderivation such that $R[x;\tau^l_\chi,\delta]$ becomes a weak Hopf algebra with*

$$\Delta(x) = \Delta(1)(g\otimes x + x\otimes 1), \qquad \epsilon(HxH) = 0, \qquad S(x) = -g^{-1}x.$$

PROOF. We first observe that $\delta(R_s) = 0$. In fact, since every element of R_s is of the form $\sum_i \lambda_i E_{ii}$, with $\lambda_i \in k$, we have $\delta\left(\sum_i \lambda_i E_{ii}\right) = \sum_i \lambda_i\delta(E_{ii}) = \sum_i \lambda_i(1-g)\alpha(E_{ii})E_{ii} = 0$. Hence, in order to obtain our result, it is enough to check the conditions of the implication $(b) \Rightarrow (a)$ of Theorem 4.4. Conditions (i) and (ii) hold by the preceding discussion and because g is an invertible central element. Also, since χS is the inverse of χ, condition (iii) follows by Corollary 2.12. Thus, the only work to do is to prove condition (iv).

We begin by noting that under our hypotheses we have $\alpha(g_{ij}E_{ij}) = -\chi(g_{ij}E_{ij})\alpha(g_{ij}^{-1}E_{ji})$. In fact, because

$$\begin{aligned} 0 &= \alpha(E_{ii}) = \alpha(g_{ij}E_{ij}g_{ij}^{-1}E_{ji}) = \alpha(g_{ij}E_{ij})\epsilon(g_{ij}^{-1}E_{ji}) + \chi(g_{ij}E_{ij})\alpha(g_{ij}^{-1}E_{ji}) \\ &= \alpha(g_{ij}E_{ij}) + \chi(g_{ij}E_{ij})\alpha(g_{ij}^{-1}E_{ji}). \end{aligned}$$

Let $a = \sum_{i,j} g_{ij}E_{ij} \in R$. Then we obtain

$$\begin{aligned} &S(g)\delta(S(\sigma(a))) \\ &= g^{-1}\sum_{i,j=1}^{n}\chi(g_{ij}E_{ij})\delta(S(g_{ij}E_{ij})) = g^{-1}\sum_{i,j=1}^{n}\chi(g_{ij}E_{ij})\delta(g_{ij}^{-1}E_{ji}) \\ &= g^{-1}\sum_{i,j=1}^{n}\chi(g_{ij}E_{ij})\delta(g_{ij}^{-1}E_{ji}) = g^{-1}\sum_{i,j=1}^{n}\chi(g_{ij}E_{ij})(1-g)\alpha(g_{ij}^{-1}E_{ji})g_{ij}^{-1}E_{ji} \\ &= (g^{-1}-1)\sum_{i,j=1}^{n}\chi(g_{ij}E_{ij})\alpha(g_{ij}^{-1}E_{ji})g_{ij}^{-1}E_{ji} = (1-g^{-1})\sum_{i,j=1}^{n}\alpha(g_{ij}E_{ij})g_{ij}^{-1}E_{ji} \\ &= \sum_{i,j=1}^{n}\alpha(g_{ij}E_{ij})(g_{ij}^{-1}E_{ji} - g^{-1}g_{ij}^{-1}E_{ji}) = \sum_{i,j=1}^{n}\alpha(g_{ij}E_{ij})S((1-g)g_{ij}E_{ij}) \\ &= \sum_{i,j=1}^{n}S((1-g)\alpha(g_{ij}E_{ij})g_{ij}E_{ij}) = S(\delta(\sum_{i,j=1}^{n}g_{ij}E_{ij})) \\ &= S(\delta(a)). \end{aligned}$$

Multiplying by g on the left yields $\delta(S(\sigma(a))) = gS(\delta(a))$ and the proof is complete. $\square$

References

[1] Gabriella Böhm, Florian Nill, and Kornél Szlachányi, *Weak Hopf algebras. I. Integral theory and C^*-structure*, J. Algebra **221** (1999), no. 2, 385–438, DOI 10.1006/jabr.1999.7984. MR1726707

[2] Gabriella Böhm and José Gómez-Torrecillas, *On the double crossed product of weak Hopf algebras*, Hopf algebras and tensor categories, Contemp. Math., vol. 585, Amer. Math. Soc., Providence, RI, 2013, pp. 153–173, DOI 10.1090/conm/585/11614. MR3077236

[3] Ken A. Brown and Ken R. Goodearl, *Lectures on algebraic quantum groups*, Advanced Courses in Mathematics. CRM Barcelona, Birkhäuser Verlag, Basel, 2002. MR1898492

[4] K. A. Brown, S. O'Hagan, J. J. Zhang, and G. Zhuang, *Connected Hopf algebras and iterated Ore extensions*, J. Pure Appl. Algebra **219** (2015), no. 6, 2405–2433, DOI 10.1016/j.jpaa.2014.09.007. MR3299738

[5] Tomasz Brzezinski and Robert Wisbauer, *Corings and comodules*, London Mathematical Society Lecture Note Series, vol. 309, Cambridge University Press, Cambridge, 2003. MR2012570

[6] Yukio Doi, *Homological coalgebra*, J. Math. Soc. Japan **33** (1981), no. 1, 31–50, DOI 10.2969/jmsj/03310031. MR597479

[7] Dmitri Nikshych, *On the structure of weak Hopf algebras*, Adv. Math. **170** (2002), no. 2, 257–286, DOI 10.1006/aima.2002.2081. MR1932332

[8] Dmitri Nikshych and Leonid Vainerman, *Finite quantum groupoids and their applications*, New directions in Hopf algebras, Math. Sci. Res. Inst. Publ., vol. 43, Cambridge Univ. Press, Cambridge, 2002, pp. 211–262. MR1913440

[9] A. N. Panov, *Ore extensions of Hopf algebras* (Russian, with Russian summary), Mat. Zametki **74** (2003), no. 3, 425–434, DOI 10.1023/A:1026115004357; English transl., Math. Notes **74** (2003), no. 3-4, 401–410. MR2022506

Department of Mathematics of the Faculty of Science and CMUP, University of Porto, Rua Campo Alegre, 687, 4169-007 Porto, Portugal
Email address: clomp@fc.up.pt

Instituto de Matemática e Estatística, Universidade Federal do Rio Grande do Sul, Av. Bento Gonçalves, 9500, 91.509-900, Porto Alegre, Brazil
Email address: alveri@mat.ufrgs.br

Instituto de Matemática e Estatística, Universidade Federal do Rio Grande do Sul, Av. Bento Gonçalves, 9500, 91.509-900, Porto Alegre, Brazil
Email address: rilesantos@gmail.com

Contemporary Mathematics
Volume **727**, 2019
https://doi.org/10.1090/conm/727/14642

Duality for (skew-)polynomial codes and a variation of Construction A

Frédérique Oggier

ABSTRACT. We consider a construction of lattices from a specific order within totally definite quaternion algebras, via a variation of the so-called Construction A, which involves (skew-)polynomial codes. We study how the notion of code duality carries over to lattice duality.

1. Introduction

Let F be a totally real number field of degree n over $\mathbb{Q}$. Let $A = (a,b)_F$ be a quaternion division F-algebra with basis $\{1, i, j, ij\}$ such that $i^2 = a,\ j^2 = b,\ ij = -ji$, for non-zero $a, b \in F$. The quaternion algebra A is endowed with a canonical involution that maps $x = x_0 + x_1 i + x_2 j + x_3 ij$ to $\bar{x} = x_0 - x_1 i - x_2 j - x_3 ij$, a reduced trace $\mathrm{tr}_{A/F}$ such that $\mathrm{tr}_{A/F}(x) = x + \bar{x} = 2x_0$ and a reduced norm $\mathrm{n}_{A/F}(x) = x\bar{x}$. There are n real embeddings $\tau_1, \ldots, \tau_n$ of F, and

$$A_{\mathbb{R}} = A \otimes_{\mathbb{Q}} \mathbb{R} \simeq s\mathbb{H} \oplus (n-s)M_2(\mathbb{R})$$

where $\mathbb{H} = (-1,-1)_{\mathbb{R}}$, $M_2(\mathbb{R})$ is the set of 2×2 matrices with real coefficients, and s is the number of real embeddings for which $A \otimes_F F_\tau \simeq \mathbb{H}$, where F_τ is the completion of F with respect to τ. We will consider the case where A is ramified at all the real places (that is $A \otimes_F F_\tau \simeq \mathbb{H}$ for all real embeddings τ thus $a < 0$, $b < 0$), in which case A is called totally definite and $A_{\mathbb{R}} \simeq n\mathbb{H}$. For $x = (x_1, \ldots, x_n) \in A_{\mathbb{R}}$, the trace $\mathrm{tr} : (x,y) \mapsto \mathrm{tr}(x\bar{y}) = \sum_{l=1}^{n} \mathrm{tr}_{\mathbb{H}/\mathbb{R}}(x_l \bar{y}_l)$ is a positive definite symmetric bilinear form since

$$\mathrm{tr}(x\bar{x}) = \sum_{l=1}^{n} \mathrm{tr}_{\mathbb{H}/\mathbb{R}}(x_l \bar{x}_l) = \sum_{l=1}^{n} 2(x_l \bar{x}_l)$$

and $x_l \bar{x}_l = x_{l0}^2 - a x_{l1}^2 - b x_{l2}^2 + ab x_{l3}^2 \geq 0$ for a totally definite quaternion F-algebra A.

Let $K = F(i)$. Then K is a maximal subfield of A, with non-trivial Galois F-automorphism $\sigma : x_0 + x_1 i \mapsto x_0 - x_1 i$, $x_0, x_1 \in F$. The quaternion algebra A can also be described as a cyclic algebra $(F(i)/F, \sigma, b) = F(i) \oplus F(i)e$, where $(x_2 + x_3 i)e = e\sigma(x_2 + x_3 i)$ and $e^2 = b$. The canonical involution is then expressed by $\overline{(x_0 + x_1 i) + (x_2 + x_3 i)e} = \sigma(x_0 + x_1 i) - e\sigma(x_2 + x_3 i) = x_0 - x_1 i - x_2 e + x_3 ei = x_0 - x_1 i - (x_2 e + x_3 i)e$. Suppose that a, b belong to $\mathcal{O}_F$, the ring of integers of F so that $\Lambda = \mathcal{O}_K \oplus \mathcal{O}_K e$ is an $\mathcal{O}_F$-order of A. Let $\mathfrak{p}$ be a prime ideal of $\mathcal{O}_F$ above p, with inertial degree f, so that $\mathfrak{p}\Lambda$ is a two-sided ideal of Λ. There is [**1**, Lemma 2.1] an $\mathbb{F}_{p^f}$-algebra isomorphism between $\Lambda/\mathfrak{p}\Lambda$ and the skew-polynomial quotient ring $(\mathcal{O}_K/\mathfrak{p}\mathcal{O}_K)[X;\sigma]/(X^2-b)$. We will denote this isomorphism by ψ. Let $\mathcal{I}$ be a left ideal of Λ. Assume that $\mathfrak{p}\subset\mathcal{I}\cap\mathcal{O}_F$. Then $\mathcal{I}/\mathfrak{p}\Lambda$ is an ideal of $\Lambda/\mathfrak{p}\Lambda$, and $\psi(\mathcal{I}/\mathfrak{p}\Lambda)$ is a left ideal of $(\mathcal{O}_K/\mathfrak{p}\mathcal{O}_K)[X;\sigma]/(X^2-b)$. For $g(X)$ a right divisor of X^2-b, the ideal $g(X)/(X^2-b)$ forms a code of length 2 over $(\mathcal{O}_K/\mathfrak{p}\mathcal{O}_K)$ [**1**, Corollary 1], consisting of codewords (c_0,c_1) that are coefficient tuples of the left multiples $c(x)=c_0+c_1X$ of $g(X)$, and $g(X)$ is called a generator polynomial of C. We set the map :

$$\rho:\Lambda\to\psi(\Lambda/\mathfrak{p}\Lambda)=(\mathcal{O}_K/\mathfrak{p}\mathcal{O}_K)[X;\sigma]/(X^2-b),$$

compositum of the canonical projection $\Lambda\to\Lambda/\mathfrak{p}\Lambda$ with ψ. Then

$$\rho^{-1}(C)=\mathcal{I}$$

is a $\mathbb{Z}$-module of rank $4[F:\mathbb{Q}]$ since $\mathcal{O}_K$ is a $\mathbb{Z}$-module of rank $2[F:\mathbb{Q}]$. The above construction of ψ and ρ holds for a cyclic division algebra over K/F, where K/F is a cyclic number field extension [**1**], for which we then obtain codes of length n. The cyclic division algebra case is obviously richer than the quaternion one from the code view point, however we will restrict us to the quaternion case here, for which we have the positive definite symmetric bilinear form tr. Thus $L=(\rho^{-1}(C),\mathrm{tr})$ forms a lattice, that is a pair formed by a free $\mathbb{Z}$-module of finite rank, and a positive definite symmetric bilinear form.

The correspondence [**1**] between the code C generated by $g(X)$, and the lattice $L=(\rho^{-1}(C),\mathrm{tr})$ falls under the general umbrella of the so-called Construction A of lattices from linear codes [**2**, **3**]. In [**3**], a map $\rho:\mathbb{Z}[\zeta_p]\times\ldots\times\mathbb{Z}[\zeta_p]\to\mathbb{F}_p\times\ldots\times\mathbb{F}_p$ for ζ_p a primitive pth root of unity is considered. Then for a linear code (vector subspace) $C\subseteq\mathbb{F}_p\times\ldots\times\mathbb{F}_p$, $\rho^{-1}(C)$ is a lattice. In [**4**], $\mathbb{Z}[\zeta_p]$ is replaced by the ring of integers of some specific number fields. Lattice properties inherited from code properties are then studied. There are numerous similar types of constructions of lattices from codes, involving for instance codes over rings (e.g. [**5**]), or quaternion algebras (e.g. [**6**]). Constructions of lattices from quaternion algebras have also been investigated from a purely number theoretic view point (e.g. [**7**]).

In [**1**], the quotient $\Lambda/\mathfrak{p}\Lambda$ of the $\mathcal{O}_F$-order Λ by $\mathfrak{p}\Lambda$ was studied, and since $\Lambda/\mathfrak{p}\Lambda)=(\mathcal{O}_K/\mathfrak{p}\mathcal{O}_K)[X;\sigma]/(X^2-b)$, left ideals of this skew-polynomial ring define codes, and it is then natural to wonder which lattices correspond to these codes. In this note, we address this question for totally definite quaternion algebras, since they are endowed with a natural positive definite symmetric bilinear form. We show that the property of inclusion of a code into its dual results in integral lattices, as expected. However, the resulting lattices seem to be heavily constrained by the basis of Λ, as we further explore for the case of F the maximal totally real subfield of a cyclotomic field, it thus would be more interesting from the lattice view point to extend the construction to other orders. However, it does give some insight on the lattice structure corresponding to this specific choice of order Λ (see [**1**, **8**] for coding motivation behind this choice of Λ, or the variation of Construction A considered).

2. Basic Properties

Given a $(2,k)$ code C (of length 2 and dimension k) with generator polynomial $g(X)\in(\mathcal{O}_K/\mathfrak{p}\mathcal{O}_K)[X;\sigma]/(X^2-b)$, since ρ is a ring homomorphism, and C is isomorphic to a submodule (and thus subgroup) of $(\mathcal{O}_K/\mathfrak{p}\mathcal{O}_K)^2$ of index

$|\mathcal{O}_K/\mathfrak{p}\mathcal{O}_K|^{2-k}$, we have that $\rho^{-1}(C)$ is a subgroup of index $|\mathcal{O}_K/\mathfrak{p}\mathcal{O}_K|^{2-k}$ of Λ, that is $|\Lambda/\mathcal{I}| = |\mathcal{O}_K/\mathfrak{p}\mathcal{O}_K|^{2-k}$.

REMARK 2.1. The case $g(X) = X^2 - b$ is the trivial code $\{0\}$, $k = 0$. The case $g(X) = 1$ gives the universe code, $k = 2$ and $\mathcal{I} = \Lambda$.

Suppose that the left ideal $\mathcal{I}$ of Λ has an $\mathcal{O}_F$-basis $\{v_1, v_2, v_3, v_4\}$, and let $\{\nu_l\}_{l=1}^n$ be an integral basis of F. Then $\{\nu_l v_k\}$ is a $\mathbb{Z}$-basis for Λ and a Gram matrix for $L = (\mathcal{I}, \mathrm{tr})$ is given by that of the trace form $(\mathrm{tr}_{A/\mathbb{Q}}(\nu_l v_k \overline{\nu_j v_i}))$. Since $\nu_i, \nu_l \in F$, we have

$$\begin{aligned}
\mathrm{tr}_{A/\mathbb{Q}}(\nu_l v_k \overline{\nu_j v_i}) &= \mathrm{tr}_{A/\mathbb{Q}}(\nu_l \nu_j v_k \bar{v}_i) \\
&= \mathrm{Tr}_{F/\mathbb{Q}}\, \mathrm{tr}_{A/F}(\nu_l \nu_j v_k \bar{v}_i) \\
&= \sum_{m=1}^{n} \tau_m(\nu_l \nu_j) \tau_m(\mathrm{tr}_{A/F}(v_k \bar{v}_i))
\end{aligned}$$

and ordering the basis elements as $\nu_l v_1$, $l = 1, \ldots, n$, $\nu_l v_2$, $l = 1, \ldots, n$, until $\nu_l v_4$, $l = 1, \ldots, n$, shows that we obtain a matrix containing 16 blocks, one for each pair (v_k, v_i), $i, k = 1, \ldots, 4$ and each of these blocks is of the form

$$\begin{bmatrix} \tau_1(\nu_1) \ldots \tau_n(\nu_1) \\ \vdots \\ \tau_1(\nu_n) \ldots \tau_n(\nu_n) \end{bmatrix} \begin{bmatrix} \tau_1(\mathrm{tr}_{A/F}(v_k \bar{v}_i)) & & 0 \\ 0 & \ddots & 0 \\ 0 & & \tau_n(\mathrm{tr}_{A/F}(v_k \bar{v}_i)) \end{bmatrix} \begin{bmatrix} \tau_1(\nu_1) \ldots \tau_1(\nu_n) \\ \vdots \\ \tau_n(\nu_1) \ldots \tau_n(\nu_n) \end{bmatrix}.$$

A Gram matrix is thus given by

$$G = BHB^T, \; B = \mathrm{diag}((\tau_m(\nu_l))_{m,l}, (\tau_m(\nu_l))_{m,l}, (\tau_m(\nu_l))_{m,l}, (\tau_m(\nu_l))_{m,l})$$

and $H = (H_{ij})_{i,j} = \mathrm{diag}(\tau_1(\mathrm{tr}_{A/F}(v_i \bar{v}_j)), \ldots, \tau_n(\mathrm{tr}_{A/F}(v_i \bar{v}_j)))$, $i, j = 1, \ldots, 4$.

Recall that by definition, the discriminant d_F of F is:

$$d_F = (\det(\tau_i(\nu_j)_{i,j}))^2$$

where τ_i, $i = 1, \ldots, n$ are the embeddings of F into $\mathbb{R}$.

PROPOSITION 1. *The squared volume* $\mathrm{vol}(L)$ *of the lattice* $L = (\mathcal{I}, \mathrm{tr})$ *is*

$$d_F^4 N_{F/\mathbb{Q}}(\det(\mathrm{tr}_{A/F}(v_i \bar{v}_j))).$$

If $\mathcal{I} = (\alpha)$ *is principal, then it simplifies to*

$$d_F^4 N_{F/\mathbb{Q}}(\mathrm{n}_{A/F}(\alpha)^4) N_{F/\mathbb{Q}}(\mathrm{disc}(\Lambda/\mathcal{O}_F)).$$

PROOF. The squared volume $\mathrm{vol}(L)$ of the lattice $(\mathcal{I}, \mathrm{tr})$ is the determinant of G:

$$\mathrm{vol}(L)^2 = \det(G) = \det(BB^T)\det(H) = d_F^4 \det(H).$$

Now permute the rows and columns so that coefficients involving τ_i appear in the ith diagonal block, and

$$\det(H) = \prod_{l=1}^{n} \tau_l(\det(\mathrm{tr}_{A/F}(v_i \bar{v}_j))) = N_{F/\mathbb{Q}} \det(\mathrm{tr}_{A/F}(v_i \bar{v}_j)).$$

For $\mathcal{I} = (\alpha)$ principal, we have $v_i = w_i \alpha$ for a left ideal, where $\{w_1, w_2, w_3, w_4\}$ is a basis of Λ, so that

$$\mathrm{tr}_{A/F}(v_i \bar{v}_j) = \mathrm{tr}_{A/F}(w_i \alpha \overline{w_j \alpha}) = \mathrm{tr}_{A/F}(\alpha w_i \bar{w}_j \bar{\alpha})$$

and

$$N_{F/\mathbb{Q}} \det(\operatorname{tr}_{A/F}(v_i\bar{v}_j)) = N_{F/\mathbb{Q}} \det(\operatorname{n}_{A/F}(\alpha) \operatorname{tr}_{A/F}(w_i\bar{w}_j))$$

to get

$$N_{F/\mathbb{Q}}(\operatorname{n}_{A/F}(\alpha))^4 N_{F/\mathbb{Q}}(\operatorname{disc}(\Lambda/\mathcal{O}_F)).$$

□

This volume computation is standard (see also [**7**] for a similar argument), here the norm of $\mathcal{I}$ is constrained by the corresponding code.

3. Duality

A lattice L is integral if L is contained in its dual $L^* = \{y \in L \otimes_{\mathbb{Z}} \mathbb{R},\ \operatorname{tr}(x\bar{y}) \in \mathbb{Z}$ for all $x \in L\}$. It is unimodular if $L = L^*$.

For a code C of length 2, the notion of dual $C^\perp$ is given by $C^\perp = \{y \in (\mathcal{O}_K/\mathfrak{p}\mathcal{O}_K)^2,\ \langle x, y\rangle = 0$ for all $x \in C\}$. If $C = C^\perp$, the code C is said to be self-dual. If C is a self-dual code of length 2, then its dimension k is $k = 1$.

We know [**1**], assuming $b^2 = 1$ in $\mathcal{O}_K/\mathfrak{p}\mathcal{O}_K$, that the dual $(g^\perp(X))/(X^2 - b)$ of the σ-constacyclic code $(g(X))/(X^2 - b)$ is a σ-constacyclic code whose generator polynomial is given by $g^\perp(X)$ (or equivalently by the monic polynomial $-bg^\perp(X)$) defined as follows. Given $g(X)$, let $h(X)$ be the parity check polynomial such that $g(X)h(X) = X^2 - b$.

- If $k = 0$, $h(X) = 1$ and $g^\perp(X) = 1$.
- If $k = 1$, $h(X) = h_0 + X$ and $g^\perp(X) = 1 + \sigma(h_0)X$.
- If $k = 2$, $h(X) = h_0 + h_1X + X^2$ and $g^\perp(X) = 1 + \sigma(h_1)X + h_0X^2$.

Since there is an automorphism σ on $\mathcal{O}_K/\mathfrak{p}\mathcal{O}_K$ inherited from the F-automorphism σ (by abuse of notation, we write σ for both cases), we could also consider a dual code with respect to the inner product $\langle x, \sigma(y)\rangle$ in which case $C^\perp = \{y \in (\mathcal{O}_K/\mathfrak{p}\mathcal{O}_K)^2,\ \langle x, \sigma(y)\rangle = 0$ for all $x \in C\}$ where σ is understood componentwise. Then for $c_0 + c_1X \in (g(X))$, $\langle (c_0, c_1), (\sigma(d_0), \sigma(d_1))\rangle = 0$ if and only if $\sigma(d_0) + \sigma(d_1)X \in (g^\perp(X))$ if and only if $d_0 + d_1X \in \sigma(g^\perp(X))$. So for $k = 1$, $\sigma(g^\perp(X)) = 1 + h_0X$. We will be interested in duality with respect to σ (sometimes, σ becomes the identity). To get a self-dual code, we thus need $g(X) = h(X)$, $b = \beta^2$ to be a square modulo $\mathfrak{p}$ and $X^2 - b = (X - \beta)^2$.

PROPOSITION 2. *Let C be a non-trivial code generated by $g(X)$ in $(\mathcal{O}_K/\mathfrak{p}\mathcal{O}_K)[X;\sigma]/(X^2 - b)$ where $b \equiv -1 \mod \mathfrak{p}$. If $C \subseteq C^\perp$ and $p\mathcal{O}_F = p$ ($\mathfrak{p} = p$) or $p\mathcal{O}_F = \mathfrak{p}^n$ then*

(1) $\frac{1}{p}\operatorname{tr}(x\bar{y}) \in \mathbb{Z}$ *for* $x, y \in \rho^{-1}(C)$, $n = [F : \mathbb{Q}]$,

(2) *we have the inclusion* $(\rho^{-1}(C^\perp), \frac{1}{p}\operatorname{tr}) \subseteq (\rho^{-1}(C), \frac{1}{p}\operatorname{tr})^*$.

PROOF. Consider the additive structure of ρ, that is, $\rho : \mathcal{O}_K \times \mathcal{O}_K \to \mathcal{O}_K/\mathfrak{p}\mathcal{O}_K \times \mathcal{O}_K/\mathfrak{p}\mathcal{O}_K$, which maps $(c_0, c_1) \mapsto (c_0 \mod \mathfrak{p}, c_1 \mod \mathfrak{p})$ where $c_0 + c_1e \in \Lambda\mathcal{I}$ and $(c_0 \mod \mathfrak{p}, c_1 \mod \mathfrak{p})$ is a codeword in C. When $C \subseteq C^\perp$, we have $c_0\sigma(d_0) + c_1\sigma(d_1) \equiv 0 \mod \mathfrak{p}\mathcal{O}_K$. Now for $c_0 + c_1e, d_0 + d_1e \in \Lambda\mathcal{I}$, $\operatorname{tr}_{A/F}((c_0 + c_1e)\overline{(d_0 + d_1e)}) = \operatorname{tr}_{A/F}(c_0\sigma(d_0) + c_1\sigma(d_1)(-b)) = \operatorname{tr}_{K/F}(c_0\sigma(d_0) + c_1\sigma(d_1)(-b))$ where $c_0\sigma(d_0) + c_1\sigma(d_1)(-b) \in \mathfrak{p}\mathcal{O}_K$. Since $\mathfrak{p} \in \mathcal{O}_F$, $\sigma(\mathfrak{p}) = \mathfrak{p}$ and $\operatorname{tr}_{K/F}(c_0\sigma(d_0) + c_1\sigma(d_1)(-b)) \in \mathfrak{p}\mathcal{O}_F$. Then if $p\mathcal{O}_F = p$ ($\mathfrak{p} = p$) or $p\mathcal{O}_F = \mathfrak{p}^n$, then $\operatorname{tr}_{F/\mathbb{Q}} \operatorname{tr}_{K/F}(c_0\sigma(d_0) + c_1\sigma(d_1)(-b)) \in p\mathbb{Z}$.

Now instead of taking $c_0 + c_1 e, d_0 + d_1 e \in \mathcal{I} = \rho^{-1}(C)$, take $c_0 + c_1 e \in \mathcal{I} = \rho^{-1}(C)$ and $d_0 + d_1 e \in \rho^{-1}(C^\perp)$. Repeating the above argument, $\mathrm{tr}_{F/\mathbb{Q}} \mathrm{tr}_{A/F}((c_0 + c_1 e)\overline{(d_0 + d_1 e)}) \in p\mathbb{Z}$. So we just showed that for $d_0 + d_1 e \in \mathcal{I} = \rho^{-1}(C^\perp)$, its inner product with any lattice point is in $p\mathbb{Z}$ and therefore $d_0 + d_1 e$ belongs to $\rho^{-1}(C)^*$ with respect to the scaled trace form $\frac{1}{p}$. □

COROLLARY 1. *Under the hypotheses of the above proposition, if* $\mathrm{vol}(\rho^{-1}(C^\perp), \frac{1}{p}\mathrm{tr}) = 1$, *then* $\rho^{-1}(C)$ *is unimodular.*

PROOF. If $C \subseteq C^\perp$, then $(\rho^{-1}(C), \frac{1}{p}\mathrm{tr})$ is an integral lattice, therefore it is included into its dual. Now having volume 1 means that the volume of the lattice and that of its dual are equal, showing that both lattices are isomorphic. □

COROLLARY 2. *The quaternion algebra* $A = (-1,-1)_{\mathbb{Q}}$ *with the above construction and* C *unimodular gives the lattice* $\rho^{-1}(C) = \mathbb{Z}^4$.

PROOF. Set $F = \mathbb{Q}$, $K = \mathbb{Q}(i)$, its non-trivial Galois automorphism is $\sigma : a + ib \mapsto a - ib$, $a, b \in \mathbb{Q}$. The quaternion algebra A is division, since $N_{K/F}(a + ib) = a^2 + b^2$, $a, b \in \mathbb{Q}$, and thus -1 cannot be a norm. We consider the $\mathbb{Z}$-order $\Lambda = \mathbb{Z}[i] \oplus \mathbb{Z}[i]e$. Take p a prime in $\mathbb{Z}$. We consider the quotient $\Lambda/p\Lambda$, and the ideal $(g_0 + X)(\mathbb{Z}[i]/p\mathbb{Z}[i])[X;\sigma]/(X^2+1)$ forms a (consta)cyclic code of dimension $k = 1$ and length $n = 2$. Then $\rho^{-1}(C) = (\alpha)$, $\rho(\alpha) = g_0 + X$ and since $\mathrm{n}(\alpha) \in \mathbb{Z}$. This shows that $(\rho^{-1}(C), \frac{1}{2\,\mathrm{n}(\alpha)}\mathrm{tr})$ is integral, with squared volume $\frac{1}{2\,\mathrm{n}(\alpha)^4}\mathrm{n}(\alpha)^4 \mathrm{disc}(\Lambda/\mathbb{Z}) = 1$ since $\mathrm{tr}(1) = \mathrm{tr}(i\bar{i}) = \mathrm{tr}(j\bar{j}) = \mathrm{tr}(ij\bar{ij}) = 2$ and $\mathrm{tr}(x\bar{y}) = 0$ for $x, y \in \{1, i, j, ij\}$, giving $\mathrm{disc}(\Lambda/\mathbb{Z}) = 2^4$. Corollary 1 tells us that the lattice is unimodular, so in dimension 4, it has to be the lattice $\mathbb{Z}^4$. □

EXAMPLE 3.1. We consider the quaternion algebra $A = (-1,-1)_{\mathbb{Q}}$ with $p = 2, 5$ and the $\mathbb{Z}$-order $\Lambda = \mathbb{Z}[i] \oplus \mathbb{Z}[i]j$. Set $p = 2$, which is ramified in $\mathbb{Q}(i)$: $(2) = (1+i)^2$. Then $\mathbb{Z}[i]/2\mathbb{Z}[i] \simeq \mathbb{F}_2 + \upsilon\mathbb{F}_2 = \{0, 1, \upsilon, \upsilon+1\}$ with $\upsilon^2 = 0$, $\Lambda/2\Lambda \simeq (\mathbb{F}_2 + \upsilon\mathbb{F}_2) \oplus (\mathbb{F}_2 + \upsilon\mathbb{F}_2)j$ so

$$\Lambda/2\Lambda \simeq (\mathbb{F}_2 + \upsilon\mathbb{F}_2)[X;\sigma]/(X^2+1) \simeq (\mathbb{F}_2 + \upsilon\mathbb{F}_2)[X]/(X^2+1).$$

In this case, the natural projection π from $\mathbb{Z}[i]$ to $\mathbb{F}_2 + \upsilon\mathbb{F}_2$ is given by $\pi : a + ib \mapsto (a+b) \mod 2 + \upsilon(b \mod 2)$ and σ becomes the identity map.

Take $\mathcal{I} = \Lambda(1+j)$, which contains 2 since the reduced norm of $(1+j)$ is $\mathrm{n}(1+j) = (1+j)(1-j) = 2$. Then

$$\psi((1+j)\mathrm{mod}2) = 1 + X.$$

The ideal $(1+X)(\mathbb{F}_2 + \upsilon\mathbb{F}_2)[X]/(X^2+1)$ forms a cyclic code of length $n = 2$ and dimension $k = 1$. Codewords are of the form $(b_0 + b_1 X)(1+X)$, that is we get the repetition code

$$(a_0, a_1) = (a_0, a_0),\ a_0 \in \mathbb{F}_2 + \upsilon\mathbb{F}_2.$$

We have $X^2 + 1 = (X+1)(X+1) \mod 2$ therefore the parity check polynomial is $h(X) = X + 1$ while the dual generator polynomial is $g^\perp(X) = 1 + \sigma(h_0)X = 1 + X = g(X)$ and the code is self-dual. We can also check that $\langle (c_0, c_0), (a_0, a_0) \rangle = 0 \mod 2$ for all a_0, c_0.

A basis for the ideal $\mathcal{I} = \Lambda(1+j)$ is $v_1 = 1+j, v_2 = i+ij, v_3 = j-1, v_4 = ij-i$. We can compute

$$\operatorname{tr}_{A/\mathbb{Q}}(v_k\bar{v}_l) = 2\begin{bmatrix} 2 & 0 & 0 & 0 \\ 0 & 2 & 0 & 0 \\ 0 & 0 & 2 & 0 \\ 0 & 0 & 0 & 2 \end{bmatrix}$$

and indeed $\frac{1}{2\,\mathrm{n}(1+j)}\operatorname{tr}(v_k\bar{v}_l)$ is the identity matrix.

Set now $p = 5$. Take $\mathcal{I} = \Lambda(2+j)$. Then $\psi((2+j) \mod 5) = 2+X$, and the ideal $(2+X)(\mathbb{F}_5\times\mathbb{F}_5)[X;\sigma]/(X^2+1)$ gives codewords of the form $(b_0+b_1X)(2+X)$, that is, $(3(a_0,a_1),(a_0,a_1))$. The action of σ on $\mathbb{F}_5\times\mathbb{F}_5$ permutes the two components, thus the dual $C^\perp$ with respect to σ is the code C itself.

For a basis for the ideal $\mathcal{I} = \Lambda(2+j)$ is $v_1 = 2+j, v_2 = 2i+ij, v_3 = 2j-1, v_4 = 2ij-i$. We can compute

$$\operatorname{tr}_{A/\mathbb{Q}}(v_i\bar{v}_j) = 2\begin{bmatrix} 5 & 0 & 0 & 0 \\ 0 & 5 & 0 & 0 \\ 0 & 0 & 5 & 0 \\ 0 & 0 & 0 & 5 \end{bmatrix}$$

and indeed $\frac{1}{2\,\mathrm{n}(2+j)}\operatorname{tr}(v_i\bar{v}_j)$ is the identity matrix.

4. The Cyclotomic Case

A lattice is l-modular for a positive integer l if it is integral and there exists an isomorphism $f: L \to L^*$ such that $l\operatorname{tr}(f(x)\bar{f(y)}) = \operatorname{tr}(x\bar{y})$ for all $x, y \in L$. The volume of an l-modular lattice $\rho^{-1}(C)$ satisfies

$$\operatorname{vol}(\rho^{-1}(C)) = \operatorname{vol}(l\rho^{-1}(C)^*) = l^{4n}\operatorname{vol}(\rho^{-1}(C)^*) = l^{4n}\operatorname{vol}(\rho^{-1}(C))^{-1}$$

and $\operatorname{vol}(\rho^{-1}(C)) = l^{2n}$. So for l-modularity, we need to show that $\frac{1}{\sqrt{l}}\rho^{-1}(C) \subseteq \rho^{-1}(C)^*$ and $\operatorname{vol}(\rho^{-1}(C)) = l^{2n}$ for $\rho^{-1}(C)$ integral.

Lemma 4.1. *For $F = \mathbb{Q}(\zeta+\zeta^{-1})$ and $\zeta = \zeta_{p^m}$ a primitive p^mth root of unity, consider the quaternion algebra $A = ((\bar{\zeta}-\zeta)^2, b)_F$, $b < 0$. Then $N_{F/\mathbb{Q}}\operatorname{disc}(\Lambda/\mathcal{O}_F) = p^2 N_{F/\mathbb{Q}}(b^2)$ for p odd and $N_{F/\mathbb{Q}}\operatorname{disc}(\Lambda/\mathcal{O}_F) = 2^4 N_{F/\mathbb{Q}}(b^2)$ for $p = 2$.*

Proof. Note that $(\bar{\zeta}-\zeta)^2 < 0$. An integral basis of $\mathbb{Z}[\zeta]$ is $\{1, \zeta, \zeta+\zeta^{-1}, (\zeta+\zeta^{-1})\zeta, \ldots, (\zeta+\zeta^{-1})^{n-1}\zeta\}$, thus an $\mathcal{O}_F$-basis of $\mathbb{Z}[\zeta]$ is $\{1,\zeta\}$ and an $\mathcal{O}_F$-basis of $\Lambda = \mathbb{Z}[\zeta]\oplus\mathbb{Z}[\zeta]e$ is $\{1,\zeta,j,\zeta j\}$. Then for $v_k, v_l \in \{1,\zeta,j,\zeta j\}$

$$\operatorname{tr}_{A/F}(v_k\bar{v}_l) = \begin{bmatrix} 2 & \operatorname{tr}(\bar{\zeta}) & 0 & 0 \\ \operatorname{tr}(\bar{\zeta}) & 2 & 0 & 0 \\ 0 & 0 & -2b & -b\operatorname{tr}(\bar{\zeta}) \\ 0 & 0 & -b\operatorname{tr}(\bar{\zeta}) & -2b \end{bmatrix}.$$

Since $i = \bar{\zeta}-\zeta$, and $\zeta = \frac{(\bar{\zeta}+\zeta)-(\bar{\zeta}-\zeta)}{2} = \frac{(\bar{\zeta}+\zeta)-i}{2}$, we have that $\operatorname{tr}(\bar{\zeta}) = \operatorname{tr}(\zeta) = \bar{\zeta}+\zeta$, and

$$\operatorname{tr}_{A/F}(v_k\bar{v}_l) = \begin{bmatrix} 2 & \bar{\zeta}+\zeta & 0 & 0 \\ \bar{\zeta}+\zeta & 2 & 0 & 0 \\ 0 & 0 & -2b & -b(\bar{\zeta}+\zeta) \\ 0 & 0 & -b(\bar{\zeta}+\zeta) & -2b \end{bmatrix}.$$

Then $\det(\mathrm{tr}(v_k\bar{v}_l)) = b^2(\bar{\zeta}-\zeta)^4$ and $N_{F/\mathbb{Q}}(b^2(\bar{\zeta}-\zeta)^4) = p^2 N_{F/\mathbb{Q}}(b^2)$. Indeed, set $K=\mathbb{Q}(\zeta_{p^m})$ and $N_{K/\mathbb{Q}}(\bar{\zeta}-\zeta)=N_{K/\mathbb{Q}}(\bar{\zeta})N_{K/\mathbb{Q}}(1-\zeta)N_{K/\mathbb{Q}}(1+\zeta)$ and $N_{K/\mathbb{Q}}(1-\zeta)=p$. Recall that $N_{K/\mathbb{Q}}(\zeta)$ is the constant term of the minimal polynomial of ζ which is $\Phi(X) = \sum_{j=0}^{p-1} X^{jp^{m-1}}$. Thus $N_{K/\mathbb{Q}}(\zeta+1)$ is the constant term of $\Phi(X-1)$ which is 1 for p odd and 2 for $p=2$. Then $N_{K/\mathbb{Q}}(\bar{\zeta}-\zeta) = N_{F/\mathbb{Q}}N_{K/F}(\bar{\zeta}-\zeta) = N_{F/\mathbb{Q}}((\bar{\zeta}-\zeta)(\zeta-\bar{\zeta})) = -N_{F/\mathbb{Q}}(\bar{\zeta}-\zeta)^2 = p$ for p odd and 4 for $p=2$. □

COROLLARY 3. *The squared volume* $\mathrm{vol}(L)$ *of the lattice* $L=(\Lambda,\mathrm{tr})$ *is*

$$p^{2(p^{m-1}(pm-m-1)-1)}p^2 N_{F/\mathbb{Q}}(b^2)$$

for p odd and

$$2^{2(2^{m-1}(m-1)-2)}2^4 N_{F/\mathbb{Q}}(b^2)$$

for $p=2$.

PROOF. We are left to compute d_F.

The discriminant of $K=\mathbb{Q}(\zeta_{p^m})$ is well known [**9**, Prop.2.1].

Set $\zeta=\zeta_{p^m}$. A $\mathbb{Q}$-basis for K is $\{1,\zeta,\dots,\zeta^{p^{m-1}(p-1)}\}$. A $\mathbb{Q}$-basis for $F = \mathbb{Q}(\zeta+\zeta^{-1})$ is $\{1,\zeta+\zeta^{-1},\dots,(\zeta+\zeta^{-1})^{p^{m-1}\frac{p-1}{2}}\}$. Also an F-basis for K is $\{1,\zeta\}$.

It is known that

$$\begin{aligned}&\mathrm{disc}_{K/\mathbb{Q}}(1,\dots,(\zeta+\zeta^{-1})^{p^{m-1}\frac{p-1}{2}},\zeta,\dots,\zeta(\zeta+\zeta^{-1})^{p^{m-1}\frac{p-1}{2}})\\ =\;&(\mathrm{disc}_{F/\mathbb{Q}}(1,\dots,(\zeta+\zeta^{-1})^{p^{m-1}\frac{p-1}{2}})^{[K:F]}N_{F/\mathbb{Q}}(\mathrm{disc}_{K/F}(1,\zeta)).\end{aligned}$$

Since $\{1,\dots,(\zeta+\zeta^{-1})^{p^{m-1}\frac{p-1}{2}},\zeta,\dots,\zeta(\zeta+\zeta^{-1})^{p^{m-1}\frac{p-1}{2}}\}$ is an integral basis of K and $\{1,\dots,(\zeta+\zeta^{-1})^{p^{m-1}\frac{p-1}{2}}\}$ is an integral basis of F, this simplifies to

$$d_K = d_F^2 N_{F/\mathbb{Q}}(\mathrm{disc}_{K/F}(1,\zeta)).$$

Then

$$\begin{aligned}\mathrm{disc}_{K/F}(1,\zeta) &= \det\left(\begin{bmatrix}1 & 1\\ \zeta & \bar{\zeta}\end{bmatrix}\begin{bmatrix}1 & \zeta\\ 1 & \bar{\zeta}\end{bmatrix}\right)\\ &= 4-(\zeta+\bar{\zeta})^2 = 2-(\zeta^2+\bar{\zeta}^2)\\ &= (\zeta-\bar{\zeta})(\bar{\zeta}-\zeta) = N_{K/F}(\zeta-\bar{\zeta})\end{aligned}$$

and

$$N_{F/\mathbb{Q}}(\mathrm{disc}_{K/F}(1,\zeta)) = N_{K/\mathbb{Q}}(\zeta-\bar{\zeta}) = N_{K/\mathbb{Q}}(\zeta)^{-1}N_{K/\mathbb{Q}}(\zeta-1)N_{K/\mathbb{Q}}(\zeta+1).$$

Thus for p odd,

$$d_K = pd_F^2 \Rightarrow |d_F| = p^{\frac{p^{m-1}(pm-m-1)-1}{2}}$$

and for $p=2$

$$d_K = 4d_F^2 \Rightarrow |d_F| = 2^{\frac{2^{m-1}(m-1)-2}{2}}.$$

□

Since $[\mathbb{Q}(\zeta+\zeta^{-1}):\mathbb{Q}] = \frac{p^{m-1}(p-1)}{2}$, the lattice (Λ,tr) has dimension $2p^{m-1}(p-1)$. Then the lattice $(\Lambda,\frac{1}{p}\mathrm{tr})$ has volume $\frac{p^{2p^{m-1}(pm-m-1)}}{p^{2p^{m-1}(p-1)}}$. For $b=-1$, a necessary condition for p-modularity is $2p^{m-1}(pm-m-1)-2p^{m-1}(p-1) = p^{m-1}(p-1)$ which gives $2m=(3p-1)/(p-1)$ and $p=3$ with $m=2$. A similar computation can be done with $p=2$ instead of 3. It turns out that 2- and 3-modular lattices respectively can be obtained, but the end results in terms of direct sums again shows the underlying basis structure of the order Λ.

EXAMPLE 4.2. Consider the quaternion algebra $A = (-1,-2)_{\mathbb{Q}(\sqrt{2})}$. Take $F = \mathbb{Q}(\sqrt{2})$ which is totally real, with embeddings $\sigma_1(\sqrt{2}) = \sqrt{2}$ and $\sigma_2(\sqrt{2}) = -\sqrt{2}$, and consider the quaternion algebra $(-1,-2)_F$. Then $K = \mathbb{Q}(\sqrt{2}, i) = \mathbb{Q}(\zeta_8)$ with $\zeta_8 = \frac{\sqrt{2}}{2}(1+i)$ a primitive 8th root of unity, and A is division since $N_{K/F}(a+ib) = a^2+b^2$, $a,b \in \mathbb{Q}(\sqrt{2})$.

We have $\mathcal{O}_K = \mathbb{Z}[\zeta_8]$, $\mathcal{O}_F = \mathbb{Z}[\sqrt{2}]$, and we consider the $\mathbb{Z}[\sqrt{2}]$-order $\Lambda = \mathcal{O}_K \oplus \mathcal{O}_K j$.

A basis of $\mathbb{Z}[\zeta_8]$ is $\{1, \zeta_8, \zeta_8+\zeta_8^{-1}, (\zeta_8+\zeta_8^{-1})\zeta_8\}$, thus an $\mathcal{O}_F$-basis of $\mathbb{Z}[\zeta_8]$ is $\{1,\zeta_8\}$ and an $\mathcal{O}_F$-basis of Λ is $\{1,\zeta_8, j, \zeta_8 j\}$.

For a basis $\{v_1,v_2,v_3,v_4\}$ of Λ

$$\mathrm{tr}_{A/F}(v_i\bar{v}_j) = \begin{bmatrix} 2 & \sqrt{2} & 0 & 0 \\ \sqrt{2} & 2 & 0 & 0 \\ 0 & 0 & 4 & 2\sqrt{2} \\ 0 & 0 & 2\sqrt{2} & 4 \end{bmatrix},$$

$\mathrm{vol}(L) = d_F^4 N_{F/\mathbb{Q}}(\det(\mathrm{tr}_{A/F}(v_i\bar{v}_j))) = 8^4 N_{F/\mathbb{Q}}(2^4)$ (see Proposition 1) and a Gram matrix is given by

$$B\begin{bmatrix} 2 & 0 & \sqrt{2} & 0 & 0 & 0 & 0 & 0 \\ 0 & 2 & 0 & -\sqrt{2} & 0 & 0 & 0 & 0 \\ \sqrt{2} & 0 & 2 & 0 & 0 & 0 & 0 & 0 \\ 0 & -\sqrt{2} & 0 & 2 & 0 & 0 & 0 & 0 \\ 0 & 0 & 0 & 0 & 4 & 0 & 2\sqrt{2} & 0 \\ 0 & 0 & 0 & 0 & 0 & 4 & 0 & -2\sqrt{2} \\ 0 & 0 & 0 & 0 & 2\sqrt{2} & 0 & 4 & 0 \\ 0 & 0 & 0 & 0 & 0 & -2\sqrt{2} & 0 & 4 \end{bmatrix}B^T$$

with

$$B = \mathrm{diag}\left(\begin{bmatrix} 1 & 1 \\ \sqrt{2} & -\sqrt{2} \end{bmatrix}, \begin{bmatrix} 1 & 1 \\ \sqrt{2} & -\sqrt{2} \end{bmatrix}, \begin{bmatrix} 1 & 1 \\ \sqrt{2} & -\sqrt{2} \end{bmatrix}, \begin{bmatrix} 1 & 1 \\ \sqrt{2} & -\sqrt{2} \end{bmatrix}\right)$$

We finally obtain

$$\begin{bmatrix} 4 & 0 & 0 & 4 & 0 & 0 & 0 & 0 \\ 0 & 8 & 4 & 0 & 0 & 0 & 0 & 0 \\ 0 & 4 & 4 & 0 & 0 & 0 & 0 & 0 \\ 4 & 0 & 0 & 8 & 0 & 0 & 0 & 0 \\ 0 & 0 & 0 & 0 & 8 & 0 & 0 & 8 \\ 0 & 0 & 0 & 0 & 0 & 16 & 8 & 0 \\ 0 & 0 & 0 & 0 & 0 & 8 & 8 & 0 \\ 0 & 0 & 0 & 0 & 8 & 0 & 0 & 16 \end{bmatrix}.$$

The lattice $\frac{1}{4}\mathrm{tr}$ is integral. Reordering the rows and columns highlights the block structure of the Gram matrix, we thus have a direct sum of four lattices:

$$G = \begin{bmatrix} 2 & 1 & 0 & 0 & 0 & 0 & 0 & 0 \\ 1 & 1 & 0 & 0 & 0 & 0 & 0 & 0 \\ 0 & 0 & 2 & 1 & 0 & 0 & 0 & 0 \\ 0 & 0 & 1 & 1 & 0 & 0 & 0 & 0 \\ 0 & 0 & 0 & 0 & 2 & 2 & 0 & 0 \\ 0 & 0 & 0 & 0 & 2 & 4 & 0 & 0 \\ 0 & 0 & 0 & 0 & 0 & 0 & 4 & 2 \\ 0 & 0 & 0 & 0 & 0 & 0 & 2 & 2 \end{bmatrix}$$

and we can check that this lattice 2-modular by computing $P2G^{-1}P^T$ with

$$P = \begin{bmatrix} 0 & 0 & 0 & 0 & -1 & 0 & 0 & 0 \\ 0 & 0 & 0 & 0 & 0 & 1 & 0 & 0 \\ 0 & 0 & 0 & 0 & 0 & 0 & 0 & -1 \\ 0 & 0 & 0 & 0 & 0 & 0 & 1 & 0 \\ -1 & 0 & 0 & 0 & 0 & 0 & 0 & 0 \\ 0 & 1 & 0 & 0 & 0 & 0 & 0 & 0 \\ 0 & 0 & 0 & -1 & 0 & 0 & 0 & 0 \\ 0 & 0 & 1 & 0 & 0 & 0 & 0 & 0 \end{bmatrix}.$$

EXAMPLE 4.3. The same computations with $p = 3$ with $m = 2$ gives a block diagonal matrix G, both blocks being (after normalization by 3):

$$\begin{bmatrix} 2 & 0 & 0 & 0 & 2 & -1 \\ 0 & 4 & -2 & 2 & -1 & 2 \\ 0 & -2 & 4 & -1 & 2 & -1 \\ 0 & 2 & -1 & 2 & 0 & 0 \\ 2 & -1 & 2 & 0 & 4 & -2 \\ -1 & 2 & -1 & 0 & -2 & 4 \end{bmatrix}.$$

The matrix Q below applied on both blocks of $3G^{-1}$ gives the transformation that shows 3-modularity:

$$Q = \begin{bmatrix} 0 & 0 & 0 & 0 & 0 & -1 \\ 0 & 0 & 0 & 0 & -1 & 0 \\ 0 & 0 & 0 & -1 & 0 & 1 \\ 0 & 0 & 1 & 0 & 0 & 0 \\ 0 & 1 & 0 & 0 & 0 & 0 \\ 1 & 0 & -1 & 0 & 0 & 0 \end{bmatrix}.$$

References

[1] J. Ducoat and F. Oggier, *On skew polynomial codes and lattices from quotients of cyclic division algebras*, Adv. Math. Commun. **10** (2016), no. 1, 79–94, DOI 10.3934/amc.2016.10.79. MR3471062

[2] J. H. Conway and N. J. A. Sloane, *Sphere packings, lattices and groups*, 3rd ed., Grundlehren der Mathematischen Wissenschaften [Fundamental Principles of Mathematical Sciences], vol. 290, Springer-Verlag, New York, 1999. With additional contributions by E. Bannai, R. E. Borcherds, J. Leech, S. P. Norton, A. M. Odlyzko, R. A. Parker, L. Queen and B. B. Venkov. MR1662447

[3] W. Ebeling, *Lattices and codes: A course partially based on lectures by F. Hirzebruch*, Advanced Lectures in Mathematics, Friedr. Vieweg & Sohn, Braunschweig, 1994. MR1280458

[4] W. Kositwattanarerk, S. S. Ong, and F. Oggier, *Construction A of lattices over number fields and block fading (wiretap) coding*, IEEE Trans. Inform. Theory **61** (2015), no. 5, 2273–2282, DOI 10.1109/TIT.2015.2416340. MR3342276

[5] Y. J. Choie, S. Dougherty, and H. Liu, *Jacobi forms and Hilbert-Siegel modular forms over totally real fields and self-dual codes over polynomial rings $\mathbb{Z}_{2m}[x]/\langle g(x)\rangle$*, Ars Combin. **107** (2012), 141–160. MR3012996

[6] C. Bachoc, *Applications of coding theory to the construction of modular lattices*, J. Combin. Theory Ser. A **78** (1997), no. 1, 92–119, DOI 10.1006/jcta.1996.2763. MR1439633

[7] X. Hou, *Construction of Arakelov-modular lattices over totally definite quaternion algebras*, Int. J. Number Theory **13** (2017), no. 7, 1855–1880, DOI 10.1142/S1793042117501020. MR3667495

[8] F. Oggier, J.-C. Belfiore, "Enabling multiplication in lattice codes via Construction A", *ITW 2013.*

[9] L. C. Washington, *Introduction to cyclotomic fields*, Graduate Texts in Mathematics, vol. 83, Springer-Verlag, New York, 1982. MR718674

Division of Mathematical Sciences, School of Physical and Mathematical Sciences, Nanyang Technological University, Singapore 639798

Email address: `frederique@ntu.edu.sg`

Contemporary Mathematics
Volume **727**, 2019
https://doi.org/10.1090/conm/727/14643

A class of semisimple Hopf algebras acting on quantum polynomial algebras

Deividi Pansera

ABSTRACT. We construct a class of non-commutative, non-cocommutative, semisimple Hopf algebras of dimension $2n^2$ and present conditions to define an inner faithful action of these Hopf algebras on quantum polynomial algebras, providing, in this way, more examples of semisimple Hopf actions which do not factor through group actions. Also, under certain condition, we classify the inner faithful Hopf actions of the Kac-Paljutkin Hopf algebra of dimension 8, H_8, on the quantum plane.

1. Introduction

Suppose that H is a finite-dimensional Hopf algebra over a field F acting on an algebra A. If I is a Hopf ideal such that $I \cdot A = 0$, we say that the action *factors through* a quotient Hopf algebra H/I. One says that the action *factors through a group action* if there exists a Hopf ideal I of H, with $I \cdot A = 0$, such that $H/I \cong F[G]$ as Hopf algebra for some group G. In this last scenario, the Hopf action can be seen, in a certain sense, as a group action. Then, the following question arises: Are there conditions on either H or A ensuring that the action factors through a group action?

The first general result appeared in [**10**]. Assuming that H is semisimple and F is algebraically closed, the authors showed that if A is a commutative domain, then the action factors thorough a group action in this setting. Cuadra, Etingof and Walton, in [**7**], showed that that is also the case for the Weyl algebra $A = A_n(F)$, i.e., they showed that any semisimple Hopf action on the Weyl algebra $A_n(F)$ must factor through a group action.

In [**14**], by using and analyzing the results obtained by Cuadra, Etingof and Walton, it was showed that any semisimple Hopf action over an algebraically closed field of characteristic zero on an skew polynomial ring of *derivation type* must factor through a group action. Although there are examples in literature of Hopf actions

2010 *Mathematics Subject Classification.* Primary 12E15, 13A35, 16T05, 16W70.

Key words and phrases. Semisimple Hopf algebras, inner faithful action, quantum polynomial rings.

I would like to thank Christian Lomp very much for many clarifications and helpful insights. The author was partially supported by CMUP (UID/MAT/00144/2013), which is funded by FCT (Portugal) with national (MEC) and European structural funds through the programs FEDER, under the partnership agreement PT2020, and also supported by CAPES, Coordination of Superior Level Staff Improvement - Brazil.

on quantum polynomial algebras that do not factor through group actions (see [**13**, 7.4-7.6]), in this paper we give some conditions to define an action of a Hopf algebra on a skew polynomial of *automorphism type* which does not factor through a group action (Theorem 3.1). In order to do that, we will construct a class of semisimple Hopf algebras H_{2n^2}, which are not group algebras, and show that there exist inner faithful actions of those algebras on the quantum polynomial algebras, in particular on the quantum plane. In a recent paper, [**9**], P. Etingof and C. Walton say that there is *no finite quantum symmetry* when the action of any finite-dimensional Hopf algebra factors through a group action. In this way, we give examples of algebras where there is quantum symmetry. Also, for the quantum plane case, under certain condition, we classify the inner faithful Hopf actions of the Kac-Paljutkin Hopf algebra of dimension 8, H_8, on it (Theorem 4.1).

Let F be a field. In this paper, all the Hopf algebras, tensor products and algebras are taken over F. In the sequel, we present the definition of *inner faithful* actions. Let H be a finite-dimensional Hopf algebra. A *representation* of H on an algebra A is an algebra homomorphism $\pi : H \to A$. The following definition was given by Banica and Bichon.

DEFINITION 1.1 ([**4**, Definition 2.7]). Let $\pi : H \to A$ be a representation of a Hopf algebra H on an algebra A. We say that π is inner faithful if $\mathrm{Ker}(\pi)$ does not contain any non-zero Hopf ideal.

Let M be a left H-module, then, if we consider the endomorphism algebra $\mathrm{End}_F(M)$, we have a representation $\pi : H \to \mathrm{End}_F(M)$ due to the fact of M be an H-module. In this case, to say that π is inner faithful is to say that $I \cdot M \neq 0$ for any non-zero Hopf ideal I of H. This leads to the definition below.

DEFINITION 1.2 ([**6**, Definition 1.2]). Let M be a left H-module. We say that M is an inner faithful H-module (or H acts inner faithfully on M) if $I \cdot M \neq 0$ for any non-zero Hopf ideal I of H. Given a Hopf action of H on an algebra A (i.e., A is a left H-module algebra), we say that this action is inner faithful if the left H-module A is inner faithful.

REMARK 1.3. Any Hopf action factors through an inner faithful Hopf action. For if H is a Hopf algebra acting and a left H-module algebra A, then we can consider the Hopf ideal $I = \sum_{J \subseteq \mathrm{Ann}_H(A)} J$, which is the largest Hopf ideal of H such that $I \cdot A = 0$. Then H/I acts inner faithfully on A.

Also, we present here a result regarding group algebras which can be found in [**11**, Lemma 4] and will be useful in the next sections.

LEMMA 1.4. *Let $H = F[G]$ be the group algebra of a finite group G and $I \subsetneq H$ be a Hopf ideal. Then there exists a normal subgroup $N \lhd G$ such that $I = (F[G])(F[N])^+$.*

2. A class of semisimple Hopf algebras

2.1. Hopf algebra structure on quotients of skew-polynomial rings. Our first main results consist in giving necessary and sufficient condition to extend the structure of a biagebra R to the skew polynomial ring of automorphism type $R[z;\sigma]$ (Theorem 2.4), and to define a Hopf algebra structure on the quotient $R[z;\sigma]/I$, for I a certain bi-ideal of $R[z;\sigma]$ (Theorem 2.7).

We start with the definition of a twist for a bialgebra.

DEFINITION 2.1. Let R be a bialgebra and J be an invertible element in $R \otimes R$. J is called a *right twist* (or a Drinfel'd twist) for R if J satisfies:

(i) $(id \otimes \Delta)(J)(1 \otimes J) = (\Delta \otimes id)(J)(J \otimes 1)$;
(ii) $(id \otimes \epsilon)(J) = 1 = (\epsilon \otimes id)(J)$.

J is called a left twist for R if it satisfies (ii) and $(1 \otimes J)(id \otimes \Delta)(J) = (J \otimes 1)(\Delta \otimes id)(J)$.

If J is a left twist for R, then J^{-1} is a right twist for R.

DEFINITION 2.2 ([**8**, 2.1]). Let R be a bialgebra. Let J be a right twist for R and $\sigma \in \operatorname{End}(R)$. We say that the pair (σ, J) is a twisted homomorphism for R if σ satisfies:

(i) $J(\sigma \otimes \sigma)\Delta(h) = \Delta(\sigma(h))J$ for all $h \in R$;
(ii) $\epsilon \circ \sigma = \epsilon$.

Note that, for any homomorphism of coalgebras $\sigma \in \operatorname{End}(R)$, the pair $(\sigma, 1 \otimes 1)$ is a twisted homomorphism for R.

REMARK 2.3. Let R be a bialgebra, (σ, J) a twist homomorphism for R and $h \in R$.

(1) If σ is an automorphism, then $(\sigma \otimes \sigma)\Delta(\sigma^{-1}(h)) = J^{-1}\Delta(h)J := \Delta^J(h)$;
(2) Since J is invertible, $(\sigma \otimes \sigma)\Delta(h) = J^{-1}\Delta(\sigma(h))J = \Delta^J(\sigma(h))$;
(3) σ is an homomorphism of coalgebras if and only if J commutes with $\Delta(\sigma(h))$, for all $h \in R$.

Now, we shall extend the bialgebra structure of a bialgebra R to its Ore extension of automorphism type $R[z;\sigma]$, for some specific $\sigma \in \operatorname{Aut}(R)$.

THEOREM 2.4. *Let R be a bialgebra and (σ, J) be a twisted homomorphism for R. Let $H = R[z;\sigma]$ be the skew polynomial ring of endomorphism type. Then the bialgebra structure of R can be extended to H such that $\Delta(z) = J(z \otimes z)$ and $\epsilon(z) = 1_F$. Conversely, if there exist an invertible element $J \in R \otimes R$ and $\sigma \in \operatorname{Aut}(R)$ such that $R[z,\sigma]$ is a bialgebra with $\Delta(z) = J(z \otimes z)$ and $\epsilon(z) = 1$, then (σ, J) is a twisted homomorphism for R.*

PROOF. Let (σ, J) be a twisted homomorphism for the bialgebra R, and let $H = R[z;\sigma]$ be the Ore extension of automorphism type of R. Since R is a bialgebra, we have a homomorphism of algebras $\Delta : R \to H \otimes H$. Consider the element $J(z \otimes z) \in H \otimes H$. Note that, for all $h \in R$, we have

$$J(z \otimes z)\Delta(h) = J(\sigma \otimes \sigma)\Delta(h)(z \otimes z) = \Delta(\sigma(h))J(z \otimes z).$$

Thus Δ satisfies the Ore condition. Hence, there exists a unique algebra homomorphism $\overline{\Delta} : H \to H \otimes H$ such that $\overline{\Delta}_{|R} = \Delta$ and $\overline{\Delta}(z) = J(z \otimes z)$. While it may be an abuse of notation, we just write $\overline{\Delta} = \Delta$.

Furthermore,

$$\begin{aligned}(id \otimes \Delta)\Delta(z) &= (id \otimes \Delta)(J)(z \otimes \Delta(z)) \\ &= (id \otimes \Delta)(J)(1 \otimes J)(z \otimes z \otimes z) \\ &= (\Delta \otimes id)(J)(J \otimes 1)(z \otimes z \otimes z) \\ &= (\Delta \otimes id)(J)(\Delta(z) \otimes z) \\ &= (\Delta \otimes id)\Delta(z);\end{aligned}$$

This implies that $\Delta : H \to H \otimes H$ is a coassociative map.

Now, since $\epsilon : R \to F$ is a homomorphism of algebras, then, for all $h \in R$, we have

$$1_F\epsilon(h) = 1_F\epsilon(\sigma(h)) = \epsilon(\sigma(h))1_F.$$

Thus ϵ satisfies the Ore condition. So, there exists a unique algebra homomorphism $\overline{\epsilon} : H \to F$ such that $\overline{\epsilon}_{|R} = \epsilon$ and $\overline{\epsilon}(z) = 1_F$. Again, while it may be an abuse of notation, we just write $\overline{\epsilon} = \epsilon$. Moreover, we have

$$(id \otimes \epsilon)\Delta(z) = (id \otimes \epsilon)(J)z = z = (\epsilon \otimes id)(J)z = (\epsilon \otimes id)\Delta(z).$$

Thus ϵ satisfies the counity property in H. Therefore, the bialgebra structure of R extends to H as stated in the lemma.

To prove the converse, suppose that there exist an invertible element $J \in R \otimes R$ and $\sigma \in \mathrm{Aut}(R)$ such that $R[z, \sigma]$ is a bialgebra with $\Delta(z) = J(z \otimes z)$ and $\epsilon(z) = 1$. Since $(id \otimes \epsilon)\Delta(z) = (\epsilon \otimes id)\Delta(z)$, we must have that $(id \otimes \epsilon)(J) = 1 = (\epsilon \otimes id)(J)$. Also,

$$(id \otimes \Delta)\Delta(z) = (id \otimes \Delta)(J)(z \otimes \Delta(z)) = (id \otimes \Delta)(J)(1 \otimes J)(z \otimes z \otimes z)$$

and

$$(\Delta \otimes id)\Delta(z) = (\Delta \otimes id)(J)(\Delta(z) \otimes z) = (\Delta \otimes id)(J)(J \otimes 1)(z \otimes z \otimes z).$$

Since $R[z, \sigma]$ is a bialgebra, we have that $(id \otimes \Delta)\Delta(z) = (\Delta \otimes id)\Delta(z)$ and hence $(id \otimes \Delta)(J)(1 \otimes J) = (\Delta \otimes id)(J)(J \otimes 1)$, that is, J is a right twist for R. Moreover, for all $h \in R$, $zh = \sigma(h)z$. This implies that

$$\epsilon(h) = \epsilon(zh) = \epsilon(\sigma(h)z) = \epsilon(\sigma(h)),$$

i.e., $\epsilon \circ \sigma = \epsilon$. Note that

$$J(\sigma \otimes \sigma)\Delta(h)(z \otimes z) = J(z \otimes z)\Delta(h) = \Delta(zh) = \Delta(\sigma(h)z) = \Delta(\sigma(h))J(z \otimes z), \quad \forall h \in R.$$

Thus, $J(\sigma \otimes \sigma)\Delta(h) = \Delta(\sigma(h))J$ and hence the pair (σ, J) is a twisted homomorphism for R. □

Given a Hopf algebra R, we will find conditions to define a Hopf algebra structure on the quotient $R[z, \sigma]/I$, for some bi-ideal I of $R[z, \sigma]$. The following lemma gives us certain conditions to find the bi-ideal on the bialgebra $R[z; \sigma]$ which will be used to define the Hopf algebra structure mentioned.

LEMMA 2.5. *Let R be a bialgebra and (σ, J) be a twisted homomorphism for R. Suppose that there exists $0 \neq t \in R$ such that $\Delta(t) = J(\sigma \otimes \sigma)(J)(t \otimes t)$. Then, for $H = R[z; \sigma]$ with the bialgebra structure as in Theorem 2.4, $I = \langle z^2 - t \rangle$ is a bi-ideal of H.*

PROOF. Let R be a bialgebra and (σ, J) be a twisted homomorphism for R. By Theorem 2.4, $H = R[z; \sigma]$ is also a bialgebra. Let $0 \neq t \in R$ as in the hypothesis and let I be the ideal in H generated by $z^2 - t$. We have to prove that $I = \langle z^2 - t \rangle$

is coideal of H. We note that t necessarily satisfies

$$\begin{aligned} t &= (\epsilon \otimes id)\Delta(t) \\ &= (\epsilon \otimes id)(J)(\epsilon \otimes id)(\sigma \otimes \sigma)(J)\epsilon(t)t \\ &= (\epsilon \otimes id)(\sigma \otimes \sigma)(J)\epsilon(t)t \\ &= (\epsilon \otimes \sigma)(J)\epsilon(t)t \\ &= \sigma((\epsilon \otimes id)(J))\epsilon(t)t \\ &= \sigma(1)\epsilon(t)t = \epsilon(t)t, \end{aligned}$$

which implies $\epsilon(t) = 1$. So, $\epsilon(z^2 - t) = \epsilon(z)^2 - \epsilon(t) = 0$. That is, $I \subseteq Ker(\epsilon)$.

Furthermore,

$$\begin{aligned} \Delta(z^2 - t) &= \Delta(z)^2 - \Delta(t) \\ &= J(z \otimes z)J(z \otimes z) - \Delta(t) \\ &= J(\sigma \otimes \sigma)(J)(z^2 \otimes z^2) - \Delta(t) \\ &= J(\sigma \otimes \sigma)(J)(z^2 - t \otimes z^2) + J(\sigma \otimes \sigma)(J)(t \otimes z^2 - t) \\ &\quad + J(\sigma \otimes \sigma)(J)(t \otimes t) - \Delta(t) \\ &= J(\sigma \otimes \sigma)(J)(z^2 - t \otimes z^2) + J(\sigma \otimes \sigma)(J)(t \otimes z^2 - t), \end{aligned}$$

which belongs to $I \otimes H + H \otimes I$. Therefore, I is a bi-ideal of H. □

REMARK 2.6. Note that, conversely, if $I = \langle z^2 - t \rangle$ is a bi-ideal of H, then $\Delta(t) - J(\sigma \otimes \sigma)(J)(t \otimes t) \in H \otimes I + I \otimes H$.

Hence, given a bialgebra R and (σ, J) a twisted homomorphism for R and an element t that satisfies the hypothesis of Lemma 2.5, we have that H/I is a bialgebra, for $H = R[z; \sigma]$ and $I = \langle z^2 - t \rangle$. The next lemma presents conditions to extend a Hopf algebra structure from R to the quotient bialgebra H/I.

THEOREM 2.7. *Let R be a Hopf algebra with antipode S, (σ, J) be a twisted homomorphism. Suppose also that $\sigma \circ S = S \circ \sigma$ and $\sigma^2 = id$. If there exists $0 \neq t \in R$, with $\Delta(t) = J(\sigma \otimes \sigma)(J)(t \otimes t)$, and such that*

(i) $tJ^1S(J^2) = 1$;
(ii) $t\sigma(S(J^1)J^2) = 1$,

where $J = J^1 \otimes J^2$ with the summation omitted, then there exists a Hopf algebra structure on H/I with $S(z) = z$. Conversely, if there exists a Hopf algebra structure on H/I, for $I = \langle z^2 - t \rangle$, with $S(z) = z$, then $tJ^1S(J^2) = 1 = t\sigma(S(J^1)J^2)$.

PROOF. Let R be a Hopf algebra with antipode S and (σ, J) a twisted homomorphism for R such that $\sigma^2 = id$. By Theorem 2.4, $R[z, \sigma]$ is a bialgebra, and by Lemma 2.5, $I = \langle z^2 - t \rangle$ is a bi-ideal of $R[z, \sigma]$.

We just write h for the element $h + I$ of H/I. And to define the antipode, we extend the antipode S of R to H/I defining $S(z) = z$. We note that $S : H/I \to H/I$ is well defined, since $S(z^2 - t) = S(z)^2 - S(t) = z^2 - t \in I$ and, using that $S \circ \sigma = \sigma \circ S$, $S(a)z = zS(\sigma(a))$ for all $a \in R$. Also, we have that

$$\begin{aligned} \mu(id \otimes S)\Delta(z) &= \mu(1 \otimes S)(J(z \otimes z)) \\ &= J^1 zS(z)S(J^2) = z^2\sigma^2(J^1)S(J^2) = tJ^1S(J^2) = 1, \end{aligned}$$

and

$$\begin{aligned}\mu(S\otimes id)\Delta(z) &= \mu(S\otimes 1)(J(z\otimes z))\\ &= S(z)S(J^1)J^2z = z^2\sigma(S(J^1))\sigma(J^2) = t\sigma(S(J^1)J^2) = 1.\end{aligned}$$

Since S is an antipode for R, the antipode property is verified for R as well. Therefore, S is an antipode of H/I and so H/I is a Hopf algebra. The converse follows from the two equations above. □

In this setting, for R a semisimple Hopf algebra, we have the following corollary.

COROLLARY 2.8. *Under the conditions of Theorem 2.7, if R is a semisimple Hopf algebra, then H/I is semisimple.*

2.2. Construction of a class of semisimple Hopf algebras. In this subsection, we shall construct semisimple Hopf algebras of dimension $2n^2$, which will be used to define actions on the quantum polynomial algebra that do not factor through group actions. From now on, we let the ground field F be algebraically closed with characteristic zero.

Let $\Gamma = \langle x \mid x^n = 1\rangle$ be the cyclic group of order $n > 1$. Let $q \in F$ be a primitive nth root of unity. For every integer j, we set

$$e_j = \frac{1}{n}\sum_{i=0}^{n-1} q^{-ij}x^i.$$

Observe that if $j \equiv j'(\text{mod } n)$, then $q^j = q^{j'}$ and $x^j = x^{j'}$, and therefore $e_j = e_{j'}$. This means that $e_0, \ldots, e_{n-1}$ lists the distinct $e_i's$. Moreover, for $0 \le j, k < n$, we have

$$e_jx^k = \frac{1}{n}\sum_{i=0}^{n-1} q^{-ij}x^{i+k} = q^{jk}\left(\frac{1}{n}\sum_{i=0}^{n-1} q^{-(i+k)j}x^{i+k}\right) = q^{jk}e_j. \tag{1}$$

LEMMA 2.9. *$\{e_0, \cdots, e_{n-1}\}$ is a complete set of orthogonal idempotents of $F[\Gamma]$.*

PROOF. Since q^{-j} is also an nth root of unity different from 1 if $j \neq 0$, we get

$$\sum_{i=0}^{n-1} e_i = \frac{1}{n}\sum_{i=0}^{n-1}\sum_{j=0}^{n-1} q^{-ij}x^j = \frac{1}{n}\sum_{j=0}^{n-1}\left(\sum_{i=0}^{n-1}\left(q^{-j}\right)^i\right)x^j = 1,$$

Also, using (1), for $0 \le l, j < n$:

$$e_je_l = \frac{1}{n}\sum_{k=0}^{n-1} q^{-lk}e_jx^k = \frac{1}{n}\sum_{k=0}^{n-1} q^{-lk+jk}e_j = \frac{1}{n}\sum_{k=0}^{n-1}\left(q^{j-l}\right)^k e_j = \begin{cases} e_j & \text{if } l = j\\ 0 & \text{if } l \neq j\end{cases}$$

Hence, $\{e_0, \cdots, e_{n-1}\}$ is a complete set of orthogonal idempotents of $F[\Gamma]$. □

We denote the elements of $G = \Gamma\times\Gamma$ by x^iy^s for $0 \le i, s < n$. Let $\{e_0, \cdots, e_{n-1}\}$ be the complete set of idempotents of $F[\Gamma]$ as in Lemma 2.9. Let $\sigma \in \text{Aut}(F[G])$ be the automorphism of $F[G]$ induced by the group isomorphism $x^iy^s \mapsto x^sy^i$, for $1 \le i, s \le n$.

Set $\overline{e_i} := \sigma(e_i)$, i.e., $\overline{e_i} = \frac{1}{n}\sum_{j=0}^{n-1} q^{-ij}y^j$. As in equation (1) one has $\overline{e_i}y^k = q^{ik}\overline{e_i}$.

Now, in $F[G]\otimes F[G]$, consider the element $J:=\sum_{i=0}^{n-1}e_i\otimes y^i$. Note that we can also write J in terms of the elements $\overline{e_i}$'s as

$$J=\frac{1}{n}\sum_{i,j=0}^{n-1}q^{-ij}x^j\otimes y^i=\sum_{i=0}^{n-1}x^i\otimes\overline{e_i}. \tag{2}$$

With this setting, we have the following lemma.

LEMMA 2.10. *The pair (σ,J) is a twisted homomorphism for $F[G]$.*

PROOF. First we note that J is invertible with inverse $J^{-1}=\sum_{j=0}^{n-1}e_j\otimes y^{-j}$. Using (1) and (2), we get

$$\begin{aligned}(\Delta\otimes 1)(J)(J\otimes 1)&=\left(\sum_{i=0}^{n-1}\Delta(x^i)\otimes\overline{e_i}\right)\left(\sum_{j=0}^{n-1}e_j\otimes y^j\otimes 1\right)\\&=\sum_{i,j=0}^{n-1}x^ie_j\otimes x^iy^j\otimes\overline{e_i}\\&=\sum_{i,j=0}^{n-1}q^{ij}e_j\otimes y^jx^i\otimes\overline{e_i}\\&=\sum_{i,j=0}^{n-1}e_j\otimes y^jx^i\otimes y^j\overline{e_i}\\&=\left(\sum_{j=0}^{n-1}e_j\otimes\Delta(y^j)\right)\left(\sum_{j=0}^{n-1}1\otimes x^i\otimes\overline{e_i}\right)=(1\otimes\Delta)(J)(1\otimes J).\end{aligned}$$

Moreover, we have

$$(1\otimes\epsilon)(J)=\sum_{i=0}^{n-1}e_i=1=\sum_{i=0}^{n-1}\overline{e_i}=(\epsilon\otimes 1)(J).$$

That is, J is a right twist as in Definition 2.1. For $x^ky^s\in G$, note that

$$(\sigma\otimes\sigma)\Delta(x^ky^s)=(\sigma\otimes\sigma)(x^ky^s\otimes x^ky^s)=x^sy^k\otimes x^sy^k=\Delta(\sigma(x^ky^s)).$$

Thus, since $F[G]$ is commutative, we have that $J(\sigma\otimes\sigma)\Delta(x^ky^s)=\Delta(\sigma(x^ky^s))J$. Moreover, clearly $\epsilon\circ\sigma=\epsilon$. Therefore, the pair (σ,J) is a twisted homomorphism for $F[G]$. □

Hence, by Theorem 2.4, $H=F[G][z;\sigma]$ is bialgebra with $\Delta(z)=J(z\otimes z)$ and $\epsilon(z)=1$. Now, consider the element $t=\sum_{i=0}^{n-1}e_iy^i$ which satisfies

$$\sigma(t)=\sum_{i=0}^{n-1}\overline{e_i}x^i=\frac{1}{n}\sum_{i=0}^{n-1}\sum_{j=0}^{n-1}q^{-ij}y^jx^i=\sum_{j=0}^{n-1}\left(\frac{1}{n}\sum_{i=0}^{n-1}q^{-ij}x^i\right)y^j=\sum_{j=0}^{n-1}e_jy^j=t.$$

Moreover, t has an inverse in $F[G]$, $t^{-1}=\sum_{i=0}^{n-1}e_iy^{-i}$.

LEMMA 2.11. *The element t satisfies $\Delta(t)=J(\sigma\otimes\sigma)(J)(t\otimes t)$.*

PROOF. Since $\sum_{k=0}^{n-1}(q^{(l-j)})^k = 0$ for $l \neq j$, note that for any i

$$\begin{aligned}\sum_{k=0}^{n-1} e_k \otimes e_{(i-k)} &= \frac{1}{n^2}\sum_{k,j,l=0}^{n-1} q^{-jk}q^{-l(i-k)}x^j \otimes x^l \\ &= \frac{1}{n^2}\sum_{k,j,l=0}^{n-1} q^{(l-j)k}q^{-li}x^j \otimes x^l \\ &= \frac{1}{n^2}\sum_{j,l=0}^{n-1}\left(\sum_{k=0}^{n-1}(q^{(l-j)})^k\right) q^{-li}x^j \otimes x^l \\ &= \frac{1}{n}\sum_{j=0}^{n-1} q^{-ij}x^j \otimes x^j = \Delta(e_i).\end{aligned}$$

So, it follows that $\Delta(t) = \sum_{i,m=0}^{n-1} e_i y^m \otimes e_{(m-i)} y^m$. Then, by (2), we get

$$\begin{aligned}J(\sigma\otimes\sigma)(J)(t\otimes t) &= \sum_{i,j=0}^{n-1} e_i y^j t \otimes y^i e_j t \\ &= \sum_{i,j,k,l=0}^{n-1} e_i e_k y^{j+k} \otimes e_j e_l y^{i+l} \\ &= \sum_{i,j=0}^{n-1} e_i y^{i+j} \otimes e_j y^{i+j} \\ &= \sum_{i,m=0}^{n-1} e_i y^m \otimes e_{(m-i)} y^m = \Delta(t).\end{aligned}$$

□

So, t satisfies the hypothesis of Lemma 2.5 and hence $I = \langle z^2 - t\rangle$ is a bi-ideal of $F[G][z;\sigma]$. Thus $H = F[G][z;\sigma]/\langle z^2 - t\rangle$ is also a bialgebra.

Note that $\sigma^2 = id$ and

$$S(t) = \frac{1}{n}\sum_{i,j=0}^{n-1} q^{-ij}x^{-i}y^{-j} = \frac{1}{n}\sum_{k,s=0}^{n-1} q^{-ks}x^k y^s = \sum_{s=0}^{n-1} e_s y^s = t.$$

Since $\sigma(S(x^k y^s)) = x^{-s}y^{-k} = S(\sigma(x^k y^s))$ and

$$tJ^{(1)}S(J^{(2)}) = \sum_{i,j=0}^{n-1} e_i e_j y^{i-j} = \sum_{i=0}^{n-1} e_i = 1,$$

and

$$\begin{aligned}t\sigma(S(J^{(1)})J^{(2)}) &= t\sigma\left(\frac{1}{n}\sum_{i,j=0}^{n-1} q^{-ij}x^{-j}y^i\right) \\ &= t\sum_{j=0}^{n-1} y^{-j}\frac{1}{n}\sum_{i=0}^{n-1} q^{-ij}x^i = t\left(\sum_{j=0}^{n-1} e_j y^{-j}\right) = tt^{-1} = 1,\end{aligned}$$

by Theorem 2.7, $H = F[G][z;\sigma]/\langle z^2 - t\rangle$ is a Hopf algebra.

Note that these Hopf algebras have dimension $2n^2$ and we shall denote them by H_{2n^2}.

Also, we note that since F has characteristic zero, $F[G]$ is semisimple, and then H_{2n^2} is also semisimple. Moreover, H_{2n^2} is non-commutative and non-cocommutative.

Before we continue, we establish the following lemma about the Hopf ideals of H_{2n^2}, which will be useful in the sequel.

LEMMA 2.12. *Let I be a Hopf ideal of H_{2n^2}. If $I \cap R = 0$, then $I = 0$.*

PROOF. Let I be a Hopf ideal of H_{2n^2} such that $I \cap R = 0$. Consider the restriction of the projection map $\pi_{|_R} : R \to H_{2n^2}/I$. Clearly, $\operatorname{Ker} \pi_{|_R} = I \cap R$. Hence, we can look at R as a Hopf subalgebra of H_{2n^2}/I. Then $\dim(R)$ divides $\dim(H_{2n^2}/I)$.

Since $\dim(H_{2n^2}) = 2\dim(R)$, we must have that $\dim(H_{2n^2}/I) = 2\dim(R)$ or $\dim(H_{2n^2}/I) = \dim(R)$. If $\dim(H_{2n^2}/I) = 2\dim(R)$, then $I = 0$.

If $\dim(H_{2n^2}/I) = \dim(R)$, then $H_{2n^2}/I \cong R$ and thus H_{2n^2}/I is commutative. Hence, we must have that $\bar{x}\bar{z} = \bar{z}\bar{x} = \bar{y}\bar{z}$, which implies that $(x - y)z \in I$. So, $(x - y)z(zt^{-1}) = (x - y) \in I \cap R = 0$, which is absurd. □

3. Inner faithful actions on quantum polynomial algebras

Let $M = (m_{ij}) \in M_{r\times r}(F^{\times})$ be a square matrix of size r such that $m_{ii} = m_{ij}m_{ji} = 1$, for all $1 \le i, j \le r$. Let $A_M = F_M[u_1, \dots, u_r]$ be the *quantum polynomial algebra* (see [**3**] and [**5**, Appendix I.14 and Chapter I.2]), i.e., the associative F-algebra generated by $u_1, \dots, u_r$ subject to the relations

$$u_i u_j = m_{ij} u_j u_i, \quad 1 \le i, j \le r.$$

Alternatively, quantum polynomial algebras can be constructed as iterated Ore extension of automorphism type

$$A_M = F[u_1][u_2, \tau_2] \cdots [u_r, \tau_r],$$

where $\tau_i(u_j) = m_{ij}u_j$ for all i, j with $1 \le j < i \le r$.

In this section, we will present inner faithful actions of H_{2n^2} on A_M. Our main result is Theorem 3.1, where we provide conditions to define inner faithful actions of H_{2n^2} on quantum polynomial algebras which do not factor through group actions.

Recall that for a nth primitive root of unity q,

$$\Delta(z) = J(z \otimes z) = \frac{1}{n} \sum_{i,j=0}^{n-1} q^{-ij} x^i z \otimes y^j z.$$

THEOREM 3.1. *Let $n > 1$ be a natural number and H_{2n^2} the Hopf algebra constructed above with respect to a primitive nth root of unity q. Let A_M be a quantum polynomial algebra in generators $u_1, \dots, u_r$ with respect to a matrix $M = (m_{ij}) \in M_{r\times r}(F)$. For any permutation $\tau \in S_r$ and elements $\lambda, \mu \in F^r$ the following holds:*

(1) *H_{2n^2} acts on A_M via $x \cdot u_i = \lambda_i u_i$, $y \cdot u_i = \mu_i u_i$ and $z \cdot u_i = u_{\tau(i)}$ if and only if there exists $b \in \mathbb{Z}_n^r$ such that*

$$\lambda_i = q^{b_i}, \qquad \mu_i = q^{b_{\tau(i)}}, \qquad m_{st} = q^{B_{\tau(s)\tau(t)}} m_{\tau(s)\tau(t)},$$

for all j, i, s, where $B_{st} = b_s b_{\tau(t)} - b_t b_{\tau(s)}$.

(2) *The action in (1) is inner faithful if and only if the map* $f : \mathbb{Z}_n^2 \to \mathbb{Z}_n^r$ *with* $f(i,j) = (ib_1 + jb_{\tau(1)}, \cdots, ib_r + jb_{\tau(r)})$ *is injective.*

(3) *In particular if there exist* s, t *such that* B_{st} *is invertible in* $\mathbb{Z}_n$*, then the action in (1) is inner faithful.*

PROOF. (1) Suppose H_{2n^2} acts on A_M via $x \cdot u_i = \lambda_i u_i$; $y \cdot u_i = \mu_i u_i$; $z \cdot u_i = u_{\tau(i)}$. Since $zy = xz$ in H_{2n^2}, we should have that $(zy) \cdot u_i = (xz) \cdot u_i$, for all $i \in \{1, \ldots, r\}$. This happens if and only if $\mu_i = \lambda_{\tau(i)}$ for all $i \in \{1, \ldots, r\}$. Since $x^n = 1$, we must have that λ_i is an nth roots of unity and hence there exist $b \in \mathbb{Z}_n^r$ such that $\lambda_i = q^{b_i}$ and $\mu_i = \lambda_{\tau(i)} = q^{b_{\tau(i)}}$ for all i.

Let us calculate $z \cdot (u_k u_l)$ for some indices k, l:

$$\begin{aligned} z \cdot (u_k u_l) &= \sum_{i,j=0}^{n-1} q^{-ij} (\lambda_{\tau(k)}^i u_{\tau(k)})(\mu_{\tau(l)}^j u_{\tau(l)}) \\ &= \left(\sum_{i,j=0}^{n-1} q^{-ij + b_{\tau(k)} i + b_{\tau(\tau(l))} j} \right) u_{\tau(k)} u_{\tau(l)} \\ &= \left(\sum_{i=0}^{n-1} q^{b_{\tau(k)} i} \sum_{j=0}^{n-1} (q^{b_{\tau(\tau(l))} - i})^j \right) u_{\tau(k)} u_{\tau(l)} = n q^{b_{\tau(k)\tau(\tau(l))}} u_{\tau(k)} u_{\tau(l)} \end{aligned}$$

Hence the relation $z \cdot (u_k u_l) = m_{kl} z \cdot (u_l u_k)$ holds if and only if:

$$m_{\tau(k)\tau(l)} q^{b_{\tau(k)} b_{\tau(\tau(l))}} = m_{kl} q^{b_{\tau(l)} b_{\tau(\tau(k))}}$$

or equivalently $m_{kl} = m_{\tau(k)\tau(l)} q^{B_\tau(k)\tau(l)} m_{\tau(k)\tau(l)}$, with $B_{\tau(k)\tau(l)} = b_{\tau(k)} b_{\tau(\tau(l))} - b_{\tau(l)} b_{\tau(\tau(k))}$.

Conversely if λ, μ and (m_{ij}) satisfy the conditions indicated above, then the action given is well-defined as $1 - x^n$, $1 - y^n$, $zx - yz$ act as zero on the u_i's and as $z \cdot (u_k u_l) = m_{kl} z \cdot (u_l u_k)$ holds.

(2) Let I be a Hopf ideal of H_{2n^2} such that $I \cdot A_M = 0$. Suppose that $I \cap R \neq 0$. Since $I \cap R$ is a Hopf ideal of the group algebra R, we can apply Lemma 1.4 and conclude that there exists a normal subgroup N of $\mathbb{Z}_n \times \mathbb{Z}_n$ such that $I \cap R = RF[N]^+$. Since $I \cap R \neq 0$ we have $N \neq \langle (0,0) \rangle$, i.e. there exist $(i,j) \neq (0,0)$, such that $x^i y^j \in N$ and hence $1 - x^i y^j \in I$. Thus, for all $s \in \{1, \ldots, r\}$, we have $(1 - q^{b_s i + b_{\tau(s)} j}) u_s = (1 - x^i y^j) \cdot u_s = 0$, which implies $ib_s + jb_{\tau(s)} = 0$ in $\mathbb{Z}_n$ for all $s \in \{1, \ldots, r\}$. Hence if f is injective, then $I \cap R = 0$, and thus, by Lemma 2.12, $I = 0$. So, the action must be inner faithful. Conversely if f is not injective and $f(i,j) = (0, \ldots, 0)$ for some $(i,j) \in \mathbb{Z}_n^2$, then the Hopf ideal generated by $1 - x^i y^j$ annihilates A_M and the action is not inner faithful.

(3) Suppose that B_{st} is invertible for some s, t. If (i,j) is a pair such that $ib_k + jb_{\tau(k)} = 0$ for all k, then in particular $b_{\tau(s)} j = -b_s i$ and hence $jB_{st} = jb_s b_{\tau(t)} - jb_{\tau(s)} b_t = -b_s b_t i + b_s b_t i = 0$. Thus $j = 0$ and therefore $ib_k = 0$ for all k. In particular $iB_{st} = 0$ and hence $i = 0$. □

Suppose F contains a primitive 6th root of unity q. Let H_{72} be the semisimple Hopf algebras with respect to q and $n = 6$. Let $A_M = F_{q^2}[u_1, u_2][u_3]$ be the quantum polynomial algebra in the generators u_1, u_2, u_3 subject to $u_2 u_1 = q^2 u_2 u_1$ and u_3 commuting with u_1 and u_2. Then there exists an inner faithful action of

H_{72} on $F_{q^2}[u_1, u_2][u_3]$ given by $b = (2, 2, 1) \in \mathbb{Z}_6^3$ and $\tau = (13) \in S_3$, i.e.

$$x \cdot u_1 = q^2 u_1, \qquad x \cdot u_2 = q^2 u_2, \qquad x \cdot u_3 = q u_3$$
$$y \cdot u_1 = q u_1, \qquad y \cdot u_2 = q^2 u_2, \qquad y \cdot u_3 = q^2 u_3$$
$$z \cdot u_1 = u_3, \qquad z \cdot u_2 = u_2, \qquad z \cdot u_3 = u_1.$$

Note that the function $f : \mathbb{Z}_6^2 \to \mathbb{Z}_6^3$ from Theorem 3.1(2) is given by $f(i,j) = (2i + j, 2(i + j), i + 2j)$ and injective. However the elements B_{st} from above are all zero divisors in $\mathbb{Z}_6$.

3.1. Inner faithful actions on the quantum plane. For a moment, let F be an arbitrary field. Let $0 \neq p \in F$ and consider the matrix

$$M = \begin{pmatrix} 1 & p^{-1} \\ p & 1 \end{pmatrix} \in M_{2\times 2}(F)$$

The *quantum plane* is the quantum polynomial algebra $A_M = F_M[u, v]$ with two generators. We denote the quantum plane $F_M[u, v]$ by $F_p[u, v]$ and we call p a *parameter*. When $p \neq 1$ the algebra $F_p[u, v]$ is non-commutative.

Consider the Hopf algebras H_{2n^2} as constructed in Section 2.2. For each n, these Hopf algebras act on $A = \mathbb{C}_p[u, v]$ with $p^2 = q$, where q is the primitive nth root of unity used to construct H_{2n^2}. The action is given by:

$$x \cdot u = qu, \qquad y \cdot u = u, \qquad z \cdot u = v,$$
$$x \cdot v = v, \qquad y \cdot v = qv, \qquad z \cdot v = u,$$

which corresponds to $\tau = (12) \in S_2$ and the vector $b = (1, 0) \in \mathbb{Z}_2^2$ in Theorem 3.1. H_{2n^2} acts on A and, since the element B_{12} is invertible in $\mathbb{Z}_2$, by Theorem 3.1, this action is inner faithful.

4. Inner faithful actions of Kac-Paljutkin Hopf algebra on the quantum plane

G.I. Kac and Paljutkin, in the 1960's, discovered a non-commutative, non-cocommutative semisimple Hopf algebra of dimension 8 [**12**]. H_8 is the algebra over F generated by x, y, and z subject to the following relations

$$x^2 = 1, \quad y^2 = 1, \quad xy = yx$$
$$z^2 = \frac{1}{2}(1 + x + y - xy), \quad zx = yz, \quad zy = xz.$$

H_8 has a coalgebra structure with

$$\Delta(x) = x \otimes x, \qquad \epsilon(x) = 1$$
$$\Delta(y) = y \otimes y, \qquad \epsilon(y) = 1$$
$$\Delta(z) = \tfrac{1}{2}(1 \otimes 1 + x \otimes 1 + 1 \otimes y - x \otimes y)(z \otimes z), \quad \epsilon(z) = 1.$$

H_8 becomes a Hopf algebra by setting $S(x) = x$, $S(y) = y$, and $S(z) = z$.

Later, in 1995, A. Masouka showed that there is only one (up to isomorphisms) semisimple Hopf algebra of dimension 8 (H_8) that is neither commutative nor cocommutative [**15**]. Although Kac and Paljutkin present H_8 as above and Masouka presents it under the perspective of biproducts and bicrossed products, in this section we present H_8 as the Hopf algebra H_{2n^2}, for $n = 2$. Also, under a certain condition, we classify the inner faithful actions of H_8 on the quantum plane.

4.1. H_8 as a quotient of an Ore extension. For the Hopf algebras constructed in that Section 2.2, we take $n = 2$ and $q = -1$. The group Γ, then, is the cyclic group of order 2, $\mathbb{Z}_2 = \{x \mid x^2 = 1\}$. In this setting, we have that the orthogonal idempotents of $F[\mathbb{Z}_2]$ as in Lemma 2.9 are given by:

$$e_0 = \frac{1}{2}(1+x) \text{ and } e_1 = \frac{1}{2}(1-x).$$

Then, for $G = \langle x, y \mid x^2 = 1 = y^2, xy = yx\rangle = \Gamma \times \Gamma$, the automorphism σ swaps x and y, i.e., $\sigma(x) = y$ and $\sigma(y) = x$. And the element J is given by

$$J = \frac{1}{2}((1+x)\otimes 1 + (1-x)\otimes y) = \frac{1}{2}(1\otimes 1 + x\otimes 1 + 1\otimes y - x\otimes y).$$

Let $R = F[G]$. So, $R[z;\sigma]$ becomes a bialgebra with

$$\Delta(z) = J(z\otimes z) = \frac{1}{2}\left(1\otimes 1 + x\otimes 1 + 1\otimes y - x\otimes y\right)(z\otimes z),$$

and $\epsilon(z) = 1$. Also, note that $zx = \sigma(x)z = yz$. Since t is given by

$$t = e_0 + e_1 y = \frac{1}{2}(1 + x + y - xy),$$

we get that $z^2 = \frac{1}{2}(1 + x + y - xy)$ in the Hopf algebra $R[z;\sigma]/\langle z^2 - t\rangle$, where $S(z) = z$.

So, H_{2n^2}, for $n = 2$, is precisely the Hopf algebra H_8. Then, from now on, every time we refer to H_8, we keep in mind its presentation as the one presented in this subsection, i.e., as a quotient of an Ore extension: $H_8 = R[z;\sigma]/\langle z^2 - t\rangle$.

4.2. Classification of the Inner faithful actions of H_8 on the quantum plane. Let $F = \mathbb{C}$ and $A = \mathbb{C}_p[u,v]$ be the quantum plane with parameter $p \in \mathbb{C}^\times$, i.e., $vu = puv$.

In the following theorem we classify the possibles inner faithful actions of H_8 on A under a certain assumption.

THEOREM 4.1. *Let $p \in \mathbb{C}^\times$. If there is a Hopf action of H_8 on the quantum plane $A = \mathbb{C}_p[u,v]$ such that $z\cdot u = v$ and $z \cdot v = u$, then this action is inner faithful and $p^2 = -1$.*

PROOF. We can assume $p \neq 1$, because if $p = 1$, then $A = \mathbb{C}[u,v]$ is the commutative polynomial ring, which is a commutative domain. Therefore, Etingof and Walton's result [**10**, Theorem 1.3] guarantee that there cannot be any inner faithful action of H_8 on A, since H_8 is not a group algebra.

If there is an action of H_8 on A, since x and y are group-like elements, they act as automorphisms of A. Hence there exist $\alpha, \beta \in \mathrm{Aut}(A)$ such that $x\cdot a = \alpha(a)$ and $y \cdot a = \beta(a)$ for all $a \in A$. Also, since $x^2 = 1 = y^2$, $\alpha^2 = id = \beta^2$.

Under the assumption that z acts by interchanging u and v, we must have $z\cdot(vu) = pz\cdot(uv)$ or equivalently

$$uv + \alpha(u)v + u\beta(v) - \alpha(u)\beta(v) = p(vu + \alpha(v)u + v\beta(u) - \alpha(v)\beta(u)). \tag{3}$$

Moreover, since $xz = zy$, it follows that $(xz)\cdot u = (zy)\cdot u$ and $(xz)\cdot v = (zy)\cdot v$, which implies that

$$\alpha(u) = z\cdot\beta(v) \quad\text{and}\quad \alpha(v) = z\cdot\beta(u). \tag{4}$$

Now, we separate the proof in cases.

CASE I: $p \neq -1$**:** In this case Alev and Chamarie showed in [**1**, 1.4.4] that the automorphisms of A are given by a torus action, i.e., x and y acts as scalars on u and v. Hence we are in the situation of Theorem 3.1 (1) with $\tau = (12) \in S_2$. Suppose

$$x \cdot u = (-1)^{b_1} u, \qquad x \cdot v = (-1)^{b_2} v, \qquad y \cdot u = (-1)^{b_{\tau(1)}} u, \qquad y \cdot v = (-1)^{b_{\tau(2)}} v,$$

and let $b = (b_1, b_2) \in \mathbb{Z}_2^2$. By Theorem 3.1, H acts on A if and only if

$$p^2 = (-1)^{b_1 b_{\tau(2)} - b_2 b_{\tau(1)}} = (-1)^{B_{12}}.$$

Since $p \neq -1$, we must have $B_{12} = 1$ and so $p^2 = -1$. Hence, by Theorem 3.1 (3), the action is inner faithful.

CASE II: $p = -1$**:** In this case Alev and Chamarie showed in [**1**, 1.4.4] that $\mathrm{Aut}(A)$ is a semidirect product of $(\mathbb{C}^\times)^2$ with the cyclic group of order 2 given by the automorphism τ that flips u and v. Hence any automorphism of A is either an element of $(\mathbb{C}^\times)^2$ or a product of an element $(\alpha_0, \alpha_1) \in (\mathbb{C}^\times)^2$ and τ. By 4, if β is just given by a torus action, then α has to be also given by a torus action, and if β is given by a torus action and τ, then α has to be also given by a torus action and τ. Therefore, these are the only two possible cases for α and β.

CASE II.a: If β is given only by a torus action, then then we are in the same situation as CASE I and hence, by what we have done for CASE I, necessarily $p^2 = -1$, which contradicts $p = -1$. So, β cannot be given by a torus action.

CASE II.b: Suppose that both α and β are compositions of a torus action and τ, then there are $(\alpha_0, \alpha_1), (\beta_0, \beta_1) \in (\mathbb{C}^\times)^2$ such that

$$x \cdot u = \alpha(u) = \alpha_0 v, \quad x \cdot v = \alpha(v) = \alpha_1 u, \quad y \cdot u = \beta(u) = \beta_0 v, \quad y \cdot v = \beta(v) = \beta_1 u.$$

Then equation (3) yields

$$uv + \alpha_0 v^2 + \beta_1 u^2 + \beta_1 \alpha_0 uv = uv - \alpha_1 u^2 - \beta_0 v^2 + \beta_0 \alpha_1 uv,$$

which is equivalent to

$$(\alpha_1 + \beta_1) u^2 + (\alpha_0 + \beta_0) v^2 + (\beta_1 \alpha_0 - \beta_0 \alpha_1) uv = 0$$

and implies $\beta_i = -\alpha_i$, for $i = 0, 1$, and $\beta_1 \alpha_0 = \beta_0 \alpha_1$. Now, since $\alpha^2 = id = \beta^2$, we must have that $\alpha_0 \alpha_1 = 1$ and $\beta_0 \beta_1 = 1$. Also, by 4, we must have that $\alpha_1 = \beta_0$ and $\alpha_0 = \beta_1$. Hence, $\alpha_0^2 = -\alpha_0 \alpha_1 = -1$. And so $\alpha_0 = \pm i$. Therefore, for $\alpha_0 \in \{-i, i\}$, the options are $\beta_1 = \alpha_0$, $\beta_0 = -\alpha_0$, $\alpha_1 = -\alpha_0$. But note that

$$xz \cdot (vu) = (vu - \alpha_0 u^2 - \alpha_0 v^2 - vu) \quad \text{and} \quad zy \cdot (vu) = (-uv - \alpha_0 u^2 - \alpha_0 v^2 - uv).$$

This leads to a contradiction, since $xz = yz$. Therefore, there can be no action at all for the case where $p = -1$. □

References

[1] J. Alev and M. Chamarie, *Dérivations et automorphismes de quelques algèbres quantiques* (French), Comm. Algebra **20** (1992), no. 6, 1787–1802, DOI 10.1080/00927879208824431. MR1162608

[2] J. M. Allman, *Actions of finite dimensional non-commutative, non-cocommutative Hopf algebras on rings* (2009).

[3] V. A. Artamonov, *Quantum polynomial algebras*, J. Math. Sci. (New York) **87** (1997), no. 3, 3441–3462, DOI 10.1007/BF02355445. Algebra, 4. MR1604198

[4] T. Banica and J. Bichon, *Hopf images and inner faithful representations*, Glasg. Math. J. **52** (2010), no. 3, 677–703, DOI 10.1017/S0017089510000510. MR2679923

[5] K. A. Brown and K. R. Goodearl, *Lectures on algebraic quantum groups*, Advanced Courses in Mathematics. CRM Barcelona, Birkhäuser Verlag, Basel, 2002. MR1898492
[6] K. Chan, C. Walton, and J. Zhang, *Hopf actions and Nakayama automorphisms*, J. Algebra **409** (2014), 26–53, DOI 10.1016/j.jalgebra.2014.04.003. MR3198834
[7] J. Cuadra, P. Etingof, and C. Walton, *Semisimple Hopf actions on Weyl algebras*, Adv. Math. **282** (2015), 47–55, DOI 10.1016/j.aim.2015.05.014. MR3374522
[8] A. Davydov, *Twisted automorphisms of Hopf algebras*, Noncommutative structures in mathematics and physics, K. Vlaam. Acad. Belgie Wet. Kunsten (KVAB), Brussels, 2010, pp. 103–130. MR2742734
[9] P. Etingof and C. Walton, *Finite dimensional Hopf actions on algebraic quantizations*, Algebra Number Theory **10** (2016), no. 10, 2287–2310, DOI 10.2140/ant.2016.10.2287. MR3582020
[10] P. Etingof and C. Walton, *Semisimple Hopf actions on commutative domains*, Adv. Math. **251** (2014), 47–61, DOI 10.1016/j.aim.2013.10.008. MR3130334
[11] R. Farnsteiner, *Burnside's theorem for Hopf algebras.* Lecture notes, available at https://www.math.uni-bielefeld.de/ sek/select/rf5.pdf.
[12] G. I. Kac and V. G. Paljutkin, *Finite ring groups* (Russian), Trudy Moskov. Mat. Obšč. **15** (1966), 224–261. MR0208401
[13] E. Kirkman, J. Kuzmanovich, and J. J. Zhang, *Gorenstein subrings of invariants under Hopf algebra actions*, J. Algebra **322** (2009), no. 10, 3640–3669, DOI 10.1016/j.jalgebra.2009.08.018. MR2568355
[14] C. Lomp and D. Pansera, *A note on a paper by Cuadra, Etingof and Walton*, Comm. Algebra **45** (2017), no. 8, 3402–3409, DOI 10.1080/00927872.2016.1236933. MR3609347
[15] A. Masuoka, *Semisimple Hopf algebras of dimension* 6, 8, Israel J. Math. **92** (1995), no. 1-3, 361–373, DOI 10.1007/BF02762089. MR1357764
[16] D. S. Passman, *The algebraic structure of group rings*, Pure and Applied Mathematics, Wiley-Interscience [John Wiley & Sons], New York-London-Sydney, 1977. MR470211

Department of Mathematics, Faculty of Science, University of Porto, Rua Campo Alegre 687, 4169-007 Porto, Portugal

Contemporary Mathematics
Volume **727**, 2019
https://doi.org/10.1090/conm/727/14644

An informal overview of triples and systems

Louis Rowen

This paper is in honor of my good friend and colleague S.K. Jain on the occasion of his 80th birthday.

ABSTRACT. We describe triples and systems, expounded as an axiomatic algebraic umbrella theory for classical algebra, tropical algebra, hyperfields, and fuzzy rings.

1. Introduction

The goal of this overview is to present an axiomatic algebraic theory which unifies, simplifies, and "explains" aspects of idempotent algebra [**2**, **14**, **25**, **26**, **64**, **65**], tropical algebra [**3**, **34**, **39**, **42**, **43**], hyperfields [**7**, **28**, **49**, **71**], and fuzzy rings [**19**, **21**, **28**] in terms of familiar algebraic concepts. It was motivated by an attempt to understand whether or not it is coincidental that basic algebraic theorems are mirrored in supertropical algebra, and was spurred by the realization that some of the same results are obtained in parallel research on hyperfields and fuzzy rings. Our objective is to hone in on the precise axioms that include these various examples, formulate the axiomatic structure, describe its uses, and review five papers [**6**, **23**, **52**, **54**, **68**] in which the theory is developed, as well as indicated where future research might be heading. The bulk of this survey concerns the long paper [**68**], in which the axiomatic framework and various consequences are laid out, since the other papers build on it.

Other treatments of tropical algebra can be found in [**15**, **30**, **57**]. Although we also deal with general categorical issues, ours is largely a "hands on" approach, emphasizing a "negation map" which exists in all of the above-mentioned examples, and which can be obtained by means of a "symmetrization" functor. The other key ingredient is a **surpassing relation** $\preceq$, to replace equality in our theorems. (In classical mathematics, $\preceq$ is just equality.)

The quadruple $(\mathcal{A}, \mathcal{T}, (-), \preceq)$ is called a $\mathcal{T}$**-system**. Although our investigation of systems has centered on semirings, having grown out of tropical considerations,

---

2010 *Mathematics Subject Classification.* Primary 08A05, 16Y60, 12K10, 13C60, 20N20; Secondary 03C05, 06F05, 13C10, 14T05.

Key words and phrases. Bipotent, category, congruence, dual basis, homology, hyperfield, linear algebra, matrix, meta-tangible, morphism, negation map, module, polynomial, prime, projective, tensor product, semifield, semigroup, semiring, split, supertropical algebra, superalgebra, surpassing relation, symmetrization, system, triple, tropical.

The author thanks the referees for valuable comments leading to improvement of the exposition and clarification of some key points in §3 and §7.1.1.

systems also can be used to develop a parallel Lie semi-algebra theory (and more generally Hopf semi-algebra theory).

1.1. Acquaintance with basic notions. One starts with a set $\mathcal{T}$ that we want to study, called the set of **tangible elements**, endowed with a partial additive algebraic structure which however is not defined on all of $\mathcal{T}$; this is resolved by embedding $\mathcal{T}$ in a larger set $\mathcal{A}$ with a fuller algebraic structure. Often $\mathcal{T}$ is a multiplicative monoid [1], a situation developed by Lorscheid [**57, 58**] when $\mathcal{A}$ is a semiring. However, there also are examples (such as Lie algebras) lacking associative multiplication. We usually denote a typical element of $\mathcal{T}$ as a, and a typical element of $\mathcal{A}$ as b.

DEFINITION 1.1. A $\mathcal{T}$-**module** over a set $\mathcal{T}$ [2] is an additive monoid $(\mathcal{A}, +, \mathbf{0}_\mathcal{A})$ together with scalar multiplication $\mathcal{T} \times \mathcal{A} \to \mathcal{A}$ satisfying distributivity over $\mathcal{T}$ [3] in the sense that

$$a(b_1 + b_2) = ab_1 + ab_2$$

for $a \in \mathcal{T}$, $b_i \in \mathcal{A}$, also stipulating that $a\mathbf{0}_\mathcal{A} = \mathbf{0}_\mathcal{A}$.

A $\mathcal{T}$-**monoid module** over a multiplicative monoid $\mathcal{T}$ is a $\mathcal{T}$-module $\mathcal{A}$ satisfying the extra conditions

$$\mathbb{1}_\mathcal{T} b = b, \qquad (a_1 a_2)b = a_1(a_2 b), \quad \forall a_i \in \mathcal{T},\ b \in \mathcal{A}.$$

For the sake of this exposition, we assume that $\mathcal{A}$ is a $\mathcal{T}$-module and $\mathcal{T} \subseteq \mathcal{A}$.

A semigroup $(\mathcal{A}, +)$ has **characteristic** $k > 0$ if $(k+1)a = a$ for all $a \in \mathcal{A}$, with $k \geq 1$ minimal. $\mathcal{A}$ has **characteristic** 0 if $\mathcal{A}$ does not have characteristic k for any $k \geq 1$.

Properties of the characteristic are described in [**68**, §6.4]. Most of our major examples have characteristic 0, but some interesting examples have characteristic 2 or more.

A **semiring**† satisfies all the axioms of ring except the existence of a $\mathbf{0}$ element and of negatives. A **semiring** [**29**] is a semiring† with $\mathbf{0}$.

1.1.1. *Brief overview.* We introduce a formal negation map $(-)$, which we describe in §1.3 after some introductory examples, such that $(-)\mathcal{T} = \mathcal{T}$ and $\mathcal{T}$ generates $\mathcal{A}$ additively, creating a $\mathcal{T}$-**triple** $(\mathcal{A}, \mathcal{T}, (-))$.

When a formal negation map is not available at the outset, we can introduce it in two ways, to be elaborated shortly:

- Declare the negation map to be the identity, as in the supertropical case, cf. §1.2.2.
- Apply symmetrization, to get the switch map, of second kind, cf. §1.2.5. Often [**4**] is applicable, where $\mathcal{T}$ could take the role of the "thin elements."

The element $b^\circ := b(-)b$ is called a **quasi-zero**, and $(-)b$ is called the **quasi-negative** of b. We write $\mathcal{A}^\circ$ for $\{b^\circ : b \in \mathcal{A}\}$.

In classical algebra, the only quasi-zero is $\mathbf{0}$ itself, and $a_1 - a_1 = \mathbf{0} = a_2 - a_2$ for all a_1, a_2. Accordingly, we call a triple "$\mathcal{T}$-classical" [**68**, Definition 3.10], when $a_1(-)a_1 = a_2(-)a_2$ for some $a_1 \neq (\pm)a_2$ in $\mathcal{T}$.

[1] Geometry over monoids has been a subject of recent interest, cf. [**16**].

[2] More generally, when $\mathcal{T}$ is not a monoid, Hopf theory could play an appropriate role.

[3] Distributivity over elements of $\mathcal{T}$ is enough to run the theory, since one can (re)define multiplication on $\mathcal{A}$ to make it distributive, as seen in [**68**, Theorem 2.5]. This rather easy result applies for instance to hyperfields such as the phase hyperfield.

Examples from classical mathematics might provide some general intuition about employing $\mathcal{A}$ to study $\mathcal{T}$. A rather trivial example: $\mathcal{T}$ is the multiplicative subgroup of a field $\mathcal{A}$. Or $\mathcal{A}$ could be a graded associative algebra, with $\mathcal{T}$ its multiplicative submonoid of homogeneous elements.

But we are mostly interested in the non-classical situation, involving semirings which are not rings. Some motivating examples: The supertropical semiring, where $\mathcal{T}$ is the set of tangible elements; the symmetrized semiring, $\mathcal{T}$ is the set of elements with one nonzero component; and the power set of a hyperfield, where $\mathcal{T}$ is the hyperfield itself. Since hyperfields are so varied, they provide a good test for this theory. Semirings in general, without negation maps, are too broad to yield the structural results that we would like, which is the reason that negation maps are introduced in the first place.

Since we need to correlate two structures ($\mathcal{T}$ and $\mathcal{A}$), as well as the negation map (which could be viewed as a unary operator), it is convenient to work in the context of universal algebra, which was designed precisely for the purpose of drawing structures together. (More recently these have been generalized to Lawvere's theories and operads and model theory, but we do not delve into these aspects.)

To round things out, we introduce the **surpassing relation** $\preceq$, to replace equality in our theorems. (In classical mathematics, $\preceq$ is just equality.) The quadruple $(\mathcal{A}, \mathcal{T}, (-), \preceq)$ is called a $\mathcal{T}$-**system**.

1.2. Motivating examples. We elaborate the main non-classical examples motivating this theory.

1.2.1. *Idempotent semirings.* Tropical geometry has assumed a prominent position in mathematics because of its ability to simplify algebraic geometry while not changing certain invariants (often involving intersection numbers of varieties), thereby simplifying difficult computations. Outstanding applications abound, including [**1, 32, 47, 61, 66**].

The main original idea, as expounded in [**33, 60**], was to take the limit of the logarithm of the absolute values of the coordinates of an affine algebraic variety as the base of the logarithm goes to ∞. The underlying algebraic structure reverted from $\mathbb{C}$ to the max-plus algebra $\mathbb{R}_{\max}$, an ordered multiplicative monoid in which one defines $a + b$ to be $\max\{a, b\}$. This is a semiring[†] and is clearly additively **bipotent** in the sense that $a + b \in \{a, b\}$. Such algebras have been studied extensively some time ago, cf. [**14, 26**].

Idempotent (in particular bipotent) semirings have characteristic 1, and their geometry has been studied intensively as "F_1-geometry," cf. [**11, 15**]. But logarithms cannot be taken over the complex numbers, and do not send sums to maxima except at the limit, so attention of tropicalists largely passed to the field of Puisseux series, which in characteristic 0 also is an algebraically closed field, but now with a natural valuation, thereby making available tools of valuation theory, cf. [**10**]. The collection [**9**] presents such a valuation theoretic approach. Furthermore, $v(a_0 + a_1)$ can be anything for a valuation v when $v(a_0) = v(a_1)$. Thus one looks for an alternative to the max-plus algebra.

1.2.2. *Supertropical semirings.* Izhakian [**34**] overcame many of the structural deficiencies of a max-plus algebra $\mathcal{T}$ by adjoining an extra copy of $\mathcal{T}$, called the **ghost** copy $\mathcal{G}$ in [**42**, Definitions 3.2, 3.4], as well as $\mathbb{0}$, and modifying addition.

More generally, a **supertropical semiring** is a semiring with ghosts $(R, \mathcal{G}, \nu) := \mathcal{T} \cup \mathcal{G}_{\mathbf{0}}$, where $\mathcal{G}_{\mathbf{0}} = \mathcal{G} \cup \{\mathbf{0}\}$, together with a projection $\nu : R \to \mathcal{G}_{\mathbf{0}}$ satisfying the extra properties:

(a) (ν-Bipotence) $a + b \in \{a, b\}$, $\forall a, b \in R$ such that $\nu(a) \neq \nu(b)$;
(b) (Supertropicality) $a + b = \nu(a)$ if $\nu(a) = \nu(b)$.

The supertropical semiring is **standard** if $\nu|_{\mathcal{T}}$ is 1:1. The supertropical semiring is called the **standard supertropical semifield** when $\mathcal{G}_{\mathbf{0}}$ is a semifield.

Mysteriously, although lacking negation, the supertropical semiring provides affine geometry and linear algebra quite parallel to the classical theory, by taking the negation map $(-)$ to be the identity, so that $a^{\circ} = a + a$, where the ghost ideal $\mathcal{G} = \{a^{\circ} : a \in \mathcal{T}\}$ takes the place of the $\mathbf{0}$ element. In every instance, the classical theorem involving equality $f = g$ is replaced by an assertion that $f = g + \text{ghost}$, called **ghost surpassing**. In particular when $g = \mathbf{0}$ this means that f itself is a ghost.

For example, an irreducible affine variety should be the set of points which when evaluated at a given set of polynomials is ghost (not necessarily $\mathbf{0}$), leading to:

- a version of the Nullstellensatz in [**42**, Theorem 7.17],
- a link between decomposition of affine varieties and (non-unique) factorization of polynomials, illustrated in one indeterminate in [**42**, Remark 8.42 and Theorem 8.46],
- a version of the resultant of polynomials that can be computed by the classical Sylvester matrix and [**45**, Theorem 4.12 and Theorem 4.19].

Matrix theory also can be developed along supertropical lines. The supertropical Cayley-Hamilton theorem [**43**, Theorem 5.2] says that the characteristic polynomial evaluated on a matrix is a ghost. A matrix is called **singular** when its permanent (the tropical replacement of determinant) is a ghost; in [**43**, Theorem 6.5] (also proved in [**5**, **40**], and [**18**] for a special case) the row rank, column rank, and submatrix rank of a matrix (in this sense) are seen to be equal. Solution of tropical equations is given in [**3**, **4**, **25**, **44**]. Supertropical singularity also gives rise to semigroup versions of the classical algebraic group SL, as illustrated in [**41**].

Supertropical valuation theory is handled in a series of papers starting with [**35**], also cf. [**51**], generalized further in [**37**] and [**38**].

Note that standard supertropical semirings "almost" are bipotent, in the sense that $a_1 + a_2 \in \{a_1, a_2\}$ for any $a_1 \neq a_2$ in $\mathcal{T}$. This turns out to be an important axiom.

1.2.3. *Hyperfields and other related constructions.* Another algebraic construction is the hyperfield [**71**], which is a multiplicative group in which sets replace elements when one takes the sum, usually denoted as $\boxplus$. Hyperfields have received considerable attention recently [**7**, **31**, **49**] in part because of their diversity, and in fact Viro's "tropical hyperfield" matches Izhakian's construction. But there are other important hyperfields. Recall the **Boolean semifield** $\mathbb{B} = \{0, 1\}$ where $0 + a = a + 0 = a$ and $1 + 1 = 1$.

(i) The Krasner hyperfield. Let $K = \{0; 1\}$ with the usual operations of $\mathbb{B}$, with the exception that $1 \boxplus 1 = \{0; 1\}$.

(ii) The hyperfield of signs. Let $S := \{0, 1, -1\}$ with the usual multiplication law, and hyperaddition given by $1 \boxplus 1 = \{1\}$, $-1 \boxplus -1 = \{-1\}$, $x \boxplus 0 = 0 \boxplus x = \{x\}$, and $1 \boxplus -1 = -1 \boxplus 1 = \{0, 1, -1\} = S$.

(iii) The phase hyperfield. Let S^1 denote the complex unit circle, together with the center $\{0\}$, and take $\mathcal{T} = S^1$. Points a and b are **antipodes** if $a = -b$. Multiplication is defined as usual (so corresponds on S^1 to addition of angles). We denote an arc from a to b of less than 180 degrees as $\overline{(a,b)}$. The hypersum is given by

$$a \boxplus b = \begin{cases} \overline{(a,b)} \text{ if } a \neq -b; \\ \{-a, 0, a\} \text{ if } a = -b \neq 0; \\ \{a\} \text{ if } b = 0. \end{cases}$$

(iv) The "triangle" hyperfield. This is defined over $\mathbb{R}^+$ by the formula

$$a \boxplus b = \{c \in \mathbb{R}^+ : |a - b| \leq c \leq a + b\}.$$

In other words, $c \in a \boxplus b$ iff there exists a Euclidean triangle with sides of lengths a, b, and c.

We also want to understand their theories along similar lines. In hypergroup theory, one can replace "zero" by the property that a given set contains 0, and then any a has a unique **hypernegative** $-a$ satisfying $0 \in a \boxplus (-a)$. An intriguing phenomenon is that linear algebra over some classes of hyperfields follows classical lines as in the supertropical case, but the hyperfield of signs provides easy counterexamples to others, as discussed in [**6**].

1.2.4. *Fuzzy rings.* Dress [**19**] introduced "fuzzy rings" a while ago in connection with matroids, and these also have been seen recently to be related to hypergroups and, more specifically, hyperfields, in [**7, 20, 21, 28, 59**].

1.2.5. *Symmetrization.* This construction uses Gaubert's "symmetrized algebras" [**3, 25, 26, 64**] (which he designed for linear algebra) as a prototype. We start with $\mathcal{T}$, take $\mathcal{A} = \widehat{\mathcal{T}} := \mathcal{T} \times \mathcal{T}$, and define the "switch map" $(-)$ by $(-)(a_0, a_1) = (a_1, a_0)$. The reader might already recognize this as the first step in constructing the integers from the natural numbers, where one identifies (a_0, a_1) with (a_0', a_1') if $a_0 + a_1' = a_0' + a_1$, but the trick here is to recognize the equivalence relation without modding it out, since everything could degenerate in the nonclassical applications. Equality $(a_0, a_1) = (b_0, b_1)$ often is replaced by the assertion $(a_0, a_1) = (b_0, b_1) + (c, c)$ for some $c \in \mathcal{T}$.

More generally, for any $\mathcal{T}$-module $\mathcal{A}$, the "symmetrized" $\mathcal{T}$-module $\widehat{\mathcal{A}} := \mathcal{A} \times \mathcal{A}$ also can be viewed as a $\widehat{\mathcal{T}}$-super-module (i.e., 2-graded), via the **twist action**

$$(a_0, a_1) \cdot_{\text{tw}} (b_0, b_1) = (a_0 b_0 + a_1 b_1, a_0 b_1 + a_1 b_0), \quad a_i \in \mathcal{T}, b_i \in \mathcal{A}. \tag{1.1}$$

The twist action is utilized in [**48**] to define and study the prime spectrum.

1.2.6. *Functions and formal polynomials.* Given some structure $\mathcal{A}$ and a set S, we can define the set of functions $\mathcal{A}^{(n)}$ from S to $\mathcal{A}$, with operators defined elementwise, i.e.,

$$\omega((f_1, \dots, f_m))(s) = \omega(f_1(s), \dots, f_m(s)).$$

For example, taking $S = \{1, \dots, n\}$, a polynomial $f(\lambda_1, \dots, \lambda_n)$ can be viewed as a function from $\mathcal{A}^{(n)}$ to $\mathcal{A}$. In other words, we take the substitution $\lambda_i \mapsto a_i$ and then send $(a_1, \dots, a_n)$ to $f(a_1, \dots, a_n)$.

On the other hand, given a semiring $\mathcal{A}$, one can define the **formal polynomial ring** $\mathcal{A}[\Lambda]_{\text{form}}$ in the usual way. We prefer to identify polynomials in terms of their values as functions on their set of definition. Then λ^2 and λ are identified as polynomials over the finite field $\mathbf{F}_2$. $\mathcal{T}_{\text{Fun}(S,\mathcal{A})}$ is the set of functions whose support has a single element, say s, and for which $f(s) \in \mathcal{T}$. Thus, the tangible polynomials are the monomials with tangible coefficients.

1.3. Negation maps, triples, and systems. These varied examples and their theories, which often mimic classical algebra, lead one to wonder whether the parallels among them are happenstance, or whether there is some straightforward axiomatic framework within which they can all be gathered and simplified. Unfortunately semirings may lack negation, so we also implement a formal negation map $(-)$ to serve as a partial replacement for negation.

DEFINITION 1.2. A **negation map** on a $\mathcal{T}$-module $(\mathcal{A}, \cdot, +)$ is a semigroup isomorphism

$$(-) : (\mathcal{A}, +) \to (\mathcal{A}, +)$$

written $a \mapsto (-)a$, together with a map $(-) : (\mathcal{T}, +) \to (\mathcal{T}, +)$, both of order ≤ 2, satisfying

$$(-)(ab) = ((-)a)b = a((-)b), \quad \forall a \in \mathcal{T},\ b \in \mathcal{A}. \tag{1.2}$$

Obvious examples of negation maps are the identity map, which might seem too trivial but in fact is the one used in supertropical algebra, the switch map $((-)(a_0, a_1) = (a_1, a_0))$ in the symmetrized algebra, the usual negation map $(-)a = -a$ in classical algebra, and the hypernegation in the definition of hyperfields. Accordingly, we say that the negation map $(-)$ is of the **first kind** if $(-)a = a$ for all $a \in \mathcal{T}$, and of the **second kind** if $(-)a \neq a$ for all $a \in \mathcal{T}$.

As indicated earlier, the quasi-zeros take the role customarily assigned to the zero element. In the supertropical theory the quasi-zeros are the "ghost" elements. In [**4**, Definition 2.6] the quasi-zeros are called "balanced elements" and have the form (a, a).

When $\mathbb{1} \in \mathcal{T}$, the element $(-)\mathbb{1}$ determines the negation map, since $(-)b = (-)(\mathbb{1}b) = ((-)\mathbb{1})b$. When $\mathcal{T} \subseteq \mathcal{A}$, several important elements then are:

$$e = \mathbb{1}^\circ = \mathbb{1}(-)\mathbb{1}, \quad e' = e + \mathbb{1}, \quad e^\circ = e(-)e = e + e = 2e. \tag{1.3}$$

The most important quasi-zero for us is e. (For fuzzy rings, $e = \mathbb{1} + \epsilon$.) But e need not absorb in multiplication; rather, in any semiring† with negation, Definition 1.1 implies

$$ea = a(-)a = a^\circ. \tag{1.4}$$

DEFINITION 1.3. A **pseudo-triple** is a collection $(\mathcal{A}, \mathcal{T}, (-))$, where $\mathcal{A}$ is a $\mathcal{T}$-module and $(-)$ is a negation map. A $\mathcal{T}$**-pseudo-triple** is a pseudo-triple in which $\mathcal{T} \subseteq \mathcal{A}$, where the negation map on $\mathcal{A}$ restricts to the negation map on $\mathcal{T}$.

A **triple** is a $\mathcal{T}$-pseudo-triple, in which $\mathcal{T}$ generates $(\mathcal{A}, +)$ and $\mathcal{T} \cap \mathcal{A}^\circ = \emptyset$, i.e., a quasi-zero cannot be tangible.

EXAMPLE 1.4. The main non-classical examples are:

- (The standard supertropical triple.) $(\mathcal{A}, \mathcal{T}, (-))$ where $\mathcal{A} = \mathcal{T} \cup \mathcal{G}$ as before and $(-)$ is the identity map.

- (The symmetrized triple.) $(\hat{\mathcal{A}}, \mathcal{T}_{\mathcal{A}}, (-))$ where $\hat{\mathcal{A}} = \mathcal{A} \times \mathcal{A}$ with componentwise addition, and $\mathcal{T}_{\mathcal{A}}, (-)) = (\mathcal{T} \times \{\mathbf{0}\}) \cup (\{\mathbf{0}\} \times \mathcal{T})$ with multiplication $\widehat{\mathcal{T}} \times \widehat{\mathcal{A}} \to \widehat{\mathcal{A}}$ given by the twist action (1.1). We take $(-)$ to be the switch map $((-)(a_0, a_1) = (a_1, a_0))$, which is of second kind.
- (The hyperfield triple and, more generally, pseudo-triple [**68**, §3.6.1].) $(\mathcal{P}(\mathcal{T}), \mathcal{T}, (-))$ where $\mathcal{T}$ is the original hyperfield, $\mathcal{P}(\mathcal{T})$ is its power set (with componentwise operations), and $(-)$ on the power set is induced from the hypernegation.

  The Krasner hyperfield triple is just the supertropicalization of the Boolean semifield $\mathbb{B}$, identifying $\{0, 1\}$ with $\nu(1)$.

  The triple arising from the hyperfield of signs is the symmetrization of $\mathbb{B}$, identifying 1 with $(1, 0)$, -1 with $(0, 1)$, and $\{0, 1, -1\}$ with $(1, 1)$.
- (The fuzzy triple [**68**, Appendix B].) For any $\mathcal{T}$-monoid module $\mathcal{A}$ with an element $\mathbb{1}' \in \mathcal{T}$ satisfying $(\mathbb{1}')^2 = \mathbb{1}$, we can define a negation map $(-)$ on $\mathcal{T}$ and $\mathcal{A}$ given by $a \mapsto \mathbb{1}'a$. In particular, $(-)\mathbb{1} = \mathbb{1}'$.
- (The polynomial triple) We take any triple $(\mathcal{A}, \mathcal{T}, (-))$ and form $(\mathcal{A}[\Lambda], \mathcal{T}_{\mathcal{A}[\Lambda]}, (-))$ where $\mathcal{T}_{\mathcal{A}[\Lambda]}$ is the set of tangible monomials. Negation is taken elementwise. From a geometric perspective it is useful to view polynomials as functions on varieties.

Although we introduced pseudo-triples since $\mathcal{T}$ need not generate $(\mathcal{A}, +)$ (for example, taking $\mathcal{A} = \mathcal{P}(\mathcal{T})$ for the phase hyperfield), we are more concerned with triples; furthermore given a pseudo-triple one can take the sub-triple generated by $\mathcal{T}$. More triples related to tropical algebra ("layered triples") are presented in [**68**, §3.3].

Structures other than monoids also are amenable to such an approach. This can all be formulated axiomatically in the context of universal algebra, as treated for example in [**46**]. Once the natural categorical setting is established, it provides the context in which tropicalization (described below) becomes a functor, thereby providing guidance to understand tropical versions of an assortment of mathematical structures.

DEFINITION 1.5. Our structure of choice, a $\mathcal{T}$-**system**, is a quadruple $(\mathcal{A}, \mathcal{T}, (-), \preceq)$, where $(\mathcal{A}, \mathcal{T}, (-))$ is a $\mathcal{T}$-triple and $\preceq$ is a "surpassing relation" ([**68**, Definition 4.5]), satisfying the crucial property that if $a + a' \succeq \mathbf{0}$ for $a, a' \in \mathcal{T}$ then $a' = (-)a$.

EXAMPLE 1.6. The main surpassing relations are:

- (for supertropical, symmetrized, and fuzzy rings) $\preceq_\circ$, defined by $b' \preceq_\circ b$ if $b = b' + c^\circ$ for some c.
- (on hyperfield triples) $\preceq$ is $\subseteq$.

The surpassing relation $\preceq$ has an important theoretical role, replacing "=" and enabling us to define a broader category than one would obtain directly from universal algebra, cf. [**68**, §5]. One major reason why $\preceq$ can replace equality in much of the theory is found in the "transfer principle" of [**4**], given in the context of systems in [**68**, Theorem 8.17].

1.3.1. *Ground triples versus module triples.* Classical structure theory involves the investigation of an algebraic structure as a small category (for example, viewing a monoid as a category with a single object whose morphisms are its elements), and homomorphisms then are functors between two of these small categories. On the other hand, one obtains classical representation theory via an abelian category, such as the class of modules over a given ring.

Analogously, there are two aspects of triples. We call a triple (resp. system) a **ground triple (resp. ground system)** when we study it as a small category with a single object in its own right, often a semiring†. Ground triples have the same flavor as Lorscheid's blueprints (albeit slightly more general, and with a negation map), whereas representation theory leads us to **module systems**, described below in §5.3 and §6.

This situation leads to a fork in the road: The first path takes us to a structure theory based on functors of ground systems, translating into homomorphic images of systems via congruences in [**54**, §6] (especially **prime systems**, in which the product of non-trivial congruences is nontrivial [**54**, §6.2]). Ground systems often are designated in terms of the structure of $\mathcal{A}$ or $\mathcal{T}$, such as "semiring systems" or "Hopf systems" or "hyperfield systems."

The paper [**6**] has a different flavor, dealing with matrices and the ensuing linear algebra over ground systems, and focusing on subtleties concerning Cramer's rule and the equality of row rank, column rank, and submatrix rank.

The second path takes us to categories of module systems. In [**54**] we also bring in tensor products and Hom functors. In [**53**] we develop the homological theory, relying on work done already by Grandis [**30**] under the name of **N-category** and **homological category** (without the negation map), and there is a parallel approach of Connes and Consani in [**15**].

2. Contents of [68]: meta-tangible triples and systems

The emphasis in [**68**] is on ground systems. One can apply the familiar constructions and concepts of classical algebra (direct powers [**68**, Definition 2.6], matrices [**68**, §6.5], involutions [**68**, §6.6], polynomials [**68**, §6.7], localization [**68**, §6.8], and tensor products [**68**, Remark 8.35]) to produce new triples and systems. The simple tensors $a_0 \otimes a_1$ where $a_i \in \mathcal{T}$, comprise the tangible elements of the tensor product. The properties of tensors and Hom are treated in much greater depth in [**54**].

Let us turn to important properties which could hold in triples.

DEFINITION 2.1. The $\mathcal{T}$-triple $(\mathcal{A}, \mathcal{T}, (-))$ has **unique quasi-negatives** if $a + b \in \mathcal{A}^\circ$ for $a, b \in \mathcal{T}$ implies $b = (-)a$.

One basic axiom for this theory, holding in all tropical situations and many related theories, is:

DEFINITION 2.2. A $\mathcal{T}$-triple $(\mathcal{A}, \mathcal{T}, (-))$ is **meta-tangible**, if the sum of two tangible elements is tangible unless they are quasi-negatives of each other.

A special case: $(\mathcal{A}, \mathcal{T}, (-))$ is $(-)$-**bipotent** if $a+a' \in \{a, a'\}$ whenever $a, a' \in \mathcal{T}$ with $a' \neq (-)a$. In other words, $a + a' \in \{a, a', a^\circ\}$ for all $a, a' \in \mathcal{T}$.

The stipulation in the definition that $a' \neq (-)a$ is of utmost importance, since otherwise all of our main examples would fail. [**68**, Proposition 7.64] shows how to view a meta-tangible triple as a hypergroup, thereby enhancing the motivation of transferring hyperfield notions to triples and systems in general.

Any meta-tangible triple has unique quasi-negatives. The supertropical triple is $(-)$-bipotent, as is the modification of the symmetrized triple described in [**68**, §3.5.2].

The Krasner hyperfield triple and the triple arising from the hyperfield of signs are $(-)$-bipotent, but the phase hyperfield triple and the triangle hyperfield triple are not even metatangible (although the latter is idempotent). But as seen in [**68**, Theorem 6.43], hyperfield triples satisfy a different property of independent interest:

DEFINITION 2.3. [**68**, Definition 6.11] A surpassing relation $\preceq$ in a system is called $\mathcal{T}$-**reversible** if $a_0 \preceq a_1 + c$ implies $a_1 \preceq a_0(-)c$ for $a_0, a_1 \in \mathcal{T}$.

The category of hyperfields as given in [**49**] can be embedded into the category of $\mathcal{T}$-reversible systems ([**68**, Theorem 6.7]). Reversibility enables one to apply systems to matroid theory, although we have not yet embarked on that endeavor in earnest.

The **height** of an element c in a triple $(\mathcal{A}, \mathcal{T}, (-))$ (sometimes called "width" in the group-theoretic literature) is the minimal t such that $c = \sum_{i=1}^{t} a_i$ with each $a_i \in \mathcal{T}$. (We say that $\mathbf{0}$ has height 0.) By definition, every element of $\mathcal{A}$ has finite height. The **height** of $\mathcal{A}$ is the maximal height of its elements, when these heights are bounded. For example, the supertropical semiring has height 2, as does the symmetrized semiring of an idempotent semifield $\mathcal{T}$.

Some unexpected examples of meta-tangible systems sneak in when they have height ≥ 3, as described in [**68**, Example 3.8, Example 4.21(iv)].

In [**68**, §7] the case is presented that meta-tangibility could be the major axiom in the theory of ground systems over a group $\mathcal{T}$, leading to a bevy of structure theorems on meta-tangible systems, starting with the observation [**68**, Lemma 5.5] that for $a, b \in \mathcal{T}$ either $a = (-)a'$, $a + a' = a$ (and thus $a^\circ + a' = a^\circ$), or $a^\circ + a' = a'$.

In [**68**, Proposition 7.18] the following assertions are seen to be equivalent for a triple $(\mathcal{A}, \mathcal{T}, (-))$ containing $\mathbb{1}$:

(i) $\mathcal{T} \cup \mathcal{T}^\circ = \mathcal{A}$,
(ii) $\mathcal{A}$ is meta-tangible of height ≤ 2,
(iii) $\mathcal{A}$ is meta-tangible with $e' \in \{\mathbb{1}, e\}$.

We obtain the following results for a meta-tangible system $(\mathcal{A}, \mathcal{T}, (-), \preceq)$:

- [**68**, Theorem 7.21] If $\mathcal{A}$ is not $(-)$-bipotent then $e' = \mathbb{1}$, with $\mathcal{A}$ of characteristic 2 when $(-)$ is of the first kind.
- [**68**, Theorem 7.28] Every element has the form a° or ma for $a \in \mathcal{T}$ and $m \neq 2$. The extent to which this presentation is unique is described in [**68**, Theorem 7.32].
- [**68**, Theorem 7.34] Distributivity follows from the other axioms.
- [**68**, Theorem 7.35] The surpassing relation $\preceq$ must "almost" be $\preceq_\circ$.
- [**68**, Theorem 7.38] A key property of fuzzy rings holds.
- [**68**, Theorem 7.44] Reversibility holds, except in one pathological situation.
- [**68**, Theorem 7.55] A criterion is given in terms of sums of squares for $(\mathcal{A}, \mathcal{T}, (-))$ to be isomorphic to a symmetrized triple.

One would want a classification theorem of meta-tangible systems that reduces to classical algebras, the standard supertropical semiring, the symmetrized semiring, layered semirings, power sets of various hyperfields, or fuzzy rings, but there are several exceptional cases. Nonetheless, [**68**, Theorem 7.56] comes close. Namely, if $(-)$ is of the first kind then either $\mathcal{A}$ has characteristic 2 and height 2 with $\mathcal{A}^\circ$ bipotent, or $(\mathcal{A}, \mathcal{T}, (-))$ is isomorphic to a layered system. If $(-)$ is of the second kind then either $\mathcal{T}$ is $(-)$-bipotent, with $\mathcal{A}$ of height 2 (except for an exceptional case), $\mathcal{A}$ is isometric to a symmetrized semiring when $\mathcal{A}$ is real, or $\mathcal{A}$ is classical. More information about the exceptions are given in [**68**, Remark 7.57].

[**68**, §9] continues with some rudiments of linear algebra over a ground triple, to be discussed in the next section.

In [**68**, §10], tropicalization is cast in terms of a functor on systems (from the classical to the nonclassical). This basic principle enables one in [**68**, §11] to define the "right" tropical versions of classical algebraic structures, including exterior algebras, Lie algebras, Lie superalgebras, and Poisson algebras.

3. Contents of [6]: Linear algebra over systems

The paper [**6**] was written with the objective of providing diverse theorems in linear algebra over semiring systems. A set of vectors $\{v_i \in \mathcal{A}^{(n)} : i \in I\}$ is called **$\mathcal{T}$-dependent** if $\sum_{i\in I'} \alpha_i v_i \in (\mathcal{A}^\circ)^{(n)}$ for some nonempty subset $I' \subseteq I$ and $\alpha_i \in \mathcal{T}$, and the **row rank** of a matrix is the maximal number of $\mathcal{T}$-independent rows. The **column rank** is analogous.

A **tangible vector** is a nonzero vector all of whose entries are in $\mathcal{T}_\mathbb{0} := \mathcal{T} \cup \{\mathbb{0}\}$. A **tangible matrix** is a nonzero matrix all of whose rows are tangible vectors.

The $(-)$-**determinant** $|A|$ of an $n \times n$ matrix $A = a_{i,j}$ is

$$\sum_{\pi \in S_n} (-)^\pi a_\pi,$$

where $a_\pi := \prod_{i=1}^n a_{i,\pi(i)}$.

DEFINITION 3.1. Write $a'_{i,j}$ for the $(-)$-determinant of the j,i minor of a matrix A. The $(-)$-**adjoint** matrix $\operatorname{adj}(A)$ is $(a'_{i,j})$.

THEOREM 3.2 (Generalized Laplace identity, [**6**, Theorem 1.56]). *Fix*

$$I = \{i_1, \dots, i_m\} \subset \{1, 2, ..., n\},$$

and for any set $J = \{j_1, \dots, j_m\} \subset \{1, 2, ..., n\}$, write $(a_{I,J})$ for the $m \times m$ minor $(a_{i,j} : i \in I, j \in J)$ and $(a'_{I,J})$ for the $(-)$-determinant of the $(n-m) \times (n-m)$ minor obtained by deleting all rows from I and all columns from J. Write $(-)^J$ for $(-)^{j_1+\dots+j_m}$. Then $|A| = \sum_{J:|J|=m} (-)^I (-)^J a'_{I,J} |a_{I,J}|$.

A matrix A is **nonsingular** if $|A| \in \mathcal{T}$. The **submatrix rank** of A is the largest size of a nonsingular square submatrix of A.

In [**68**, Corollary 9.5] we see that the submatrix rank of a matrix over a $\mathcal{T}$-cancellative meta-tangible system is less than or equal to both the row rank and the column rank. This is improved in [**6**] with a version of "Cramer's rule":

THEOREM 3.3 ([**6**, Theorem 4.8(i)]). *Let $(\mathcal{A}, \mathcal{T}, (-), \preceq_\circ)$ be a system. For any vector v, the vector $y = \operatorname{adj}(A)v$ satisfies $|A|v \preceq_\circ Ay$. In particular, if $|A|$ is invertible in $\mathcal{T}$, then $x := \frac{1}{|A|}\operatorname{adj}(A)v$ satisfies $v \preceq_\circ Ax$.*

In [**6**, Theorem 4.7(ii)], uniqueness of x is obtained using a property called "strong balance elimination." The existence of a tangible such x is subtler. One considers valuations of systems $\nu : (\mathcal{A}, +) \to (\mathcal{G}, +)$, and their **fibers** $\{a \in \mathcal{T} : \nu(a) = g\}$ for $g \in \mathcal{G}$; we call the system $\mathcal{T}$-**Noetherian** if any ascending chain of fibers stabilizes. The property needed to prove the "Jacobi decomposition" of [**4**, Proposition 5.18] is:

(3.1) $$\text{If } a \in \mathcal{T} \text{ and } \nu(a) \le \nu(c), \text{ then } \mathcal{F}(\nu(c)) \text{ contains some } b \in \mathcal{T} \text{ with } \nu(a) \le \nu(b) = \nu(c).$$

THEOREM 3.4 (The Jacobi algorithm, [**6**, Theorem 4.26]). *The "Jacobi algorithm" of* [**4**] *applies to any $\mathcal{T}$-Noetherian $\mathcal{G}$-valued system $\mathcal{A}$ satisfying* (3.1), *and a matrix $A \in M_n(\mathcal{A})$ having a "dominant diagonal," in the sense that $\nu(a_{1,1} \cdots a_{n,n})$ is invertible in A° and $\nu(a_{1,1} \cdots a_{n,n}) \ge \nu(|A|)$. Explicitly, let $A = D + N$ be a Jacobi decomposition. Then*

(i) *One can construct an ascending sequence $x_1, x_2, \dots,$ for $x_k \in \mathcal{T}_{\mathbf{0}}^{(n)}$ (with respect to the "natural order" $\le$, satisfying $Dx_{k+1}(-)(Nx_k + v) \succeq \mathbf{0}$, with $\nu(Dx_{k+1}) = \nu(Nx_k + v)$.*

(ii) *When $\mathcal{A}$ satisfies* (3.1), *the sequence $\nu(x_1), \nu(x_2), \dots,$ stabilizes at n, i.e., $x_k = x_{k+1}$ for all $k \ge n$, and moreover $\nu(x_n) = \nu(|A|)^{-1}\nu(\mathrm{adj}(A)v)$ when $\nu(|A|)$ is invertible.*

(iii) *When $\mathcal{A}$ satisfies* [**4**, Property 5.2] *and is $\mathcal{T}$-Noetherian, the sequence $x_1, x_2, \dots,$ stabilizes, i.e., there is m such that $\nu(x_k) = \nu(x_{k+1})$ for all $k \ge m$ and, moreover, the limit x_m is a solution of $Ax(-)v \succeq \mathbf{0}$.*

(iv) *When fibers have length 1 (*[**4**, Property 5.4]*), one can take $m = n$ in (iii).*

After translating some more of the concepts of [**4**] into the language of systems, we turn to the question, raised privately for hyperfields by Baker:

Question A. When does the submatrix rank equal the row rank?

Our initial hope was that this would always be the case, in analogy to the supertropical situation. However, Gaubert observed that a (nonsquare) rebuttal to Question A already can be found in [**3**], and the underlying system even is meta-tangible. Here the kind of negation map is critical: A rather general counterexample for triples of the second kind is given in [**6**, Proposition 3.3]; the essence of the example already exists in the "sign hyperfield." Although the counterexample as given is a nonsquare (3×4) matrix, it can be modified to an $n \times n$ matrix for any $n \ge 4$. This counterexample is minimal in the sense that Question A has a positive answer for $n \le 2$ and for 3×3 matrices under a mild assumption, cf. [**6**, Proposition 5.5 and Theorem 6.1].

Nevertheless, general positive results are available. A positive answer for Question A along the lines of [**4**, Theorems 5.11, 5.20, 6.9] is given in [**6**, Theorem 5.10] for systems satisfying certain technical conditions. In [**6**, Theorem 6.2 and Theorem 6.3], we show that Question A has a positive answer for various triples of first kind of height 2, and this seems to be the "correct" framework in which we can lift theorems from classical algebra. A positive answer for all rectangular matrices is given in [**6**, Theorem 6.5], but with restrictive hypotheses that essentially reduce to the supertropical situation.

4. Contents of [23]: Grassmann semialgebras

This paper unifies classical and tropical theory. Ironically, even though a vector space over a semifield† need not be provided with a negation map, the elements of degree ≥ 2 in the tensor semi-algebra do have a negation map given by

$$(-)(v \otimes w) = w \otimes v,$$

so these can be viewed in terms of triples and systems. The version given in [**23**, Theorem 2.6 and Definition 2.12] (over a free module V over an arbitrary semifield) is the construction which seems to "work" for the Grassmann semialgebra. This is obtained by taking a given base $\{b_0, b_1, \dots, b_{n-1}\}$ of V, defining $(-)b_i \wedge b_j = b_j \wedge b_i$ and $b_i \wedge b_i = \mathbf{0}$ for each $0 \leq i < j \leq n-1$, and extending to all of the Grassmann semialgebra. This does not imply $v \wedge v = \mathbf{0}$ for arbitrary $v \in V$; for example, taking $v = b_0 + b_1$ yields $v \wedge v = b_0 \wedge b_1 + b_1 \wedge b_0$ which need not be $\mathbf{0}$, but it acts like $\mathbf{0}$.

Thus the tensor semialgebra of a free module V has a natural negation map on all homogeneous vectors with the ironic exception of V itself, obtained by switching two tensor components. This provides us "enough" negation to carry out the theory. (But we do not mod out all elements $v \otimes v$ for arbitrary $v \in V$!)

Our main theorem, [**23**, Theorem 3.17], describes the relation between a "Hasse-Schmidt derivation" $D\{z\}$ and its "quasi-inverse" $\overline{D}\{z\}$ defined in such a way to yield:

THEOREM 4.1. $\overline{D}\{z\}(D\{z\}u \wedge v) \succeq u \wedge \overline{D}\{z\}v.$

In the classical case, one gets equality as shown in [**23**, Remark 3.19], so we recover [**24**].

In the process we investigate Hasse-Schmidt derivations on Grassmann exterior systems and use these results to provide a generalization of the Cayley-Hamilton theorem to semi-algebras in [**23**, Theorem 3.18], yielding

COROLLARY 4.2 ([**23**, Corollary 3.20]).

$$((D_n u + e_1 D_{n-1} u + \cdots + e_n u) \wedge v)\,(-)\,((D_n u + e_1' D_{n-1} u + \cdots + e_n' u) \wedge v) \succeq 0$$

for all $u \in \bigwedge^{>0} V_n$, *where* e_i, e_i' *are given by* $\overline{D}_i \zeta = e_i \zeta + e_i' \zeta'$.

5. Contents of [54]: Basic categorical considerations

The paper [**54**] elaborates on the categorical aspects of systems, with emphasis on important functors. The convolution triple $\mathcal{C}^S$ [**54**, Proposition 3.7] embraces important constructions including the symmetrized triple and polynomial triples (via the usual convolution product given before [**54**, Definition 3.4]).

5.1. Prime systems. In order to pave the way towards geometry, we consider "prime" systems and congruences, proving some basic results about polynomial systems:

PROPOSITION 5.1 ([**54**, Proposition 6.19]). *For every $\mathcal{T}$-congruence Φ on a commutative $\mathcal{T}$-semiring system, $\sqrt{\Phi}$ is an intersection of prime $\mathcal{T}$-congruences.*

DEFINITION 5.2. A **$\preceq$-root** of a polynomial $f \in \mathcal{A}[\lambda_1, \dots, \lambda_n]$ is some n-tuple $(a_1, \dots, a_n)$ such that $f(a_1, \dots, a_n) \succeq \mathbf{0}$.

THEOREM 5.3 ([**54**, Theorem 6.29]). *Over a prime commutative triple, any nonzero polynomial $f \in \mathcal{T}[\lambda]$ of degree n cannot have $n+1$ distinct $\preceq$-roots in $\mathcal{T}$.*

COROLLARY 5.4 ([**54**, Corollary 6.30]). *If $(\mathcal{A}, \mathcal{T}, (-))$ is a prime commutative triple with $\mathcal{T}$ infinite, then so is $(\mathcal{A}[\lambda], \mathcal{T}, (-))$.*

As in [**57, 58**], the emphasis on [**54**] is for $\mathcal{T}$ to be a cancellative multiplicative monoid (even a group), which encompasses many major applications but slights the Lie theory. In fact one may also consider Hopf systems. Motivation can be found in [**69**].

5.2. Morphisms. An issue that must be confronted is the proper definition of morphism, cf. [**54**, Definitions 4.1, 4.3]. In categories arising from universal algebra, one's intuition would be to take the homomorphisms, i.e., those maps which preserve equality in the operators. However, this approach loses some major examples of hyperfields. Applications in tropical mathematics and hyperfields (cf. [**49**, Definition 2.3]) tend to depend on the surpassing relation $\preceq$, so we are led to a broader definition called **$\preceq$-category** in [**54**, §4]. [**54**, Proposition 4.37ff.] gives a way of verifying that some $\preceq$-morphisms automatically are morphisms.

5.3. Module systems. The situation is stricter for module systems over ground triples [**54**, §5, 7, 8]. The sticky point here is that the semigroups of morphisms $\mathrm{Hom}(A, B)$ in our $\mathcal{T}$-module categories are not necessarily groups, so the traditional notion of abelian category has to be replaced by "semi-abelian," [**54**, Theorem 7.16], and these lack some of the fundamental properties of abelian categories. The tensor product is only functorial when we restrict our attention to the stricter definition of morphisms, [**54**, Proposition 5.8]!

In both cases, in the theory of semirings and their modules, homomorphisms are described in terms of congruences, so congruences should be a focus of the theory. The null congruences contain the diagonal, and not necessarily zero, and lead us to null morphisms, [**54**, Definition 2.2]. In [**54**, §8], "Hom" is studied together with its dual. In this way, the categorical view is to be carried out in [**53**] for homology, again at times with a Hopfian flavor.

The functors between the various categories arising in this theory are described in [**54**, §9], also with an eye towards valuations of systems.

6. Contents of [52]: $\preceq$-Projective module systems

To develop the algebraic theory of systems, it is crucial to have the "right" version of projective module, which is the subject of this paper, in the last stages of revision. Here $\preceq$-plays a key role, where the traditional definitions are replaced by $\preceq$-morphisms. We start with a $\preceq$ version of split epics (weaker than the classical definition):

DEFINITION 6.1. An epic $\pi : \mathcal{M} \to \mathcal{N}$ **$\preceq$-splits** if there is an N-monic $\nu : \mathcal{N} \to \mathcal{M}$ such that $\mathbb{1}_{\mathcal{N}} \preceq \pi\nu$. In this case, we also say that (π, ν) **$\preceq$-splits**, and $\mathcal{N}$ is a **$\preceq$-retract** of $\mathcal{M}$.

This leads to projective module systems.

DEFINITION 6.2. A $\mathcal{T}$-module system $\mathcal{P}$ is **$\preceq$-projective** if for any ($\mathcal{T}$-module system) $\preceq$-onto morphism $h : \mathcal{M} \to \mathcal{M}'$, every $\preceq$-morphism $f : \mathcal{P} \to \mathcal{M}'$ $\preceq$-lifts to a $\preceq$-morphism $\tilde{f} : \mathcal{P} \to \mathcal{M}$, in the sense that $f \preceq h\tilde{f}$.

Their fundamental properties are then obtained, including a $\preceq$-Dual Basis Lemma [**52**, Proposition 4.16], leading to a $\preceq$-Schanuel's Lemma [**52**, Theorem 4.18 and Corollary 4.19]. [**52**, § 4.4] also contains a $\preceq$-Morita theory.

Built on $\preceq$-projective modules, homology is work in progress [**53**], as is parallel work in geometry. One obtains a homology theory in the context of homological categories [**30**] and derived functors, in connection to the recent work of Connes and Consani [**15**].

7. Interface between systems and tropical mathematics

We conclude by relating systems to other approaches taken in tropical mathematics, [**10**, **11**, **33**, **60**]. The interface with other approaches to tropical algebra is given briefly in [**54**, §10].

7.1. The "standard" (idempotent) tropical approach. One often works in the polynomial semiring $\mathbb{R}_{\max}[\Lambda]$, although here we replace $\mathbb{R}_{\max}$ by any partially ordered semigroup $(\Gamma, \cdot)$, with $\Gamma_0 := \Gamma \cup \{\mathbf{0}\}$ where $\mathbf{0}a = \mathbf{0}$ for all $a \in \Gamma$. For $\mathbf{i} = (i_1, \dots, i_n) \in \mathbb{N}^{(n)}$, we write $\Lambda^{\mathbf{i}}$ for $\lambda_1^{i_1} \cdots \lambda_n^{i_n}$. A tropical hypersurface of a tropical polynomial $f = \sum_{\mathbf{i}} \alpha_{\mathbf{i}} \Lambda^{\mathbf{i}} \in \Gamma[\Lambda]$ is defined as the set of points in which two monomials take on the same dominant value, which is the same thing as the supertropical value of f being a ghost.

Definition 7.1 ([**27**, Definition 5.1.1]). Given a polynomial $f = \sum \alpha_{\mathbf{i}} \Lambda^{\mathbf{i}}$, define $\operatorname{supp}(f)$ to be all the tuples $\mathbf{i} = (i_1 \cdots i_n)$ for which the monomial in f has nonzero coefficient $\alpha_{\mathbf{i}}$, and for any such monomial h, write $f_{\hat{h}}$ for the polynomial obtained by deleting h from the summation.

The **bend congruence** (with respect to a tropical hypersurface V defined by f) is generated by the **bend relation** given by all

$$\{f \equiv_{\text{bend}} f_{\hat{h}=\alpha_{\mathbf{i}}\Lambda^{\mathbf{i}}} : \mathbf{i} \in \operatorname{supp}(f)\}.$$

The **bend congruence** (with respect to a tropical affine variety V defined by an ideal $\mathcal{I}$) is generated by the **bend relation** given by all

$$\{f \equiv_{\text{bend}} f_{\hat{h}=\alpha_{\mathbf{i}}\Lambda^{\mathbf{i}}} : f \in \mathcal{I},\ \mathbf{i} \in \operatorname{supp}(f)\}.$$

The bend relation reflects when some monomial in a tropical polynomial f evaluated on a variety V is superfluous. This happens for example when f has two monomials (not necessarily the same throughout) taking equal dominant values at each point of V. This treatment in [**27**] uses formal polynomials.

7.1.1. *Tropicalization and tropical ideals.* Next, one relates tropical algebra to Puiseux series via the following tropicalization map.

Definition 7.2. For any additive group $\mathcal{M}$, one can define the group $\mathcal{M}\{\{t\}\}$ of Puiseux series on the variable t, which is the set of formal series of the form $p = \sum_{j=\ell}^{\infty} c_j t^{j/N}$ where $N \in \mathbb{N}$, $\ell \in \mathbb{Z}$, and $c_j \in \mathcal{A}$.

Suppose from now on that $\mathcal{M}$ is a field K, and $\Lambda = \{\lambda_i : i \in I\}$ is a set of commuting indeterminates.

Remark 7.3. $K\{\{t\}\}$ is also a field under the usual convolution product, and is algebraically closed when $\mathcal{M} = \mathbb{C}$.

Definition 7.4. One has the **Puiseux valuation** $val : \widehat{K\{\{t\}\}} \setminus \{0\} \to \Gamma$ defined by

$$val(p) = \min_{c_j \neq 0}\{j/N\}, \tag{7.1}$$

This induces a map

$$\text{trop} : K\{\{t\}\}[\Lambda] \to \Gamma[\Lambda],$$

called **tropicalization**, sending $p(\lambda_1 \cdots \lambda_n) := \sum p_{\mathbf{i}} \lambda_1 \cdots \lambda_n$ to $\sum \operatorname{val}(p_{\mathbf{i}}) \lambda_1 \cdots \lambda_n$.

We can **normalize** a Puiseux series $p = \sum p_{\mathbf{i}}\lambda_1^{i_1}\cdots\lambda_n^{i_n}$ at any given $\mathbf{i} \in \operatorname{supp}(p)$ by dividing through by $p_{\mathbf{i}}$; then the normalized coefficient is $\mathbb{1}$. Given two Puiseux series $p = \sum p_{\mathbf{i}}\lambda_1^{i_1}\cdots\lambda_n^{i_n}$, $q = \sum q_{\mathbf{i}}\lambda_1^{i_1}\cdots\lambda_n^{i_n}$ having a common monomial $\Lambda^{\mathbf{i}} = \lambda_1^{i_1}\cdots\lambda_n^{i_n}$ in their support, one can normalize both and assume that $p_{\mathbf{i}} = q_{\mathbf{i}} = \mathbb{1}$, and remove this monomial from their difference $p - q$, i.e., the coefficient of $\Lambda^{\mathbf{i}}$ in $\operatorname{val}(p-q)$ is $\mathbf{0}$ $(= -\infty)$.

Accordingly, a **tropical ideal** of $\Gamma[\Lambda]$ is an ideal $\mathcal{I}$ such that for any two polynomials $f = \sum f_{\mathbf{j}}\lambda_1^{j_1}\cdots\lambda_n^{j_n}$, $g = \sum g_{\mathbf{j}}\lambda_1^{j_1}\cdots\lambda_n^{j_n} \in \mathcal{I}$ having a common monomial $f_{\mathbf{i}}\lambda_1^{i_1}\cdots\lambda_n^{i_n} = g_{\mathbf{i}}\lambda_1^{i_1}\cdots\lambda_n^{i_n}$, there is $h = \sum h_{\mathbf{j}}\lambda_1^{j_1}\cdots\lambda_n^{j_n} \in \mathcal{I}$ whose coefficient of $\Lambda^{\mathbf{i}}$ is $\mathbf{0}$, and for which $h_{\mathbf{j}} = \min\{f_{\mathbf{j}}, g_{\mathbf{j}}\}$, $\forall \mathbf{j}$.

For any tropical ideal $\mathcal{I}$, the sets of minimal indices of supports constitutes the set of circuits of a matroid. This can be formulated in terms of valuated matroids, defined in [**20**, Definition 1.1] as follows:

A **valuated matroid** of **rank** m is a pair (E, v) where E is a set and $v : E^{(m)} \to \Gamma$ is a map satisfying the following properties:

(i) There exist $e_1, \dots, e_m \in E$ with $v(e_1, \dots, e_m) \neq 0$.

(ii) $v(e_1, \dots, e_m) = v(e_{\pi(1)}, \dots, e_{\pi(m)})$. for each $e_1, \dots, e_m \in E$ and every permutation π. Furthermore, $v(e_1, \dots, e_m) = \mathbf{0}$ in case some $e_i = e_j$.

(iii) For $(e_0, \dots, e_m, e_2', \dots, e_m') \in E$ there exists some i with $1 \leq i \leq m$ and

$$v(e_1, \dots, e_m)v(e_0, e_2', \dots, e_m') \leq v(e_0, \dots, e_{i-1}, e_{i+1}, \dots, e_m)v(e_i, e_2', \dots, e_m').$$

This information is encapsulated in the following result.

[**59**, Theorem 1.1] Let K be a field with a valuation $\operatorname{val} : K \to \Gamma_0$, and let Y be a closed subvariety of $K^{\times(n)}$ defined by an ideal $\mathcal{I} \triangleleft K[\lambda_1^{\pm 1}, \dots, \lambda_n^{\pm 1}]$. Then any of the following three objects determines the others:

(i) The bend congruence on $\{\operatorname{trop}(f) : f \in \mathcal{I}\}$.

(ii) The ideal $\operatorname{trop}(\mathcal{I})$ in the semiring $\mathcal{S} := \Gamma_0[\lambda_1^{\pm 1}, \dots, \lambda_n^{\pm 1}]$ of tropical Laurent polynomials.

(iii) The set of valuated matroids of the vector spaces $\mathcal{I}_h^d$, where $\mathcal{I}_h^d$ is the degree d part of the homogenization of the tropical ideal $\mathcal{I}$.

7.2. The supertropical approach. In supertropical mathematics the definitions run somewhat more smoothly. V was defined in terms of $\mathcal{T}^\circ$ so for $f, g \in \mathcal{T}[\Lambda]$ we define the **$\circ$-equivalence** on $\mathcal{T}[\Lambda]$: $f \equiv_\circ g$ if and only if $f^\circ = g^\circ$ on V, i.e. $f(a)^\circ = g(a)^\circ$ for each $a \in V$.

PROPOSITION 7.5. *The bend relation (on V) implies $\circ$-equivalence, in the sense that if $f \equiv_{\text{bend}} g$ then $f \equiv_\circ g$, for any polynomials in $f, g \in \mathcal{T}[\Lambda]$.*

PROOF. The bend relation is obtained from a sequence of steps, each removing or adding on a monomial which takes on the same value of some polynomial f on V. Thus the defining relations of the bend relation are all $\circ$-relations. $\square$

Conversely, given a $\circ$-relation $f^\circ = g^\circ$, where $f = \sum f_i$ and $g = \sum g_j$ for monomials f_i, g_j, we have

$$f \equiv_{\text{bend}} g_1 + f \equiv_{\text{bend}} g_1 + g_2 + f \equiv_{\text{bend}} \cdots$$
$$\cdots \equiv_{\text{bend}} g + f \equiv_{\text{bend}} g + f_{\hat{h}} \equiv_{\text{bend}} \cdots \equiv_{\text{bend}} g,$$

implying $f \equiv_{\text{bend}} g$ when the intermediate polynomials are in the tropical ideal, but we must take into account the discrepancy between formal polynomials and their functions. (The same sort of argument is given in the proof of [**63**, Proposition 2.6].)

In supertropical algebra, given a polynomial $f = \sum \alpha_{\mathbf{i}}\Lambda^{\mathbf{i}}$, define ν-$\operatorname{supp}(f)$ to be the set of tuples $\mathbf{i} = (i_1 \cdots i_n)$ for which the corresponding monomial in f has coefficient $\alpha_{\mathbf{i}} \in \mathcal{T}$.

The supertropical version of "tropical ideal" is that if $f, g \in \mathcal{I}$ and

$$\mathbf{i} \in \nu\text{-}\operatorname{supp}(f) \cap \nu\text{-}\operatorname{supp}(g),$$

then, by normalizing, taking a_f, b_g such that $\mathbf{i} \notin \nu\text{-}\operatorname{supp}(a_f f + b_g g)$, there is $h \in \mathcal{I}$ whose $\mathbf{i}$ coefficient is 0, for which $\nu\text{-}\operatorname{supp}(h) \subseteq \nu\text{-}\operatorname{supp}(f) \cup \nu\text{-}\operatorname{supp}(g)$ but

$$a_f f + b_g g + h = a_f f + b_g g + \text{ghost}.$$

Supertropical "d-bases" over a super-semifield F are treated in [**36**], where vectors are defined to be independent when no tangible linear combination is a ghost. If V is defined as the set

$$\{v \in F^{(n)} : f_j(v) \in \nu(F),\ \forall j \in J\}$$

for a set $\{f_j : j \in J\}$ of homogeneous polynomials of degree m, then taking

$$\mathcal{I} = \{f : f(V) \in \nu(F)\}$$

and $\mathcal{I}_m$ to be its polynomials of degree m, one sees that the d-bases of I_m of cardinality m comprise a matroid (whose circuits are those polynomials of minimal support), by [**36**, Lemma 4.10]. On the other hand, submodules of free modules can fail to satisfy Steinitz' exchange property ([**36**, Examples 4.8, 4.9]), so there is considerable room for further investigation.

"Supertropicalization" strop : $K[\lambda_1^{\pm 1}, \dots, \lambda_n^{\pm 1}] \to (\Gamma \cup \mathcal{G}_{\mathbf{0}})[\lambda_1^{\pm 1}, \dots, \lambda_n^{\pm 1}]$ then is the same tropicalization map as trop , now taken to the standard supertropical semifield (where $\mathcal{T} = \Gamma$).

7.3. The systemic approach. The supertropical approach can be generalized directly to the systemic approach, which also includes hyperfields and fuzzy rings. We assume $(\mathcal{A}, \mathcal{T}, (-), \preceq)$ is a system.

DEFINITION 7.6. The $\circ$-**equivalence** on $\operatorname{Fun}(S, \mathcal{A})$ is defined by, $f \equiv_\circ g$ if and only if $f^\circ = g^\circ$, i.e. $f(s)^\circ = g(s)^\circ$ for each $s \in S$.

This matches the supertropical definition. Given a polynomial $f = \sum \alpha_{\mathbf{i}}\Lambda^{\mathbf{i}}$, define $\mathcal{T}$-supp(f) to be the set of tuples $\mathbf{i} = (i_1 \cdots i_n)$ for which the corresponding monomial in f has coefficient $\alpha_{\mathbf{i}} \in \mathcal{T}$.

The systemic version of "tropical ideal" is:

If $f, g \in \mathcal{I}$ and $s \in \mathcal{T}$-supp$(f) \cap \mathcal{T}$-supp(g), then there are $a_f, b_g \in \mathcal{T}$ and $h \in \mathcal{I}$ such that $s \notin \mathcal{T}$-supp(h), and $a_f f(-) b_g g + h \succeq a_f f(-) b_g g$.

In this context, one can view tropicalization as a functor, as in [**68**, §10].

8. Suggested areas for further research

8.1. Geometry. $\preceq$-Roots lead naturally to affine $\preceq$-varieties (as common $\preceq$-roots of a set of polynomials), and algebraic geometry. An alternative approach is through Hopf semi-algebras.

8.2. Valuated matroids over systems. Viewing tropicalization as a functor, define the appropriate valuated matroid continuing the approach of §7.3. Then one should be able to address the recent work on matroids and valuated matroids, and formulate them over systems in analogy to [**21**]. Presumably, as in [**6**], in the presence of various assumptions, one could carry out the proofs of many of these assertions.

References

[1] K. Adiprasito, J. Huh, and E. Katz, *Hodge theory of matroids*, Notices Amer. Math. Soc. **64** (2017), no. 1, 26–30, DOI 10.1090/noti1463. MR3586249

[2] M. Akian, R. Bapat, and S. Gaubert, *Max-plus algebra*, In: Hogben, L., Brualdi, R., Greenbaum, A., Mathias, R. (eds.) Handbook of Linear Algebra, Chapman and Hall, London, 2006.

[3] M. Akian, S. Gaubert, and A. Guterman, *Linear independence over tropical semirings and beyond*, Tropical and idempotent mathematics, Contemp. Math., vol. 495, Amer. Math. Soc., Providence, RI, 2009, pp. 1–38, DOI 10.1090/conm/495/09689. MR2581511

[4] M. Akian, S. Gaubert, and A. Guterman, *Tropical Cramer determinants revisited*, Tropical and idempotent mathematics and applications, Contemp. Math., vol. 616, Amer. Math. Soc., Providence, RI, 2014, pp. 1–45, DOI 10.1090/conm/616/12324. MR3221324

[5] M. Akian, S. Gaubert, and A. Guterman, *Tropical polyhedra are equivalent to mean payoff games*, Internat. J. Algebra Comput. **22** (2012), no. 1, 1250001, 43, DOI 10.1142/S0218196711006674. MR2900854

[6] M. Akian, S. Gaubert, and L. Rowen, *Linear algebra over systems*, preprint, 2019.

[7] M. Baker and N. Bowler, *Matroids over hyperfields*, arXiv:1601.01204v4, 2016.

[8] M. Baker and N. Bowler, *Matroids over partial hyperstructures*, preprint, 2017.

[9] M. Baker and S. Payne (eds.), *Nonarchimedean and tropical geometry*, Simons Symposia, Springer, [Cham], 2016. Papers based on two Simons Symposia held in St. John, April 2013 and in Puerto Rico, February 2015. MR3645464

[10] V. Berkovich, *Non-Archimedean analytic geometry: first steps*, *p*-adic geometry, Univ. Lecture Ser., vol. 45, Amer. Math. Soc., Providence, RI, 2008, pp. 1–7, DOI 10.1090/ulect/045/01. MR2482344

[11] V. Berkovich, *Algebraic and analytic geometry over the field with one element*, preprint, 2014.

[12] A. Bertram and R. Easton, *The tropical Nullstellensatz for congruences*, Adv. Math. **308** (2017), 36–82, DOI 10.1016/j.aim.2016.12.004. MR3600055

[13] N. Bourbaki, *Commutative Algebra,* Paris and Reading, 1972.

[14] P. Butkovič, *Max-algebra: the linear algebra of combinatorics?*, Linear Algebra Appl. **367** (2003), 313–335, DOI 10.1016/S0024-3795(02)00655-9. MR1976928

[15] A. Connes and C. Consani, *Homological algebra in characteristic one*, arXiv:1703.02325v1, 2017.

[16] G. Cortiñas, C. Haesemeyer, M. E. Walker, and C. Weibel, *Toric varieties, monoid schemes and cdh descent*, J. Reine Angew. Math. **698** (2015), 1–54, DOI 10.1515/crelle-2012-0123. MR3294649

[17] A. Almeida Costa, *Sur la théorie générale des demi-anneaux* (French), Publ. Math. Debrecen **10** (1963), 14–29. MR0168608

[18] M. Develin, F. Santos, and B. Sturmfels, *On the rank of a tropical matrix*, Combinatorial and computational geometry, Math. Sci. Res. Inst. Publ., vol. 52, Cambridge Univ. Press, Cambridge, 2005, pp. 213–242. MR2178322

[19] A. W. M. Dress, *Duality theory for finite and infinite matroids with coefficients*, Adv. in Math. **59** (1986), no. 2, 97–123, DOI 10.1016/0001-8708(86)90047-2. MR834224

[20] A. W. M. Dress and W. Wenzel, *Valuated matroids*, Adv. Math. **93** (1992), no. 2, 214–250, DOI 10.1016/0001-8708(92)90028-J. MR1164708

[21] A. Dress and W. Wenzel, *Algebraic, tropical, and fuzzy geometry*, Beitr. Algebra Geom. **52** (2011), no. 2, 431–461, DOI 10.1007/s13366-011-0017-y. MR2842641

[22] P. Etingof, S. Gelaki, D. Nikshych, and V. Ostrik, *Tensor categories*, Mathematical Surveys and Monographs, vol. 205, American Mathematical Society, Providence, RI, 2015. MR3242743

[23] L. Gatto and L. Rowen, *Grassman semialgebras and the Cayley-Hamilton theorem*, arXiv:1803.08093v1.

[24] L. Gatto and I. Scherbak, *Remarks on the Cayley-Hamilton Theorem*, arXiv:1510.03022.

[25] S. Gaubert, *Théorie des systèmes linéaires dans les diodes,* Thèse, École des Mines de Paris, 1992.

[26] S. Gaubert and M. Plus, *Methods and applications of* (max, +) *linear algebra*, STACS 97 (Lübeck), Lecture Notes in Comput. Sci., vol. 1200, Springer, Berlin, 1997, pp. 261–282, DOI 10.1007/BFb0023465. MR1473780

[27] J. Giansiracusa and N. Giansiracusa, *Equations of tropical varieties*, Duke Math. J. **165** (2016), no. 18, 3379–3433, DOI 10.1215/00127094-3645544. MR3577368

[28] J. Giansiracusa, J. Jun, and O. Lorscheid, *On the relation between hyperrings and fuzzy rings*, Beitr. Algebra Geom. **58** (2017), no. 4, 735–764, DOI 10.1007/s13366-017-0347-5. MR3702974

[29] J. Golan, *Semirings and their Applications*, Springer-Science + Business, Dordrecht, 1999. (Previously published by Longman, 1192, and Kluwer Acad. Publ., 1999.)

[30] M. Grandis, *Homological algebra: In strongly non-abelian settings*, World Scientific Publishing Co. Pte. Ltd., Hackensack, NJ, 2013. MR3155611

[31] S. Henry, *Symmetrization of monoïds as hypergroups*, arXiv:1309.1963, 2013.

[32] I. Itenberg, V. Kharlamov, and E. Shustin, *Welschinger invariants of real del Pezzo surfaces of degree* $\geq$ 3, Math. Ann. **355** (2013), no. 3, 849–878, DOI 10.1007/s00208-012-0801-5. MR3020146

[33] I. Itenberg, G. Mikhalkin, and E. Shustin, *Tropical algebraic geometry*, Oberwolfach Seminars, vol. 35, Birkhäuser Verlag, Basel, 2007. MR2292729

[34] Z. Izhakian, *Tropical arithmetic and matrix algebra*, Comm. Algebra **37** (2009), no. 4, 1445–1468, DOI 10.1080/00927870802466967. MR2510993

[35] Z. Izhakian, M. Knebusch, and L. Rowen, *Supertropical semirings and supervaluations*, J. Pure Appl. Algebra **215** (2011), no. 10, 2431–2463, DOI 10.1016/j.jpaa.2011.01.002. MR2793947

[36] Z. Izhakian, M. Knebusch, and L. Rowen, *Supertropical linear algebra*, Pacific J. Math. **266** (2013), no. 1, 43–75, DOI 10.2140/pjm.2013.266.43. MR3105777

[37] Z. Izhakian, M. Knebusch, and L. Rowen, *Layered tropical mathematics*, J. Algebra **416** (2014), 200–273, DOI 10.1016/j.jalgebra.2014.05.019. MR3232800

[38] Z. Izhakian, M. Knebusch, and L. Rowen, *Categories of layered semirings*, Comm. Algebra **43** (2015), no. 5, 1807–1836, DOI 10.1080/00927872.2013.878838. MR3316824

[39] Z. Izhakian, M. Knebusch, and L. Rowen, *Algebraic structures of tropical mathematics*, Tropical and idempotent mathematics and applications, Contemp. Math., vol. 616, Amer. Math. Soc., Providence, RI, 2014, pp. 125–150, DOI 10.1090/conm/616/12312. MR3221330

[40] Z. Izhakian and L. Rowen, *The tropical rank of a tropical matrix*, Comm. Algebra **37** (2009), no. 11, 3912–3927, DOI 10.1080/00927870902828793. MR2573227

[41] Z. Izhakian, A. Niv, and L. Rowen, *Supertropical* SL_n, Linear Multilinear Algebra **66** (2018), no. 7, 1461–1483, DOI 10.1080/03081087.2017.1363148. MR3795818

[42] Z. Izhakian and L. Rowen, *Supertropical algebra*, Adv. Math. **225** (2010), no. 4, 2222–2286, DOI 10.1016/j.aim.2010.04.007. MR2680203

[43] Z. Izhakian and L. Rowen, *Supertropical matrix algebra*, Israel J. Math. **182** (2011), 383–424, DOI 10.1007/s11856-011-0036-2. MR2783978

[44] Z. Izhakian and L. Rowen, *Supertropical matrix algebra II: solving tropical equations*, Israel J. Math. **186** (2011), 69–96, DOI 10.1007/s11856-011-0133-2. MR2852317

[45] Z. Izhakian and L. Rowen, *Supertropical polynomials and resultants*, J. Algebra **324** (2010), no. 8, 1860–1886, DOI 10.1016/j.jalgebra.2010.05.019. MR2678825

[46] N. Jacobson, *Basic algebra. II*, W. H. Freeman and Co., San Francisco, Calif., 1980. MR571884

[47] D. Jensen and S. Payne, *Combinatorial and inductive methods for the tropical maximal rank conjecture*, J. Combin. Theory Ser. A **152** (2017), 138–158, DOI 10.1016/j.jcta.2017.06.003. MR3682730

[48] D. Joó and K. Mincheva, *Prime congruences of additively idempotent semirings and a Nullstellensatz for tropical polynomials*, Selecta Math. (N.S.) **24** (2018), no. 3, 2207–2233, DOI 10.1007/s00029-017-0322-x. MR3816503

[49] J. Jun, *Algebraic geometry over hyperrings*, Adv. Math. **323** (2018), 142–192, DOI 10.1016/j.aim.2017.10.043. MR3725876

[50] J. Jun, *Čech cohomology of semiring schemes*, J. Algebra **483** (2017), 306–328, DOI 10.1016/j.jalgebra.2017.04.001. MR3649822

[51] J. Jun, *Valuations of semirings*, J. Pure Appl. Algebra **222** (2018), no. 8, 2063–2088, DOI 10.1016/j.jpaa.2017.08.020. MR3771848

[52] J. Jun, K. Mincheva, and L. Rowen, *Projective module systems*, arXiv:1809.01996, 2018.

[53] J. Jun, K. Mincheva, and L. Rowen, *Homology of systems*, in preparation.

[54] J. Jun and L. Rowen, *Categories with negation*, arXiv:1709.0318, Contemporary Mathematics (2019), to appear.

[55] Y. Katsov, *Tensor products of functors*, Siberian J. Math. **19** (1978), 222-229, trans. from Sirbiskii Mathematischekii Zhurnal **19** (1978), no. 2, 318–327.

[56] Y. Katsov, *Toward homological characterization of semirings: Serre's conjecture and Bass's perfectness in a semiring context*, Algebra Universalis **52** (2004), no. 2-3, 197–214, DOI 10.1007/s00012-004-1866-0. MR2161650

[57] O. Lorscheid, *The geometry of blueprints: Part I: Algebraic background and scheme theory*, Adv. Math. **229** (2012), no. 3, 1804–1846, DOI 10.1016/j.aim.2011.12.018. MR2871157

[58] O. Lorscheid, *A blueprint view on $\mathbb{F}_1$-geometry*, Absolute arithmetic and $\mathbb{F}_1$-geometry, Eur. Math. Soc., Zürich, 2016, pp. 161–219. MR3701901

[59] D. Maclagan and F. Rincón, *Tropical schemes, tropical cycles, and valuated matroids*, arXiv:1401.4654v1.

[60] D. Maclagan and B. Sturmfels, *Introduction to tropical geometry*, Graduate Studies in Mathematics, vol. 161, American Mathematical Society, Providence, RI, 2015. MR3287221

[61] G. Mikhalkin, *Enumerative tropical algebraic geometry in $\mathbb{R}^2$*, J. Amer. Math. Soc. **18** (2005), no. 2, 313–377, DOI 10.1090/S0894-0347-05-00477-7. MR2137980

[62] A. Patchkoria, *On exactness of long sequences of homology semimodules*, J. Homotopy Relat. Struct. **1** (2006), no. 1, 229–243. MR2264745

[63] T. Perri and L. H. Rowen, *Kernels in tropical geometry and a Jordan-Hölder theorem*, J. Algebra Appl. **17** (2018), no. 4, 1850066, 99, DOI 10.1142/S0219498818500664. MR3786745

[64] M. Plus, *L'algèbre* $(\max, +)$ *et sa symétrisation ou l'algèbre des équilibres* (French, with English summary), C. R. Acad. Sci. Paris Sér. II Méc. Phys. Chim. Sci. Univers Sci. Terre **311** (1990), no. 4, 443–448. MR1074073

[65] C. Reutenauer and H. Straubing, *Inversion of matrices over a commutative semiring*, J. Algebra **88** (1984), no. 2, 350–360, DOI 10.1016/0021-8693(84)90070-X. MR747520

[66] Q. Ren, K. Shaw, and B. Sturmfels, *Tropicalization of del Pezzo surfaces*, Adv. Math. **300** (2016), 156–189, DOI 10.1016/j.aim.2016.03.017. MR3534831

[67] L.H. Rowen, *Symmetries in tropical algebra*, arXiv:1602.00353.

[68] L.H. Rowen, *Algebras with a negation map*, arXiv:1602.00353.

[69] Y. Sommerhäuser, *Hopf algebras and associated group actions*, slides of lecture at ACC conference Combinatorics of group actions, Saint John's, Newfoundland, August 29, 2017.

[70] B. Toën and M. Vaquié, *Au-dessous de* $\operatorname{Spec}\mathbb{Z}$ (French, with English and French summaries), J. K-Theory **3** (2009), no. 3, 437–500, DOI 10.1017/is008004027jkt048. MR2507727

[71] O.Y. Viro, *Hyperfields for tropical geometry I. Hyperfields and dequantization*, arXiv AG/1006.3034, 2010.

Department of Mathematics, Bar-Ilan University, Ramat-Gan 52900, Israel
Email address: `rowen@math.biu.ac.il`

Contemporary Mathematics
Volume **727**, 2019
https://doi.org/10.1090/conm/727/14645

On dual automorphism-invariant and superfluous ADS-modules

Truong Cong Quynh and Serap Şahinkaya

ABSTRACT. Let M and N be two right R-modules. The module M is called dual automorphism-N-invariant if whenever K_1 is a small submodule of M and K_2 is a small submodule of N, then any epimorphism $p : M/K_1 \to N/K_2$ with small kernel lifts to a homomorphism $\varphi : M \to N$. Let $\pi_1 : P_1 \to M$ and $\pi_2 : P_2 \to N$ be projective covers. M is called an s-ADS-module if for every decomposition $M = S \oplus T$ of M and every supplement T' of S with $T' + T = M$ and $(T \cap T') \ll M$, we have $M = S \oplus T'$. It is shown that an amply supplemented R-module M is s-ADS if and only if for each decomposition $M = A \oplus B$, A and B are relatively dual automorphism invariant.

Introduction

Lee and Zhou introduced the concept of automorphism invariant modules in [**5**]: A module M is called *automorphism invariant* if it is invariant under automorphisms of its injective hull, equivalently if every isomorphism between two essential submodules of M extends to an automorphism of M. Quasi-injective modules are automorphism-invariant. A module M is called *pseudo-injective* if, for any submodule A of M, every monomorphism from A to M can be extended to some element of $End(M)$. In [**5**], the authors showed that pseudo-injective modules are automorphism-invariant, and asked that "Is an automorphism-invariant module pseudo-injective?". In [**3**], Er, Singh and Srivastava have shown that pseudo-injective modules and automorphism-invariant modules coincide.

Recall that a submodule N of a module M is called small in M (denoted as $N \ll M$) if $N + K = M$ for any submodule K of M then $K = M$. In [**8**], Singh and Srivastava introduced the concept of dual automorphism-invariant modules. A right R-module M is called *dual automorphism-invariant* if whenever K_1 and K_2 are small submodules of M, then any epimorphism $\eta : M/K_1 \to M/K_2$ with small kernel lifts to an endomorphism φ of M. A module M is called a *pseudo-projective module* if for every submodule N of M, any epimorphism from M to M/N can be lifted to a homomorphism from M to M. Any pseudo-projective module is dual automorphism-invariant, see [**8**, Proposition 7].

2010 *Mathematics Subject Classification.* Primary 16D40, 16E50; Secondary 16N20.
Key words and phrases. Dual automorphism-invariant module, pseudo-projective module, ADS-modules.
The first author has been partially founded by Funds for Science and Technology Development of the University of Danang under project number B2017-DN03-08.

In the literature there are many generalizations of (quasi) injectivity and (quasi) projectivity. It is well known that pseudo injectivity (respectively, pseudo projectivity) is a generalization of quasi injectivity (respectively, quasi projectivity). As we mentioned in the previous paragraphs, the class of (dual) automorphism-invariant modules have been characterized by pseudo injectivity (projectivity). The purpose of this paper is to initiate the study of dual automorphism N-invariant modules via N-pseudo-projective modules. We call M *dual automorphism N-invariant* if, whenever K_1 is a small submodule of M and K_2 is a small submodule of N, then any epimorphism $p : M/K_1 \to N/K_2$ with small kernel lifts to a homomorphism $\varphi : M \to N$. A module M is called *N-pseudo-projective*(or *pseudo-projective relative to* M) if for every submodule A of M, any epimorphism $\alpha : N \to M/A$ can be lifted to a homomorphism $\beta : N \to M$. Moreover M is called *pseudo-projective* if N is M-pseudo-projective. Note that M is a dual automorphism invariant module if and only if M is dual automorphism M-invariant, and the class of dual automorphism M-invariant modules is shown to be closed under direct summands.

One of the main result of the present paper is the following theorem which generalizes [**8**, Theorem 27] including an equivalent condition for a module with projective cover to be a dual automorphism-invariant module:

" Let $\pi_1 : P_1 \to M$ be a projective cover and $\pi_2 : P_2 \to N$ be a projective cover. Then M is dual automorphism N-invariant if and only if $\sigma(Ker(\pi_1)) \leq Ker(\pi_2)$ for any isomorphism $\sigma : P_1 \to P_2$."

In [**4**], Koşan, Quynh and Zemlicka introduced the concept of e-ADS modules. A module M is called *e-ADS* if, for every decomposition $M = S \oplus T$ and every complement T' of S with $T' \cap T = 0$ and $S \cap (T' \oplus T) \leq^e S$, we have $M = S \oplus T'$. Assume that an amply supplemented R-module X has a decomposition $X = M \oplus N$ for some R-modules M and N. In Theorem 2.1, we prove that M is dual automorphism N-invariant if and only if, for any supplement K of N in X with $K + M = X$ and $(K \cap M) \ll X$, the module X has a decomposition $X = K \oplus N$ if and only if, for each submodule K of X such that K is a supplement of N in X and M is a supplement of K in X, we have $X = K \oplus N$. In light of this observation, we can define *s-ADS*-modules, as modules M such that or every decomposition $M = S \oplus T$ and every supplement T' of S with $T' + T = M$ and $(T \cap T') \ll M$, we have $M = S \oplus T'$. This definition generalizes in natural way quasi-discrete, ADS^* ([**9**]) as well as dual automorphism-invariant modules. We show that an amply supplemented R-module M is s-ADS if and only if for each decomposition $M = A \oplus B$, A and B are relatively dual automorphism invariant.

Throughout this paper, rings are associative with unity and modules are unital right R-modules, where R denotes such a ring and M denotes such a module. We write Rad and $End(M)$ for the radical of the module M and the endomorphism ring of of the module M, respectively. We also write $N \ll M$ and $N \leq^{\oplus} M$ to indicate that N is a small submodule of M and a direct summand of M, respectively. For any term not defined here the reader is referred to [**2**] and [**6**].

1. The Results

A right R-module M is called *dual automorphism-invariant* if whenever K_1 and K_2 are small submodules of M, then any epimorphism $\eta : M/K_1 \to M/K_2$

with small kernel lifts to an endomorphism φ of M. Regular modules or pseudo-projective modules are dual automorphism-invariant:

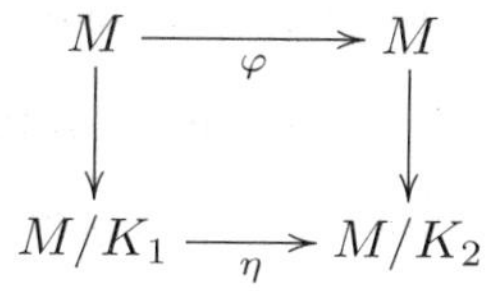

In [**8**, Lemma 1], Singh and Srivastava proved that the endomorphism φ in the diagram is an automorphism of M.

The module M is called *dual automorphism N-invariant* if, whenever K_1 is a small submodule of M and K_2 is a small submodule of N, then any epimorphism $p: M/K_1 \to N/K_2$ with small kernel lifts to a homomorphism $\varphi: M \to N$:

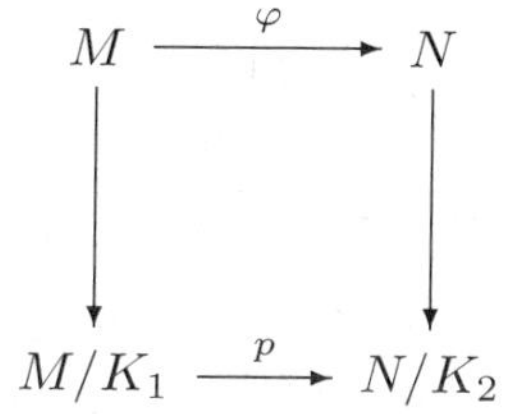

LEMMA 1.1. *Let M be a dual automorphism N-invariant module. If $f: N \to M$ is an epimorphism with small kernel, then f is an isomorphism.*

PROOF. Let $K = Ker(f)$. Then f induces an isomorphism $\overline{f}: N/K \to M$. Consider $\bar{f}^{-1}: M \to N/K$. Since M is a dual automorphism N-invariant module, we obtain that $\bar{f}^{-1}$ lifts to a homomorphism $\lambda: M \to N$. Hence we have $\lambda(M) + K = N$. As $K \ll N$, we get $\lambda(M) = N$. Thus λ is an epimorphism. Then for any $x \in M$, $\bar{f}^{-1}(x) = \lambda(x) + K$. Now $x = \bar{f}\bar{f}^{-1}(x) = \bar{f}(\lambda(x) + K) = f\lambda(x)$. This proves that $f\lambda = 1_M$. Thus $f^{-1} = \lambda$ and hence f is an isomorphism. $\square$

As a consequence, we have the following;

THEOREM 1.2. *Let M and N be two right R-modules. Then the following conditions are equivalent:*

(1) *M is dual automorphism N-invariant.*
(2) *For any small submodule K_1 of M and small submodule K_2 of N, every epimorphism $p: M/K_1 \to N/K_2$ with small kernel lifts to an epimorphism $\varphi: M \to N$.*
(3) *For any small submodule K_2 of N, every epimorphism $f: M \to N/K_2$ with small kernel lifts to a homomorphism $\varphi: M \to N$.*

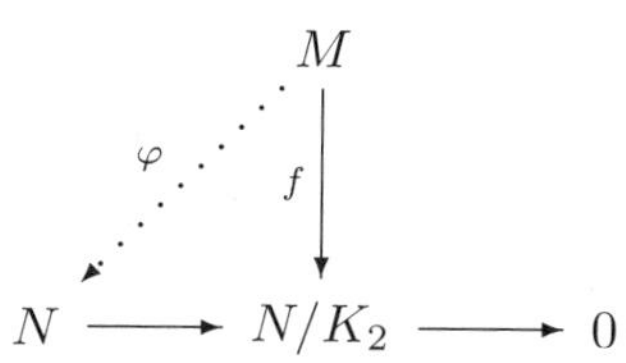

PROOF. (1) $\Rightarrow$ (2) It follows from the definition.
(2) $\Rightarrow$ (3) Clear.
(3) $\Rightarrow$ (1) Let $K_1 \ll M$ and $K_2 \ll N$. Let $p: M/K_1 \to N/K_2$ be an epimorphism with small kernel. We consider the natural mappings $\pi_1: M \to M/K_1$ and

$\pi_2 : N \to N/K_2$. By (3), $p\pi_2$ lifts to a homomorphism, say φ. Then $p\pi_2 = \pi_1\varphi$. Therefore $0 = p\pi_2(K_1) = \pi_1\varphi(K_1)$ gives $\varphi(K_1) \subseteq K_2$. Hence φ is a lifting of p. This proves that M is dual automorphism N-invariant. □

Letting $M = R$, Theorem 1.2 yields the following corollary:

COROLLARY 1.3. *The following conditions are equivalent for a right R-module M:*

(1) *M is dual automorphism invariant.*
(2) *For any small submodule K of M, every epimorphism $f : M \to M/K$ with small kernel lifts to an endomorphism of M.*

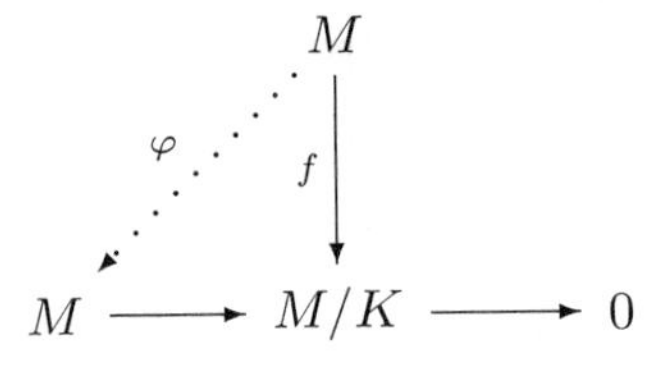

Let N and L be submodules of M. The module N is called a *supplement* of L in M if $N + L = M$ and N is minimal with respect to this property, equivalently, $M = N + L$ and $N \cap L \ll N$. M is called *supplemented* if every submodule of M has a supplement in M.

PROPOSITION 1.4. *Let M and N be modules and $X = M \oplus N$. The following conditions are equivalent:*

(1) *M is dual automorphism N-invariant.*
(2) *For each submodule K of X such that N is a supplement of K in X, there exists $C \leq K$ such that $N \oplus C = X$.*

PROOF. (1) ⇒ (2) Define $f : M \to N/(N \cap K)$ by $f(m) = n + (M \cap K)$ for all $m = k + n \in N$ with $k \in K, n \in N$. Clearly the R-homomorphism f is an epimorphism and $Ker(f) = K \cap M \ll M$. Since M is dual automorphism N-invariant, there exists $h : M \to N$ such that $\pi h = f$, where $\pi : N \to N/(N \cap K)$ is the natural projection. Let $C = \{m - h(m) | \ m \in M\}$. Now it is easy to see that $C \leq K$ and $N \oplus C = X$.
(2) ⇒ (1) Let $A \ll N$, $f : M \to N/A$ be a small epimorphism and $\pi : N \to N/A$ the natural projection. Let $K = \{n + m | \ f(m) = -\pi(n)\}$. It is easy to see that

$$K + M = K + N = X$$

$$K \cap M = Ker(f) \ll M$$

and

$$K \cap N = A \ll N.$$

By (2), there exists $C \leq K$ such that $N \oplus C = X$. Let $p : N \oplus C \to N$ denote the canonical projection. Then $\pi p|_M = f$, which implies that M is dual automorphism N-invariant. □

A module M is called *hollow* if every proper submodule of M is small in M, and it is said to be *local* if it is hollow and has a unique maximal submodule. The following observation was proved for local modules in [**8**, Proposition 13]. For the direct sum of hollow modules, we have the following.

PROPOSITION 1.5. *Assume that M_1, M_2 are two hollow modules. If M_1 is dual automorphism M_2-invariant, then M_1 is M_2-projective.*

PROOF. We consider the following diagram

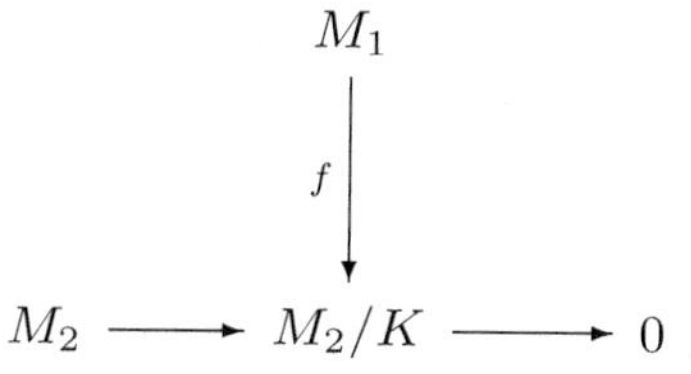

with $K \leq M_2, K \neq M_2$ and $K \ll M_2$. We show that there exists a homomorphism $\varphi : M_1 \to M_2$ such that the above diagram is commutative. If $f = 0$, then the claim was already proved. Otherwise, if $f \neq 0$ then $Ker(f) \neq M_1$ and so $Ker(f) \ll M_1$ by the hypothesis. Since M_1 is dual automorphism M_2-invariant, there exists $\varphi : M_1 \to M_2$ such that $\pi\varphi = f$ where $\pi : M_2 \to M_2/K$ the canonical projection. Thus M_1 is M_2-projective. □

In [**8**, Proposition 11 and Remark 12], Singh and Srivastava proved that the class of dual automorphism-invariant modules is closed under taking direct summands but not submodules and the direct sum.

THEOREM 1.6. *Let M and N be two R-modules.*

(1) *Every direct summand of a dual automorphism M-invariant module is also dual automorphism M-invariant.*
(2) *M is dual automorphism N-invariant if and only if any isomorphism $f : M/B \to N/A$ with $B \ll M$ and $A \ll N$ lifts to a homomorphism from M to N.*
(3) *If M is a dual automorphism N-invariant module and $K \cong N$, then M is dual automorphism K-invariant.*
(4) *Assume that $N = A \oplus B$ and $M = C \oplus D$ such that there exists a small epimorphism from D to B. If M is dual automorphism N-invariant, then C is dual automorphism A-invariant.*

PROOF. (1) Assume that N is dual automorphism M-invariant and $N = A \oplus B$. Let X be a small submodule of M and $f : A \to M/X$ be an epimorphism. Define the epimorphism $g : A \oplus B \to M/X$ with $g = f\phi_A$, where ϕ_A is the natural projection of N onto A. Then $g(a, b) = f(a)$, where $a \in A$ and $b \in B$. Since N is dual automorphism M-invariant, there is $g^* : A \oplus B \to M$ which is lifting of g. Then $f^* = g^*|_A$ is a homomorphism which lifts to f. Therefore A is dual automorphism M-invariant.
(2) Assume that M is a dual automorphism N-invariant module. Let $B \ll M$, $A \ll N$ and let $f : M/B \to N/A$ be any isomorphism. Since $B \ll M$ we have a natural small epimorphism $g : M \to M/B$. Note that the composition map $fg : M \to N/A$ is an epimorphism. Also $Kerfg \ll M$ because $Kerfg = Kerf$. Since M is dual automorphism N-invariant we get the lifting homomorphism from M to N by Theorem 1.2.

Taking $B = 0$, the converse yields.
(3) Let $L \ll K$ and $g : K \to N$ be an isomorphism. Clearly $g(L) \ll N$. Let $f : M \to K/L$ be a small epimorphism. There exists a small epimorphism $\pi\iota f : M \to N/g(L)$ where $\pi : N \to N/g(L)$ is a natural projection and $\iota : K/L \to N$ is an

isomorphism (There exists such a unique ι because $K \to K/L$ natural projection, $g : K \to N$ is an isomorphism and $Kerg \subseteq L$). Since M is dual automorphism N-invariant, the composition map $\pi\iota f$ lifts to $h : M \to N$. So $g^{-1}h : M \to K$ is desired homomorphism.
(4) Assume that $N = A \oplus B$, $M = C \oplus D$ and $\alpha : D \to B$ is a small epimorphism. Let $f : C \to A/H$ be a small epimorphism with $H \ll A$. We note that

$$B \cong B/(H \cap B) \cong (B + H)/H,$$

and so $\overline{\alpha} : D \to (B + H)/H$ is a small epimorphism. Hence $f \oplus \overline{\alpha} : M \to N/H$ is a small epimorphism. Also it is clear $H \ll N$ because $H \ll A$. Since M is dual automorphism N-invariant it lifts to some $g \in Hom(M, N)$. For the inclusion $\iota : C \to M$ and the projection $\pi : N \to A$, the map $g\pi\iota : C \to A$ is a lifting of f. □

Recall that a *projective cover* of a module M is an epimorphism $f : P \to A$ such that P is a projective module and $Ker(f) \ll P$.

LEMMA 1.7 ([**8**, Lemma 26]). *Let M and N be two R-modules. Assume that $\nu : M \to N$ is a lifting of an epimorphism $\delta : M/K_1 \to N/K_2$, where $K_1 \ll M$ and $K_2 \ll N$. Then the following conditions are equivalent.*

(1) *ν is an epimorphism.*
(2) *If δ is an isomorphism and $Ker(\nu)$ is a direct summand of M, then ν is an isomorphism.*

The following theorem extends Singh and Srivastava [**8**, Theorem 27].

THEOREM 1.8. *Let $\pi_1 : P_1 \to M$ and $\pi_2 : P_2 \to N$ be projective covers. Then the following conditions are equivalent.*

(1) *M is dual automorphism N-invariant.*
(2) *$\sigma(Ker(\pi_1)) \leq Ker(\pi_2)$ for any isomorphism $\sigma : P_1 \to P_2$.*

PROOF. (1) $\Rightarrow$ (2) Assume that M is a dual automorphism N-invariant module. Let $K_1 = Ker(\pi_1)$ and $K_2 = Ker(\pi_2)$. Note that $M \cong P_1/Ker(\pi_1)$, $N \cong P_2/Ker(\pi_2)$ and P_1/K_1 is dual automorphism P_2/K_2-invariant. Let $\sigma : P_1 \to P_2$ be an isomorphism. The map σ induces an epimorphism

$$\overline{\sigma} : P_1/K_1 \to P_2/(\sigma(K_1) + K_2)$$

given by

$$\overline{\sigma}(x + K_1) = \sigma(x) + \sigma(K_1) + K_2.$$

Since $Ker(\overline{\sigma}) = (\sigma^{-1}(K_2) + K_1)/K_1$ is small in P_1/K_1, we obtain $\overline{\sigma}$ lifts to an isomorphism $\eta : P_1/K_1 \to P_2/K_2$ and $\eta^{-1}(K_2 + \sigma(K_1)/K_2) = (\sigma^{-1}(K_2) + K_1)/K_1$. Consider the following diagram.

$$\begin{array}{ccc} P_1/K_1 & \xrightarrow{\ \eta\ } & P_2/K_2 \\ \downarrow & & \downarrow \\ P_1/K_1 & \xrightarrow{\ \overline{\sigma}\ } & P_2/(\sigma(K_1) + K_2) \end{array}$$

Now η lifts to a homomorphism, say λ from P_1 to P_2. Now we have the following diagram.

$$\begin{array}{ccc} P_1 & \xrightarrow{\lambda} & P_2 \\ \downarrow & & \downarrow \\ P_1/K_1 & \xrightarrow{\eta} & P_2/K_2 \end{array}$$

By Lemma 1.7, the homomorphism λ is an isomorphism. Hence $\lambda(K_1) \leq K_2$. Since λ is the lifting of the isomorphism η, we obtain $\lambda^{-1}(K_2) \leq K_1$ or $K_2 \leq \lambda(K_1)$. Thus $\lambda(K_1) = K_2$. Note that

$$\eta(\sigma^{-1}(K_2) + K_1)/K_1) = [K_2 + \sigma(K_1)]/K_2,$$

which implies that

$$\lambda(\sigma^{-1}(K_2) + K_1) = K_2 + \sigma(K_1).$$

Put $C := \sigma^{-1}(K_2)$. Clearly, $C \ll P_1$. Now consider the homomorphisms $\overline{\lambda}$ and $\overline{\eta}$ given as follows:

$$\overline{\lambda} : P_1 \to P_2/K_2$$

such that $\overline{\lambda}(x) = \lambda(x) + K_2$, and

$$\overline{\mu} : P_1 \to P_2/K_2$$

such that $\overline{\mu}(x) = \sigma(x) + K_2$. Clearly, $\overline{\mu}(C) = 0$. Now

$$\begin{aligned} \eta[(\sigma^{-1}(K_2) + K_1)/K_1] &= [\lambda(\sigma^{-1}(K_2)) + \lambda(K_1)]/K_2 \\ &= [\lambda(\sigma^{-1}(K_2)) + K_2]/K_2 \\ &= (\lambda(C) + K_2)/K_2 \end{aligned}$$

Therefore $\overline{\lambda}(C) = (\sigma(K_1) + K_2)/K_2$. Let $\overline{P}_2 := P_2/K_2$. Then

$$\overline{P}_2/\overline{\lambda}(C) = P_2/(\sigma(K_1) + K_2).$$

Consider the a natural homomorphism $\pi : P_2/K_2 \to P_2/(\sigma(K_1)+K_2)$. Set $\overline{\lambda'} := \pi\overline{\lambda}$ and $\overline{\mu'} := \pi\overline{\mu}$. Let $x \in P_1$. Then

$$\overline{\lambda'}(x) = \pi(\lambda(x) + K_2) = \pi\eta(x + K_1).$$

Now $\eta(x + K_1) = y + K_2$ for some $y \in P_2$. It follows that

$$\begin{aligned} \overline{\lambda'}(x) &= y + \sigma(K_1) + K_2 \\ &= \overline{\sigma}(x + K_1) \\ &= \sigma(x) + \sigma(K_1) + K_2 \\ &= \overline{\mu'}(x). \end{aligned}$$

Therefore $\overline{\lambda'} = \overline{\mu'}$. By [**8**, Lemma 25], we conclude that $\overline{\lambda} = \overline{\mu}$ which implies that $\overline{\lambda}(K_1) = \overline{\mu}(K_1) = 0$. Since $\lambda(K_1) = K_2$, we get $[\sigma(K_1) + K_2]/K_2 = 0$ or $\sigma(K_1) \leq K_2$.

(2) $\Rightarrow$ (1) For any small submodule K of N and any epimorphism $f : M \to N/K$ with small kernel, assume that there exists an epimorphism $\sigma : P_1 \to P_2$ such that $f\pi_1 = p_K\pi_2\sigma$ where $\pi_K : N \to N/K$ is the canonical projection. It follows that σ is an isomorphism. By hypothesis, $\sigma(Ker(\pi_1)) \leq Ker(\pi_2)$, there exists $\overline{f} : M \to N$ such that $p_K\overline{f} = f$. $\square$

Letting $M = N$, Theorem 1.8 yields the following corollary:

COROLLARY 1.9. *Let $\pi : P \to M$ be a projective cover. Then M is dual automorphism-invariant if and only if $\sigma(Ker(\pi)) \leq Ker(\pi)$ for any isomorphism $\sigma : P \to P$.*

A module N is called *M-pseudo-projective*(or *pseudo-projective relative to M*) if for every submodule A of M, any epimorphism $\alpha : N \to M/A$ can be lifted to a homomorphism $\beta : N \to M$. Moreover N is called *pseudo-projective* if N is N-pseudo-projective.

Clearly, every quasi-projective module is pseudo-projective and, by [**8**, Proposition 7], any pseudo-projective module is dual automorphism-invariant. In [**8**, Remark 10], the authors also pointed out that dual automorphism-invariant modules need not be pseudo-projective.

PROPOSITION 1.10. *Any N-pseudo-projective module is dual automorphism N-invariant.*

PROOF. Suppose M is an N-pseudo-projective module. Let L_1 be a small submodule of M and L_2 be a small submodule of N. Let $\sigma : M/L_1 \to N/L_2$ be an epimorphism and let $\pi_1 : M \to M/L_1$ be the natural homomorphism. Since M is N-pseudo-projective, we have $\sigma\pi_1$ lifts to a homomorphism $\eta : M \to N$. Let $\pi_2 : N \to N/L_2$ be the natural homomorphism. Then $\pi_2\eta = \sigma\pi_1$. Now $\pi_2\eta(L_1) = \sigma\pi_1(L_1) = 0$ which gives that $\eta(L_1) \subseteq L_2$. Hence η is a lifting of σ. This proves that M is dual automorphism N-invariant. □

Any two modules N and M are called *mutually pseudo-projective* if N is M-pseudo-projective and M is N-pseudo-projective, and M and N are called *mutually dual automorphism invariant* if M is dual automorphism N-invariant and N is dual automorphism M-invariant.

THEOREM 1.11. *Let M and N be mutually dual automorphism invariant modules and $\pi_1 : P_1 \to M$ and $\pi_2 : P_2 \to N$ be projective covers. If $P_1 \cong P_2$, then every isomorphism $\sigma : P_1 \to P_2$ reduces an isomorphism from $Ker(\pi_1)$ to $Ker(\pi_2)$.*

PROOF. Let $\sigma : P_1 \to P_2$ be an isomorphism. By Theorem 1.8, we have $\sigma(Ker(\pi_1)) \subseteq Ker(\pi_2)$. Similarly $\sigma^{-1}(Ker(\pi_2)) \subseteq Ker(\pi_1)$. Thus $Ker(\pi_2) = \sigma\sigma^{-1}(Ker(\pi_2)) \subseteq \sigma(Ker(\pi_1))$, i.e., $\sigma(Ker(\pi_1)) = Ker(\pi_2)$. Therefore $\sigma|_{Ker(\pi_1)} : Ker(\pi_1) \to Ker(\pi_2)$ is an isomorphism. □

2. s-ADS modules

Any module M is called *amply supplemented* if B contains a supplement of A in M whenever $M = A + B$.

THEOREM 2.1. *Assume that an amply supplemented R-module X has a decomposition $X = M \oplus N$ for some R-modules M and N. Then the following conditions are equivalent:*

(1) *M is dual automorphism N-invariant.*
(2) *For any supplement K of N in X with $K + M = X$ and $(K \cap M) \ll X$, the module X has a decomposition $X = K \oplus N$.*
(3) *For each submodule K of X such that K is a supplement of N in X and M is a supplement of K in X, we have $X = K \oplus N$.*

PROOF. (1) $\Rightarrow$ (2) Let K be a supplement of N in X with $K+M=X$ and $(K\cap M)\ll X$. It follows that M and N are supplement of K in M. By Lemma 1.4, there exists $T\leq K$ such that $X=T\oplus N$. Since K contains T and K is a supplement of N we have $X=K\oplus N$.
(2) $\Rightarrow$ (3) Since M is a direct summand of X and M is a supplement of K in X, the result is clear.
(3) $\Rightarrow$ (1) Let $A\ll N$, $f:M\to N/A$ a small epimorphism and $\pi:N\to N/A$ the natural projection. Let $K=\{n+m|\ f(m)=-\pi(n)\}$. It is easy to see that $K+M=K+N=X$ and $K\cap M=Ker(f)\ll M$. Since X is an amply supplemented module, there exists $H\leq K$ such that H is a supplement of N in X and so K is a supplement of N in X. By (3), we have $N\oplus K=X$. Let $p:N\oplus K\to N$ be the canonical projection. Clearly $\pi p|_M=f$. Thus M is dual automorphism N-invariant. □

A module M is called an *esential -ADS module* (shortly, e-ADS module) if, for every decomposition $M=S\oplus T$ and every complement T' of S with $T'\cap T=0$ and $S\cap(T'\oplus T)\leq^e S$, we have $M=S\oplus T'$ (see [**4**]).

We call M an *superfluous -ADS-module* (shortly, s-ADS module) if, for every decomposition $M=S\oplus T$ of M and every supplement T' of S with $T'+T=M$ and $(T\cap T')\ll M$, we have $M=S\oplus T'$.

LEMMA 2.2. *The class of s-ADS modules is closed under taking direct summands.*

PROOF. Assume that M is an s-ADS module. Let $M=M_1\oplus M_2$ for some submodules M_1 and M_2 of M. Now we show that M_1 is an s-ADS module. Assume that M_1 has a decomposition $M_1=S\oplus T$ where T' is a supplement of S in M_1 and T is a supplement of T' in M_1. It is easy to see that

$$S\cap(T'\oplus M_2)\leq(S\cap T')\oplus M_2$$

and

$$(T\oplus M_2)\cap(T'\oplus M_2)\leq(T\cap T')\oplus M_2.$$

Then M has a decomposition $M=S\oplus(T\oplus M_2)$. Since M is an s-ADS module, we have $M=S\oplus(T'\oplus M_2)$. Therefore, by the modularity law, we obtain $M_1=S\oplus T'$ which implies that M_1 is an s-ADS module. □

Let us formulate several equivalent conditions for a module to be s-ADS.

THEOREM 2.3. *The following conditions are equivalent for a module M:*

(1) *M is s-ADS.*
(2) *For every decomposition $M=S\oplus T$, if T' is a supplement of S in M and T is a supplement of T' in M, then $M=S\oplus T'$.*

PROOF. (1) $\Rightarrow$ (2) Suppose that the module M has a decomposition $M=S\oplus T$ such that T' is a supplement of S and T is a supplement of T' in M. Then $T+T'=M$ and $T\cap T'\ll T$ which implies $T\cap T'\ll M$ since T is a supplement of T' in M. By (1), we have $M=S\oplus T'$.
(2) $\Rightarrow$ (1) Let $M=S\oplus T$ of M and T' be a supplement of S such that $T'+T=M$ and $(T\cap T')\ll M$. Since T is a direct summand of M we have $(T\cap T')\ll T$. Then by (2), we have $M=S\oplus T'$. □

A right R-module M is said to be ADS^* if for every decomposition $M = S \oplus T$ and for every supplement T' of S, we have $M = S \oplus T'$ (see [**9**]). It is clear from definition that every ADS^* module is s-ADS.

In [**9**, Theorem 2.1], it is shown that an amply supplemented R-module M is ADS^* if and only if If $M = A \oplus B$, then A and B are mutually projective. The following result extends this observation.

THEOREM 2.4. *An amply supplemented R-module M is s-ADS if and only if for each decomposition $M = A \oplus B$, A and B are relatively dual automorphism invariant.*

PROOF. It is clear from Theorem 2.1. □

As an immediate consequence of Theorem 2.4 , we obtain the following corollary.

COROLLARY 2.5. *Every amply supplemented dual automorphism-invariant module is s-ADS.*

Recall that a ring R is called a right *V-ring* if every simple right R-module is injective. It is well-known that if R is a right V-ring then for every right R-module M, $Rad(M) = 0$, and so M has no nonzero small submodule.

LEMMA 2.6. *Let R be a right V-ring. Then every right R-module is ADS^*.*

PROOF. See [**9**, Example 2.2]. □

THEOREM 2.7. *The following conditions are equivalent for a ring R:*

(1) *R is a right V-ring.*
(2) *Every 2-generated right R-module is ADS^*.*
(3) *Every 2-generated right R-module is s-ADS.*

PROOF. (1) $\Rightarrow$ (2) by Lemma 2.6.
(2) $\Rightarrow$ (3) Clear.
(3) $\Rightarrow$ (1) Suppose every 2-generated right R-module is s-ADS. Assume that R is not a right V-ring. Then there exists a simple right R-module S such that which is not injective. Let $E(S)$ be the injective hull of S. Then $E(S) \neq S$. Choose any $x \in E(S) \setminus S$. Then S is small in xR. Let $X = xR \oplus xR/S$. Then X is s-ADS. Now we put $A = \{(y, y + S) | y \in xR\}$. It is easy to see that $X = A + (xR \oplus 0)$ and $A \cap (xR \oplus 0) = S \oplus 0$. Let $(S \oplus 0) + A' = A$ and $H = \{t \in xR| \ (t, t + S) \in A'\}$. Then H is a submodule of xR. Since $(S \oplus 0) + A' = A$, we have $H + S = xR$. Then $H = xR$ and so $A' = A$. Thus A is a supplement of $xR \oplus 0$ in X. On the other hand, $A \oplus (0 \oplus xR/S) = X$, that means $0 \oplus xR/S$ is a supplement of A in X. But $A \cap (xR \oplus 0) \neq 0$, a contradiction. □

REMARK 2.8. We know that every cyclic module over a commutative ring is quasi-projective, by [**10**, 41.14(2)]. By [**8**, Remark 6], every quasi-projective over a commutative ring is ADS^* and so every cyclic module over a commutative ring is ADS^*. If we consider R to be a commutative ring which is not von Neumann regular, then every cyclic module over R is ADS^* but R is not a V-ring.

References

[1] A. Alahmadi, S. K. Jain, and A. Leroy, *ADS modules*, J. Algebra **352** (2012), 215–222, DOI 10.1016/j.jalgebra.2011.10.035. MR2862183

[2] F. W. Anderson and K. R. Fuller, *Rings and categories of modules*, Springer-Verlag, New York-Heidelberg, 1974. Graduate Texts in Mathematics, Vol. 13. MR0417223

[3] N. Er, S. Singh, and A. K. Srivastava, *Rings and modules which are stable under automorphisms of their injective hulls*, J. Algebra **379** (2013), 223–229, DOI 10.1016/j.jalgebra.2013.01.021. MR3019253

[4] M.T. Koşan, T.C. Quynh, J. Žemlička, *e-ADS modules and rings*, preprint.

[5] T.-K. Lee and Y. Zhou, *Modules which are invariant under automorphisms of their injective hulls*, J. Algebra Appl. **12** (2013), no. 2, 1250159, 9, DOI 10.1142/S0219498812501599. MR3005608

[6] S. H. Mohamed and B. J. Müller, *Continuous and discrete modules*, London Mathematical Society Lecture Note Series, vol. 147, Cambridge University Press, Cambridge, 1990. MR1084376

[7] T. C. Quynh and M. T. Koşan, *On automorphism-invariant modules*, J. Algebra Appl. **14** (2015), no. 5, 1550074, 11, DOI 10.1142/S0219498815500747. MR3323335

[8] S. Singh and A. K. Srivastava, *Dual automorphism-invariant modules*, J. Algebra **371** (2012), 262–275, DOI 10.1016/j.jalgebra.2012.08.012. MR2975396

[9] D. Keskin Tütüncü, *A note on ADS* modules*, Bull. Math. Sci. **2** (2012), no. 2, 359–363, DOI 10.1007/s13373-012-0020-0. MR2994209

[10] R. Wisbauer, *Foundations of module and ring theory*, Revised and translated from the 1988 German edition, Algebra, Logic and Applications, vol. 3, Gordon and Breach Science Publishers, Philadelphia, PA, 1991. A handbook for study and research. MR1144522

Department of Mathematics, Danang University, 459 Ton Duc Thang, DaNang city, Vietnam

Email address: tcquynh@live.com

Department of Mathematics, Gebze Institute of Technology, Çayirova Campus, 41400, Gebze/ Kocaeli, Turkey

Email address: ssahinkaya@gtu.edu.tr

Contemporary Mathematics
Volume **727**, 2019
https://doi.org/10.1090/conm/727/14646

When every endomorphism of a Σ-injective module is a sum of two commuting automorphisms

Feroz Siddique

ABSTRACT. It is well known that if M is an injective module then each endomorphism of M is a sum of two automorphisms if and only if $End(M)$ has no homomorphic image isomorphic to the field of two elements. In this paper we show that if M is a Σ-injective module such that each homomorphism of M is a sum of two commuting automorphisms then M must be directly finite.

All rings in this paper are associative with the identity element 1 and denoted by R unless specified. All modules are unital right R-modules and all endomorphisms act on the left unless specified. We denote the Jacobson radical of a ring R by $J(R)$.

1. Introduction

The question of when the automorphism group of a module additively generates its endomorphism ring has been a topic of interest for many years and has been studied by many authors. It has a rich history, the earliest instances of which can be found in the investigations of Dieudonné on Galois theory of simple and semisimple rings [**Di**], published in 1948. In 1950, G. Hochschild showed that in any simple algebra over a field, every element is a sum of units [**Ho**]. The study received a significant contribution when D. Zelinsky[**Ze**] and W.G Wolfson[**WoZe**] independently showed that every linear transformation of a right vector space V over a division ring D is a sum of two invertible linear transformations except when V is one dimensional over $\mathbb{F}_2$. We can come across several conditions on endomorphism rings of certain classes of modules, that allows each endomorphism to be expressed as a sum of automorphisms. In 1963, L. Fuchs [**Fu**] raised the following question: *For which abelian group G, does the automorphism group of G generate the endomorphism ring on G?.* Several authors including R.S Pierce [**Pi**], R.W Stringall [**St**], H. Freedman [**Fr**], G.S. Monk [**Mo**] and F. Castgan [**Cas**] gave results that explore the affirmation and negation of the above statement for different classes of modules.

This study of additive representations of elements in rings spilled beyond invertible elements and generated quite an interest when Nicholson in his study of lifting idempotents and exchange rings defined rings whose elements were characterized to

2010 *Mathematics Subject Classification.* Primary 16U60, 16D50, 16D70.
Key words and phrases. Units, Σ-injective modules, injective modules, directly finite.

be a sum of invertible elements and idempotents. Such rings were defined as clean rings. A ring R is strongly clean if each element in R is a sum of an idempotent and invertible element which commute. As strongly $\pi-$regular rings are directly finite, and strongly π-regular rings are strongly clean, Nicholson asked the question whether every strongly clean ring is directly finite.

This was partially answered in the affirmative by Camillo, Dorsey and Nielsen [**CDN**] when they showed that if a module is quasi-continuous and strongly clean, then it is directly finite.

Recall that a module M is called directly finite (Dedekind Finite, Von Neumann Finite) if M is not isomorphic to any proper direct summand of itself, or equivalently for any $f, g \in End(M)$, $fg = 1 \implies gf = 1$. For more results on direct finiteness, the reader is referred to Goodearl [**Go**]. A ring R is *2-good* (strongly *2-good*) if each element in R can be expressed as a sum of two invertible elements (that commute).

Since injective modules are quasi-continuous, it is quite natural to move tangentially from the notion of strongly clean property to strongly 2-good property and first explore the possibilities when each endomorphism on an injective module is a sum of two commuting automorphisms. We already know from Khurana and Srivastava [**KS**] that a right-self injective ring R is *2-good* if and only if R has no factor ring isomorphic to $\mathbb{F}_2$. The questions remains as to what characterizes a right self-injective ring to be strongly 2-good.

Tang and Zhou studied this question and characterized finite size matrix rings over division rings in terms of strongly 2-good (2-sum) property.

THEOREM 1.1 (Tang, Zhou [**TaZh**]). *Let V be a vector space space over a division ring D. Then each linear transformation is a sum of two commuting invertible linear transformations if and only if $|D| \geq 3$ and $dim(V) < \infty$.*

Since endomorphism rings of right vector spaces over division rings are right self-injective, it is quite natural to wonder whether Tang and Zhou's result extends to all right self injective rings that have no factor isomorphic to $M_n(\mathbb{F}_2)$ for any n.

We first look at the case of Σ-injective modules. Recall that Σ-injective modules were first introduced by Faith [**Fa**] where he defines an injective module M to be Σ-injective if an infinite direct sum of copies of M is injective.

In this paper we partially answer the question by proving that if E_R is a Σ-injective module and every endomorphism on E_R can be expressed as a sum of two commuting automorphisms then E must be directly finite.

2. Main Results

A classical result independently due to Matlis [**Ma**] and Papp [**Pa**] says that if ring R is left-Noetherian then every injective left R-module is a direct sum decomposition of indecomposable (injective) left R-modules. Later this was generalized by Cailleau [**Ca**] and Faith [**Fa**] in the next Lemma.

LEMMA 2.1. *An injective R-module is Σ-injective if and only if it has a direct sum decomposition of Σ-injective indecomposable modules.*

We now proceed with our main result.

THEOREM 2.2. *Let E_R be a Σ-injective module. If every element in $R = End(E_R)$ can be expressed as a sum of two commuting units then E_R is directly finite.*

PROOF. Let E be Σ-injective module.

Then by Lemma 2.1, E can be expressed as a direct sum of indecomposable Σ-injective modules, say $E = \bigoplus_{i \in I} P_i$.

Let us assume that E is not directly finite. Then there exists an infinite subfamily $\{P_i\}_{I'}$ of $\{P_i\}_I$ in which all the indecomposables are isomorphic among each other. Take an infinite countable subset $I'' \subseteq I'$, thereof identifying I'' with $\mathbb{N}$ and calling $Q = \bigoplus_{i \notin I''} P_i$, we can write $E \cong \bigoplus_{i=1}^{\infty} P_i \bigoplus Q = M$ (say) where $P_i \cong P_{i+1}$ for each positive integer i.

Let $\phi_0 : P_1 \to P_1$ be the identity morphism and $\phi_i : P_i \to P_{i+1}$ be an isomorphism from P_i to P_{i+1} for each $i \in \mathbb{N}$.

Choose any element $a \in \bigoplus_{i=1}^{\infty} P_i$.

Then we can find a n such that $a = \phi_0(x_0) + \sum\limits_{i=1}^{n} \phi_i(x_i) = \sum\limits_{i=0}^{n} \phi_i(x_i)$ with each $x_i \in P_i$.

Define $f\colon \bigoplus_{i=1}^{\infty} P_i \longrightarrow \bigoplus_{i=1}^{\infty} P_i$ by $f(a) = f(\sum\limits_{i=0}^{n} \phi_i(x_i))$

$$= \sum_{i=0}^{n} (\phi_{i+1} \circ \phi_i)(x_i)$$

Clearly f is a well-defined endomorphism on $\bigoplus_{i=1}^{\infty} P_i$.

We now consider the endomorphism $\sigma = f \oplus 1_Q$ on $M = \bigoplus_{i=1}^{\infty} P_i \bigoplus Q$ where 1_Q is the identity map on Q.

Let us check that Q is invariant under any endomorphism on M that commutes with σ. In other words, if $h\colon M \to M$ such that $\sigma h = h\sigma$ then $h(Q) \subseteq Q$.

In fact, take an arbitrary $a \in h(Q)$. Then $a = h(x)$ for some $x \in Q$.

As $h(x) \in M$ we can write $h(x) = \sum\limits_{i=0}^{n} \phi_i(x_i) + w$ where $x_i \in P_i$ and $w \in Q$.

Since $\sigma|_Q$ is an identity and therefore $\sigma(x) = x \in Q$, we have $\sum\limits_{i=0}^{n} \phi_i(x_i) + w =$

$h(x) = h(\sigma(x)) = \sigma(h(x)) = f(\sum\limits_{i=0}^{n} \phi_i(x_i)) + 1_Q(w) = \sum\limits_{i=0}^{n} (\phi_{i+1} \circ \phi_i)(x_i) + w$

$\implies \phi_0(x_0) = 0.$

Applying this step repeatedly gives $\phi_i(x_i) = 0$, and since each ϕ_i is an isomorphism we have $a = h(x) = w \in Q$.

Therefore $h(Q) \subseteq Q$.

By our hypothesis, every endomorphism on E can be expressed as a sum of two commuting automorphisms and $E \cong M$. Therefore, $\sigma\colon M \mapsto M$ can be expressed as a sum of 2 commuting automorphisms.

Let $\sigma = \alpha + \beta$ where $\alpha\beta = \beta\alpha$. Then $\alpha, \beta, \alpha^{-1}, \beta^{-1}$ commute with σ.

Choose any $x \in M = \bigoplus_{i=1}^{\infty} P_i \bigoplus Q$.

Then $x = \sum\limits_{i=0}^{n} \phi_i(x_i) + y$ where $y \in Q$ and $\phi_0(x_0) \in P_0$.

Then $\alpha(\phi_0(x_0)) \notin Q$ and $\alpha^{-1}(\phi_0(x_0)) \notin Q$.

Because if $\alpha(\phi_0(x_0)) \in Q$, then from above, as $\alpha^{-1}\colon M \to M$ and $\alpha^{-1}\sigma = \sigma\alpha^{-1}$, we will have $\alpha^{-1}(\alpha(\phi_0(x_0))) = \phi_0(x_0) \in Q$ which is a contradiction.

Similarly, $\alpha^{-1}(\phi_0(x_0)) \notin Q$.

Without loss of generality, let $\alpha(\phi_0(x_0)) = \sum\limits_{i=0}^{n} \phi_i(y_i) + w$ where $\phi_n(y_n) \neq 0 \implies y_n \neq 0$ as ϕ_n is a homomorphism.

Then $\phi_0(x_0) = \alpha^{-1}(\alpha(\phi_0(x_0))$

$= \alpha^{-1}(\sum_{i=0}^{n} \phi_i(y_i) + w)$

$= \alpha^{-1}(\phi_0(y_0) + \phi_1(y_1) + \phi_2(y_2) + ... + \phi_n(y_n)) + \alpha^{-1}(w)$

$= \alpha^{-1}(\phi_0(y_0)) + \alpha^{-1}(\phi_1(y_1)) + \alpha^{-1}(\phi_2(y_2)) + \alpha^{-1}(\phi_3(y_3)) + ...$

$+ ...\alpha^{-1}(\phi_n(y_n)) + \alpha^{-1}(w)$

As each ϕ_i is an isomorphism, we have

$\phi_0(x_0) = \alpha^{-1}(\phi_0(y_0)) + \alpha^{-1}(\phi_1(\phi_0(y_1^1)) + \alpha^{-1}(\phi_2(\phi_1(y_2^1))$
$+ ... + \alpha^{-1}(\phi_n(\phi_{n-1}(y_{n-1}^1)) + \alpha^{-1}(w)$ where $\phi_0(y_1^1) = y_1, \phi_1(y_2^1) = y_2, \ . \ . \ . \ , \phi_{n-1}(y_n^1) = y_n$

Observe that in each term, we will express the isomorphism ϕ_i as a composition of $\phi_i \circ \phi_{i-1} \circ \phi_{i-2} ... \circ \phi_0$ where the last element in the composition is ϕ_0. So $\phi_i(y_i) = \phi_i(\phi_{i-1}(y_i^1)$ where $\phi_{i-1}(y_i^1) = y_i$ as ϕ_i is an isomorphism on P_i. Then we write $\phi_i(y_i) = \phi_i(\phi_{i-1}(y_i^1)) = \phi_i(\phi_{i-1}(\phi_{i-2}(y_i^2)))$ where $\phi_{i-2}(y_i^2) = y_i^1$ and so forth.

Therefore, continuing in this way, we get,

$\phi_0(x_0) = \alpha^{-1}(\phi_0(y_0)) + \alpha^{-1}(\phi_1(\phi_0(y_1^1)) + \alpha^{-1}(\phi_2(\phi_1(\phi_0(y_2^2)))$
$+ \alpha^{-1}(\phi_3(\phi_2(\phi_1(\phi_0(y_3^3))) + ... +$
$\alpha^{-1}(\phi_n(\phi_{n-1}(\phi_{n-2}...(\phi_0(y_n^n))) + \alpha^{-1}(w)$
where $\phi_0(y_3^3) = y_3^2$,..., $\phi_0(y_n^n) = y_n^{n-1}$

And by the very definition of σ,

$= \alpha^{-1}(\phi_0(y_0) + \alpha^{-1}(\sigma(\phi_0(y_1^1)) + \alpha^{-1}(\sigma(\sigma(\phi_0(y_2^2)))$
$+ \alpha^{-1}(\sigma(\sigma(\sigma(\phi_0(y_3^3))) + ... +$
$\alpha^{-1}(\sigma(\sigma(\sigma...(\phi_0(y_n^n)) + \alpha^{-1}(w)$
$= \alpha^{-1}(\phi_0(y_0) + \alpha^{-1}(\sigma(\phi_0(y_1^1)) + \alpha^{-1}(\sigma^2(\phi_0(y_2^2)))$
$+ \alpha^{-1}(\sigma^3(\phi_0(y_3^3))) + ... + \alpha^{-1}(\sigma^n(\phi_0(y_n^n)) + \alpha^{-1}(w)$
$= \sum_{i=0}^{n} \alpha^{-1}(\sigma^i(\phi_0(y_i^i)) + \alpha^{-1}(w)$ where $y_0^0 = y_0$

Since σ commutes with α^{-1} we have,

$\phi_0(x_0) = \sum_{i=0}^{n} \alpha^{-1}(\sigma^i(\phi_0(y_i^i)) + \alpha^{-1}(w).$
$= \sum_{i=0}^{n} \sigma^i(\alpha^{-1}(\phi_0(y_i^i)) + \alpha^{-1}(w).$

As we have assumed in the beginning that $\phi_n(y_n) \neq 0 \implies y_n \neq 0$. And since $y_n = \phi_{n-1}(y_n^1) = \phi_{n-1}(\phi_{n-2}(y_n^2) = ...\phi_{n-1}(\phi_{n-2}(...\phi_0(y_n^n)))$ and $\phi_{n-1}, \phi_{n-2}, ...\phi_0$ are all automorphisms, we have $y_n^n \neq 0$ and $y_n^n \in P_1$.

Therefore,

$\phi_0(x_0) = \sum_{i=0}^{n-1} \sigma^i(\alpha^{-1}(\phi_0(y_i^i)) + \sigma^n(\alpha^{-1}(\phi_0(y_n^n)) + \alpha^{-1}(w)$ where $\alpha^{-1}(\phi_0(y_n^n) \neq 0$.

Note that $\alpha^{-1}(\phi_0(y_n^n))) \notin Q$.

Because if $\alpha^{-1}(\phi_0(y_n^n)) \in Q$, then from above, as $\alpha\colon M \to M$ and $\alpha\sigma = \sigma\alpha$, we will have $\alpha(\alpha^{-1}(\phi_0(y_n^n))) = \phi_0(y_n^n) \in Q$ which is a contradiction.

Without loss of generality, let $\alpha^{-1}(\phi_0(y_n^n)) = \sum_{i=0}^{m} \phi_i(z_i) + k$ where $\phi_m(z_m) \neq 0$ and $k \in Q$. Then $z_m \neq 0$. Thereof, $\alpha^{-1}(\phi_0(y_n^n)) = \sum_{i=0}^{m-1} \phi_i(z_i) + \phi_m(z_m) + k$

Going back we see that,

$$\begin{aligned}\phi_0(x_0) &= \sum_{i=0}^{n-1} \sigma^i(\alpha^{-1}(\phi_0(y_i^i)) + \sigma^n(\alpha^{-1}(\phi_0(y_n^n)) + \alpha^{-1}(w) \\ &= \sum_{i=0}^{n-1} \sigma^i(\alpha^{-1}(\phi_0(y_i^i)) + \sigma^n(\sum_{i=0}^{m-1} \phi_i(z_i) + \phi_m(z_m) + k) + \alpha^{-1}(w) \\ &= \sum_{i=0}^{n-1} \sigma^i(\alpha^{-1}(\phi_0(y_i^i)) + \sigma^n(\sum_{i=0}^{m-1} \phi_i(z_i) + k) + \sigma^n(\phi_m(z_m) + \alpha^{-1}(w) \\ &= \sum_{i=0}^{n-1} \sigma^i(\alpha^{-1}(\phi_0(y_i^i)) + \sigma^n(\sum_{i=0}^{m-1} \phi_i(z_i)) + (\phi_{m+n} \circ \phi_{m+n-1} \circ \ldots \circ \phi_{m+1} \\ &\quad \circ \phi_m)(z_m) + k + \alpha^{-1}(w) \text{ where}\end{aligned}$$

$(\phi_{m+n} \circ \phi_{m+n-1} \circ \ldots \phi_{m+1} \circ \phi_m(z_m) \neq 0$ as $z_m \neq 0$ and $\sigma|_Q$ is an identity.

Thus observing the left hand side term which is in P_1 and the term $(\phi_{m+n} \circ \phi_{m+n-1} \circ \ldots \phi_{m+1} \circ \phi_m(z_m)$ which is in P_{n+m+1} we conclude that $m + n = 0$ where m, n are non-negative integers.

Therefore $m = n = 0$.

Thus $\alpha(\phi_0(x_0)) = \phi_0(y_0) + w$.

Similarly, $\beta(\phi_0(x_0)) = \phi_0(s_0) + w'$ for some $s_0 \in P_1$ and $w' \in Q$.

Therefore $(\alpha + \beta)(\phi_0(x_0)) = \phi_0(y_0 + s_0) + w + w'$.

$\Longrightarrow \sigma(\phi_0(x_0)) = \phi_0(y_0 + s_0) + w + w'$.

$\Longrightarrow \phi_1(\phi_0(x_0)) = \phi_0(y_0+s_0)+w+w'$ where $\phi_1(\phi_0(x_0)) \in P_2$ and $\phi_0(y_0+s_0) \in P_1$.

Therefore $\phi_1(\phi_0(x_0)) = 0$ and since ϕ_0, ϕ_1 are isomorphisms, we have $x_0 = 0$.

Now since $x = \sum_{i=0}^{n} \phi_i(x_i) + y$ was chosen to be arbitrary, we can always assume $x_0 \neq 0$ (equivalent to assuming $\phi_0(x_0) \neq 0$).

Thus we have a contradiction.

Therefore E is directly finite. □

REMARK 2.3. Note that the result is no longer true if each endomorphism on E is a sum of two units that do not necessarily commute with one another. For example consider the endomorphism ring of a countably infinite dimensional right vector space V over a division ring D. This ring is not directly finite but it is known that every element in the endomorphism ring $End(V_D)$ is a sum of two units.

If V_D is a finite-dimensional, right vector space over a division ring D such that $|D| \geq 3$, then from [**TaZh**] every element in $End(V_D)$ is a sum of two commuting automorphisms.

As a consequence, we obtain part of Theorem 5 in [**TaZh**] by Tang and Zhou.

COROLLARY 2.4. *If V_D is a right vector space over a division ring D such that $|D| \geq 3$. Then the following are equivalent.*

(1) Every element in $R = End(V_D)$ can be expressed as a sum of two commuting automorphisms.

(2) V_D is finite dimensional.

Recall that a ring R is unit-regular if for each element $x \in R$ there exists a invertible element $u \in R$ such that $xux = x$. It can be easily seen that unit-regular rings are directly finite.

As a consequence of the above theorem, we have the following corollary.

Corollary 2.5. *Let E_R be a Σ-injective module. If every element in $End(E_R)$ can be expressed as a sum of two commuting automorphisms, then $End(E_R)/J(End(E_R))$ is unit-regular.*

Proof. Since every element in $End(E_R)$ can be expressed as a sum of two commuting automorphisms, by *Theorem* 3, E_R is directly finite. In particular, since Σ-injective modules are automorphism invariant modules and E_R is directly finite, we have that $End(E_R)/J(End(E_R))$ is unit regular from Guil Asensio and Srivastava[**GuSr**]. □

In view of our main theorem on Σ-injective modules and the above corollaries, we would like to propose the following question.

Question 2.6. If E_R is an injective module such that $End(E_R)$ has no homomorphic image isomorphic to $M_n(\mathbb{F}_2)$ for any n, then the following are equivalent.

(1) E_R is directly finite.
(2) Each endomorphism on E_R can be expressed as a sum of two commuting automorphisms.
(3) $End(E_R)/J(End(E_R))$ is unit-regular.

Acknowledgement

I would like to thank Dr. Pedro A. Guil Asensio and Dr. Ashish K Srivastava for all the discussions and encouragement received.

References

[BA] H. Bass, *Injective dimension in Noetherian rings*, Trans. Amer. Math. Soc. **102** (1962), 18–29, DOI 10.2307/1993878. MR0138644

[BJS08] K. I. Beidar, S. K. Jain, and A. K. Srivastava, *New characterization of Σ-injective modules*, Proc. Amer. Math. Soc. **136** (2008), no. 10, 3461–3466, DOI 10.1090/S0002-9939-08-09420-3. MR2415029

[Ca] A. Cailleau, *Une caractérisation des modules Σ-injectifs* (French), C. R. Acad. Sci. Paris Sér. A-B **269** (1969), A997–A999. MR0260782

[CDN] V. Camillo, T. J. Dorsey, and P. P. Nielsen, *Dedekind-finite strongly clean rings*, Comm. Algebra **42** (2014), no. 4, 1619–1629, DOI 10.1080/00927872.2012.746976. MR3169656

[CE56] H. Cartan and S. Eilenberg, *Homological algebra*, Princeton University Press, Princeton, N. J., 1956. MR0077480

[Cas] F. Castgan, *Sums of automorphisms of a primary abelian group* , Pac. J. Math 27(3), 1968.

[Fa] C. Faith, *Rings with ascending condition on annihilators*, Nagoya Math. J. **27** (1966), 179–191. MR0193107

[Di] J. Dieudonné, *La théorie de Galois des anneaux simples et semi-simples* (French), Comment. Math. Helv. **21** (1948), 154–184, DOI 10.1007/BF02568032. MR0024892

[Fa] C. Faith, *Lectures on injective modules and quotient rings*, Lecture Notes in Mathematics, No. 49, Springer-Verlag, Berlin-New York, 1967. MR0227206

[Fr] H. Freedman, *On endomorphisms of primary Abelian groups*, J. London Math. Soc. **43** (1968), 305–307, DOI 10.1112/jlms/s1-43.1.305. MR0225880

[Fu] L. Fuchs, *Recent results and problems on abelian groups*, Topics in Abelian Groups (Proc. Sympos., New Mexico State Univ., 1962), Scott, Foresman and Co., Chicago, Ill., 1963, pp. 9–40. MR0169906

[Go] K. R. Goodearl, *Ring theory: Nonsingular rings and modules*, Pure and Applied Mathematics, vol. 33, Marcel Dekker, Inc., New York-Basel, 1976. MR0429962

[GuSr] P. A. Guil Asensio and A. K. Srivastava, *Automorphism-invariant modules satisfy the exchange property*, J. Algebra **388** (2013), 101–106, DOI 10.1016/j.jalgebra.2013.05.003. MR3061680

[Ho] G. Hochschild, *Automorphisms of simple algebras*, Trans. Amer. Math. Soc. **69** (1950), 292–301, DOI 10.2307/1990359. MR0037834

[KS] D. Khurana and A. K. Srivastava, *Right self-injective rings in which every element is a sum of two units*, J. Algebra Appl. **6** (2007), no. 2, 281–286, DOI 10.1142/S0219498807002181. MR2316422

[La] T. Y. Lam, *Lectures on modules and rings*, Graduate Texts in Mathematics, vol. 189, Springer-Verlag, New York, 1999. MR1653294

[Ma] E. Matlis, *Injective modules over Noetherian rings*, Pacific J. Math. **8** (1958), 511–528. MR0099360

[Mo] G. S. Monk, *On the endomorphism ring of an abelian p-group, and of a large subgroup*, Pacific J. Math. **41** (1972), 183–193. MR0323923

[Pa] Z. Papp, *On algebraically closed modules*, Publ. Math. Debrecen **6** (1959), 311–327. MR0121390

[Pi] R. S. Pierce, *Homomorphisms of primary abelian groups*, Topics in Abelian Groups (Proc. Sympos., New Mexico State Univ., 1962), Scott, Foresman and Co., Chicago, Ill., 1963, pp. 215–310. MR0177035

[St] R. W. Stringall, *Endomorphism rings of Abelian groups generated by automorphism groups*, Acta Math. Acad. Sci. Hungar **18** (1967), 401–404, DOI 10.1007/BF02280299. MR0214660

[TaZh] G. Tang and Y. Zhou, *When is every linear transformation a sum of two commuting invertible ones?*, Linear Algebra Appl. **439** (2013), no. 11, 3615–3619, DOI 10.1016/j.laa.2013.09.038. MR3119876

[WoZe] K. G. Wolfson, *An ideal-theoretic characterization of the ring of all linear transformations*, Amer. J. Math. **75** (1953), 358–386, DOI 10.2307/2372458. MR0053080

[Ze] D. Zelinsky, *Every linear transformation is a sum of nonsingular ones*, Proc. Amer. Math. Soc. **5** (1954), 627–630, DOI 10.2307/2032048. MR0062728

Department of Mathematics, University of Wisconsin Colleges, Wisconsin

Current address: Department of Mathematics, University of Wisconsin Eau Claire-Barron County, Rice Lake, Wisconsin 54868

Email address: `feroz.siddique@uwc.edu`

Selected Published Titles in This Series

727 **André Leroy, Christian Lomp, Sergio López-Permouth, and Frédérique Oggier, Editors,** Rings, Modules and Codes, 2019

722 **Yves Aubry, Everett W. Howe, and Christophe Ritzenthaler, Editors,** Arithmetic Geometry: Computation and Applications, 2019

721 **Petr Vojtěchovský, Murray R. Bremner, J. Scott Carter, Anthony B. Evans, John Huerta, Michael K. Kinyon, G. Eric Moorhouse, and Jonathan D. H. Smith, Editors,** Nonassociative Mathematics and its Applications, 2019

720 **Alexandre Girouard, Editor,** Spectral Theory and Applications, 2018

719 **Florian Sobieczky, Editor,** Unimodularity in Randomly Generated Graphs, 2018

718 **David Ayala, Daniel S. Freed, and Ryan E. Grady, Editors,** Topology and Quantum Theory in Interaction, 2018

717 **Federico Bonetto, David Borthwick, Evans Harrell, and Michael Loss, Editors,** Mathematical Problems in Quantum Physics, 2018

716 **Alex Martsinkovsky, Kiyoshi Igusa, and Gordana Todorov, Editors,** Surveys in Representation Theory of Algebras, 2018

715 **Sergio R. López-Permouth, Jae Keol Park, S. Tariq Rizvi, and Cosmin S. Roman, Editors,** Advances in Rings and Modules, 2018

714 **Jens Gerlach Christensen, Susanna Dann, and Matthew Dawson, Editors,** Representation Theory and Harmonic Analysis on Symmetric Spaces, 2018

713 **Naihuan Jing and Kailash C. Misra, Editors,** Representations of Lie Algebras, Quantum Groups and Related Topics, 2018

712 **Nero Budur, Tommaso de Fernex, Roi Docampo, and Kevin Tucker, Editors,** Local and Global Methods in Algebraic Geometry, 2018

711 **Thomas Creutzig and Andrew R. Linshaw, Editors,** Vertex Algebras and Geometry, 2018

710 **Raphaël Danchin, Reinhard Farwig, Jiří Neustupa, and Patrick Penel, Editors,** Mathematical Analysis in Fluid Mechanics, 2018

709 **Fernando Galaz-García, Juan Carlos Pardo Millán, and Pedro Solórzano, Editors,** Contributions of Mexican Mathematicians Abroad in Pure and Applied Mathematics, 2018

708 **Christian Ausoni, Kathryn Hess, Brenda Johnson, Ieke Moerdijk, and Jérôme Scherer, Editors,** An Alpine Bouquet of Algebraic Topology, 2018

707 **Nitya Kitchloo, Mona Merling, Jack Morava, Emily Riehl, and W. Stephen Wilson, Editors,** New Directions in Homotopy Theory, 2018

706 **Yeonhyang Kim, Sivaram K. Narayan, Gabriel Picioroaga, and Eric S. Weber, Editors,** Frames and Harmonic Analysis, 2018

705 **Graham J. Leuschke, Frauke Bleher, Ralf Schiffler, and Dan Zacharia, Editors,** Representations of Algebras, 2018

704 **Alain Escassut, Cristina Perez-Garcia, and Khodr Shamseddine, Editors,** Advances in Ultrametric Analysis, 2018

703 **Andreas Malmendier and Tony Shaska, Editors,** Higher Genus Curves in Mathematical Physics and Arithmetic Geometry, 2018

702 **Mark Grant, Gregory Lupton, and Lucile Vandembroucq, Editors,** Topological Complexity and Related Topics, 2018

701 **Joan-Carles Lario and V. Kumar Murty, Editors,** Number Theory Related to Modular Curves, 2018

700 **Alexandre Girouard, Dmitry Jakobson, Michael Levitin, Nilima Nigam, Iosif Polterovich, and Frédéric Rochon, Editors,** Geometric and Computational Spectral Theory, 2017